Contents

Preface

A HELPFUL WORD TO THE STUDENT:

The primary purpose of this book is to help you review for the New York State Regents examination in physics. You can also use it effectively to review for classroom, midterm, and final examinations. This book contains a number of special features that are designed to aid you in achieving good grades on your examinations. These features include the following sections:

- *How to Use this Book*
- *Test-taking Tips*
- *What to Expect on the Regents Examination in Physics*
- A topic outline of the New York State Regents physics syllabus
- A question index that keys the Regents examination questions found in this book to the topic outline
- *Reference Tables for Physics*
- *Explanation of Equations Used in New York State Physics Course*
- A glossary of important terms
- Recent Regents examinations
- A detailed explanation of the answers and a self-analysis chart for each examination

HOW TO STUDY FOR CLASSROOM TESTS

It is assumed that you do your homework daily. Before you start on the new assignment, review the day's classroom notes and your notes of the previous day's work. After you have done the assignment, close your notebook and in your mind go over what you have read and written. This will help you remember the important things.

When the test is announced, review all your notes for the topics to be covered on the test. If your notes are inadequate, you may find a review book useful.

After reviewing the subject matter, turn to the Topic Outline in this book, and test yourself by means of the questions arranged by topic. Use the more recent Regents examinations first. Check your answers and study the explanations; the explanations also review material needed for answering many related questions.

HOW TO STUDY FOR REGENTS AND OTHER UNIFORM EXAMINATIONS

Use the examinations in this book as a study guide. The questions will indicate to you what is important. They will also help you find your weak spots. Review from notes and test material will then be more purposeful and productive.

Take the most recent examination under test conditions. Mark the examination and note your weak points. Study the answers to the questions you could not answer or which you answered inadequately. Repeat this procedure, and watch your knowledge grow.

This book is dedicated to all physics students who want to do better and to my helpers: Matty, Paul, Jonathan, George, and Zoë.

Herman Gewirtz

I dedicate this book to all students at the Bronx High School of Science and to my special students, Dana and Rebecca, and to my son, Jay, for his invaluable assistance.

Stanley Blumenstein

This book is dedicated to Albert S. Tarendash with gratitude for all the teaching and encouragement that made physics fun to learn, and for all his work in preparing the introductory material in this book, and to my chem/phys students at Stuyvesant, who made physics fun to teach.

Miriam A. Lazar

How to Use This Book

1. Read the section *Test-taking Tips* to learn how to prepare properly for an examination and how to take the examination with maximum efficiency.
2. Read the section *What to Expect on the Regents Examination in Physics* in order to familiarize yourself with the structure and contents of this examination.
3. Read the section *Reference Tables for Physics* to familiarize yourself with the contents and use of these tables and equations.

   Note: On occasion, the New York State Education Department may change the content or format of the Regents examination and/or the reference tables. Your classroom teacher is your best source of information about such changes.
4. Take the first Regents examination in this book, answering *all of the questions*.
5. Refer to the *Glossary* to learn the meanings of words and terms you do not understand.
6. Check your answers and complete the self-analysis chart found at the end of the examination. This will help you pinpoint the areas of your strengths and weaknesses.
7. Review the detailed explanations of *all of the questions*, and pay closest attention to the questions you answered incorrectly. Occasionally, the *wrong choices* are explained, and these explanations may help you understand *why* you chose an incorrect answer.
8. When you have determined your areas of weakness, refer to the Regents *Topic Outline and Question Index* to locate similar questions found on other recent examinations. (You can also use the outline and index to determine which areas have been stressed in recent years.)
9. Repeat steps 4–8 for the other examinations, *with the exception of the most recent examination.*

10. Take the most recent examination *under strict examination conditions* when you have completed your studying, but no more than 1–2 days before the actual examination.
11. After you have checked your answers to this last examination, you will have a rough idea of how you will perform on the actual examination.

Barron's Regents Exams and Answers

Physics

HERMAN GEWIRTZ
Former Chairman, Physical Sciences Department
Bronx High School of Science,
New York City

STANLEY L. BLUMENSTEIN
Assistant Principal, Physical Sciences Department
Bronx High School of Science,
New York City

MIRIAM A. LAZAR
Formerly, Department of Chemistry and Physics
Stuyvesant High School,
New York City

Barron's Educational Series, Inc.

The section on how to answer Part III questions and the glossary were taken from *Let's Review: Physics* by Miriam A. Lazar and Albert S. Tarendash, published by Barron's Educational Series, Inc., 1996.

All inquiries should be addressed to:
Barron's Educational Series, Inc.
250 Wireless Boulevard
Hauppauge, New York 11788
http://www.barronseduc.com

ISBN 0-8120-3349-3
ISSN 0190-5767

PRINTED IN THE UNITED STATES OF AMERICA

9 8 7 6 5 4 3 2 1

Test-taking Tips

The following pages contain several tips to help you achieve a good grade on the Physics Regents exam.

GENERAL HELPFUL TIPS

TIP 1

Be confident and prepared.

SUGGESTIONS

- Review previous tests.
- Use a clock or watch, and take previous exams at home under examination conditions, (i.e., don't have the radio or television on.)
- Get a review book. (One review book is Barron's *Let's Review: Physics*.)
- Talk over the answers to questions on these tests with someone else, such as another student in your class or someone at home.
- Finish all your homework assignments.
- Look over classroom exams that your teacher gave during the term.
- Take class notes carefully.
- Practice good study habits.
- Know that there is an answer for every question.
- Be aware that the people who made up the Regents exam want you to pass.
- Remember that thousands of students over the last few years have taken and passed a Physics Regents. You can pass too!

- Complete your study, and review at least one day before the examination. Last-minute cramming does not help and may hurt your performance.
- On the night prior to the exam day: lay out all the things you will need, such as clothing, pens, and admission cards.
- Go to bed early; eat wisely.
- Bring the required materials to the examination. This generally means a pen, two sharpened pencils, and a good quality eraser. In addition, Part III may require the use of a ruler and a protractor. If your school does not supply a calculator, be certain to bring one to the examination. Some schools also require a signed Regents admission card for identification.
- Good advice: Assume your school will *not* supply you with any materials!
- Once you are in the exam room, arrange things, get comfortable, be relaxed, attend to personal needs (the bathroom).
- Keep your eyes on your own paper; do not let them wander over to anyone else's paper.
- Be polite in making any reasonable requests of the exam room proctor, such as changing your seat or having window shades raised or lowered.

TIP 2

Read test instructions carefully.

SUGGESTIONS

- Be familiar with the format of the examination. Know which parts must be answered by *all* students and how many optional groups you will be required to answer. Make sure you are familiar with the *content* of each optional group.
- Decide upon the task(s) that you have to complete.
- Know how the test will be graded.
- If your school supplies an electronic scoring sheet, be certain you are familiar with the additional directions for recording and changing answers.
- If you decide to change an answer, be certain that you erase your original response completely.
- Any stray marks on your answer sheet should be erased completely.
- Leave all *unanswered* groups on Part II blank.
- Be familiar with the directions for Part III (free-response) questions. Answer each question completely. Explanations must be written as

whole sentences and substitutions into equations must include *units*. Be certain that your answers are clearly labeled and well organized. Place a box around numerical answers. Be neat!

- Ask for assistance from the exam room proctor if you do not understand the directions.

TIP 3

Read each question carefully and read each choice before you record your answer.

SUGGESTIONS

- Be sure you understand *what* the question is asking.
- Try to recognize information that is *given* in the question.
- Will a physics formula help you find the answer to the question?
- Some choices may look appealing yet be incorrect. (These traps are known as *distractors*.)
- Try to eliminate those choices that are *obviously* incorrect.

TIP 4

Budget your test time (3 hours).

SUGGESTIONS

- Bring a watch or clock to the test.
- The Regents examination is designed to be completed in 1½ to 2 hours.
- If you are absolutely uncertain of the answer to a question, mark your question booklet and move on to the next question.
- If you persist in trying to answer every difficult question *immediately*, you may find yourself rushing or unable to finish the remainder of the examination.
- When you have finished the examination, return to those unanswered questions.
- Good advice: If at all possible, reread the *entire* examination—and your responses—at least one more time. (This will help you eliminate those errors that result from misreading questions.)

TIP 5

Use your reasoning skills.

SUGGESTIONS

- Answer *all* questions.
- Relate (connect) the question to anything that you studied, wrote in your notebook, or heard your teacher say in class.
- Relate (connect) the question to any film, demonstration, or experiment you saw in class, any project you did, or to anything you may have learned from newspapers, magazines, or television.
- Look over the entire test to see whether one part of it can help you answer another part.

TIP 6

Use your reference tables and refer frequently to the "Summary of Equations."

SUGGESTIONS

- You should be familiar with the *content* of each table.
- Frequently, the answers to questions can be found from information contained within the table.
- The equations are grouped according to where they appear in the syllabus.
- The definition of each symbol in the equation is also provided.
- These equations will aid you in answering many of the questions on the examination.

TIP 7

Don't be afraid to guess.

SUGGESTIONS

- In general, go with your first answer choice.
- Eliminate obvious incorrect choices.
- If still unsure of an answer, make an educated guess.
- There is no penalty for guessing; therefore, answer ALL questions. An omitted answer gets no credit.

TIP 8

Sign the declaration found on your answer sheet.

SUGGESTION

- Unless this declaration is signed, your paper cannot be scored.

SUMMARY OF TIPS

1. Be confident and prepared.
2. Read test instructions carefully.
3. Read each question carefully and read each choice before you record your answer.
4. Budget your test time (3 hours).
5. Use your reasoning skills.
6. Use your reference tables and refer frequently to the "Summary of Equations."
7. Don't be afraid to guess.
8. Sign the declaration found on your answer sheet.

HOW TO ANSWER PART III QUESTIONS

A *free-response question* is an examination question that requires the test taker to do more than to choose among several responses or to fill in a blank. You may need to perform numerical calculations, draw and interpret graphs, and provide extended written responses to a question or problem.

Part III of the New York State Regents examination in physics contains free-response questions. This section is designed to provide you with a number of general guidelines for answering them.

SOLVING PROBLEMS INVOLVING NUMERICAL CALCULATIONS

To receive full credit you must:

- Provide the appropriate equation(s).
- Substitute values and units into the equation(s).
- Display the answer, with appropriate units.
- If the answer is a vector quantity, include its direction.

Although SI units are used on the Regents examination, you are expected to have some familiarity with other metric units such as the gram and the kilometer.

You will *not* be penalized if your answer has an incorrect number of significant digits. However, it is always good practice to pay attention to this detail.

You should write as legibly as possible. Teachers are human, and nothing irks them more than trying to decipher a careless, messy scrawl. It is also a good idea to identify your answer clearly, either by placing it in a box or by writing the word "answer" next to it.

A final word: If you provide the correct answer but do not show any work, you will not receive any credit for the problem!

The following is a sample problem and its model solution.

Problem

A 5.0-kilogram object has a velocity of 10. meters per second [east]. Calculate the momentum of this object.

Solution

$$\mathbf{p} = m\mathbf{v}$$

$$\mathbf{p} = (5.0 \text{ kg})(10. \text{ m/s [east]})$$

$$\boxed{\mathbf{p} = 50. \text{ kg·m/s [east]}}$$

GRAPHING EXPERIMENTAL DATA

To receive full credit you must:

- Label both axes with the appropriate variables and units.
- Divide the axes so that the data ranges fill the graph as nearly as possible.
- Plot all data points accurately.
- Draw a best-fit line carefully with a straightedge. The line should pass through the origin *only if the data warrant it.*
- If a part of the question requires that the slope be calculated, calculate the slope *from the line,* not from individual data points.

Generally a graph should have a title, and the *independent variable* is usually drawn along the *x*-axis. However, you will not be penalized on the Regents examination if you do not follow these conventions.

The following is a sample problem and its model solution.

Problem

The weights of various masses, measured on Planet *X*, are given in the table below.

Mass (kg)	Weight (N)
15	21
20.	32
25	35
30.	48
35	56

1. Draw a graph that illustrates these data.
2. Use the *graph* to calculate the acceleration due to gravity on Planet *X*.

Solution

1. The graph shown below incorporates the essential items that were listed in the table.

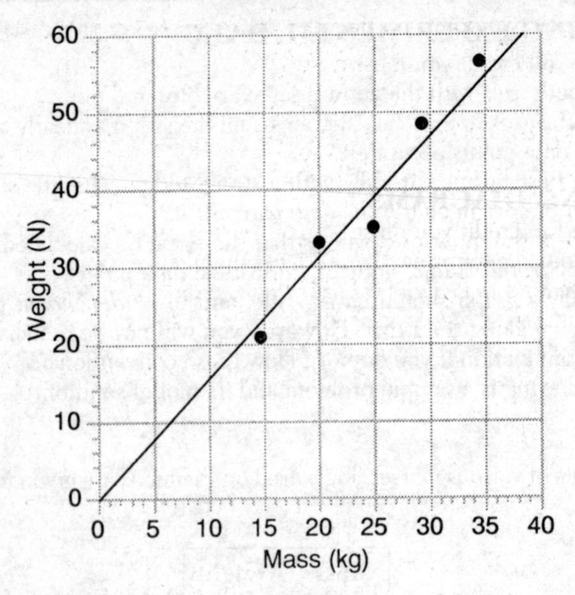

2. Since the magnitude of the gravitational acceleration can be determined by calculating the ratio of weight to mass ($g = W/m$), we can calculate the value of g from the slope of the graph.

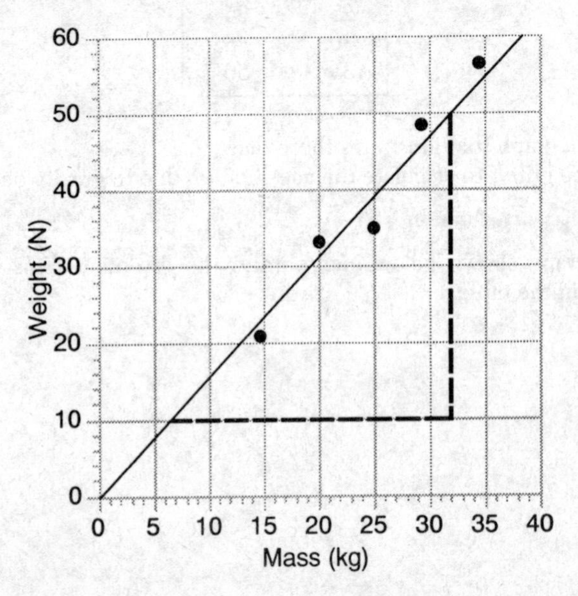

We choose two points on the line; we do not use the data points themselves:

$$g = \frac{\Delta W}{\Delta m} = \frac{50.\ \text{N} - 10.\ \text{N}}{32\ \text{kg} - 6\ \text{kg}} = 1.5\frac{\text{N}}{\text{kg}} = 1.5\ \text{m/s}^2$$

DRAWING DIAGRAMS

To receive full credit you must:

- Draw your diagrams neatly, and label them clearly.
- Draw vectors *to scale* and *in the correct direction*. If you are given a scale, you must draw your vectors to that scale.
- Bring a straightedge and a protractor with you so that you can draw neat, accurate diagrams.

WRITING A FREE-RESPONSE ANSWER

To receive full credit you must:

- Use complete, clear sentences that make sense to the reader.
- Use correct physics in your explanations.

A sample question and acceptable and unacceptable answers are given below.

Question

If the data from two different photoemissive materials are graphed, which characteristic of the two graphs will be the same?

Acceptable Answers

- The characteristic that will be the same is the slope of the line.
- It is the slope.
- The two lines have the same slope.

The following answers are unacceptable:

- The slope. (Incomplete sentence)
- Both lines will have the same y-intercept. (Incorrect physics)

What to Expect on the Regents Examination in Physics

The Regents examination in physics is divided into three parts.

PART I

Part I consists of 55 multiple-choice questions that count for 65 points. All students must answer every question in this part of the examination. An "adjustment" is built into the grading of Part I, as shown in the table below.

No. Right	Credits	No. Right	Credits
55	**65**	34	**48**
54	**64**	33	**47**
53	**63**	32	**46**
52	**63**	31	**45**
51	**62**	30	**44**
50	**61**	29	**44**
49	**60**	28	**43**
48	**59**	27	**42**
47	**58**	26	**41**
46	**58**	25	**40**
45	**57**	24	**40**
44	**56**	23	**39**
43	**55**	22	**38**
42	**54**	21	**37**
41	**54**	20	**36**
40	**53**	19	**35**
39	**52**	18	**35**
38	**51**	17	**34**
37	**50**	16	**33**
36	**49**	15	**32**
35	**49**	14	**31**

No. Right	Credits	No. Right	Credits
13	**31**	6	**14**
12	**29**	5	**12**
11	**26**	4	**10**
10	**24**	3	**7**
9	**21**	2	**5**
8	**19**	1	**2**
7	**17**	0	**0**

This table was taken from a recent Regents physics examination. Note that the adjustment (which varies slightly from year to year) works in your favor. For example, a student who answers 40 questions correctly receives 53 credits. Without this adjustment, a student would receive only 47 credits.

The questions for Part I are drawn from five units of the New York State physics syllabus:

Unit I—Mechanics
Unit II—Energy
Unit III—Electricity and Magnetism
Unit IV—Wave Phenomena
Unit V—Modern Physics

Questions on Part I generally appear in the order of the units listed above, with the exception of a few questions found at the end of Part I that are reserved for "decreases-increases-remains the same" items.

PART II

Part II consists of six groups that contain ten multiple-choice questions each. A student must choose only two of these groups and answer every question in each of the two groups he or she has chosen. Part II counts for 20 points. The questions for Part II are drawn from the following units of the New York State physics syllabus:

Group 1
Unit VI—Motion in a Plane

Group 2
Unit VII—Internal Energy

Group 3
Unit VIII—Electromagnetic Applications

Group 4
Unit IX—Geometrical Optics

Group 5
Unit X—Solid-State Physics

Group 6
Unit XI—Nuclear Energy

Students always wonder *which* groups they should choose from on Part II. The answer depends on how comfortable you feel with the material. When you have finished answering the sample examinations in this book, you will have a fairly good idea about which groups will be best for you. When you take the actual examination, however, be certain to inspect *every* group before you make your final decision. You may find that a group you had intended to ignore contains material that is quite familiar to you.

You should refer to the detailed Regents Topic Outline found on page 18 for a complete listing of the topics covered in each unit of the New York State physics syllabus.

PART III

Part III is a free-response section that counts for 15 points. All students must answer every question in this part of the examination. You may be asked to draw a graph or a diagram, solve a mathematical problem, or provide explanations for physical phenomena.

Every part of the Regents examination in physics contains questions designed to test your ability to:

- Recall factual information
- Interpret diagrams, tables, and graphs correctly
- Solve problems involving mathematical relationships
- Compare the relative magnitudes of various quantities
- Apply the physics concepts that were developed during the year

A sampling and analysis of examination questions drawn from Parts I and III is given below.

1. A person travels 6 meters north, 4 meters east, and 6 meters south. What is the total displacement?
   (1) 16 m east (2) 6 m north (3) 6 m south (4) 4 m east
   In this question, you are being asked to add vector quantities and calculate the resultant vector. You need to recognize that the 6-meter displacements negate each other (since they are in opposite directions), leaving a result of 4 m east.

2. If an object's velocity changes from 25 meters per second to 15 meters per second in 2.0 seconds, the magnitude of the object's acceleration is
(1) 5.0 m/s^2 (2) 7.5 m/s^2 (3) 13 m/s^2 (4) 20. m/s^2
In this question, you need to refer to the "Summary of Equations" to find and use the relationship $a = \frac{\Delta v}{\Delta t}$ in order to solve the problem. The answer is choice (1).
3. What is the approximate thickness of this piece of paper?
(1) 10^1 m (2) 10^0 m (3) 10^{-2} m (4) 10^{-4} m
This question tests your familiarity with the *size* of metric units. Since a meter is approximately three feet, the only reasonable answer is choice (4).
4. A glass rod is given a positive charge by rubbing it with silk. The rod has become positive by
(1) gaining electrons (2) gaining protons (3) losing electrons
(4) losing protons
In this question, you are expected to know that objects acquire electric charges by gaining or losing electrons. Therefore, a rod becomes positive by *losing electrons*.
5. The wavelength of the periodic wave shown in the diagram below is 4.0 meters. What is the distance from point B to point C?

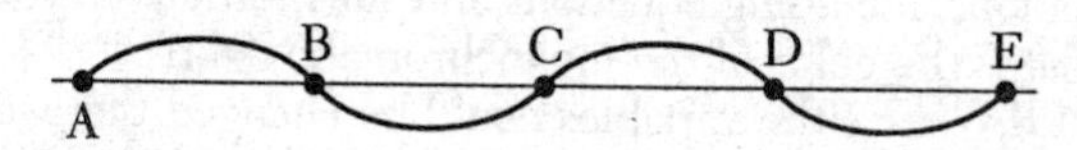

(1) 1.0 m (2) 2.0 m (3) 3.0 m (4) 4.0 m
In this question, you are expected to recognize that one wavelength includes one crest and one trough (e.g., the distance from point A to point C). Since the distance specified in the question includes *only* one trough, it represents one-half of a wavelength, or 2.0 m.
6. The wavelength of photon A is greater than the wavelength of photon B. Compared with the energy of photon A, the energy of photon B is
(1) less (2) greater (3) the same
This question asks for a comparison, but you must know the mathematical relationships between wavelength, frequency, and energy for a photon. The "Summary of Equations" informs us that $E = hf$ and $c = f\lambda$, where c is the speed of a photon in space. We can combine these two relationships to yield $E = \frac{hc}{\lambda}$. We can conclude from this relationship that the smaller the wavelength of a photon, the greater is its energy. Since photon B has a smaller wavelength than photon A, photon B has greater energy.

7. (Free-response) Monochromatic light is incident on a two-slit apparatus as shown in the diagram below. The distance between the slits is 1.0×10^{-3} meter, the distance from the two-slit apparatus to a screen displaying the interference pattern is 4.0 meters. The distance between the central maximum and the first-order maximum is 2.4×10^{-3} meter.

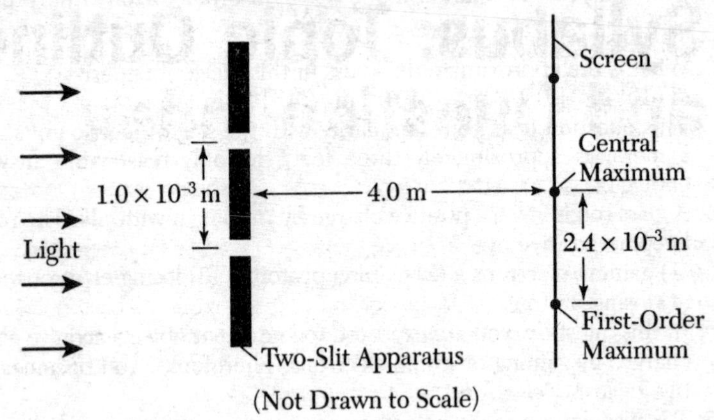

(Not Drawn to Scale)

a What is the wavelength of the monochromatic light? (Show all calculations, including equations and substitutions with units.)

b What is the color of the monochromatic light?

c List *two* ways the variables could be changed that would cause the distance between the central maximum and the first-order maximum to increase.

MODEL SOLUTIONS TO FREE-RESPONSE QUESTION 7:

a $\lambda = \frac{dx}{L} = \frac{(1.0\times10^{-3}\text{m})(2.4\times10^{-3}\text{m})}{4.0\text{m}} = 6.0\times10^{-7}\text{m}$

b The color of the monochromatic light is *orange*.

c The equation in part *a* can be rearranged to yield: $x = \frac{\lambda L}{d}$. In order to increase the distance between the central maximum and the first-order maximum (x), one could either *increase* the wavelength of the monochromatic light (λ) or *increase* the distance between the screen and the two-slit apparatus (L).

(Note that *c* also has a third possibility: The distance between the slits (d) could be decreased.)

A FINAL WORD: Practice really makes perfect! The more questions you answer, the better the likelihood that you will achieve a good grade on the Regents examination.

New York State Regents Physics Syllabus: Topic Outline and Question Index

The topic outline on pages 18–27 is taken from the Regents Physics Syllabus of the New York State Education Department. Regents physics courses and the Regents Examination are based upon this syllabus.

This topic outline also serves as an index to relevant questions in the examinations contained and explained in this book. Note: Questions marked with an asterisk (*) embrace more than one topic.

Topic Outline and Question Index

Examination Questions	1998	1997	1996	1995	1994	1993
UNIT ONE – MECHANICS						
I. KINEMATICS						
A. Distance and displacement	1	1	8, 116	117	1, 55	
B. The meter	2	3	1, 117	1		2
C. Velocity and speed	5, 7, 120	2	3–7	2, 118		1, 4
D. Acceleration		6, 54		116	4, 8, 9	
F. Final velocity of an object at constant acceleration					6	
G. Distance traveled by an object at constant acceleration		4, 9			10	3
H. Freely falling objects	3			3, 4	3	6
II. STATICS						
A. Force	83°, 116	19	10, 20	5	19	117
B. Vector addition of concurrent forces	4°, 22°	5, 116–119	2	6, 7	5	9, 116
C. Resolution of forces	6, 12°	8	9	9, 52	7	7
D. Equilibrium	4°			7	2, 16, 58	8, 10
III. DYNAMICS						
A. Newton's first law of motion	12°	7, 120	13			
B. Newton's second law of motion	9, 117	11, 12	11, 12	11, 14°	11, 14°	11
C. Newton's universal law of gravitation	10	15		54	12	14, 15
D. Gravitational field strength	11		14	8	13	16
E. Friction	12°	13	16	10, 12		17

NOTE: Questions with an asterisk (°) embrace more than one topic

Examination Questions	1998	1997	1996	1995	1994	1993
F. Momentum	13–15, 54	10, 14, 55	17	13, 14*, 53	14*	
G. Newton's third law of motion	8, 16		15		15	
UNIT TWO – ENERGY						
I. WORK AND ENERGY	121–124	17				
A. Work	18, 21	16	118	15	17	18, 19
B. Power	17	20	19	16, 29	13	
C. Kinetic Energy	20	21	22, 119	18	21	13
D. Potential Energy	19	18	18, 21	17	20	12
E. Work–energy relationship				19		20
F. Conservation of energy	63*			26	22	24–26
UNIT THREE – ELECTRICITY AND MAGNETISM						
I. STATIC ELECTRICITY						
A. Microstructure of matter					23	
B. Charged objects	22*	22	24	21		21
C. Conservation of charge		23	25	22		22
D. Elementary charges	23	26		83	81	
E. Quantity of charge			82	25		
F. Coulomb's law	24		26, 28	23, 119, 120	24, 25	23
G. Electric fields	25, 83*, 119*	24, 30			26	
H. Potential difference	26, 119*	25	27	24, 26	27	27, 28
II. ELECTRIC CURRENT			32			
A. Conductivity in solids						
B. Conditions necessary for an electric current						

Topic Outline and Question Index

Examination Questions	1998	1997	1996	1995	1994	1993
C. Unit of current	27°, 29°, 118	27, 28	31	29	28	
D. Resistance	53	29	29	27	29	29, 30
E. Ohm's law	27°, 85°		30°	26		
F. Series circuits	28, 32°		30°	28	30°, 33	31, 54°
G. Parallel circuits	29°, 32°	124, 125, 126			30°	54°
H. Power and energy in electric circuits	30, 31	31, 32	36	30	31	32, 53
III. MAGNETISM						
A. Magnetic force		33				
B. Magnetic field	33, 34, 36	34	33, 37	31, 32, 35	32, 34	34, 36
C. Force on a moving charge-carrier in a magnetic field		35	34, 83, 85°	36, 80–82	35, 78, 79	35, 79, 83
IV. ELECTROMAGNETIC INDUCTION						
A. Induced EMF						
B. Electromagnetic radiation			35		36	
UNIT FOUR – WAVE PHENOMENA						
I. INTRODUCTION TO WAVES						
A. Transfer of energy	35		39		37	
B. Pulses and periodic waves		36				
C. Types of wave motion				37	38	38
II. CHARACTERISTICS OF PERIODIC WAVES						
A. Frequency	38		40°	34, 44		41

Examination Questions	1998	1997	1996	1995	1994	1993
B. Period	42	37	40*			
C. Amplitude		38, 122			43	39
D. Phase			38, 44	38		
E. Wavelength	41	40, 123	121	39	39	
F. Speed	37		122	33	40	40, 42
G. Doppler effect	47*	41	55	41	41	37
H. Wave fronts					42	
III. PERIODIC WAVE PHENOMENA						
A. Interference	39	121	42, 46	42		43
B. Standing waves						44
C. Resonance	40					
IV. LIGHT						
A. Speed					44	
B. Reflection	43, 48*		123, 124	40	45	
C. Refraction	48*	42		45, 121, 122		55
D. Absolute index of refraction	44*, 48*	39	43, 47			
E. Snell's law	45*, 48*	45	41		49	45
F. Critical angle and total internal reflection	52	44	54	123	46	47
G. Dispersion	46		45	47, 55		46
H. Interference and diffraction		43, 47				118
I. Polarization		46	50		47	48
J. Electromagnetic spectrum				43	54	
K. Coherent light		48				

Topic Outline and Question Index

Examination Questions	1998	1997	1996	1995	1994	1993
UNIT FIVE – MODERN PHYSICS						
I. DUAL NATURE OF LIGHT						
A. Wave phenomena						
B. Particle phenomena	47*		48		48	49, 119
II. THE QUANTUM THEORY						
A. The quantum of energy			52	46		
B. The photon						
C. The photoelectric effect	50, 55	49, 51	49	48	52	50
D. Photon-particle collisions	51				50	
E. Photon momentum		50				52
F. Matter waves						
III. MODELS OF THE ATOM			120			
A. The Rutherford model	49	52	53	49	51	51
B. The Bohr model of the hydrogen atom	45		51	50	53	
C. Atomic spectra		53				
D. The cloud model				51		
UNIT SIX – MOTION IN A PLANE (Examination Group 1 of Part II)						
I. TWO DIMENSIONAL MOTION AND TRAJECTORIES						
A. A projectile fired horizontally	56, 57	64	65	56–58	62, 63	56, 57, 59
B. A projectile fired at an angle	61, 62	57, 61, 62	56–58	64	64	58

Examination Questions	1998	1997	1996	1995	1994	1993
II. UNIFORM CIRCULAR MOTION	59		62	59		
A. Centripetal acceleration	58	59	59	60, 61	57, 59	62
B. Centripetal force	60	58, 60	60, 61	62	56, 60	61, 63, 64
III. KEPLER'S LAWS						
A. Kepler's first law			63		61	60
B. Kepler's second law	63*, 64	65	64		65	65
C. Kepler's third law	65	63		65		
IV. SATELLITE MOTION						
A. Geosynchronous orbits		56				
B. Artificial satellites				63		
UNIT SEVEN – INTERNAL ENERGY (Examination Group 2 of Part II)						
I. TEMPERATURE			66			75
A. Absolute temperature	66	66		66	66, 75	
B. Temperature scales	67		72		67	66
II. INTERNAL ENERGY AND HEAT			70			73
A. Specific heat		68, 69	69	68, 69	69, 70	70
B. Exchange of internal energy	71*	67			68	
C. Change of phase	70, 73	71	68	67, 75		
D. Heat of fusion	72	73	67	70	71	
E. Heat of vaporization	69					67, 71
F. Factors affecting freezing and boiling points	74	75	71	71	74	72, 74

Topic Outline and Question Index

Examination Questions	1998	1997	1996	1995	1994	1993
III. KINETIC THEORY OF GASES		70		72		
A. Pressure			74			
B. Gas laws		72	73, 75	73	73	69
IV. THE LAWS OF THERMODYNAMICS						
A. The first law	75					
B. The second law	68, 71*	74		74	72	68
C. The third law						
UNIT EIGHT – ELECTROMAGNETIC APPLICATIONS (Examination Group 3 of Part II)						
I. TORQUE ON A CURRENT-CARRYING LOOP	78				77	
A. Galvanometer				76		
B. Ammeter	79					
C. Voltmeter		76, 77	76			80
D. Motor (DC)			78	77		
E. Back EMF	80			78		
II. ELECTRON BEAMS						
A. Thermionic emission	81		85*			
B. Control of electron beams by electric and magnetic fields	76, 77	78, 79, 85	80, 81	79	76, 80	81, 84

Examination Questions	1998	1997	1996	1995	1994	1993
III. ION BEAMS						
A. The mass spectrometer	82					
B. Particle accelerators						
IV. INDUCED POTENTIAL DIFFERENCE						
A. Magnitude and direction of induced EMF	84	81	79	84		82
B. Lenz's law						
C. Generator (AC)		83	77		83–85	
D. Transformers	85*	80, 84	84	85	82	76–78
E. Induction coils		82				
V. THE LASER						85
UNIT NINE – GEOMETRIC OPTICS (Examination Group 4 of Part II)						
I. IMAGES						
A. Real images	87*, 90*					
B. Virtual images						
II. IMAGES FORMED BY REFLECTION						
A. Images formed by a plane mirror	93	86	86, 87	86, 95	86	86, 95
B. Images formed by spherical convex mirrors	89, 94	87		90	88	89
C. Images formed by spherical concave mirrors	86, 87*, 88	88–91	89, 90	87, 88	87, 89, 90	87, 88
III. IMAGES FORMED BY REFRACTION			95			
A. Images formed by converging lenses	90*–92	92–94	91–94	89, 91, 92	91–94	90–92, 94
B. Images formed by diverging lenses		95		93		
C. Lens defects	95		88	94	95	93

Topic Outline and Question Index

Examination Questions	1998	1997	1996	1995	1994	1993
UNIT TEN – SOLID STATE PHYSICS (Examination Group 5 of Part II)						
I. CONDUCTION IN SOLIDS		96			97	99
A. Conductors	101				96	
B. Insulators						
C. Semiconductors			96			97
D. The electron–sea model of conduction						
E. The band model of conduction	102		97	97		96
II. EXTRINSIC SEMICONDUCTORS						
A. Doping	104*		99, 102		105	
B. N-type semiconductors	99		100	98*	99*	
C. P-type semiconductors	104*	98, 105		96, 98*, 99	98, 99*	101
III. SEMICONDUCTOR DEVICES						
A. The junction diode	96, 103	99–101	103*, 105	100, 101	100, 101*	98
B. Forward and reverse biasing	97, 98	102, 103	101, 103*		104	100, 102
C. N-P-N and P-N-P (bipolar) transistors	100, 105	97, 104	98, 104	102–104	101*–103	103, 105
D. Integrated circuits				105		104
UNIT ELEVEN – NUCLEAR ENERGY (Examination Group 6 of Part II)						
I. THE NUCLEUS						
A. Nucleons	106			25, 106		106
B. Atomic number		108				

Examination Questions	1998	1997	1996	1995	1994	1993
C. Mass number					106	
D. Nuclear force	108		108	107		
E. The atomic mass unit						
F. Mass-energy relationship			106	108	107	107
G. Nuclear mass and binding energy		107	107	115		
H. Isotopes			109*	109	108	
I. Nuclides			110			
J. Particle accelerators		109				
K. Detection devices						
L. Subatomic particles	109					
II. NUCLEAR REACTIONS	107	106, 111				
A. Natural radioactivity	110–112	110	112, 113	110, 111	110, 111	109
B. Half-life	113	112	115	112	112	110
C. Artificial transmutation					113	112
D. Nuclear fission	114	113, 114			115	113, 114
E. Fission reactors			114	113		
F. Nuclear fusion	115	115	109*	114	114	115

Reference Tables for Physics

LIST OF PHYSICAL CONSTANTS

Name	*Symbol*	*Value(s)*
Gravitational constant	G	6.7×10^{-11} N•m^2/kg^2
Acceleration due to gravity (up to 16 km altitude)	g	9.8 m/s^2
Speed of light in a vacuum	c	3.0×10^8 m/s
Speed of sound at STP		3.3×10^2 m/s
Mass-energy relationship		1 u (amu) = 9.3×10^2 MeV
Mass of the Earth		6.0×10^{24} kg
Mass of the Moon		7.4×10^{22} kg
Mean radius of the Earth		6.4×10^6 m
Mean radius of the Moon		1.7×10^6 m
Mean distance from Earth to Moon		3.8×10^8 m
Electrostatic constant	k	9.0×10^9 N•m^2/C^2
Charge of the electron (1 elementary charge)		1.6×10^{-19} C
One coulomb	C	6.3×10^{18} elementary charges
Electronvolt	eV	1.6×10^{-19} J
Planck's constant	h	6.6×10^{-34} J•s
Rest mass of the electron	m_e	9.1×10^{-31} kg
Rest mass of the proton	m_p	1.7×10^{-27} kg
Rest mass of the neutron	m_n	1.7×10^{-27} kg

ABSOLUTE INDICES OF REFRACTION

($\lambda = 5.9 \times 10^{-7}$ m)

Air	1.00
Alcohol	1.36
Canada Balsam	1.53
Corn Oil	1.47
Diamond	2.42
Glass, Crown	1.52
Glass, Flint	1.61
Glycerol	1.47
Lucite	1.50
Quartz, Fused	1.46
Water	1.33

WAVELENGTHS OF LIGHT IN A VACUUM

Color	Wavelength
Violet	$4.0 - 4.2 \times 10^{-7}$ m
Blue	$4.2 - 4.9 \times 10^{-7}$ m
Green	$4.9 - 5.7 \times 10^{-7}$ m
Yellow	$5.7 - 5.9 \times 10^{-7}$ m
Orange	$5.9 - 6.5 \times 10^{-7}$ m
Red	$6.5 - 7.0 \times 10^{-7}$ m

HEAT CONSTANTS

	Specific Heat (average) (kJ/kg•C°)	*Melting Point* (°C)	*Boiling Point* (°C)	*Heat of Fusion* (kJ/kg)	*Heat of Vaporization* (kJ/kg)
Alcohol (ethyl)	2.43 (liq.)	−117	79	109	855
Aluminum	0.90 (sol.)	660	2467	396	10500
Ammonia	4.71 (liq.)	−78	−33	332	1370
Copper	0.39 (sol.)	1083	2567	205	4790
Iron	0.45 (sol.)	1535	2750	267	6290
Lead	0.13 (sol.)	328	1740	25	866
Mercury	0.14 (liq.)	−39	357	11	295
Platinum	0.13 (sol.)	1772	3827	101	229
Silver	0.24 (sol.)	962	2212	105	2370
Tungsten	0.13 (sol.)	3410	5660	192	4350
Water: ice	2.05 (sol.)	0	—	334	—
Water: water	4.19 (liq.)	—	100	—	2260
Water: steam	2.01 (gas)	—	—	—	—
Zinc	0.39 (sol.)	420	907	113	1770

ENERGY LEVEL DIAGRAMS FOR MERCURY AND HYDROGEN

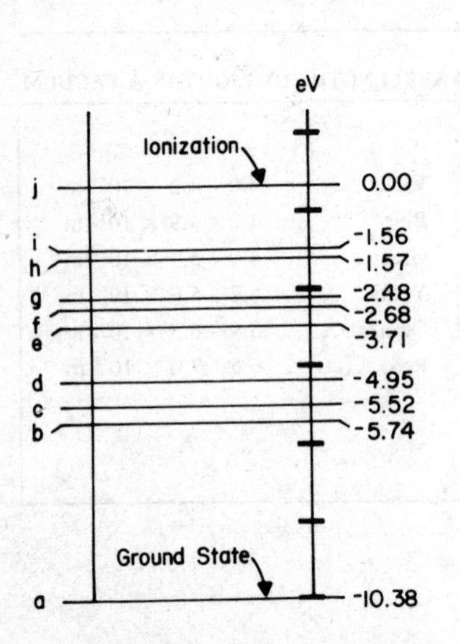

A few energy levels for the mercury atom

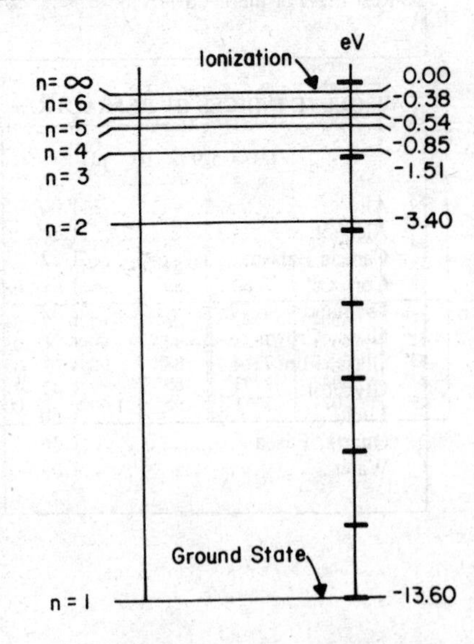

Energy levels for the hydrogen atom

VALUES OF TRIGONOMETRIC FUNCTIONS

Angle	Sine	Cosine	Angle	Sine	Cosine
1°	.0175	.9998	46°	.7193	.6947
2°	.0349	.9994	47°	.7314	.6820
3°	.0523	.9986	48°	.7431	.6691
4°	.0698	.9976	49°	.7547	.6561
5°	.0872	.9962	50°	.7660	.6428
6°	.1045	.9945	51°	.7771	.6293
7°	.1219	.9925	52°	.7880	.6157
8°	.1392	.9903	53°	.7986	.6018
9°	.1564	.9877	54°	.8090	.5878
10°	.1736	.9848	55°	.8192	.5736
11°	.1908	.9816	56°	.8290	.5592
12°	.2079	.9781	57°	.8387	.5446
13°	.2250	.9744	58°	.8480	.5299
14°	.2419	.9703	59°	.8572	.5150
15°	.2588	.9659	60°	.8660	.5000
16°	.2756	.9613	61°	.8746	.4848
17°	.2924	.9563	62°	.8829	.4695
18°	.3090	.9511	63°	.8910	.4540
19°	.3256	.9455	64°	.8988	.4384
20°	.3420	.9397	65°	.9063	.4226
21°	.3584	.9336	66°	.9135	.4067
22°	.3746	.9272	67°	.9205	.3907
23°	.3907	.9205	68°	.9272	.3745
24°	.4067	.9135	69°	.9336	.3584
25°	.4226	.9063	70°	.9397	.3420
26°	.4384	.8988	71°	.9455	.3256
27°	.4540	.8910	72°	.9511	.3090
28°	.4695	.8829	73°	.9563	.2924
29°	.4848	.8746	74°	.9613	.2756
30°	.5000	.8660	75°	.9659	.2588
31°	.5120	.8572	76°	.9703	.2419
32°	.5299	.8480	77°	.9744	.2250
33°	.5446	.8387	78°	.9781	.2079
34°	.5592	.8290	79°	.9816	.1908
35°	.5736	.8192	80°	.9848	.1736
36°	.5878	.8090	81°	.9877	.1564
37°	.6018	.7986	82°	.9903	.1392
38°	.6157	.7880	83°	.9925	.1219
39°	.6293	.7771	84°	.9945	.1045
40°	.6428	.7660	85°	.9962	.0872
41°	.6561	.7547	86°	.9976	.0698
42°	.6691	.7431	87°	.9986	.0523
43°	.6820	.7314	88°	.9994	.0349
44°	.6947	.7193	89°	.9998	.0175
45°	.7071	.7071	90°	1.0000	.0000

URANIUM DISINTEGRATION SERIES

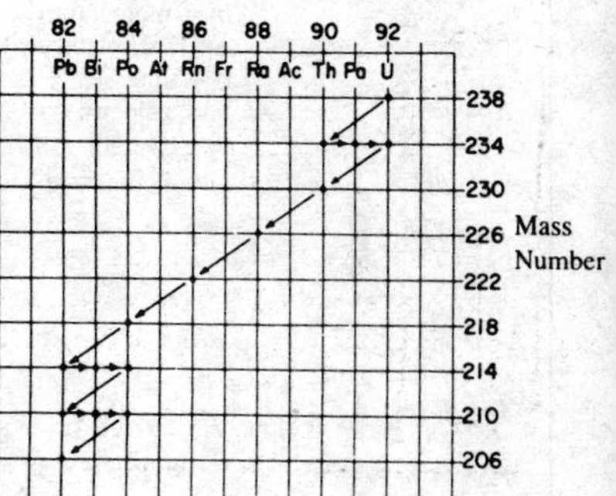

SUMMARY OF EQUATIONS

MECHANICS

$\bar{v} = \frac{\Delta s}{\Delta t}$

$\bar{v} = \frac{v_f + v_i}{2}$

$\bar{a} = \frac{\Delta v}{\Delta t}$

$\Delta s = v_i \Delta t + \frac{1}{2}a(\Delta t)^2$

$v_f^2 = v_i^2 + 2a\Delta s$

$F = ma$

$w = mg$

$F = \frac{Gm_1m_2}{r^2}$

$p = mv$

$J = F\Delta t$

$F\Delta t = m\Delta v$

a = acceleration
r = distance between centers
F = force
g = acceleration due to gravity
G = universal gravitation constant
J = impulse
m = mass
p = momentum
Δs = displacement
t = time
v = velocity
w = weight

ENERGY

$W = F\Delta s$

$P = \frac{W}{\Delta t} = \frac{F\Delta s}{\Delta t} = F\bar{v}$

$\Delta PE = mg\Delta h$

$KE = \frac{1}{2}mv^2$

$F = kx$

$PE_s = \frac{1}{2}kx^2$

F = force
g = acceleration due to gravity
h = height
k = spring constant
KE = kinetic energy
m = mass
P = power
PE = potential energy
PE_s = potential energy stored in a spring
Δs = displacement
t = time
v = velocity
W = work
x = change in spring length from the equilibrium position

ELECTRICITY AND MAGNETISM

$F = \frac{kq_1q_2}{r^2}$

$E = \frac{F}{q}$

$V = \frac{W}{q}$

$E = \frac{V}{d}$

$I = \frac{\Delta q}{\Delta t}$

$R = \frac{V}{I}$

$P = VI = I^2R = \frac{V^2}{R}$

$W = Pt = VIt = I^2Rt$

d = separation of parallel plates
r = distance between centers
E = electric field intensity
F = force
I = current
k = electrostatic constant
ℓ = length of conductor
P = power
q = charge
R = resistance
t = time
V = electric potential difference
W = energy

Series Circuits:

$I_t = I_1 = I_2 = I_3 = \ldots$

$V_t = V_1 + V_2 + V_3 + \ldots$

$R_t = R_1 + R_2 + R_3 + \ldots$

Parallel Circuits:

$I_t = I_1 + I_2 + I_3 + \ldots$

$V_t = V_1 = V_2 = V_3 = \ldots$

$\frac{1}{R_t} = \frac{1}{R_1} + \frac{1}{R_2} + \frac{1}{R_3} + \ldots$

INTERNAL ENERGY

$Q = mc\Delta T_c$

$Q_f = mH_f$

$Q_v = mH_v$

c = specific heat
H_f = heat of fusion
H_v = heat of vaporation
m = mass
Q = amount of heat
T_c = Celsius temperature

WAVE PHENOMENA

$T = \frac{1}{f}$

$v = f\lambda$

$n = \frac{c}{v}$

$\sin \theta_c = \frac{1}{n}$

$n_1 \sin \theta_1 = n_2 \sin \theta_2$

$n_1 v_1 = n_2 v_2$

$\frac{\lambda}{d} = \frac{x}{L}$

c = speed of light in a vacuum
d = distance between slits
f = frequency
L = distance from slit to screen
n = index of absolute refraction
T = period
v = speed
x = distance from central maximum to first-order maximum
λ = wavelength
θ = angle
θ_c = critical angle of incidence relative to air

MODERN PHYSICS

$W_o = hf_o$

$E_{photon} = hf$

$KE_{max} = hf - W_o$

$p = \frac{h}{\lambda}$

$E_{photon} = E_i - E_f$

c = speed of light in a vacuum
E = energy
f = frequency
f_o = threshold frequency
h = Planck's constant
KE = kinetic energy
p = momentum
W_o = work function
λ = wavelength

MOTION IN A PLANE

$v_{iy} = v_i \sin \theta$

$v_{ix} = v_i \cos \theta$

$a_c = \frac{v^2}{r}$

$F_c = \frac{mv^2}{r}$

a_c = centripetal acceleration
F_c = centripetal force
m = mass
r = radius
v = velocity
θ = angle

ELECTROMAGNETIC APPLICATIONS

$F = qvB$

$\frac{N_p}{N_s} = \frac{V_p}{V_s}$

$V_p I_p = V_s I_s$ (ideal)

% Efficiency = $\frac{V_s I_s}{V_p I_p} \times 100$

$V = B\ell v$

B = flux density
F = force
I_p = current in primary coil
I_s = current in secondary coil
N_p = number of turns of primary coil
N_s = number of turns of secondary coil
q = charge
v = velocity
V_p = voltage of primary coil
V_s = voltage of secondary coil
ℓ = length of conductor
V = electric potential difference

GEOMETRIC OPTICS

$\frac{1}{d_o} + \frac{1}{d_i} = \frac{1}{f}$

$\frac{S_o}{S_i} = \frac{d_o}{d_i}$

d_i = image distance
d_o = object distance
f = focal length
S_i = image size
S_o = object size

NUCLEAR ENERGY

$E = mc^2$

$m_f = \frac{m_i}{2^n}$

c = speed of light in a vacuum
E = energy
m = mass
n = number of half-lives

Explanation of Equations Used in New York State Physics Course*

UNIT I—MECHANICS

1. $\bar{v} = \dfrac{\Delta s}{\Delta t}$

The average speed at which an object travels is equal to the distance traveled divided by the time it took to travel that distance.

EXAMPLE: It takes a car 0.20 hour to travel a distance of 12 kilometers.

$$\bar{v} = \frac{\Delta s}{\Delta t}$$
$$= \frac{12.\ \text{km}}{0.20\ \text{h}}$$
$$= 60.\ \text{km/h}$$

2. $\bar{v} = \dfrac{v_f + v_i}{2}$

The average speed of an object moving with constant acceleration is equal to one-half of the sum of its final and its initial speeds.

EXAMPLE: An object starting from rest is accelerated uniformly for 4.0 seconds attaining a maximum speed of 10. meters per second. Since the initial speed $v_i = 0$, then

$$\text{average speed } \bar{v} = \frac{1}{2}v_f$$
$$= \frac{1}{2}(10.\ \text{m/s})$$
$$= 5.0\ \text{m/s}$$

*The numerical values of constants can be found, if needed, in the preceding Physics Reference Tables under List of Physical Constants and Heat Constants.

3. $a = \dfrac{\Delta v}{\Delta t}$

The acceleration of an object is equal to the rate of change of its velocity. The average acceleration over a given time interval equals the change in velocity divided by the time in which the change takes place.

EXAMPLE: An object is uniformly accelerated from rest to a speed of 25. meters per second in 10. seconds.

$$\text{acceleration } a = \frac{(25. - 0)\text{ m/s}}{10.\text{s}}$$
$$= 2.5\text{ m/s}^2$$

°4. $\Delta s = \dfrac{1}{2} a \Delta t^2 + v_i \Delta t$

For motion having an initial speed with constant acceleration, the distance traveled is equal to one-half the product of its acceleration and the square of the time of travel, plus the product of its initial velocity and the time of travel.

EXAMPLE: A cart with an initial velocity of 6.00 meters per second rolls down an inclined plane with an acceleration of 4.00 meters per second squared. What distance will it travel in 3.00 seconds?

$$s = \frac{1}{2}(4.00\text{ m/s}^2)(3.00\text{ s})^2 + (6.00\text{ m/s})(3.00\text{ s})$$
$$= 36.0\text{ m}$$

°5. $v_f^{\,2} = 2\,a\,\Delta s + v_i^{\,2}$

For an object moving with an initial speed and then accelerating uniformly, the square of its speed at any instant is equal to the sum of twice the product of its acceleration and its displacement plus the square of its initial speed.

EXAMPLE: A car moving at a constant speed of 4.10 meters per second then accelerates uniformly at 3.20 meters per second squared. What will be its speed when it has traveled a distance of 40.0 meters?

$$v_f^{\,2} = (2)(3.20\text{ m/s}^2)(40.0\text{ m}) + (4.10\text{ m/s})^2$$
$$= 256\text{ m}^2/\text{s}^2 + 16.8\text{ m}^2/\text{s}^2$$
$$v_f = 16.5\text{ m/s}$$

6. $F = ma$

The net force acting on an object is equal to the product of the object's mass and the resulting acceleration.

°Constant acceleration along straight lines.

EXAMPLE: A 3.0-kilogram mass is being moved along a horizontal surface by a force of 6.0 newtons. If the surface is frictionless, what is the acceleration produced by the 6.0-newton force?

$$F = ma$$
$$6.0 \text{ N} = (3.0 \text{ kg})\ (a)$$
$$a = 2.0 \text{ m/s}^2$$

7. $w = mg$

The weight of an object is equal to the product of its mass and the acceleration due to gravity of a freely falling object.

EXAMPLE: The acceleration due to gravity on planet A is 20. meters per second squared. On this planet what is the weight of an object whose mass is 2.0 kilgrams?

$$\begin{aligned} w &= mg \\ &= (2.0 \text{ kg})\ (20. \text{ m/s}^2) \\ &= 40\text{N} \end{aligned}$$

8. $F = \dfrac{Gm_1m_2}{d^2}$

The gravitational force between two objects is given by the product of the gravitational constant (G) and the masses of the two objects, divided by the square of the distance between the two objects.

EXAMPLES:

a. Two masses of 10.0 kg and 1.0 kg, respectively, are located 1.0 meter apart. How large a gravitational force does each mass exert on the other?

$$F = (6.67 \times 10^{-11} \text{ N} \cdot \text{m}^2/\text{kg}^2)\ (10. \text{ kg})\ \frac{(1.0 \text{ kg})}{(1.0 \text{ m})^2}$$

$$= 6.67 \times 10^{-10} \text{ N}$$

The value of the gravitational constant is the first item in the Physics Reference Tables.

b. Two small objects of equal mass are a fixed distance apart. How would the gravitational force between them change if the mass of each object were three times as great?

According to the formula given, the gravitational force between two objects is proportional to the product of their masses. If the mass of only one object is tripled, the product of their masses is tripled and the gravitational force is tripled. If the mass of each object is tripled, the product of their masses becomes 9 times as great, and the gravitational force between them becomes 9 times as great.

9. $p = mv$

The momentum of an object is equal to the product of its mass and its velocity. It is a vector quantity.

EXAMPLE: A cart whose mass is 20. kg moves with a velocity of 5.0 meters per second eastward. What is its momentum?

$$
\begin{aligned}
p &= mv \\
&= (20.\text{ kg})(5.0\text{ m/s}) \\
&= 100\text{ kg}\cdot\text{m/s in the direction of the velocity.}
\end{aligned}
$$

10. $J = F\,\Delta t = \Delta mv;\ m_1v_1 = m_2v_2$

a. $F\,\Delta t = \Delta mv$

The product of the net force acting on an object and the length of time that it acts is equal to the product of the object's mass and its change in velocity. (Note that the first product is called *impulse*, and the second product is called *change of momentum*.)

EXAMPLE: A spring exerts a force of 50.0 newtons on a cart located on a frictionless plane. The cart has a mass of 2.0 kilograms and the force acts for 0.20 second.

$$
\begin{aligned}
\text{J} = \text{Impulse } F\,\Delta t &= (50.0\text{ N})(0.20\text{ s}) \\
&= 10.\text{ N}\cdot\text{s}
\end{aligned}
$$

The change of momentum of the cart is also 10. N · s, and that is the same thing as 10. kilogram-meters per second.

$$
\begin{aligned}
\Delta mv &= 10.\text{ kg}\cdot\text{m/s} \\
(2.0\text{ kg})(\Delta v) &= 10.\text{ kg}\cdot\text{m/s} \\
\Delta v &= 5.0\text{ m/s}
\end{aligned}
$$

b. $m_1v_1 = m_2v_2$

In an explosion of a stationary object that produces two pieces going off in opposite directions, the magnitude of the product of the mass and velocity of one piece is equal to that of the mass and velocity of the other piece.

EXAMPLE: The spring of the above example is actually between the cart in the above example and another cart, pushing with the same force of 50.0 newtons on each cart but in opposite directions. The second cart has a mass of 1.0 kilogram (and gets a change of momentum in the opposite direction).

$$
\begin{aligned}
m_1v_1 &= m_2v_2 \\
(2.0\text{ kg})(5.0\text{ m/s}) &= (1.0\text{ kg})(v_2) \\
v_2 &= 10.\text{ m/s}
\end{aligned}
$$

UNIT II—ENERGY

1. $W = F\Delta s$

When an object moves in the direction of the force applied to it, the work done by the force is equal to the product of the applied force and the distance moved by the object. If they are not in the same direction, we use the component of the force in the direction of the object's motion.

EXAMPLE: If a horizontal force of 30 newtons is used to push an object 40 meters along a horizontal surface, the work done on the object

$$= F\Delta s$$
$$= (30 \text{ N})(40 \text{ m})$$
$$= 1200 \text{ joules}$$

Note: 1 newton-meter = 1 joule

2. $P = \dfrac{W}{\Delta t} = \dfrac{F\Delta s}{\Delta t} = F\bar{v}$

Power is the rate of doing work.

EXAMPLE: An object whose mass is 3.0 kilograms is moved at constant speed by a force of 5.0 newtons. If the object is moved 40. meters in 20. seconds, then

$$\text{the required power} = \frac{\text{work}}{\text{time}}.$$

But the work which is done is equal to the product of the applied force and the distance that the object is moved:

$$W = F\Delta s$$
$$= (5.0 \text{ N})(40. \text{ m})$$
$$= 200 \text{ joules.}$$

$$\text{power} = \frac{\text{work done}}{\text{time required}}$$
$$= \frac{200 \text{ joules}}{20. \text{ s}}$$
$$= 10. \text{ watts.}$$

$$\text{Also, power} = \frac{\text{force} \times \text{distance moved}}{\text{time required}}$$
$$= \frac{5.0 \text{ N} \times 40. \text{ m}}{20. \text{ s}}$$
$$= 10. \text{ watts}$$

(Also note that speed, $\bar{v}$, is equal to the distance moved divided by the time required to move that distance.)

3. $\Delta E_p = mg\Delta h$ ($= \Delta PE$)

The gain (or loss) in gravitational potential energy of an object (E_p) is equal to the product of its mass (m), the value of the acceleration due to gravity (g), and the change in its vertical height above the surface (h).

EXAMPLE: A 3.0-kilogram mass is raised 4.0 meters from a surface. Calculate its gain in gravitational potential energy.

$$E_p = (3.0 \text{ kg})(9.8 \text{ m/s}^2)(4.0 \text{ m})$$
$$= 120 \text{ joules}$$

4. $\Delta E_p = -\Delta E_k$

As an object falls freely, its loss in gravitational potential energy is equal to its gain in kinetic energy.

EXAMPLE: Calculate the kinetic energy gained by a 10.-kilogram mass as its falls freely through a distance of 6.0 meters near the Earth's surface.

$$mg\Delta h = \Delta E_k$$
$$(10.\text{ kg})(9.8 \text{ m/sec}^2)(6.0 \text{ m}) = \Delta E_k$$
$$+\Delta E_k = 590 \text{ joules}$$

5. $KE = \frac{1}{2}mv^2$

The kinetic energy of an object is equal to one-half the product of its mass and the square of its speed.

EXAMPLE: A car with a mass of 1000 kilograms travels with a speed of 20 meters per second.

$$E_k = \frac{1}{2}(1000 \text{ kg})(20 \text{ m/s})^2$$
$$= 200{,}000 \text{ joules}$$

6. $W = \Delta PE + \Delta KE + W_f$ [Not required.]

The work done in raising an object along an incline is equal to the sum of the gain in potential energy, the gain in kinetic energy, and the work done against friction.

EXAMPLE: A block of mass 20. kg is raised from the bottom to the top of an inclined plane 5.0 meters long and 3.0 meters above the ground at the top. How much work must be done by a force parallel to the incline to move the block to the top if the coefficient of friction is 0.20?

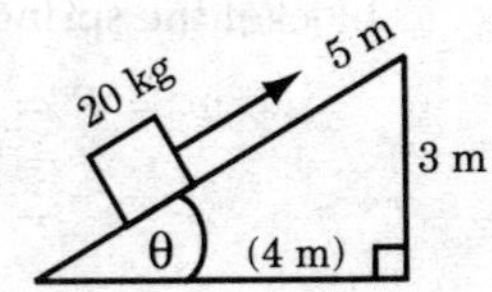

a. The gain in gravitational potential energy is equal to the product of the weight of the object and the vertical height through which it was raised:

$$\begin{aligned} \Delta PE &= \mathrm{mg}\Delta \mathrm{h} \\ &= (20.\ \mathrm{kg})\ (9.8\ \mathrm{m/s^2})\ (3.0\ \mathrm{m}) \\ &= 5.8 \times 10^2\ \text{joules} \end{aligned}$$

b. The change in kinetic energy is equal to one-half the product of the mass of the object and the square of its speed: ($\Delta v = 2.0$ m/s)

$$\begin{aligned} KE &= \frac{1}{2}(20.\ \mathrm{kg})\ (2.0\ \mathrm{m/s})^2 \\ &= 40.\ \text{joules} \end{aligned}$$

c. The work done against friction is equal to the product of the friction and the distance moved. Friction is equal to the product of the coefficient of friction and the normal force between the two surfaces:

$$\begin{aligned} \text{friction} &= mg \cos \theta \\ &= 0.20\ (20.\ \mathrm{kg})\ (9.8\ \mathrm{m/s^2})\ (4/5) \\ &= 31.4\ \text{newtons} \end{aligned}$$

$$\begin{aligned} W_f &= \text{friction} \times \text{distance moved} \\ &= (31.4\ \mathrm{N})\ (5.0\ \mathrm{m}) \\ &= 1.57 \times 10^2\ \text{joules} \end{aligned}$$

d. $$\begin{aligned} \text{Total work} &= (5.8 + 1.6 + 0.4) \times 10^2\ \text{joules} \\ &= 7.8 \times 10^2\ \text{joules} \end{aligned}$$

7. $F = kx$

The change in length of a spring is proportional to the stretching force. (Hooke's Law)

EXAMPLE: A block is suspended from a spring which has a spring constant, k, of 200. newtons per meter. What is the weight of the block if the spring is stretched 0.20 meter?

$$\begin{aligned} w = F &= kx \\ &= (200.\ \mathrm{N/m})\ (0.20\ \mathrm{m}) \\ &= 40.\ \text{newtons} \end{aligned}$$

8. $PE_s = \dfrac{kx^2}{2}$

The potential energy stored in a spring is equal to one-half the product of its spring constant and the square of the distance that the spring is stretched.

EXAMPLE: A block is suspended from a spring and as a result the spring is stretched 0.20 meter. The spring constant is 200. newtons per meter. How much potential energy is stored in the spring?

$$PE_s = \frac{kx^2}{2}$$

$$= (200.\ \text{N/m})\frac{(0.20\ \text{m})^2}{2}$$

$$= 8.0\ \text{J}$$

UNIT III—ELECTRICITY AND MAGNETISM

1. $\dfrac{F = kq_1q_2}{d^2}$

The force between two small charged spheres is proportional to the product of the two charges and inversely proportional to the square of the distance between them, where k is the electrostatic constant.

EXAMPLE: Two small charged spheres are 3.0 meters apart, and each sphere has a charge of 2.0×10^{-6} coulomb. Calculate the magnitude of the force exerted by one sphere on the other.

$$F = \frac{(9.00 \times 10^9\ \text{N}\bullet\text{m}^2/\text{coul}^2)\ (2.0 \times 10^{-6}\ \text{coul})^2}{(3.0\ \text{m})^2}$$

$$= 4.0 \times 10^{-3}\ \text{N}$$

2. $E = \dfrac{F}{q}$

The magnitude of the electric field intensity at a certain point is the force per unit charge at that point.

EXAMPLE: When a charge of 0.04 coulomb is placed at a point in the electric field, the force on the charge is 100 newtons. What is the magnitude of the electric field at that point?

$$E = \frac{100\ \text{N}}{0.04\ \text{coul}}$$

$$= 2500\ \text{N/C}$$

3. $V = \dfrac{W}{q}$

The potential difference between two points is the work required to move a unit charge between the two points.

EXAMPLE: The work required to move a charge of 0.04 coulomb from one point to another in an electric field is 200 joules.

$$\text{potential difference} = \frac{200 \text{ joules}}{0.04 \text{ coulomb}} = 5000 \text{ volts}$$

4. $E = \dfrac{V}{d}$

The magnitude of the electric field intensity between two parallel charged plates is equal to the ratio of the potential difference between the plates and the distance between the plates.

EXAMPLE: The potential difference between two large parallel metal plates is 100. volts, and the distance between the plates is 4.0×10^{-3} meter. What is the magnitude of the electric field intensity?

$$E = \frac{V}{d}$$
$$= \frac{100. \text{ volts}}{4.0 \times 10^{-3} \text{ m}}$$
$$= 25 \times 10^{3} \text{ N/C}$$
$$= 2.5 \times 10^{4} \text{ N/C}$$

Note: 1 volt/meter = 1 newton/coulomb.

5. $V = IR$

The potential difference across a resistor is equal to the product of the current in it and its resistance.

EXAMPLE: A resistor has a resistance of 10. ohms and a current in it of 0.50 ampere. What is the potential difference across the resistor?

$$V = (0.50 \text{ A})(10. \text{ ohms})$$
$$= 5.0 \text{ volts}$$

6. $P = VI = I^2R = \dfrac{V^2}{R}$

a. $P = VI$

The power used by a resistor (or a circuit) is equal to the product of the potential difference across it and the current in it.

EXAMPLE: The potential difference across a resistor is 15. volts and the current in it is 0.50 ampere.

$$\text{The power used by it} = (15.\ \text{volts})\ (0.50\ \text{amperes}) = 7.5\ \text{watts}$$

b. $P = I^2R$

The power used by a resistor (or a circuit) is equal to the product of the square of the current in it and its resistance.

EXAMPLE: The current in a resistor is 3.0 ampere, and its resistance is 10. ohms.

$$\text{The power used in it} = (3.0\ \text{amp})^2\ (10.\ \text{ohms}) = 90.\ \text{watts}$$

c. $P = \dfrac{V^2}{R}$

The power used by a resistor (or a circuit) is equal to the ratio of the square of the potential difference across it and its resistance.

EXAMPLE: The potential difference across a resistor is 100. volts, and its resistance is 50.0 ohms.

$$\text{The power used by it} = \frac{(100.\ \text{volts})^2}{50.0\ \text{ohms}} = 200.\ \text{watts}$$

7. $W = Pt$

The energy used by a device is equal to the product of the power it uses and the length of time that the power is supplied to it.

EXAMPLE: An electric light bulb uses 100 watts of power for 20 minutes. How much energy does it use?

$$\begin{aligned} \text{Energy} &= (\text{power})\ (\text{time}) \\ &= (100\ \text{watts})\ (20\ \text{minutes}) \\ &= 2000\ \text{watt-minutes} \\ &= 120{,}000\ \text{watt-seconds} \\ &= 120{,}000\ \text{joules} \end{aligned}$$

8. $W = Pt = VIt = I^2Rt$

One definition of power is: the rate of using or supplying energy; $P = \frac{W}{t}$. We combine this with the previous equations (item 7) and get $W = Pt$, etc.

a. $W = Pt$

EXAMPLE: An electric toaster rated at 1500 watts is used for 40 seconds. How much energy does it use?

$$W = (1500 \text{ joules per second})(40 \text{ s})$$
$$= 6.0 \times 10^4 \text{ joules}$$

b. $W = VIt$

EXAMPLE: The potential difference across a resistor is 15. volts and the current in it is 0.50 ampere. How much energy is used by the resistor in 5.0 minutes when operating at the rated power?

$$W = (15 \text{ volts})(0.50 \text{ ampere})(300 \text{ seconds})$$
$$= 2.3 \times 10^3 \text{ joules}$$

c. $W = I^2Rt$

EXAMPLE: The current in a resistor is 3.0 ampere, and its resistance is 10. ohms. How much energy is used by the resistor in 5.0 minutes?

$$W = (3.0 \text{ amp})^2 (10. \text{ ohms})(300 \text{ seconds})$$
$$= 2.7 \times 10^4 \text{ joules}$$

9. a. Series Circuit: $I_t = I_1 = I_2 = I_3$

In a series circuit the current supplied by the source of power (or the total current) is the same as the current in each resistor in the circuit.

EXAMPLE: In a series circuit containing three resistors having resistances of 5, 10, and 15 ohms, respectively, we know that the current in the 15-ohm resistor is 0.5 ampere. This tells us that the current is also 0.5 ampere in the 5-ohm and 10-ohm resistors as well as in the generator connected to the circuit.

b. Series Circuit: $V_t = V_1 + V_2 + V_3$

In a series circuit the total (or source) potential difference is equal to the sum of the potential differences (or voltages) across the individual resistors in the circuit.

EXAMPLE: In the circuit described in the previous question, if we know that the voltages across the three resistors are 2.5, 5.0, and 7.5 volts, respectively, then the voltage supplied by the source is the sum of these three values, namely 15. volts.

c. Series Circuit: $R_t = R_1 + R_2 + R_3$

In a series circuit the total resistance of the circuit is equal to the sum of the individual resistances.

EXAMPLE: In the circuit described in the previous two questions, we can add the individual resistances to get the total resistance.

$$\begin{aligned}\text{total resistance} &= (5 + 10 + 15)\text{ ohms}\\ &= 30\text{ ohms}\end{aligned}$$

10. a. Parallel Circuit: $I_t = I_1 + I_2 + I_3$

In a parallel circuit the total current supplied by the source is equal to the sum of the branch currents.

EXAMPLE: A parallel circuit has two branches, a 10-ohm resistor connected parallel to a 20-ohm resistor. If the current in the 10-ohm resistor is 6.0 amperes, and the current in the 20-ohm resistor is 3.0 amperes, how much current is supplied by the source?

$$\begin{aligned}I_t &= 6.0\text{ amp} + 3.0\text{ amp}\\ &= 9.0\text{ amp}\end{aligned}$$

b. Parallel Circuit: $V_t = V_1 = V_2 = V_3$

In a parallel circuit the potential difference of the source is equal to the potential difference across each of the branches.

EXAMPLE: If the potential difference across the 20-ohm resistor is 60. volts, what is the potential difference supplied by the source?

$$\begin{aligned}V_t &= V_1\\ &= 60.\text{ volts}\end{aligned}$$

c. Parallel Circuit: $\frac{1}{R_t} = \frac{1}{R_1} + \frac{1}{R_2} + \frac{1}{R_3}$

In a parallel circuit the reciprocal of the total resistance (or equivalent resistance) is equal to the sum of the reciprocals of the individual resistances.

EXAMPLE: In the circuit described in the previous two questions, what is the total resistance of the circuit?

$$\frac{1}{R_t} = \frac{1}{10 \text{ ohms}} + \frac{1}{20 \text{ ohms}}$$

$$\frac{1}{R_t} = \frac{3}{20 \text{ ohms}}$$

$$R_t = 6.7 \text{ ohms}$$

UNIT IV—WAVE PHENOMENA

1. $T = \dfrac{1}{f}$

The period of a wave is equal to the reciprocal of its frequency. (It is the time it takes for one cycle to occur.)

EXAMPLE: If the frequency of a wave is 0.25 cycle per second, what is the period of the wave?

$$T = \frac{1}{f} = \frac{1}{0.25 \text{ Hz}} = 4.0 \text{ s}$$

2. $v = f\lambda$

The speed of a wave is equal to the product of its frequency and its wavelength.

EXAMPLE: The frequency of a wave is 2.0 cycles per second, and its wavelength is 0.03 meter.

$$\text{Its speed} = (2.0 \text{ Hz})\cdot(0.03 \text{ m}) = 0.06 \text{ m/s}$$

3. $v = \dfrac{c}{n}$; $n = \dfrac{c}{v}$

The speed of light in a material medium is equal to the ratio of the speed of light in a vacuum to the index of refraction of the material.

EXAMPLE: A ray of monochromatic light travels from air into a material whose index of refraction is 1.4. Calculate the speed of light in this material.

The speed of light in air is very close to its speed in a vacuum. Therefore,

$$v = \frac{3.0 \times 10^8 \text{ m/s}}{1.4}$$
$$= 2.1 \times 10^8 \text{ m/s}$$

4. $\sin \theta_c = \dfrac{1}{n}$

The sine of the critical angle of incidence is equal to the reciprocal of the index of refraction.

EXAMPLE: Light travels from benzene into air. Calculate the sine of the critical angle for benzene.

$$\sin \theta_c = \frac{1}{1.50}$$
$$= 0.67$$

5. $\dfrac{\sin \theta_1}{\sin \theta_2} = \dfrac{n_2}{n_1}$ *or* $n_1 \sin \theta_1 = n_2 \sin \theta_2$

When light goes obliquely from medium 1 to medium 2, the ratio of the sine of the angle of incidence to the sine of the angle of refraction is equal to the ratio of the index of refraction of medium 2 to the index of refraction of medium 1.

EXAMPLE: If the angle of incidence in air is 30°, what is the sine of the angle of refraction in crown glass?

$$\frac{\text{sine of angle in air}}{\text{sine of angle in crown glass}} = \frac{\text{index of refraction of crown glass}}{\text{index of refraction of air}}$$

$$\frac{\sin 30°}{\sin \theta_2} = \frac{1.52}{1.00}$$

$$\frac{0.50}{\sin \theta_2} = 1.52$$

$$\sin \theta_2 = 0.33$$

6. $\dfrac{n_2}{n_1} = \dfrac{v_1}{v_2}$ *or* $n_1 v_1 = n_2 v_2$

The speed of light in two media varies inversely as their index of refraction.

EXAMPLE: If the speed of light in a medium is 1.5×10^8 meters per second, what is the index of refraction of the medium?

In vacuum (or air), whose index of refraction (n_1) is taken as 1.00, the speed of light is 3.0×10^8 meters per second.

$$\frac{n_2}{n_1} = \frac{v_1}{v_2}$$

where n_1 is the index of refraction in medium 1, and v_1 is the speed of light in the same medium. Substituting, we get

$$\frac{n_2}{1.0} = \frac{3.0 \times 10^8 \text{ m/s}}{1.5 \times 10^8 \text{ m/s}}$$

$$n_2 = 2.0$$

7. $\lambda = \frac{dx}{L}$ *or* $\frac{\lambda}{d} = \frac{x}{L}$

This applies to the double-slit experiment where λ is the wave length of the light, d is the distance between the slits, x is the distance of the first bright fringe from the central maximum, and L is the distance from the double-slit barrier to the screen.

EXAMPLE: In a double-slit experiment the distance between the two slits is 1.0×10^{-3} meter, the distance of the first bright fringe from the central maximum is 1.2×10^{-3} meter, and the distance from the slits to the screen is 2.0 meters. Calculate the wavelength of the light used in the experiment.

$$\lambda = \frac{(1.0 \times 10^{-3} \text{ m})(1.2 \times 10^{-3} \text{ m})}{2.0 \text{ m}}$$

$$= 6.0 \times 10^{-7} \text{ m}$$

UNIT V—MODERN PHYSICS

1. $W_o = hf_o$

The work function of a material is equal to the product of Planck's constant and its threshold frequency.

EXAMPLE: When the frequency of monochromatic light which shines on a certain metal is raised from 2.0×10^{15} hertz to 2.47×10^{15} hertz, electrons are barely emitted by the metal. What is the work function of the metal?

$$W_o = hf_o$$

$$= (6.6 \times 10^{-34} \text{ J} \cdot \text{s})(2.47 \times 10^{15} \text{ Hz})$$

$$= 1.6 \times 10^{-18} \text{ joules}$$

2. $E_{photon} = hf$

The energy of a photon is equal to the product of Planck's constant (h) and the frequency of the light (f).

EXAMPLE: If the frequency of a monochromatic light beam is 1.2 $\times$ 10^{15} cycles per second, what is the energy of each of its photons?

$$\begin{aligned} E &= (6.6 \times 10^{-34} \text{ joule-sec})(1.2 \times 10^{15} \text{ Hz}) \\ &= 7.9 \times 10^{-19} \text{ joule.} \end{aligned}$$

Note: $E_{photon} = hc/\lambda$

The energy of a photon is equal to the ratio of the product of Planck's constant and the speed of light to the wavelength of the light.

EXAMPLE: Calculate the energy of a photon of monochromatic orange light whose wavelength is 6.0×10^{-7} m.

$$\begin{aligned} E_{photon} &= (6.63 \times 10^{-34} \text{ J} \cdot \text{s})(3.00 \times 10^{8} \text{ m/s})/6.0 \times 10^{-7} \text{ m} \\ &= 3.3 \times 10^{-19} \text{ joule} \end{aligned}$$

3. $KE_{max} = hf - W_o$

When light shines on a photoemissive material, the maximum kinetic energy of the emitted electron (E_k) is equal to the difference between the energy of the absorbed photon (hf) and the work function of the material W_o.

EXAMPLE: When a photoemissive material with a work function of 1.3×10^{-19} joule is exposed to incident photons with an energy of 3.3 $\times$ 10^{-19} joule, what is the maximum kinetic energy of an ejected photoelectron?

$$\begin{aligned} \text{maximum kinetic energy} &= (3.3 \times 10^{-19} \text{ joule}) - (1.3 \times 10^{-19} \text{ joule}) \\ &= 2.0 \times 10^{-19} \text{ joule} \end{aligned}$$

4. $p = h/\lambda$

The momentum of a photon is equal to the ratio of Planck's constant to the wavelength of the photon.

EXAMPLE: Calculate the momentum of the photon of orange light described in Example 2, Note.

$$\begin{aligned} p &= h/\lambda \\ &= (6.63 \times 10^{-34} \text{ J} \cdot \text{s})/6.0 \times 10^{-7} \text{ m} \\ &= 1.1 \times 10^{-27} \text{ kg} \cdot \text{m/s} \end{aligned}$$

5. $E_{photon} = E_i - E_f$

When an atom jumps from one energy level to another, it emits (or absorbs) a photon whose energy is equal to the difference between the energies of the two levels.

EXAMPLE: A hydrogen atom jumps from an energy level of −0.85 eV to an energy level of −13.6 eV. What is the energy of the photon emitted as a result of this transition?

$$\begin{aligned} E_{photon} &= (-0.85\text{ eV}) - (-13.6\text{ eV}) \\ &= 12.75\text{ eV} \end{aligned}$$

UNIT VI—MOTION IN A PLANE (GROUP 1)

1. $v_{iy} = v_i \sin \theta_i$

2. $v_{ix} = v_i \cos \theta_i$

These two equations can be used for projectile motion. The first one gives the vertical component of the initial velocity, and the second one gives the horizontal component. The horizontal component remains constant throughout the flight of the projectile if friction is negligible. The vertical component is affected by the downward acceleration due to gravity: $v_y = v_{iy} + gt$.

EXAMPLE: A projectile is fired in the horizontal direction with an initial velocity of 100. meters per second. If friction is neglected, (a) what is the speed of the projectile in the horizontal direction initially and after two seconds of flight? (b) Calculate the same for the vertical component.

The angle is zero degrees. The cosine of 0° = 1 ; the sine of 0° = 0.

(a) $v_{ix} = v_i = 100$ m/s throughout the flight.

(b) $v_{iy} = (100\text{ m/s})(0) = 0$, at the beginning. After 2 seconds,

$$\begin{aligned} v_{iy} &= 0 + gt \\ &= (9.8\text{ m/s}^2)(2\text{ s}) \\ &= 19.6\text{ m/s after two seconds of flight.} \end{aligned}$$

3. $a_c = \dfrac{v^2}{r}$

4. $F_c = ma_c = \dfrac{mv^2}{r}$

The centripetal force acting on an object moving at constant speed around a circle is equal to the product of the object's mass and its centripetal acceleration; this acceleration is equal to the ratio of the square of the object's speed and the radius of the circle.

EXAMPLE: An object moves around a circle whose radius is 2.0 meters. The constant speed of the object is 6.0 meters per second and the mass of the object is 0.20 kilogram.

$$\text{centripetal acceleration } a_c = \frac{v^2}{r}$$
$$= \frac{(6.0 \text{ m/s})^2}{2.0 \text{ m}}$$
$$= 18. \text{ m/s}^2$$

$$\text{centripetal force } F_c = ma_c \left(= \frac{mv^2}{r} \right)$$
$$= (0.20 \text{ kg})(18. \text{ m/s}^2)$$
$$= 3.6 \text{ newtons}$$

5. $\frac{R_1^3}{T_1^2} = \frac{R_2^3}{T_2^2}$ [Not required.]

This is Kepler's Third Law which states that the periods or times (T) required for any two planets of the sun to complete their elliptical paths are related to their average or mean distances from the sun (R) by the above proportion.

EXAMPLE: Imagine a solar system which has two planets. The more distant one from its sun, on the average, is four times as far as the nearer one, which takes two of our years to revolve once around its sun. How long is the period of the more distant planet?

$$\frac{R_1^3}{T_1^2} = \frac{R_2^3}{T_2^2}$$
$$\frac{4^3}{T_1^2} = \frac{1^3}{2^2}$$
$$T_1 = 16 \text{ years}$$

UNIT VII—INTERNAL ENERGY (GROUP 2)

1. $\Delta Q = mc\Delta T_c$

The heat required to raise the temperature of a substance without change of phase is equal to the product of its mass, specific heat, and change in temperature.

EXAMPLE: A large container contains 20. kilograms of water at

20°C. How much heat is required to bring the water to its boiling point? (Neglect heat absorbed by the container.)

$$\Delta Q = (20 \text{ kg})(4.19 \times 10^3 \text{ J/kg} \cdot {}^\circ\text{C})(100^\circ - 20^\circ)$$
$$= 6.7 \times 10^6 \text{ J}$$

2. $Q_f = mH_f$

The heat required to melt or fuse a substance at its melting point is equal to the product of its mass and its heat of fusion.

EXAMPLE: How much heat is required to change 10. kilograms of solid tungsten at its melting point (3370°C) to liquid tungsten?

$$Q_f = (10. \text{ kg})(192 \times 10^3 \text{ J/kg})$$
$$= 1.9 \times 10^6 \text{ J}$$

3. $Q_v = mH_v$

The heat required to vaporize a substance at its boiling point is equal to the product of the mass vaporized and its heat of vaporization.

EXAMPLE: How much heat is required to vaporize 20. kilograms of water at its boiling point (100°C)?

$$Q_v = (20 \text{ kg})(2260 \times 10^3 \text{ J/kg})$$
$$= 4.5 \times 10^7 \text{ J}$$

UNIT VIII—ELECTROMAGNETIC APPLICATIONS (GROUP 3)

1. $F = qvB$

When an electric charge moves at right angles through a magnetic field, the magnetic force acting on it has a magnitude equal to the product of the magnitude of the charge (q), its speed of motion, and the magnetic flux density.

EXAMPLE: A beam of electrons moves with a speed of 3.0×10^6 meters per second at right angles through a magnetic field whose flux density is 5.0×10^{-5} weber per square meter. What is the magnitude of the force acting on each electron?

$$F = qvB$$
$$= (1.6 \times 10^{-19} \text{ C})(3.0 \times 10^6 \text{ m/s})(5.0 \times 10^{-5} \text{ Wb/m}^2)$$
$$= 2.4 \times 10^{-17} \text{ newtons}$$

(**Note:** one *tesla* (T) = one weber/m^2)

2. $\frac{N_p}{N_s} = \frac{V_p}{V_s}$

In a transformer, the ratio of the number of turns in the primary coil to the number of turns in the secondary coil is equal to the ratio of the voltage applied to the primary coil to the voltage induced in the secondary coil.

EXAMPLE: A transformer has 60 turns on the primary winding and 300 turns on the secondary winding. If 120 volts AC are supplied to the primary winding, what is the voltage induced in the secondary winding?

$$\frac{(N_p)}{(N_s)} = \frac{(V_p)}{(V_s)}$$
$$\frac{60}{300} = \frac{(120\ V)}{V_s}$$
$$V_s = 600 \text{ volts}$$

3. $V_sI_s = V_pI_p$

EXAMPLE: In the above transformer, if its efficiency is 100% and its power output is 100 watts, what is the current in the primary coil?

$$V_sI_s = V_pI_p$$
$$100\ W = (120\ V)\ I_p$$
$$I_p = 0.83 \text{ ampere}$$

4. $(V_sI_s)/V_pI_p) = \text{efficiency}$

EXAMPLE: In the previous example, if the efficiency of the transformer is 90%, what is the power supplied to the primary?

$$(100 \text{ watts})/V_pI_p = 0.90$$
$$V_pI_p = 110 \text{ watts}$$

5. $V = Blv$

If a wire is moved at right angles to the direction of a magnetic field, an electric potential difference is induced in the wire that is of a magnitude equal to the product of the magnetic flux density (B), the length of wire in the field (1), and the speed with which the wire moves (v).

EXAMPLE: If a wire whose length is 1.0 meter is moved with a speed of 10. meters per second at right angles to a magnetic field whose flux density is 8.0 webers per square meter, what is the potential difference induced in the wire?

$$\begin{aligned} V &= Blv \\ &= (8.0\ \text{Wb/m}^2)\,(1.0\ \text{m})\,(10.\ \text{m/s}) \\ &= 80.\ \text{volts} \end{aligned}$$

UNIT IX—GEOMETRIC OPTICS (GROUP 4)

1. $\dfrac{1}{d_o} + \dfrac{1}{d_i} = \dfrac{1}{f}$

For a lens,

$$\frac{1}{\text{object distance}} + \frac{1}{\text{image distance}} = \frac{1}{\text{focal length}}$$

EXAMPLE: The object is located 3.0 meters from a converging lens. The image is formed on the other side of the lens, 1.5 meters away from it. Calculate the focal length of the lens.

$$\frac{1}{3.0\ \text{m}} + \frac{1}{1.5\ \text{m}} = \frac{1}{f}$$

$$f = 1.0\ \text{m}$$

2. $\dfrac{s_i}{s_o} = \dfrac{d_i}{d_o}$

For any lens,

$$\frac{\text{size of image}}{\text{size of object}} = \frac{\text{distance of image from lens}}{\text{dist. of object from lens}}$$

EXAMPLE: An object placed 0.40 meter from the lens produces an image located 1.20 meters from the lens. If the size of the object is 0.05 meter, calculate the size of the image.

$$\frac{\text{size of image}}{0.05\ \text{m}} = \frac{1.20\ \text{m}}{0.40\ \text{m}}$$

$$\text{size of image} = 0.15\ \text{m}$$

3. Note that the equations used above also apply to spherical mirrors except that the calculations for the convex lens are applicable to the concave mirror, and those for the concave lens are applicable to the convex mirror. Also note that for the spherical mirror, the focal length is equal to one-half of the sphere's radius.

EXAMPLE: An object is located 6.0 meters from a concave spherical mirror whose focal length is 2.0 meters. Where will the image be?

$$\frac{1}{6.0 \text{ m}} + \frac{1}{d_i} = \frac{1}{2.0 \text{ m}}$$

$$d_i = 3.0 \text{ meters}$$

The image is located 3.0 meters from the mirror on the same side from the mirror as the object. Also note that the radius of the spherical surface is equal to $2f$, or 4.0 meters.

UNIT X—SOLID-STATE PHYSICS (GROUP 5)

There are no equations for this unit.

UNIT XI—NUCLEAR ENERGY (GROUP 6)

1. $E = mc^2$

The energy equivalence (E) of a given amount of mass is equal to the product of this mass (m) and the square of the speed of light (c).

EXAMPLE: What is the energy equivalence, in joules, of 1.0×10^{-3} kilogram of matter?

$$\begin{aligned} E &= (1.0 \times 10^{-3} \text{ kg})(3.0 \times 10^{8} \text{ m/s})^2 \\ &= 9.0 \times 10^{13} \text{ joules} \end{aligned}$$

(In some respects this conversion is similar to expressing a given amount of heat energy, in kilocalories, to an equivalent amount of mechanical energy, in joules.)

2. $m_f = \dfrac{1}{2^n} m_i$

A radioactive isotope decays at such a rate that the mass of isotope that is left after n half-lives (m_f) is equal to the product of the initial mass (m_i) and one-half raised to the nth power.

EXAMPLE: The half-life of a certain radioactive element is 15 minutes. What fraction of a given sample of this element will be left after 4 half-lives (1 hour)?

$$\begin{aligned} m_f &= \left(\frac{1}{2}\right)^4 m_i \\ &= \text{one-sixteenth of the initial mass} \end{aligned}$$

Practice Questions and Answers for Part III

1. In a laboratory exercise, a student collected the following data as the unbalanced force applied to a body of mass M was changed.

Data Table

Force (newtons)	Acceleration (meters per second^2)
4.0	2.1
8.0	4.0
12.0	6.0
16.0	7.9
20.0	10.0

a Label the axes of the graph with the appropriate values for force and acceleration.

b Plot an acceleration versus force graph for the laboratory data provided.

c Using the data or your graph, determine the mass, M, of the body. [Show all calculations.]

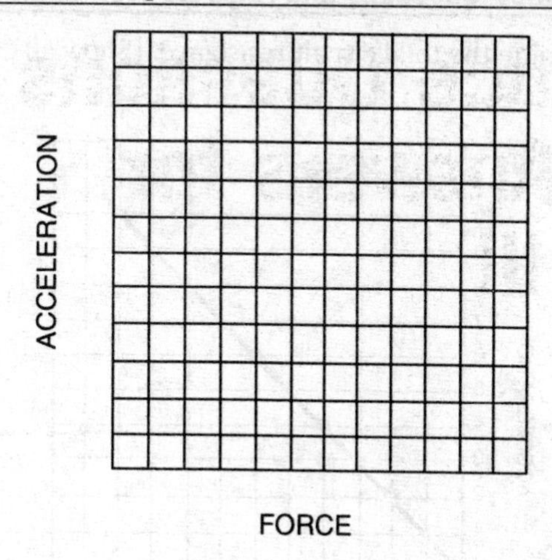

2. *a* *In the space provided*, draw a circuit diagram showing the following elements connected in parallel:

Elements

One 12.0-volt battery
One 2.0-ohm resistor
One 3.0-ohm resistor

Place an ammeter in the circuit to read the total current. Use the symbols shown below. [Assume availability of any number of wires of negligible resistance.]

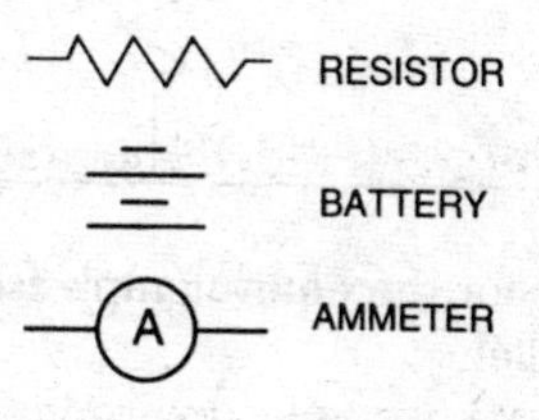

b Determine the total circuit resistance. [Show all calculations.]

1. *a* and *b*

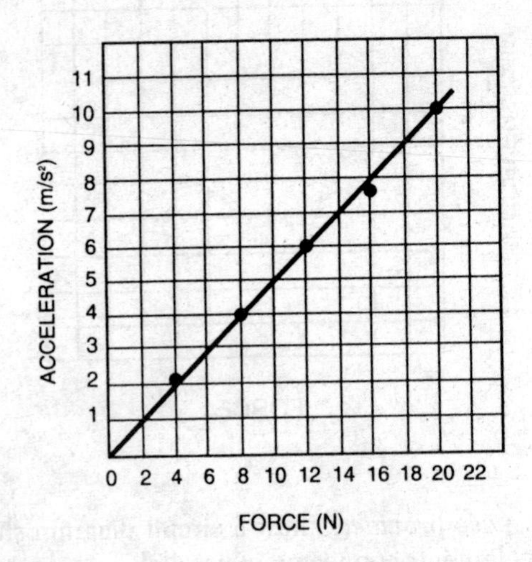

c Applying Newton's second law ($F = ma$):

$$
\begin{aligned}
M &= (F/a) \\
&= (20.0\ \text{N})/(10.0\ \text{m/s}^2) \\
&= 2.0\ \text{kg}
\end{aligned}
$$

2. *a*

A

12.0 V

2.0 Ω

3.0 Ω

b A convenient rule for the equivalent resistance of two resistors connected in parallel:

$$
\begin{aligned}
R_T &= (R_1 R_2)/(R_1 + R_2) \\
&= (2.0\ \text{ohms} \times 3.0\ \text{ohms})\ /\ (2.0\ \text{ohms} + 3.0\ \text{ohms}) \\
&= 1.2\ \text{ohms}
\end{aligned}
$$

Glossary of Terms

absolute index of refraction The ratio of the speed of light in a vacuum to the speed of light in a medium.
absolute temperature The temperature as measured on the Kelvin scale; a measure of the average kinetic energy of the molecules of a body.
absolute zero The temperature at which the internal energy of an object is at a minimum (0 K or –273°C).
absorption spectrum A series of dark spectral lines or bands formed by the absorption of specific wavelengths of light by atoms or molecules.
acceleration The time rate of change in velocity. The SI unit is meters per second^2.
acceptor material A substance used to dope a semiconductor in order to increase the number of positive holes in the semiconductor.
accuracy The agreement of a measured value with an accepted standard.
alpha decay A natural radioactive process that results in the emission of an alpha particle from a nuclide.
alpha particle A helium nucleus; a particle consisting of two protons and two neutrons.
alternating current An electric current that varies in magnitude and alternates in direction.
ammeter A device used to measure electric current. It is constructed by placing a low-resistance shunt across the coil of a galvanometer.
ampere (A) The SI unit of electric current, equivalent to the unit coulomb per second.
amplitude The maximum displacement in periodic phenomena such as wave motion, pendulum motion, and spring oscillation.
angle of incidence The angle made by the incident wave with the surface of a medium; the angle made by the incident ray with the normal to the surface of the medium.

angle of reflection The angle made by the reflected wave with the surface of a medium; the angle made by the reflected ray with the normal to the surface of the medium.

angle of refraction The angle made by the refracted wave with the surface of a medium; the angle made by the refracted ray with the normal to the surface of the medium.

anode The positive terminal of a DC source of potential difference.

antimatter One or more atoms composed entirely of antiparticles.

antinode The point or locus of points on an interference pattern (such as a standing wave or double slit pattern) that results in maximum constructive interference.

antiparticle The counterpart of a subatomic particle. An antiparticle has the same mass as its companion particle, but its electric charge is opposite in sign.

atomic mass unit (u) A unit of mass defined as one-twelfth the mass of an atom of carbon-12.

atomic number The number of protons in the nucleus of an atom. The atomic number defines the element.

avalanche The breakdown of a semiconductor diode caused by placing an excessively large potential difference across it when it is reverse biased.

back emf The potential difference that develops in a circuit that opposes the potential difference of the source. A back emf arises as a consequence of Lenz's law.

Balmer series The visible-ultraviolet line spectrum of atomic hydrogen. It is the result of electrons falling from higher levels to the $n = 2$ state.

base The thin middle portion of a bipolar transistor.

battery A combination of two or more electric cells.

beta (–) decay A natural radioactive process that results in the emission of a beta (–) particle from a nuclide.

beta (–) particle An electron formed in the nucleus by the disintegration of a neutron.

beta (+) decay A natural radioactive process that results in the emission of a beta (+) particle from a nuclide.

beta (+) particle A positron, the antiparticle of the electron, formed in the nucleus by the disintegration of a proton.

binding energy The energy equivalent of the mass defect of a nucleus.

boiling The condition in which the liquid and gaseous phases of matter are in equilibrium. A synonym for boiling is *vaporization*.

Boyle's law The volume of an ideal gas is inversely proportional to the pressure at constant temperature.

breeder reactor A fission reactor that produces additional fissionable fuel as a result of neutron bombardment of uranium-238.

cathode The negative terminal of a DC source of potential difference.

cathode ray tube A device for visualizing an electron beam. It consists of an evacuated tube with a source of electrons at one end and a fluorescent screen at the other end. The electron beam is controlled by electric and magnetic fields.

celsius scale (°C) The temperature scale that fixes the (atmospheric) freezing point of water at 0° and the boiling point of water at 100°.

center of curvature In spherical mirrors and lenses, the point on the principal axis that is located a distance of one radius from the center of the mirror or lens.

centripetal acceleration The acceleration that is directed along the radius and toward the center of a curved path in which an object is moving.

centripetal force The force that causes centripetal acceleration. It is responsible for changing an object's direction, not its speed.

chain reaction In nuclear fission, a continuing series of reactions that results from bombardment by neutrons that are, themselves, fission products.

Charles's law The volume occupied by an ideal gas is directly proportional to the Kelvin temperature at constant pressure.

chromatic aberration A lens defect in which different colors of light are focused at different points.

circuit A closed loop formed by a source of potential difference connected to one or more resistances.

cloud chamber A device that tracks the path of a charged particle by condensing an undercooled gas into a liquid.

coefficient of kinetic friction The ratio of the force of kinetic friction on an object to the normal force on it.

coherent light A series of light waves that have a fixed phase relationship; the type of light produced by a laser. Lasers produce beams of monochromatic coherent light.

collector The outer layer in a bipolar transistor, which receives the majority of charge carriers.

commutator A split ring that is connected to the armature coil of a DC motor or generator, causing the current to reverse direction with each half-turn.

component One of the two or more vectors into which a given vector may be resolved.

concave lens A lens that is thinner in the middle than at the edges.

concave mirror A curved mirror whose reflecting surface is the inner surface of the curve.

concurrent forces Two or more forces acting at the same point.

conduction band The energy band that contains the electrons that are able to carry an electric current.

conductivity The reciprocal of a material's resistivity. The SI unit of conductivity is the mho, which is equivalent to the $ohm^{-1} \cdot meter^{-1}$.
conductor A material that allows electrons to flow through it freely. Metals such as copper and silver are conductors.
constructive interference The combination of two in-phase wave disturbances to produce a single wave disturbance whose amplitude is the sum of the amplitudes of the individual disturbances.
control rod A device used to regulate the rate of a nuclear chain reaction by absorbing neutrons.
converging lens A lens that focuses its transmitted light to a point. Generally, convex lenses are converging lenses.
converging mirror A mirror that focuses its reflected light to a point. Generally, concave mirrors are converging mirrors.
convex lens A lens that is thicker in the middle than at the edges.
convex mirror A curved mirror whose reflecting surface is the outer surface of the curve.
coolant A circulating fluid that removes and transfers the heat energy generated by devices such as fission reactors.
core The part of a fission reactor that is the focus of the fission reaction.
coulomb (C) The SI unit of electric charge, approximately equal to 6.25×10^{18} elementary charges.
Coulomb's law The electrostatic force between two point charges is directly proportional to the product of the charges and inversely proportional to the square of the distance between the charges.
critical angle The angle of incidence for which the corresponding angle of refraction is 90°.
critical mass The minimum amount of a fissionable nuclide needed to sustain a chain reaction.
cycle One complete repetition of the pattern in any periodic phenomenon.
cyclotron One of a number of devices that accelerate charged nuclear particles by using electric and magnetic fields.

de Broglie wavelength The wavelength of a matter wave.
destructive interference The combination of two out-of-phase wave disturbances to produce a single wave disturbance whose amplitude is the difference of the amplitudes of the individual disturbances.
deuterium An isotope of hydrogen that contains one proton and one neutron in its nucleus.
diode A device that permits charge to flow in one direction only. A semiconductor diode consists of a *P*-type semiconductor joined to an *N*-type semiconductor.
diffraction The bending of a wave around a barrier.
diffuse reflection The reflection of parallel light rays by irregular surfaces.

direct current An electric current that flows in one direction only.
dispersion The separation of polychromatic light into its individual colors.
dispersive medium A medium in which the speed of a wave depends on its frequency.
displacement A change of position in a specific direction.
diverging lens A lens that spreads its transmitted light outward. Generally, concave lenses are diverging lenses.
diverging mirror A mirror that spreads its reflected light outward. Generally, convex mirrors are diverging mirrors.
donor material A substance used to dope a semiconductor in order to increase the number of electrons in the semiconductor.
doping The process of inserting small amounts of certain impurities into a semiconductor in order to increase the number of electrons or positive holes that it contains.
Doppler effect An apparent change in frequency that results when a wave source and an observer are in relative motion with respect to each other.

efficiency The degree to which a device such as a transformer or machine compares with an ideal device. Efficiency is usually expressed as a percentage.
Einstein's postulates of special relativity
(1) The laws of physics are valid in all inertial frames of reference.
(2) The speed of light has the same value in all frames of reference.
elastic potential energy The energy stored in a spring when it is compressed or stretched.
electric current The time rate of flow of charged particles. The SI unit of electric current is the ampere (A).
electric field The region of space around a charged object that affects other charges.
electric field barrier The region in a semiconductor, established by the combination of holes and electrons, that prevents further migration of charge carriers across the *P-N* junction.
electric field intensity The ratio of the force that an electric field exerts on a charge to the magnitude of the charge.
electric motor A device that converts electrical energy into mechanical energy.
electric potential The total work done by an electric field in bringing 1 coulomb of positive charge from infinity to a specific point. The potential is a positive number if the charge is repelled by the field and a negative number if the charge is attracted by the field. At infinity, the potential is taken to be zero. Electric potential is measured in volts.
electromagnet A solenoid whose magnetic field is intensified by the insertion of certain ferromagnetic materials.

electromagnetic induction The process by which the magnetic field and the mechanical energy are used to generate a potential difference.
electromagnetic radiation The propagation of electromagnetic waves in space.
electromagnetic spectrum The entire range of electromagnetic waves from the lowest to the highest frequencies.
electromagnetic wave A periodic wave, consisting of mutually perpendicular electric and magnetic fields, that is radiated away from the vicinity of an accelerating charge.
electromotive force The potential difference produced as a result of the conversion of other forms of energy into electrical energy.
electron A fundamental, negatively charged, subatomic particle.
electron capture A radioactive process in which a nucleus absorbs one of an atom's innermost electrons.
electron cloud In quantum theory, the region of space where an electron is most likely to be found.
electron-volt (eV) A unit of energy equal to the work needed to move an elementary charge across a potential difference of 1 volt.
electroscope A device used to detect the presence of electric charges.
elementary charge The magnitude of charge present on a proton or an electron. An elementary charge is approximately equal to 1.6×10^{-19} coulomb.
emission spectrum A series of bright spectral lines or bands formed by the emission of certain wavelengths of light by excited atoms falling to lower energy states.
emitter One of the outer layers of a bipolar transistor, which supplies the charge carriers to the rest of the transistor.
energy A quantity related to work.
entropy A measure of the disorder or randomness present in a system.
equilibrant A single balancing force that maintains the static equilibrium of an object.
equivalent resistance A single resistance that can be substituted for a group of resistances in series or in parallel.
ether A hypothetical medium whose existence was proposed as the carrier of all electromagnetic waves.
excited state A condition in which the energy of an atom is greater than its lowest energy state.
extrinsic semiconductor Silicon or germanium that has been made semiconductive by doping.

farad (F) The SI unit of capacitance, equivalent to the unit coulomb per volt.
ferromagnetic Referring to a material, such as iron, that has the ability to strengthen greatly the magnetic field of a current-carrying coil.

field lines A series of lines used to represent the magnitude and direction of a field.

first law of thermodynamics A change in the internal energy of a system is equal to the difference of the heat energy absorbed by the system and the work done by it. It is a statement of the law of conservation of energy.

fission The process of splitting a heavy nucleus, such as uranium-235, into lighter fragments. Fission is accompanied by the release of large quantities of energy.

flux density The number of magnetic field lines per unit area. The flux density is one way of measuring the strength of a magnetic field.

focal length The distance between the center of a lens or mirror and its focal point.

focal point The point of a lens or mirror where incoming parallel light rays meet. The focal point is also known as the *principal focus*.

force A push or a pull on an object. If the force is unbalanced, an acceleration will result.

forward bias A potential difference applied across a *P-N* junction in a direction that facilitates both electron and hole flow across the junction.

frame of reference A coordinate grid and a set of synchronized clocks that can be used to determine the position and time of an event.

free fall A motion in the Earth's gravitational field without regard to air resistance.

frequency The number of repetitions produced per unit time by periodic phenomena.

friction The force present as the result of contact between two surfaces. The direction of a frictional force is opposite to the direction of motion.

fuel rods Rods packed with nuclear fuel pellets and placed in the core of a fission reactor, which is the source of energy from the fission reaction.

fusion (1) The process of uniting lighter nuclei, such as deuterium, into a heavier nucleus. Fusion is accompanied by the release of large quantities of energy. (2) In the study of heat and thermodynamics, a synonym for *melting*.

Gallilean-Newtonian relativity principle The laws of mechanics are valid in all inertial frames of reference.

galvanometer A device, consisting of a coil-shaped wire placed between the opposite poles of a permanent magnet, that is used to detect small amounts of electric current.

gamma radiation Very high energy photons of electromagnetic radiation. Gamma photons have the highest frequencies in the electromagnetic spectrum.

geiger counter A device that detects charged nuclear particles.

generator A device that uses a magnetic field and mechanical energy to induce a source of electromotive force.

geosynchronous orbit An orbit in which the period of a satellite is equal to the period of the Earth's rotation (approximately 24 hours). A satellite in a geosynchronous orbit remains continually in the same position above the Earth's surface.

gravitational field The region of space around a mass that affects other masses.

gravitational field intensity The ratio of the force that a gravitational field exerts on a mass to the magnitude of the mass, numerically equal to the acceleration due to gravity.

gravitational force The universal attraction between two masses.

gravitational potential energy The energy that an object acquires as a result of the work done in moving the object against a gravitational field.

ground An extremely large source or reservoir of electrons, which can supply or accept electrons as the need arises.

ground state The lowest energy state of an atom.

half-life The time needed reduce the number of radioactive nuclei present in a sample to one-half its initial value.

heat energy The energy that is transferred from warmer objects to cooler ones because of a temperature difference between them.

heat of fusion The quantity of heat energy needed to convert a unit mass of a solid entirely into a liquid at its melting temperature.

heat of vaporization The quantity of heat energy needed to convert a unit mass of a liquid entirely into a gas at its boiling temperature.

hertz (Hz) The SI unit of frequency, equivalent to the unit $second^{-1}$.

hole An electron vacancy in a semiconductor. In an electric field, holes behave like positive charges.

ideal gas A simplified model of a gas in which the molecules have mass, but no volume, and do not exert forces except during collisions. Real gases may approximate ideal behavior at low pressure and high temperature.

image An optical reproduction of an object.

impulse The product of the net force acting on an object and the time during which the force acts. The impulse delivered to an object is equal to its change in momentum. The direction of the impulse is the direction of the force. The SI unit of impulse is the newton · second, which is equivalent to the kilogram · meter per second.

incident ray A ray of a wave impinging on a surface.

incident wave A wave impinging on a surface.

induced current An electric current that is the result of an induced electromotive force.

induced emf A potential difference created when a magnetic field is interrupted over a time period.
induction (1) A method of charging a neutral object by using a charged object and a ground. The induced charge is always opposite to the charge on the charged object. (2) See *induced current* and *induced emf.*
inertia The property of matter that resists changes in motion. Mass is the quantitative measure of inertia.
inertial frame of reference A frame of reference in which Newton's first law of motion is valid.
instantaneous velocity The ratio of displacement to time at any given instant; the slope of a line tangent to a displacement-time graph at any given point.
insulator A material that is a very poor conductor because it has few conduction electrons. Wood and glass are examples of insulators.
integrated circuit A miniaturized circuit containing an array of semiconductor devices.
interference pattern Regions of constructive and destructive interference that are present in a medium as a result of the combination of two or more waves.
internal energy The total kinetic and potential energy associated with the atoms and molecules of an object.
ionization energy See *ionization potential*.
ionization potential The quantity of energy needed to remove a single electron from an atom or ion.
isotopes Atoms with identical atomic numbers but different mass numbers. Two isotopes of the same element have identical numbers of protons but different numbers of neutrons.

joule (J) The SI unit of work and energy, equivalent to the unit newton · meter.

k-capture Electron capture.
Kelvin (K) The SI unit of temperature.
Kelvin scale (K) The absolute temperature scale. The single fixed point on the Kelvin scale is the triple point of water, which is set at 273.16K.
Kepler's first law The orbits of all the planets are elliptical, with the Sun placed at one focus of the ellipse.
Kepler's second law A line from the Sun to a planet sweeps out equal areas in equal periods of time.
Kepler's third law The ratio of the cube of the mean radius of a planet's orbit to the square of its period of revolution about the Sun is the same for all planetary bodies in the solar system.
kilogram (kg) The SI unit of mass; a fundamental unit.

kinetic energy The energy that an object possesses because of its motion.

kinetic-molecular theory of gases A simplified model of gas behavior based on the properties of an ideal gas.

laser An acronym for light amplification by the stimulated emission of radiation. A laser is a device that emits extremely intense, monochromatic, coherent light.

length contraction A consequence of Einstein's theory of special relativity. The length measured in a moving frame of reference is shorter than the length measured in a stationary frame of reference (the *proper length*).

linear accelerator A device that accelerates charged particles continuously in an electric field along a linear path.

longitudinal wave A wave in which the disturbance is parallel to the direction of the wave's motion. Sound waves are longitudinal.

magnet Any material that aligns itself, when free to do so, in an approximate north-south direction. Magnets exert forces on one another and on charged particles in motion.

magnetic field The region of space around a magnet or charge in motion that exerts a force on magnets or other moving charges.

magnetic induction The force that a magnetic field exerts on a 1-meter-long wire in the field when the wire carries a current of 1 ampere. The unit of magnetic induction is the tesla.

magnification The ratio of image size to object size.

mass (1) The measure of an object's ability to obey Newton's second law of motion. (2) The measure of an object's ability to obey Newton's law of universal gravitation. The SI unit of mass is the kilogram.

mass defect The mass lost by a nucleus when it is assembled from its nucleons. (See also *binding energy*.)

mass number The sum of the number of protons and neutrons in a nucleus; the number of nucleons the nucleus contains.

mass spectrometer A device that uses a magnetic field to separate charged particles of differing masses.

matter waves According to quantum theory, the waves associated with moving particles.

mechanical advantage In a simple machine, the ratio of the input force (*resistance*) to the output force (*effort*).

medium A material through which a disturbance, such as a wave, travels.

moderators Materials used in fission reactors to slow neutrons so that they can be absorbed by the fuel nuclei.

momentum The product of mass and velocity. The direction of an object's momentum is the direction of its velocity. The SI unit of momentum is the kilogram · meter per second.

natural frequency A specific frequency with which an elastic body may vibrate if disturbed.

net force The unbalanced force present on an object; the accelerating force.

neutrino A subatomic particle with no charge and questionable mass. It and its antiparticle are products of beta-decay reactions.

Newton (N) The SI unit of force, equivalent to the unit kilogram · meter per second^2.

Newton's first law of motion Objects remain in a state of uniform motion unless acted upon by an unbalanced force.

Newton's law of universal gravitation Any two bodies in the universe are attracted to each other with a force that is directly proportional to their masses and inversely proportional to the square of the distance between them.

Newton's second law of motion The unbalanced force on an object is equal to the product of its mass and acceleration.

Newton's third law of motion If object *A* exerts a force on object *B*, then object *B* exerts an equal and opposite force on object *A*.

node The point or locus of points on an interference pattern, such as a standing wave or double-slit pattern, that results in total destructive interference.

nondispersive medium A medium in which the speed of a wave does *not* depend on its frequency.

normal A line perpendicular to a surface.

normal force The force that keeps two surfaces in contact. If an object is on a *horizontal* surface, the normal force on the object is equal to its weight.

nuclear force The attractive, short-range force responsible for binding protons and neutrons in the nucleus.

nucleon A proton or a neutron.

nucleus The dense, positively charged core of an atom.

nuclide An atomic nucleus.

ohm (Ω) The SI unit of electrical resistance, equivalent to the unit volt per ampere.

Ohm's law A relationship in which the ratio of the potential difference across certain conductors to the current in them is constant at constant temperature.

parallel circuit An electric circuit with more than one current path.

particle accelerator A device used to accelerate charged nuclear particles.

pascal (Pa) The SI unit of pressure, equivalent to the unit newton per meter^2.

period The time for one complete repetition of a periodic phenomenon. The SI unit of period is the second.

periodic wave A regularly repeating series of waves.

phase (1) A form in which matter can exist, including liquid, solid, gas, and plasma. (2) In wave motion, the points on the wave that have specific time and space relationships.

photoelectric effect A phenomenon in which light causes electrons to be ejected from certain materials. (See also *photoemissive*.)

photoelectrons Electrons that have been emitted as a result of the photoelectric effect.

photoemissive Referring to materials whose surfaces can eject electrons on exposure to light.

photon The fundamental particle of electromagnetic radiation.

Planck's constant A universal constant (h) relating the energy of a photon to its frequency; its approximate value is 6.62×10^{-34} joule · second.

P-N junction The narrow region in a semiconductor diode in the vicinity of the boundary between the *P*-type and *N*-type materials.

point charge A charge with negligible physical dimensions.

polarization A process that produces transverse waves that vibrate in only one plane. Polarization is limited to transverse waves: light can be polarized; sound cannot.

polychromatic Referring to light waves of different colors (frequencies).

positron See *Beta (+) particle*.

potential difference The ratio of the work required to move a test charge between two points in an electric field to the magnitude of the test charge. The unit of potential difference is the volt.

potential energy The energy that a system has because of its relative position or condition.

power The time rate at which work is done or energy is expended. The SI unit of power is the watt, which is equivalent to the unit joule per second.

precision The limit of the ability of a measuring device to reproduce a measurement.

pressure The force on a surface per unit area. The SI unit of pressure is the pascal.

primary coil The wire coil of a transformer that is connected to a source of alternating current.

principal axis A line passing through the center of curvature and the center of a curved mirror or lens.

principal quantum number The integer that defines the main energy level of an atom.

proton A positively charged subatomic particle with a charge equal in magnitude to that of the electron.

pulse A nonperiodic disturbance in a medium.

quantum A discrete quantity of energy.
quarks The particles of which protons, neutrons, and certain other subatomic particles are composed. Quarks carry a charge of either one-third or two-thirds of an elementary charge and come in six "flavors": top, bottom, up, down, charm, and strange.

radioactive decay A spontaneous change in the nucleus of an atom.
radioactivity Changes in the nucleus of an atom that produce the emission of subatomic particles or photons.
radius of curvature The distance from the center of curvature to the curved surface of a curved mirror or lens.
ray A straight line indicating the direction of travel of a wave.
real image An image created by the actual convergence of light waves. Real images from single mirrors and single lenses are inverted and can be projected on a screen.
rectifier A device (usually a diode) that converts alternating current into pulsed direct current.
refraction The change in the direction of a wave when it passes obliquely from one medium to another in which it moves at different speed.
regular reflection The reflection of parallel light rays incident on a smooth plane surface.
resistance The opposition of a material to the flow of electrons through it; the ratio of potential difference to current.
resistivity A quantity that allows the resistance of substances to be compared. Numerically, it is the resistance of a 1-meter conductor with a cross-sectional area of 1 square meter. The SI unit of resistivity is the ohm·meter.
resistor A device that supplies resistance to a circuit.
resolution The process of determining the magnitude and direction of the components of a vector.
resonance The spontaneous vibration of an object at a frequency equal to that of the wave that initiates the resonant vibration.
rest energy A quantity derived from the special theory of relativity; the energy equivalent of mass.
resultant A vector sum.
reverse bias A potential difference applied across a *P-N* junction in a direction that inhibits both electron and hole flow across the junction.

satellite A body that revolves around a larger body as a result of a gravitational force.

scalar quantity A quantity, such as mass or work, that has magnitude but not direction.

scintillation counter A device that detects nuclear particles emitted from atoms by converting their kinetic energy into photons of light.

secondary coil In a transformer, the coil that is not connected to the primary source of potential difference.

second law of thermodynamics Heat energy will not flow spontaneously from a cooler body to a warmer body unless work is done. Alternatively, it is impossible to convert heat energy entirely into work.

semiconductor A metalloid whose conductivity can be increased either by raising its temperature or by adding certain impurities. Silicon and germanium are examples of semiconductors.

series circuit A circuit with only one current path.

shunt A low-resistance device for diverting electric current, used to convert a galvanometer into an ammeter.

significant digits The digits that are part of any measurement.

simultaneity The occurrence of two or more events at the same time. According to special relativity, events that are simultaneous in one frame of reference need not be simultaneous in another.

solenoid A coil of wire wound as a helix. When a current is passed through the solenoid, it becomes an electromagnet.

specific heat The amount of heat energy absorbed or liberated by a unit mass of a substance as it changes its temperature by one unit.

speed The time rate of change of distance; the magnitude of velocity. The SI unit of speed is the meter per second.

spherical aberration The failure of mirrors and lenses with spherical surfaces to bring light to a single focus.

spherical lens A lens in which at least one of the refracting surfaces is a portion of a sphere.

spherical mirror A mirror having a reflecting surface that is a portion of a sphere.

spring constant The ratio of the force required to stretch or compress a spring to the magnitude of the stretch or compression.

standard pressure One atmosphere; approximately 1.01×10^5 pascals.

standard temperature Approximately 273 kelvins.

standing wave A wave pattern created by the continual interference of an incident wave with its reflected counterpart. The standing wave does not travel, but oscillates about an equilibrium position.

static equilibrium The condition of a body when a net force of zero is acting on it.

stationary state A condition in which an atom is neither absorbing nor releasing energy.

step-down transformer A transformer in which the potential difference across the secondary coil is less than that across the primary coil.
step-up transformer A transformer in which the potential difference across the secondary coil is greater than that across the primary coil.
superconductor A material with no electrical resistance.

temperature The "hotness" of an object, measured with respect to a chosen standard.
tesla (T) The SI unit of flux density, equal to the units weber per square meter and newton per ampere · meter.
thermal equilibrium The point at which materials in contact reach the same temperature.
thermal neutrons Neutrons able to initiate fission reactions because they have speeds low enough to permit their absorption by fuel nuclei.
thermionic emission The emission of electrons from substances such as metallic filaments when these substances are heated.
thermodynamics The study of the interrelationships among heat energy, work, and other forms of energy.
third law of thermodynamics It is impossible to reach absolute zero in a finite number of operations.
threshold frequency The lowest frequency at which photons striking a specific surface can produce a photoelectric effect.
time dilation A consequence of the theory of special relativity. The time interval measured in a moving frame of reference is longer than the time interval measured in a stationary frame of reference (the *proper time*).
torque A force, applied perpendicularly to a designated line, that tends to produce rotational motion.
total internal reflection The reflection of a wave inside a relatively dense medium produced when the angle of the wave with the boundary exceeds the critical angle.
total mechanical energy The sum of the potential and kinetic energies of a mechanical system.
transformer A device that uses a changing magnetic field to increase or decrease the potential difference between a primary and a secondary circuit.
transistor (bipolar) A semiconducting device in which one type of semiconductor is sandwiched between two semiconductors of the opposite type. Transistors serve as amplification devices.
transmutation The change of one radioactive nuclide into another, either by decay or by bombardment.
transverse wave A wave in which the disturbance is perpendicular to the direction of the wave's motion. Light waves are transverse.

uniform In the study of motion, a term that is equivalent to *constant*.
uranium disintegration series The sequence of alpha and beta decays by which uranium-238 is transformed into lead-206.

valence band The energy band that contains electrons from the outermost energy levels of a group of atoms in a crystal.
vaporization See *boiling*.
vector A representation of a vector quantity; an arrow in which the length represents the magnitude of the quantity and the arrowhead points in the direction of its orientation.
vector quantity A quantity, such as force or velocity, that has both magnitude and direction.
velocity The time rate of change of displacement.
virtual focus The point where incident light waves appear to originate after they are refracted by a diverging lens or reflected by a diverging mirror.
virtual image An image formed by projecting diverging light behind a mirror or a lens.
volt (V) The SI unit of potential difference, equivalent to the unit joule per coulomb.
voltage Another term for *potential difference*.
voltmeter A device used to measure potential difference and constructed by placing a large resistor in series with the coil of a galvanometer.

watt (W) The SI unit of power, equivalent to the unit joule per second.
wave A series of periodic oscillations of a particle or a field both in time and in space.
wave front All points on a wave that are in phase with each other.
wavelength The length of one complete wave cycle.
weber (Wb) The SI unit of magnetic flux, equivalent to the unit joule per ampere.
weight The gravitational force present on an object.
work The product of the force on an object and its displacement. The SI unit of work is the joule.
work function The minimum radiant energy required to remove an electron from a photoemissive surface.

Regents Examinations, Answers, and Self-Analysis Charts

Examination June 1993

Physics

PART I

Answer all 55 questions in this part. [70]

Directions (1–55): For *each* statement or question, select the word or expression that, of those given, best completes the statement or answers the question. Record the answers to these questions in the spaces provided.

1 A car travels a distance of 98 meters in 10. seconds. What is the average speed of the car during this 10.-second interval?

(1) 4.9 m/s
(2) 9.8 m/s
(3) 49 m/s
(4) 98 m/s

1 2

2 Which measurement of an average classroom door is closest to 1 meter?

1 thickness
2 width
3 height
4 surface area

2 2

3 A boat initially traveling at 10. meters per second accelerates uniformly at the rate of 5.0 meters per second^2 for 10. seconds. How far does the boat travel during this time?

(1) 50. m
(2) 250 m
(3) 350 m
(4) 500 m

3 3

4 The graph below represents the relationship between distance and time for an object.

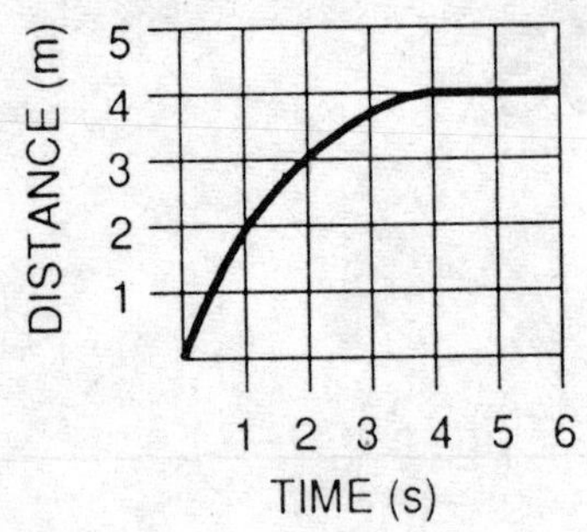

What is the instantaneous speed of the object at t = 5.0 seconds?

(1) 0 m/s (2) 2.0 m/s (3) 5.0 m/s (4) 4.0 m/s 4 1

5 An object accelerates uniformly from rest to a speed of 50. meters per second in 5.0 seconds. The average speed of the object during the 5.0-second interval is

(1) 5.0 m/s (2) 10. m/s (3) 25 m/s (4) 50. m/s 5 2

6 A 5-newton ball and a 10-newton ball are released simultaniously from a point 50 meters above the surface of the Earth. Neglecting air resistance, which statement is true?

1 The 5-N ball will have a greater acceleration than the 10-N ball.
2 The 10-N ball will have a greater acceleration than the 5-N ball.
3 At the end of 3 seconds of free-fall, the 10-N ball will have a greater momentum than the 5-N ball.
4 At the end of 3 seconds of free-fall, the 5-N ball will have a greater momentum than the 10-N ball. 6 3

7 In the diagram below, the weight of a box on a plane inclined at 30°, is represented by the vector W.

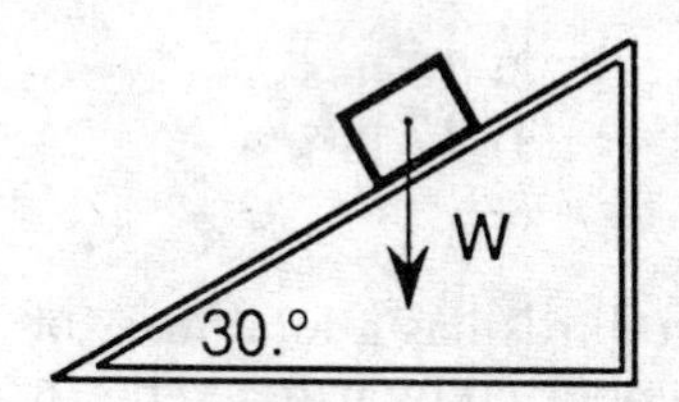

What is the magnitude of the component of the weight (W) that acts parallel to the incline?

(1) W
(2) $0.50W$
(3) $0.87W$
(4) $1.5W$

7 ____

8 The diagram at the right represents a force acting at point P. Which pair of concurrent forces would produce equilibrium when added to the force acting at point P?

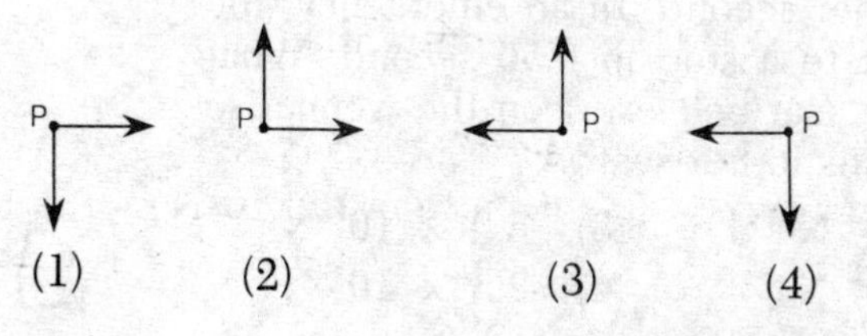

8 

9 A boat heads directly eastward across a river at 12 meters per second. If the current in the river is flowing at 5.0 meters per second due south, what is the magnitude of the boat's resultant velocity?

(1) 7.0 m/s
(2) 8.5 m/s
(3) 13 m/s
(4) 17 m/s

9 ____

10 A bird feeder with two birds has a total mass of 2.0 kilograms and is supported by wire as shown in the diagram below.

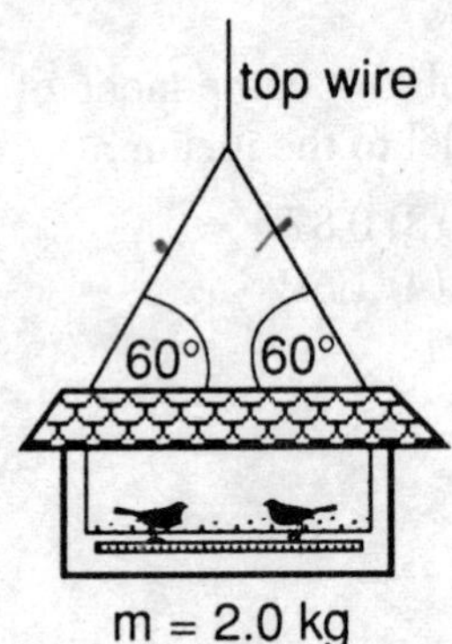

The force in the top wire is approximately

(1) 10. N
(2) 14 N
(3) 20. N
(4) 39 N

10 ____

11 A 50.-kilogram woman wearing a seat belt is traveling in a car that is moving with a velocity of +10. meters per second. In an emergency, the car is brought to a stop in 0.50 second. What force does the seat belt exert on the woman so that she remains in her seat?

(1) -1.0×10^3 N
(2) -5.0×10^2 N
(3) -5.0×10^1 N
(4) -2.5×10^1 N

11 ____

12 A 0.10-kilogram ball dropped vertically from a height of 1.0 meter above the floor bounces back to a height of 0.80 meter. The mechanical energy lost by the ball as it bounces is approximately

(1) 0.080 J
(2) 0.20 J
(3) 0.30 J
(4) 0.78 J

12 ____

13 A student rides a bicycle up a 30.° hill at a constant speed of 6.00 meters per second. The combined mass of the student and bicycle is 70.0 kilograms. What is the kinetic energy of the student-bicycle system during this ride?

(1) 210. J
(2) 420. J
(3) 1,260 J
(4) 2,520 J

13 ____

Base your answers to questions 14 and 15 on the information and diagram below.

Spacecraft *S* is traveling from planet P_1 toward planet P_2. At the position shown, the magnitude of the gravitational force of planet P_1 on the spacecraft is equal to the magnitude of the gravitational force of planet P_2 on the spacecraft.

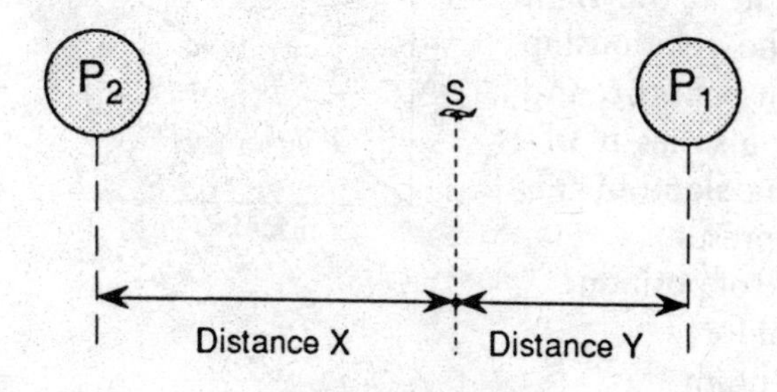

(not drawn to scale)

Note that questions 14 and 15 have only three choices.

14 If distance X is greater than distance Y, then the mass of P_1 must be
1 less than the mass of P_2
2 greater than the mass of P_2
3 equal to the mass of P_2

14 2

15 As the spacecraft moves from the position shown toward planet P_2, the ratio of the gravitational force of P_2 on the spacecraft to the gravitational force of P_1 on the spacecraft will
1 decrease
2 increase
3 remain the same

15 2

---

16 The graph at the right shows the relationship between weight and mass for a series of objects. The slope of this graph represents

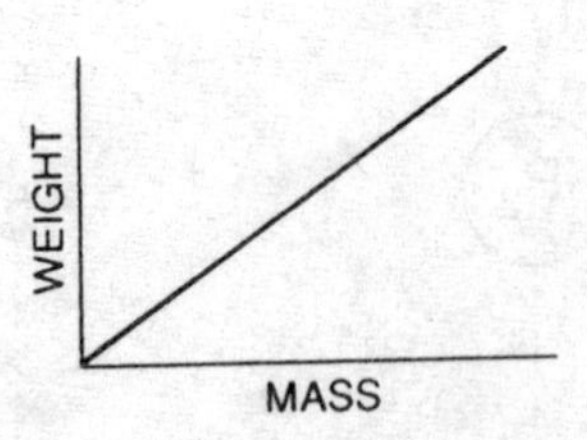

1 change of position
2 normal force
3 momentum
4 acceleration due to gravity

16 2

17 Each diagram below shows a different block being pushed by a force across a surface at a constant velocity.

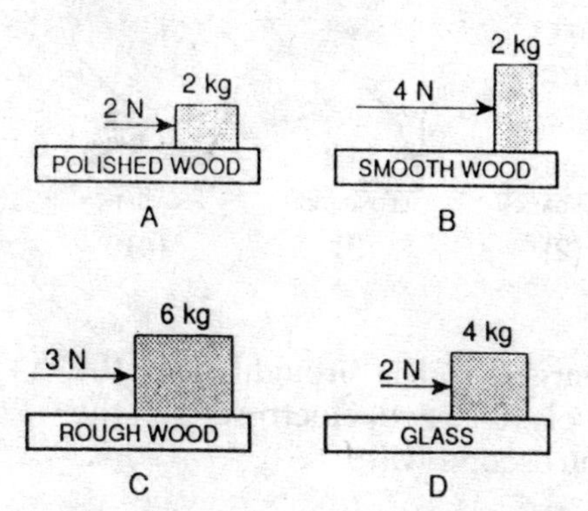

In which two diagrams is the force of friction the same?

(1) *A* and *B*
(2) *B* and *D*
(3) *A* and *D*
(4) *C* and *D*

17 4

18 A student running up a flight of stairs increases her speed at a constant rate. Which graph best represents the relationship between work and time for the student's run up the stairs?

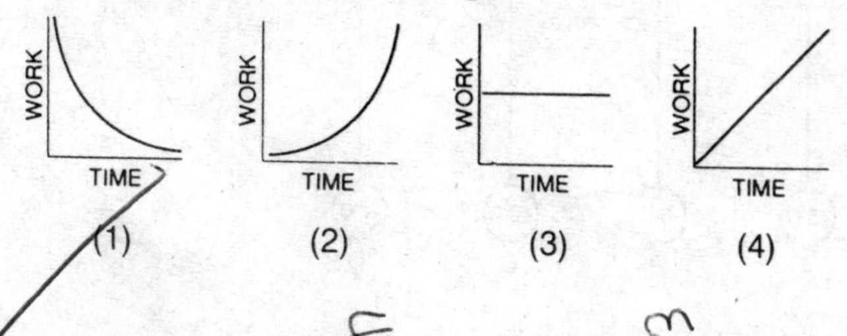

18 2

19 A net force of 5.0 newtons moves a 2.0-kilogram object a distance of 3.0 meters in 3.0 seconds. How much work is done on the object?

(1) 1.0 J
(2) 10. J
(3) 15 J
(4) 30. J

19 3

20 Which graph best represents the relationship between the elongation of a spring whose elastic limit has not been reached and the force applied to it?

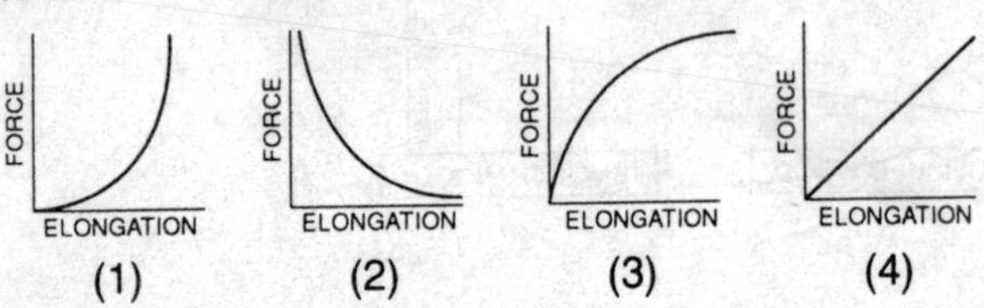

20 4

21 If a positively charged rod is brought near the knob of a positively charged electroscope, the leaves of the electroscope will

1 converge, only
2 diverge, only
3 first diverge, then converge
4 first converge, then diverge

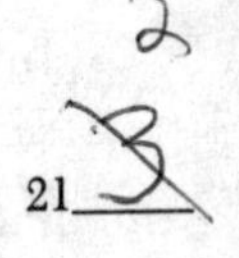

21 3

22 The diagram below shows four charged metal spheres suspended by strings. The charge of each sphere is indicated.

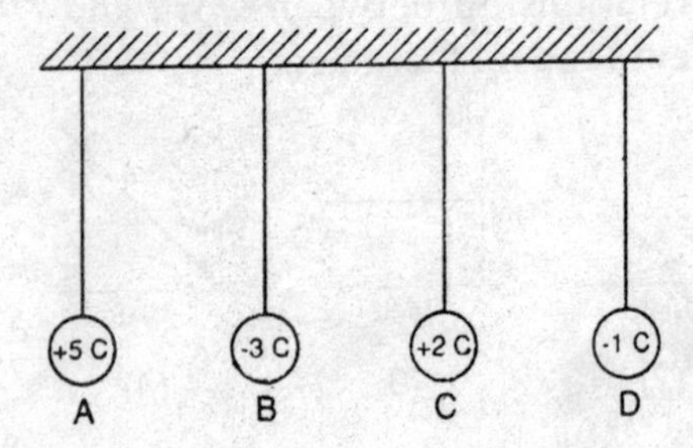

If spheres *A*, *B*, *C*, and *D* simultaneously come into contact, the net charge on the four spheres will be

(1) +1 C
(2) +2 C
(3) +3 C
(4) +4 C

22 3

23 The diagram below represents the electric field lines in the vicinity of two isolated electrical charges, *A* and *B*.

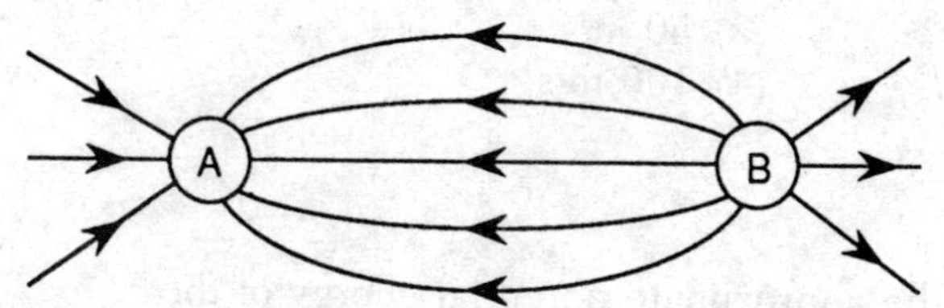

Which statement identifies the charges of *A* and *B*?

(1) *A* is negative and *B* is positive.
(2) *A* is positive and *B* is negative.
(3) *A* and *B* are both positive.
(4) *A* and *B* are both negative.

23 1

Base your answers to questions 24 through 26 on the diagram below which represents a frictionless track. A 10-kilogram block starts from rest at point *A* and slides along the track.

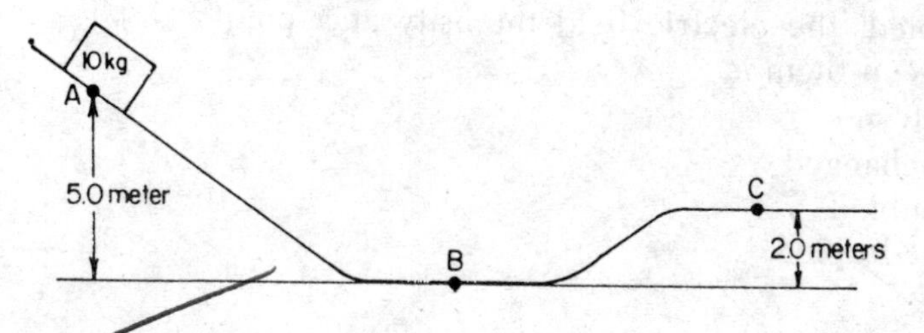

24 As the block moves from point *A* to point *B*, the total amount of gravitational potential energy changed to kinetic energy is approximately

(1) 5 J
(2) 20 J
(3) 50 J
(4) 500 J

24 4

25 What is the approximate speed of the block at point *B*?

(1) 1 m/s
(2) 10 m/s
(3) 50 m/s
(4) 100 m/s

25____

26 What is the approximate potential energy of the block at point *C*?

(1) 20 J
(2) 200 J
(3) 300 J
(4) 500 J

26____

---

27 If the potential difference between two oppositely charged parallel metal plates is doubled, the electric field intensity at a point between them is

1 halved
2 unchanged
3 doubled
4 quadrupled

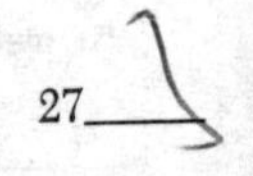
27____

28 Moving a point charge of 3.2×10^{-19} coulomb between points *A* and *B* in an electric field requires 4.8×10^{-19} joule of energy. What is the potential difference between these two points?

(1) 0.67 V
(2) 2.0 V
(3) 3.0 V
(4) 1.5 V

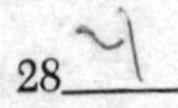
28____

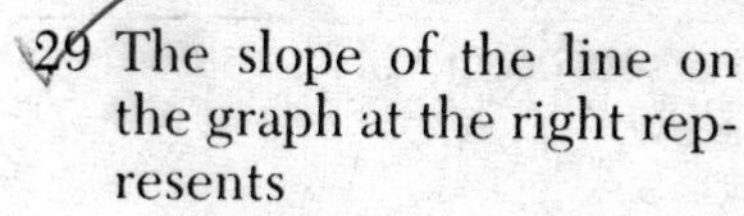

29 The slope of the line on the graph at the right represents

1 resistance of a material
2 electric field intensity
3 power dissipated in a resistor
4 electrical energy

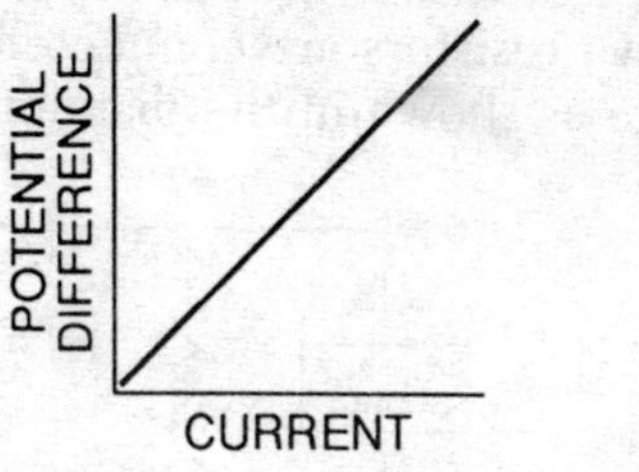

29_____

30 In the diagrams below, ℓ represents a unit length of copper wire and A represents a unit cross-sectional area. Which copper wire has the *smallest* resistance at room temperature?

(1) A, ℓ

(2) A, 2ℓ

(3) 2A, ℓ

(4) 2A, 2ℓ

30_____

31 Two resistors are connected to a source of voltage as shown in the diagram below.

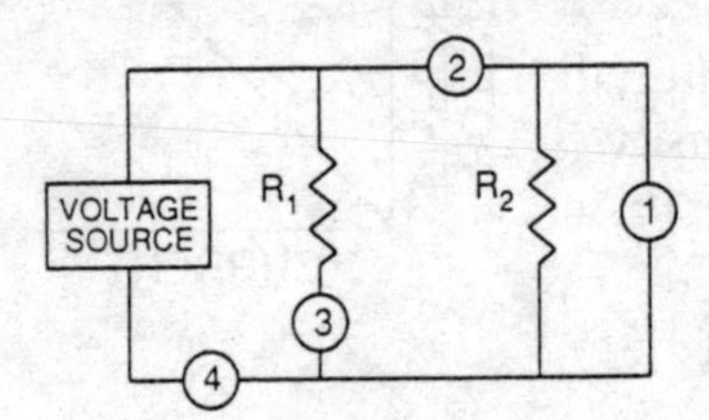

At which position should an ammeter be placed to measure the current passing only through resistor R_1?

(1) 1
(2) 2
(3) 3
(4) 4

31 ____

32 A toaster dissipates 1,500 watts of power in 90. seconds. The amount of electric energy used by the toaster is approximately

(1) 1.4×10^5 J
(2) 1.7×10^1 J
(3) 5.2×10^8 J
(4) 6.0×10^{-2} J

32 ____

33 In the diagram below, a steel paper clip is attached to a string, which is attached to a table. The clip remains suspended beneath a magnet.

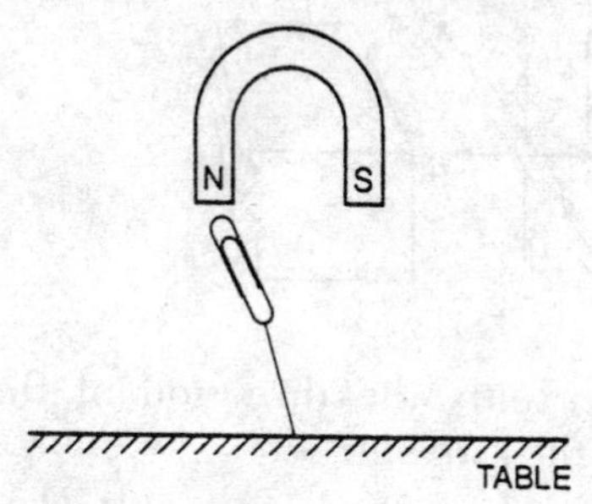

As the magnet is lifted, the paper clip begins to fall as a result of

1 an increase in the potential energy of the clip
2 an increase in the gravitational field strength near the magnet
3 a decrease in the magnetic properties of the clip
4 a decrease in the magnetic field strength near the clip

33 4

34 The diagram below shows the magnetic field that results when a piece of iron is placed between unlike magnetic poles.

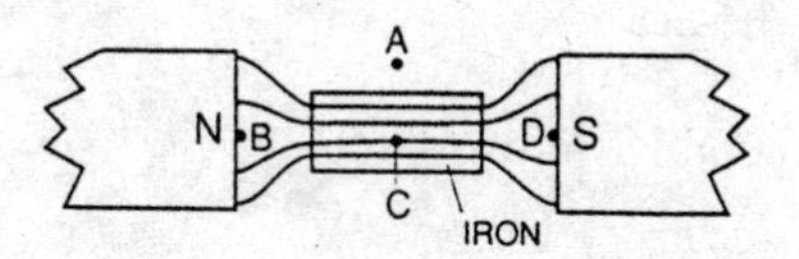

At which point is the magnetic field strength greatest?

(1) *A* (2) *B* (3) *C* (4) *D*

34 3

35 A wire carrying an electron current (e^-) is placed between the poles of a magnet, as shown in the diagram below.

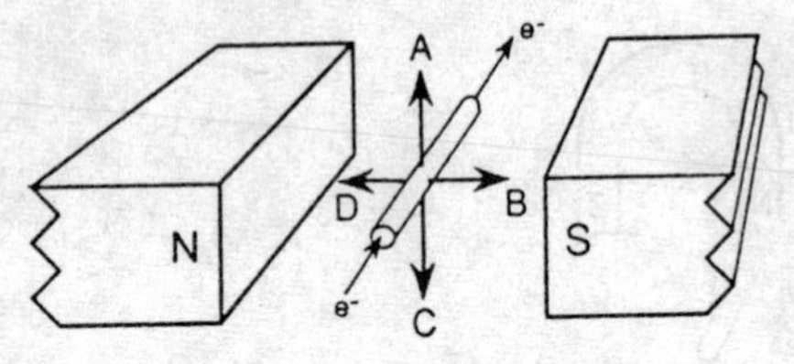

Which arrow represents the direction of the magnetic force on the current?

(1) *A* (2) *B* (3) *C* (4) *D*

35 2

36 The diagram below shows a coil of wire connected to a battery.

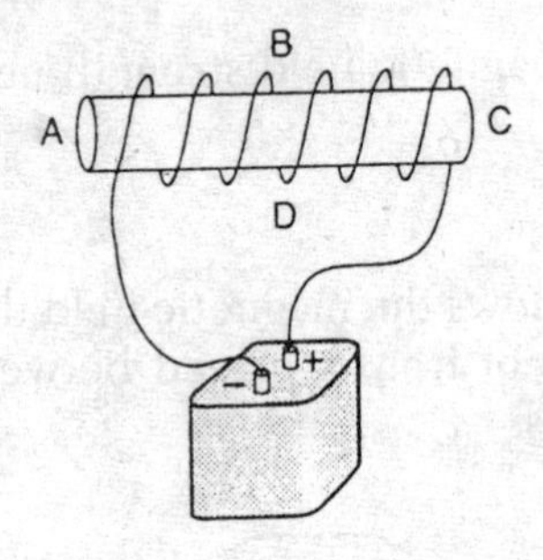

The *N*-pole of this coil is closest to

(1) *A*
(2) *B*
(3) *C*
(4) *D*

36 3

37 The diagram below shows radar waves being emitted from a stationary police car and reflected by a moving car back to the police car.

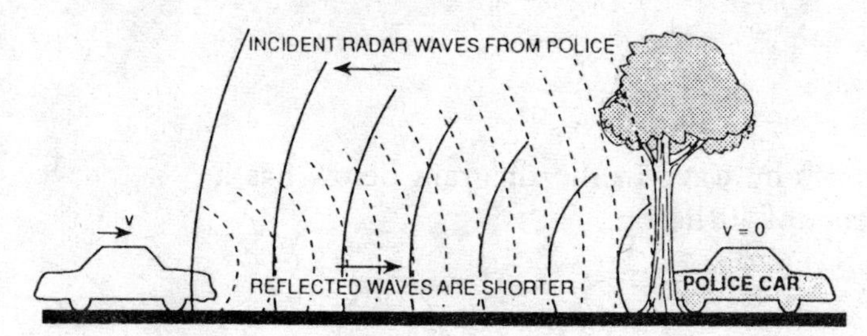

The difference in apparent frequency between the incident and reflected waves is an example of

1 constructive interference
2 refraction
3 the Doppler effect
4 total internal reflection

37____

38 The diagram below shows a transverse pulse moving to the right in a string.

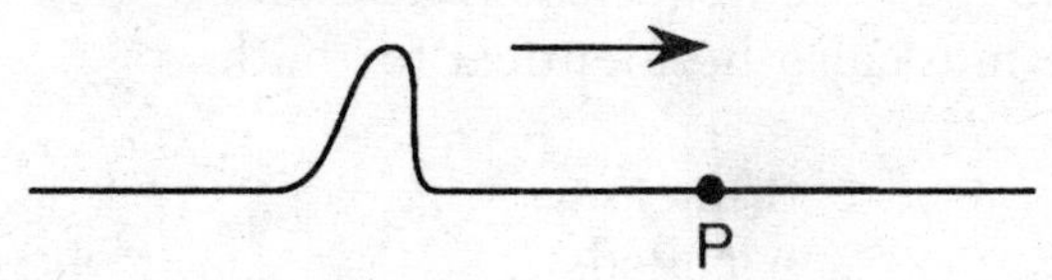

Which diagram best represents the motion of point *P* as the pulse passes point *P*?

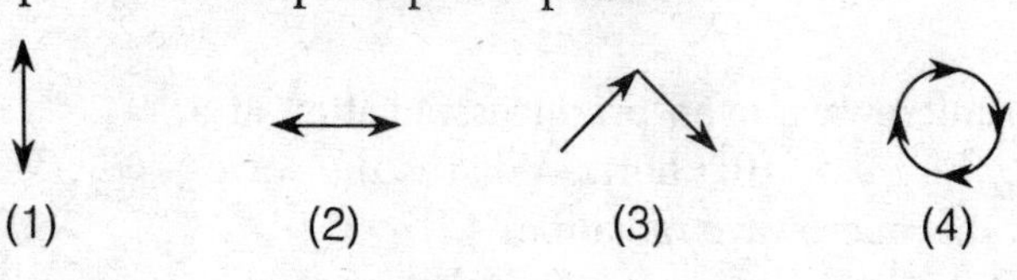

38____

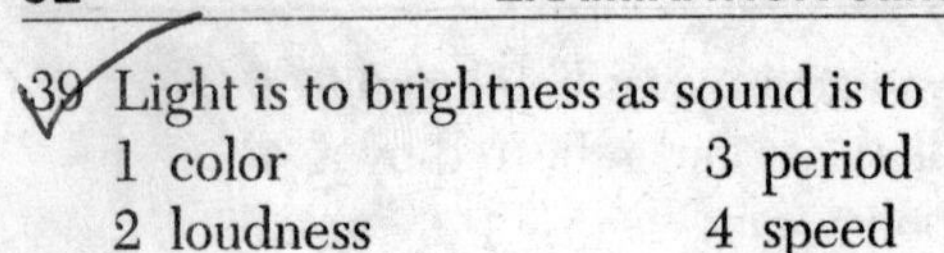

39 Light is to brightness as sound is to

1 color
2 loudness
3 period
4 speed

39 ____

40 The periodic wave in the diagram below has a frequency of 40. hertz.

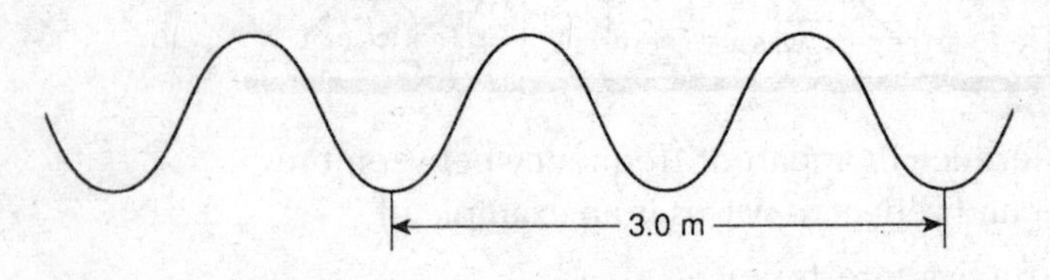

What is the speed of the wave?

(1) 13 m/s
(2) 27 m/s
(3) 60. m/s
(4) 120 m/s

40 ____

41 Two waves have the same frequency. Which wave characteristic must also be identical for both waves?

1 phase
2 amplitude
3 intensity
4 period

41 ____

42 A typical microwave oven produces radiation at a frequency of 1.0×10^{10} hertz. What is the wavelength of this microwave radiation?

(1) 3.0×10^{-1} m
(2) 3.0×10^{-2} m
(3) 3.0×10^{10} m
(4) 3.0×10^{18} m

42 ____

43 Two wave sources operating in phase in the same medium produce the circular wave patterns shown in the diagram below. The solid lines represent wave crests and the dashed lines represent wave troughs.

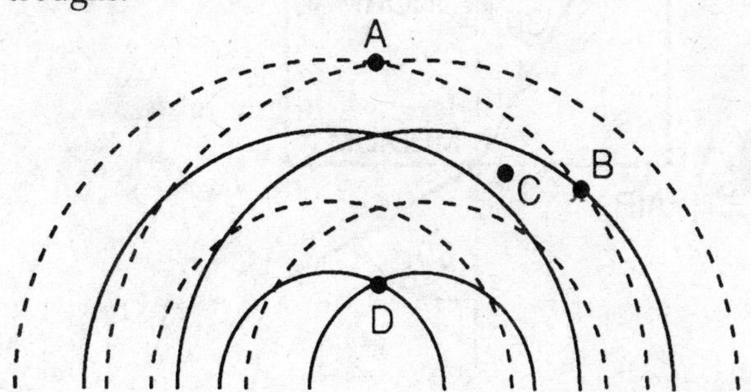

Which point is at a position of maximum destructive interference?

(1) *A* (3) *C*
(2) *B* (4) *D*

43 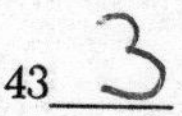

44 The distance between successive antinodes in the standing wave pattern shown at the right is equal to

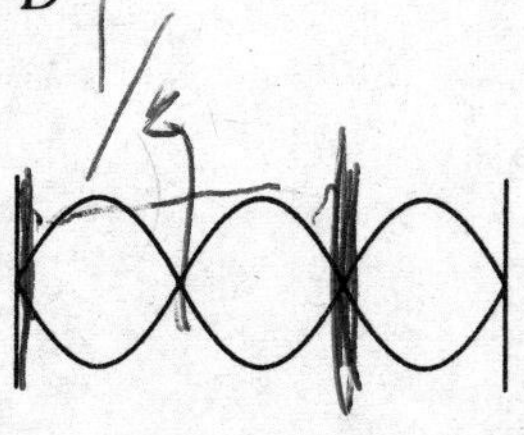

(1) 1 wavelength
(2) 2 wavelengths
(3) $\frac{1}{2}$ wavelength
(4) $\frac{1}{3}$ wavelength

44 4

45 The diagram below shows a ray of light passing from medium *X* into air.

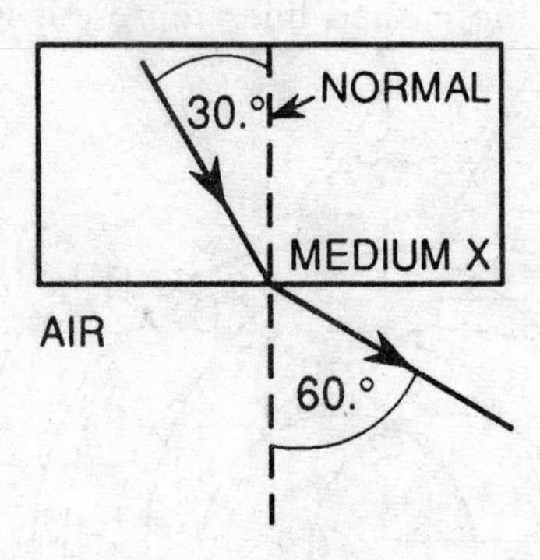

What is the absolute index of refraction of medium *X*?

(1) 0.50
(2) 2.0
(3) 1.7
(4) 0.58

45 3

46 In the diagram below, a ray of monochromatic light (*A*) and a ray of polychromatic light (*B*) are both incident upon an air-glass interface.

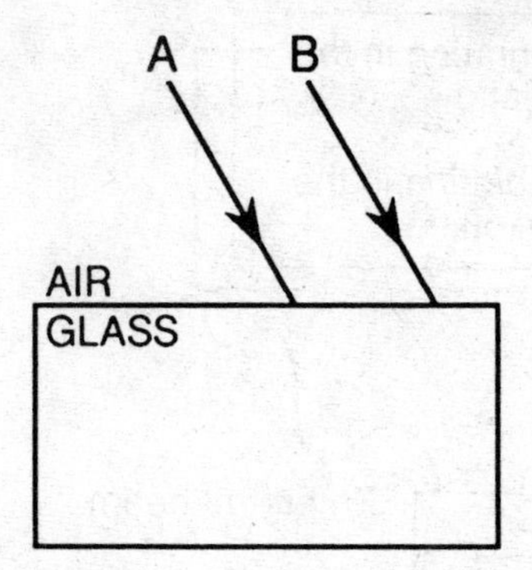

Which phenomenon could occur with ray *B*, but *not* with ray *A*?

1 dispersion
2 reflection
3 polarization
4 refraction

46__1__

47 If the critical angle for a substance is 44°, the index of refraction of the substance is equal to

(1) 1.0
(2) 0.69
(3) 1.4
(4) 0.023

47__2__

48 The diagram below shows a beam of light entering and leaving a "black box."

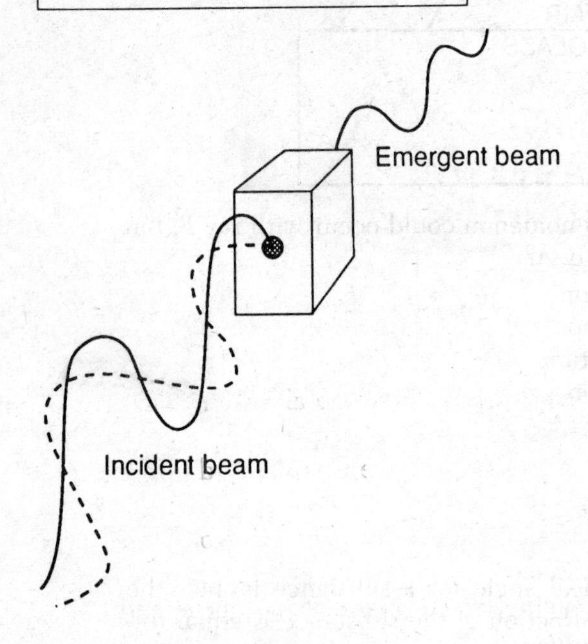

The box most likely contains a

1 prism
2 converging lens
3 double slit
4 polarizer

48 4

49 Which graph best represents the relationship between the intensity of light that falls on a photoemissive surface and the number of photoelectrons that the surface emits?

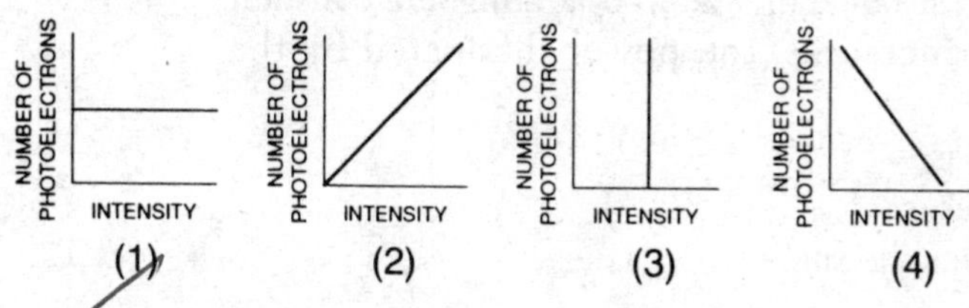

49 ____

50 The work function of a certain photoemissive material is 2.0 electronvolts. If 5.0-electronvolt photons are incident on the material, the maximum kinetic energy of the ejected photoelectrons will be

(1) 7.0 eV
(2) 5.0 eV
(3) 3.0 eV
(4) 2.5 eV

50 ____

51 Alpha particles fired at thin metal foil are scattered in hyperbolic paths due to the

1 attraction between the electrons and alpha particles
2 magnetic repulsion between the electrons and alpha particles
3 gravitational attraction between the nuclei and alpha particles
4 repulsive forces between the nuclei and alpha particles

51 ____

52 The momentum of a photon with a wavelength of 5.9×10^{-7} meter is

(1) 8.9×10^{26} kg•m/s
(2) 1.6×10^{-19} kg•m/s
(3) 1.1×10^{-27} kg•m/s
(4) 3.9×10^{-40} kg•m/s

52 ____

Note that questions 53 through 55 have only three choices.

53 As the resistance of a lamp operating at a constant voltage increases, the power dissipated by the lamp
1 decreases
2 increases
3 remains the same

53 ____

54 Circuit *A* and circuit *B* are shown below.

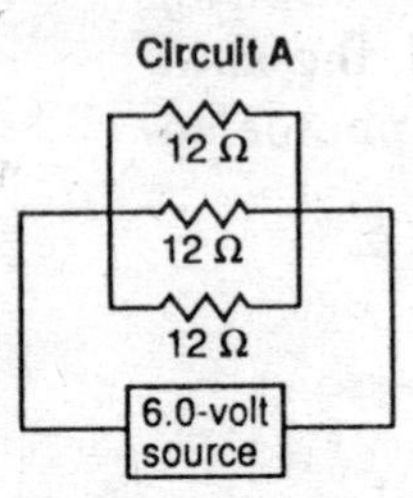

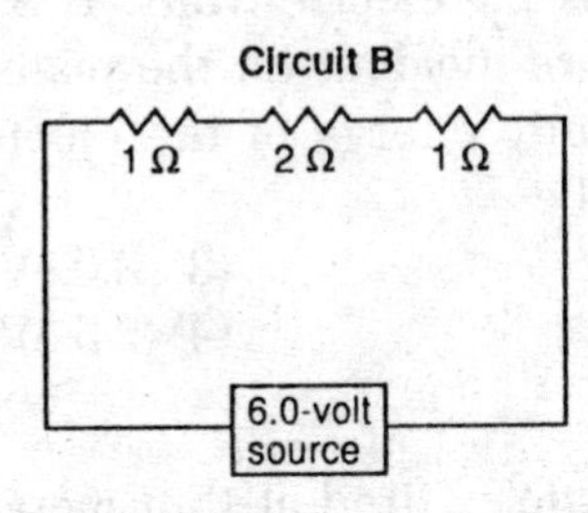

Compared to the total resistance of circuit *A*, the total resistance of circuit *B* is
1 less
2 greater
3 the same

54 ____

55 The diagram at the right represents the path of periodic waves passing from medium *A* into medium *B*. As the waves enter medium *B*, their speed
1 decreases
2 increases
3 remains the same

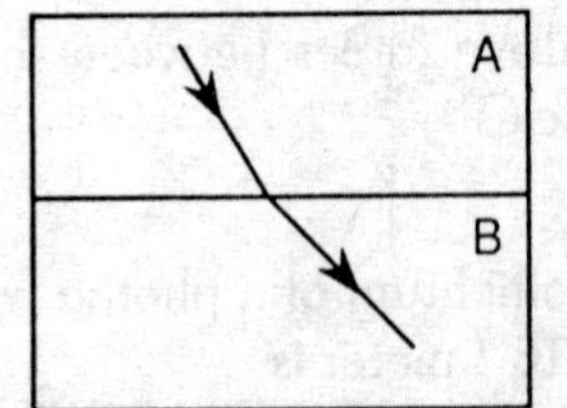

55 ____

PART II

This part consists of six groups, each containing ten questions. Each group tests an optional area of the course. Choose two of these six groups. Be sure that you answer all ten questions in each group chosen. Record the answers to these questions in the spaces provided. [20]

GROUP 1—**Motion in a Plane**

If you choose this group, be sure to answer questions **56–65.**

56 A ball is thrown horizontally at a speed of 20. meters per second from the top of a cliff. How long does the ball take to fall 19.6 meters to the ground?

(1) 1.0 s (2) 2.0 s (3) 9.8 s (4) 4.0 s 56_____

57 A book is pushed with an initial horizontal velocity of 5.0 meters per second off the top of a desk. What is the initial vertical velocity of the book?

(1) 0 m/s (3) 5.0 m/s
(2) 2.5 m/s (4) 10. m/s 57_____

58 The diagram below shows a baseball being hit with a bat. Angle θ represents the angle between the horizontal and the ball's initial direction of motion.

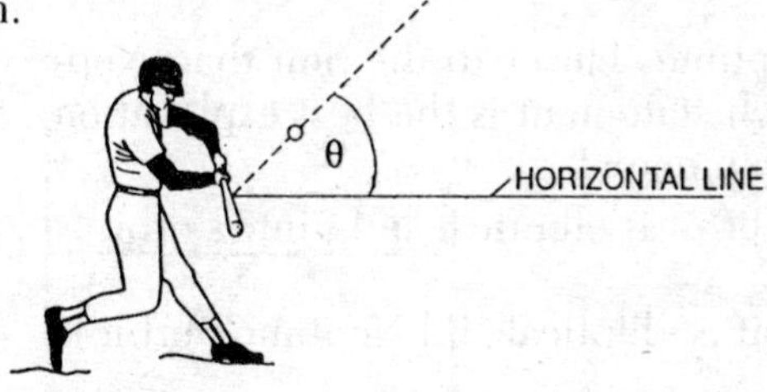

Which value of θ would result in the ball traveling the longest horizontal distance? [Neglect air resistance.]

(1) 25° (2) 45° (3) 60° (4) 90° 58_____

59 The diagram below represents a bicycle and rider traveling to the right at a constant speed. A ball is dropped from the hand of the cyclist.

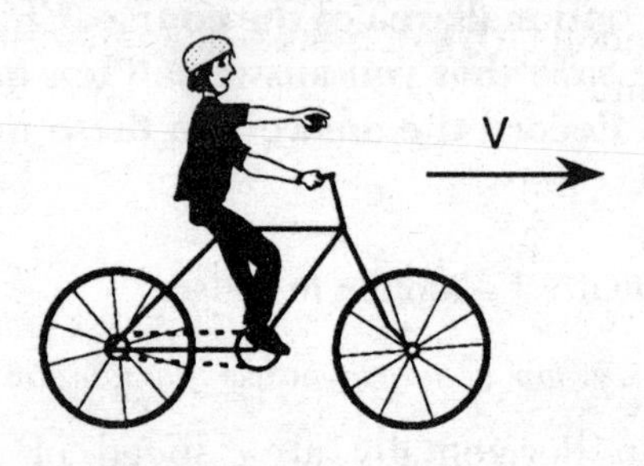

Which set of graphs best represents the horizontal motion of the ball relative to the ground? [Neglect air resistance.]

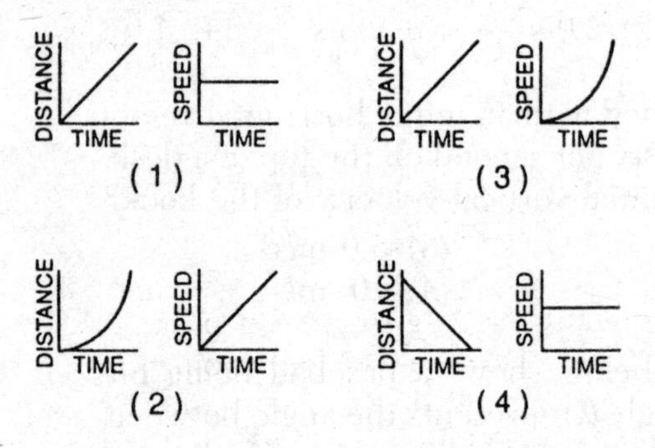

59_____

60 Pluto is sometimes closer to the Sun than Neptune is. Which statement is the best explanation for this phenomenon?

1 Neptune's orbit is elliptical and Pluto's orbit is circular.
2 Pluto's orbit is elliptical and Neptune's orbit is circular.
3 Pluto and Neptune have circular orbits that overlap.
4 Pluto and Neptune have elliptical orbits that overlap.

60_____

Base your answers to questions 61 through 63 on the diagram below which shows a 2.0-kilogram model airplane attached to a wire. The airplane is flying clockwise in a horizontal circle of radius 20. meters at 30. meters per second.

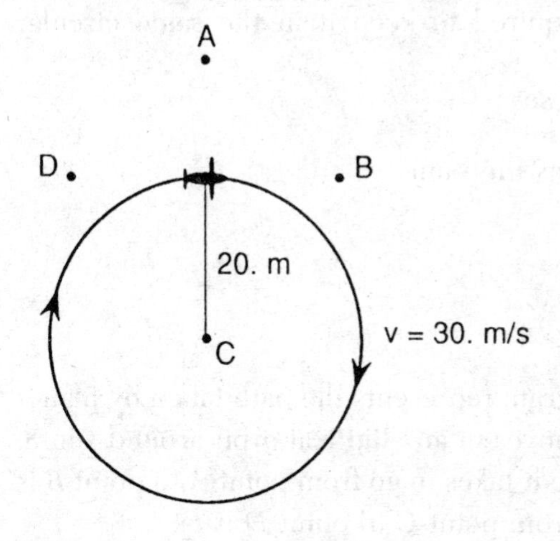

61 The centripetal force acting on the airplane at the position shown is directed toward point

(1) *A*
(2) *B*
(3) *C*
(4) *D*

61_____

62 What is the magnitude of the centripetal acceleration of the airplane?

(1) 0 m/s^2
(2) 1.5 m/s^2
(3) 45 m/s^2
(4) 90. m/s^2

62_____

63 If the wire breaks when the airplane is at the position shown, the airplane will move toward point

(1) *A*
(2) *B*
(3) *C*
(4) *D*

63_____

Note that questions 64 and 65 have only three choices.

64 A motorcycle travels around a flat circular track. If the speed of the motorcycle is increased, the force required to keep it in the same circular path

1 decreases
2 increases
3 remains the same

64_____

65 The diagram represents the path taken by planet *P* as it moves in an elliptical orbit around sun *S*. The time it takes to go from point *A* to point *B* is t_1, and from point *C* to point *D* is t_2.

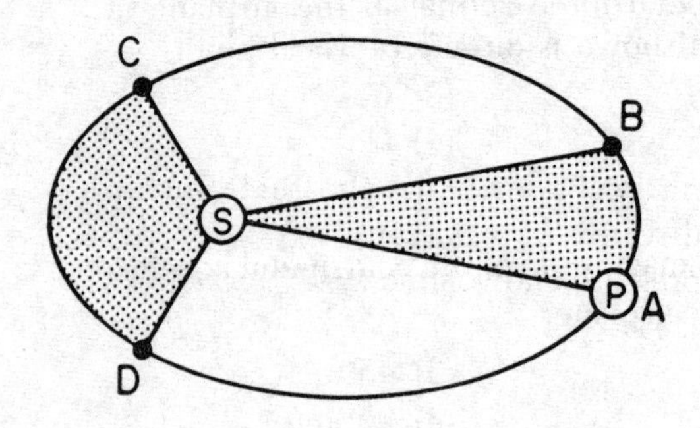

If the two shaded areas are equal, then t_1 is

1 less than t_2
2 greater than t_2
3 the same as t_2

65_____

GROUP 2—**Internal Energy**

If you choose this group, be sure to answer questions **66–75**.

66 What is the difference between the melting point and boiling point of ethyl alcohol on the Kelvin scale?

(1) 38 (3) 352
(2) 196 (4) 469 66____

67 A kilogram of each of the substances below is condensed from a gas to a liquid. Which substance releases the most energy?

1 alcohol 3 water
2 mercury 4 silver 67____

68 Which sample of metal will gain net internal energy when placed in contact with a block of lead at 100°C?

1 platinum at 60°C 3 leat at 125°C
2 iron at 100°C 4 silver at 200°C 68____

69 Which graph best represents the relationship between absolute temperature (T) and the product of pressure and volume ($P \bullet V$) for a given mass of ideal gas?

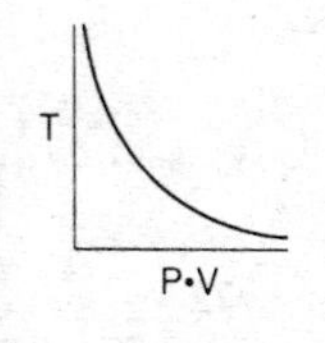

(1)

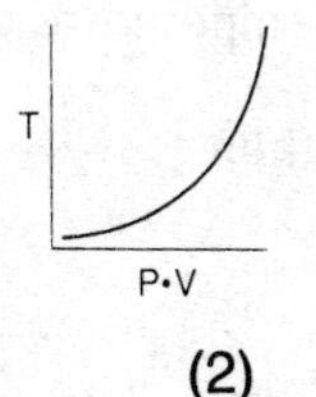

(2)

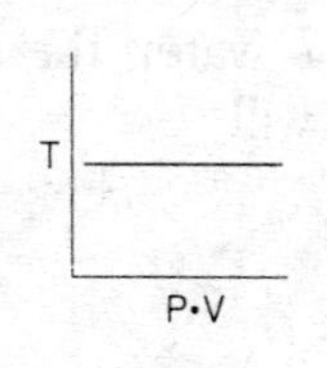

(3)

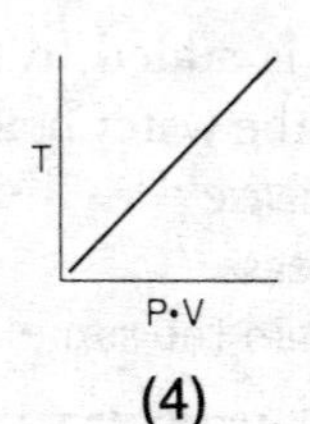

(4)

69____

Base your answers to questions 70 through 72 on the information below.

Ten kilograms of water initially at 20°C is heated to its boiling point (100°C). Then 5.0 kilograms of the water is converted into steam at 100°C.

70 What was the approximate amount of heat energy needed to raise the temperature of the water to its boiling point?
(1) 840 kJ
(2) 3,400 kJ
(3) 4,200 kJ
(4) 6,300 kJ

70_____

71 The amount of heat energy needed to convert the 5.0 kilograms of water at 100°C into steam at 100°C is approximately
(1) 1,700 kJ
(2) 2,100 kJ
(3) 5,500 kJ
(4) 11,000 kJ

71_____

Note that question 72 has only three choices.

72 If salt is added to the water, the temperature at which the water boils will
1 decrease
2 increase
3 remain the same

72_____

73 The graph below shows temperature versus time for 1.0 kilogram of a substance at constant pressure as heat is added at a constant rate of 100 kilojoules per minute. The substance is a solid at 20°C.

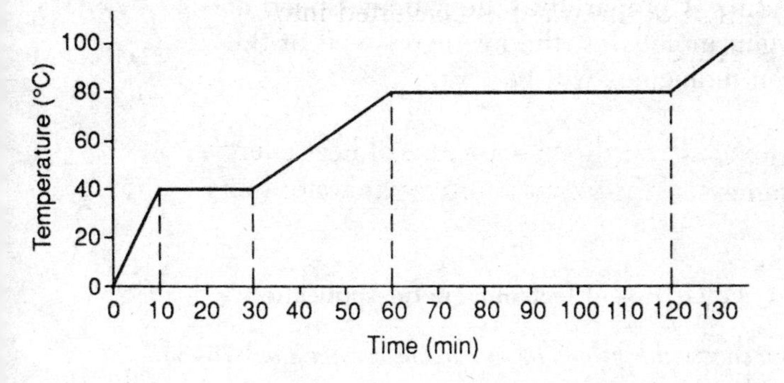

How much heat was added to change the substance from a liquid at its melting point to a vapor at its boiling point?

(1) 3,000 kJ (3) 9,000 kJ
(2) 6,000 kJ (4) 11,000 kJ 73____

Note that questions 74 and 75 have only three choices.

74 As pressure is applied to a snowball, the melting point of the snow

1 decreases
2 increases
3 remains the same 74____

75 Oxygen molecules are about 16 times more massive than hydrogen molecules. An oxygen gas sample is in a closed container and a hydrogen gas sample is in a second closed container of different size. Both samples are at room temperature. Compared to the average speed of the oxygen molecules, the average speed of the hydrogen molecules will be

1 less
2 greater
3 the same

75_____

GROUP 3—**Electromagnetic Applications**

If you choose this group, be sure to answer questions **76–85.**

Base your answers to questions 76 through 78 on the information and data table below.

During a laboratory investigation of transformers, a group of students obtained the following data during four trials, using a different pair of coils in each trial.

	Primary Coil		Secondary Coil	
	V_p (volts)	I_p (amperes)	V_s (volts)	I_s (amperes)
Trial 1	3.0	12.0	16.0	2.0
Trial 2	6.0	3.0	8.0	2.2
Trial 3	9.0	4.3	54.0	0.7
Trial 4	12.0	2.5	5.0	9.0

76 What is the efficiency of the transformer in Trial 1?

(1) 75%
(2) 89%
(3) 100%
(4) 113%

76_____

77 What is the ratio of the number of turns in the primary coil to the number of turns in the secondary coil in trial 3?

(1) 1:6 (3) 6:1
(2) 1:9 (4) 9:1

77_____

78 In which trial was an error most likely made in recording the data?

(1) 1 (3) 3
(2) 2 (4) 4

78_____

79 A wire 0.50 meter long cuts across a magnetic field with a magnetic flux density of 20. teslas. The wire moves at a speed of 4.0 meters per second and travels in a direction perpendicular to the magnetic flux lines. What is the maximum potential difference induced between the ends of the wire?

(1) 2.5 V
(2) 10. V
(3) 40. V
(4) 160 V

79_____

80 Compared to the resistance of the circuit being measured, the internal resistance of a voltmeter is designed to be very high so that the meter will draw

1 no current from the circuit
2 little current from the circuit
3 most of the current from the circuit
4 all the current from the circuit

80_____

81 A proton and an electron traveling with the same velocity enter a uniform electric field. Compared to the acceleration of the proton, the acceleration of the electron is

1 less, and in the same direction
2 less, but in the opposite direction
3 greater, and in the same direction
4 greater, but in the opposite direction

81_____

82 The diagram below shows an end view of a straight conducting wire, *W*, moving with constant speed in uniform magnetic field *B*.

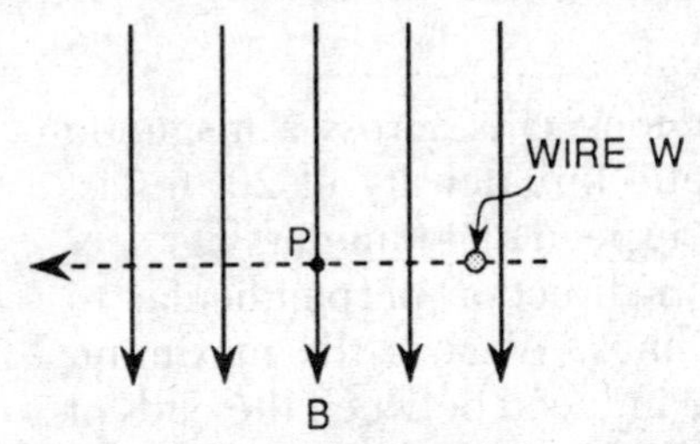

As the conductor moves through position *P*, the electron current induced in the wire is directed

1 toward the bottom of the page
2 toward the top of the page
3 into the page
4 out of the page

82_____

83 An electron moves at 3.0×10^7 meters per second perpendicularly to a magnetic field that has a flux density of 2.0 teslas. What is the magnitude of the force on the electron?

(1) 9.6×10^{-19} N
(2) 3.2×10^{-19} N
(3) 9.6×10^{-12} N
(4) 4.8×10^{-12} N

83_____

84 In each diagram below, an electron travels to the right between points *A* and *B*. In which diagram would the electron be deflected toward the bottom of the page?

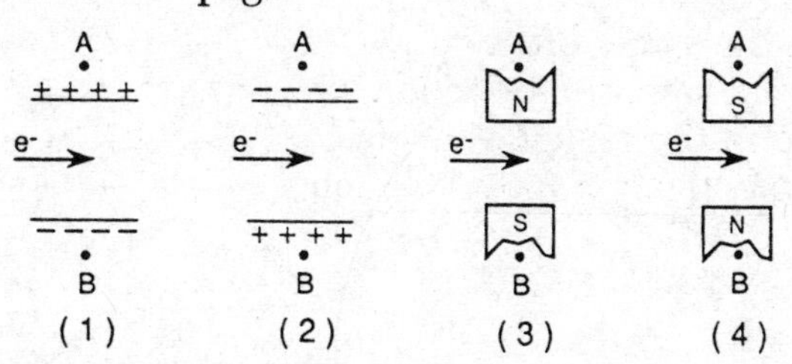

84_____

85 What is one characteristic of a light beam produced by a monochromatic laser?

(1) It consists of coherent waves.
(2) It can be dispersed into a complete continuous spectrum.
(3) It cannot be reflected or refracted.
(4) It does not exhibit any wave properties.

85_____

GROUP 4—**Geometric Optics**

If you choose this group, be sure to answer questions **86–95.**

86 An object is placed in front of a plane mirror as shown in the diagram at the right. Which diagram below best represents the image that is formed?

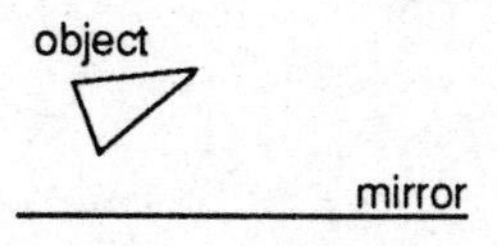

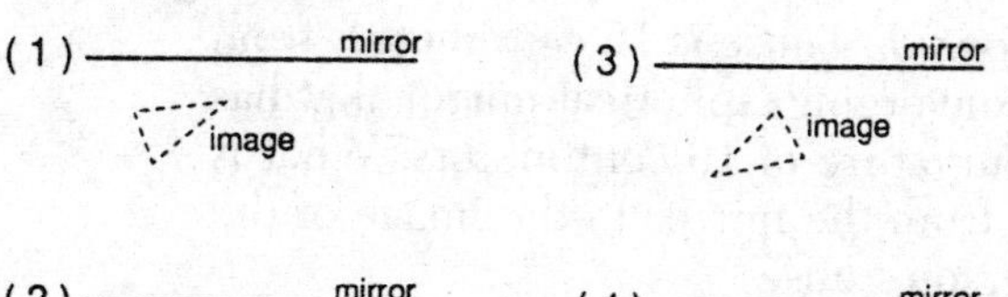

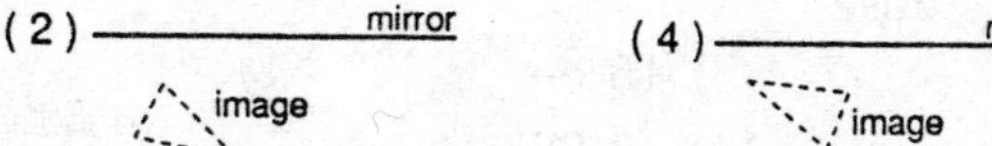

86_____

87 The diagram below shows light ray R parallel to the principal axis of a spherical concave (converging) mirror. Point F is the focal point of the mirror and C is the center of curvature.

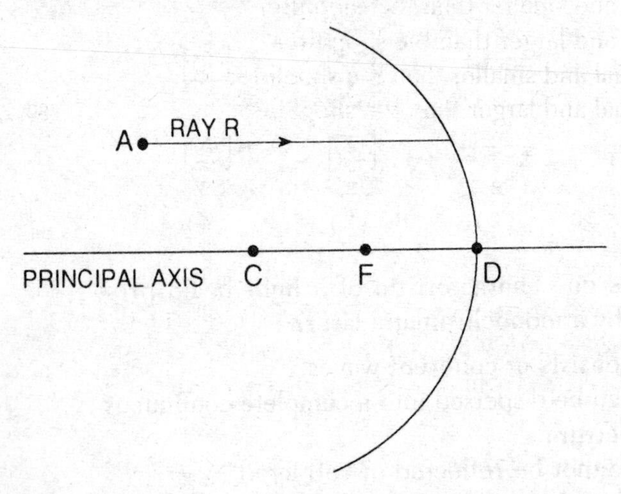

After reflecting, the light ray will pass through point

(1) A
(2) F
(3) C
(4) D

87_____

88 The tip of a person's nose is 12 centimeters from a concave (converging) spherical mirror that has a radius of curvature of 16 centimeters. What is the distance from the mirror to the image of the tip of the person's nose?

(1) 8.0 cm
(2) 12 cm
(3) 16 cm
(4) 24 cm

88_____

89 The image of a shoplifter in a department store is viewed on a convex (diverging) mirror. The image is

1 real and smaller than the shoplifter
2 real and larger than the shoplifter
3 virtual and smaller than the shoplifter
4 virtual and larger than the shoplifter

89 ____

90 When light rays pass through the film in a movie projector, an image of the film is produced on a screen. In order to produce the image on the screen, what type of lens does the projector use and how far from the lens must the film be placed?

1 converging lens, at a distance greater than the focal length
2 converging lens, at a distance less than the focal length
3 diverging lens, at a distance greater than the focal length
4 diverging lens, at a distance less than the focal length

90 ____

91 Two light rays from a common point are refracted by a lens. A real image is formed when these two refracted rays

1 converge to a single point
2 diverge and appear to come from a single point
3 travel in parallel paths
4 totally reflect inside the lens

91 ____

92 The diagram below represents a convex (converging) lens with focal point F.

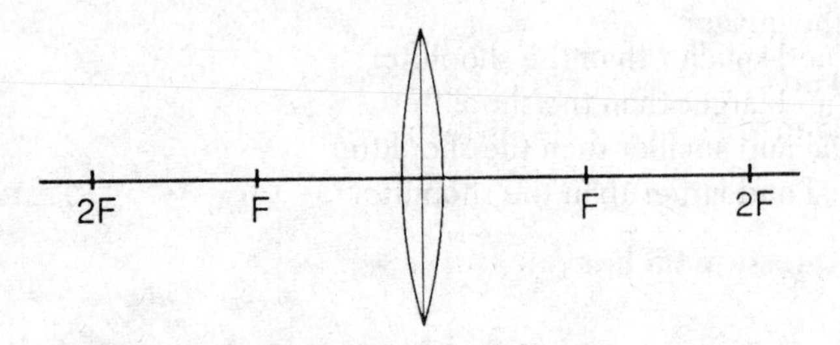

If an object is placed at $2F$, the image will be

1 virtual, erect, and smaller than the object
2 real, inverted, and the same size as the object
3 real, inverted, and larger than the object
4 virtual, erect, and the same size as the object 92 ____

93 The diagram below shows the refraction of the blue and red components of a white light beam.

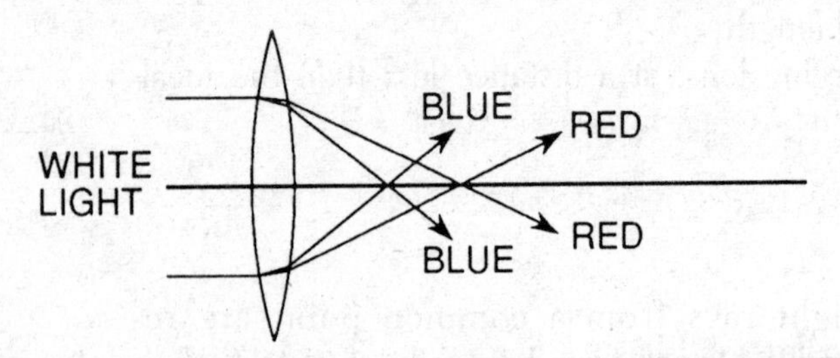

Which phenomenon does the diagram illustrate?

1 total internal reflection
2 critical angle reflection
3 spherical aberration
4 chromatic aberration 93____

94 When a 2.0-meter-tall object is placed 4.0 meters in front of a lens, an image is formed on a screen located 0.050 meter behind the lens. What is the size of the image?

(1) 0.10 m (2) 2.5 m
(3) 0.025 m (4) 0.40 m

94____

Note that question 95 has only three choices.

95 As the distance between a man and a plane mirror increases, the size of the image of the man produced by the mirror

1 decreases
2 increases
3 remains the same

95____

GROUP 5—**Solid State**

If you choose this group, be sure to answer questions **96–105**.

96 A particular solid has a small energy gap between its valence and conduction bands. This solid is most likely classified as

1 a good conductor
2 a semiconductor
3 a type of glass
4 an insulator

96____

97 The diagram below shows an electron moving through a semiconductor.

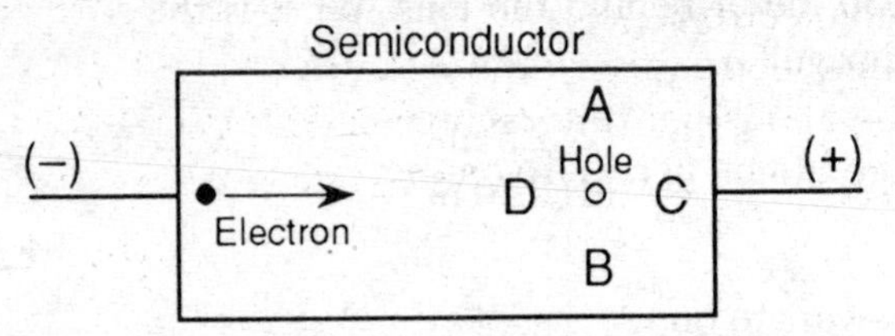

Toward which letter will the hole move?

(1) *A*
(2) *B*
(3) *C*
(4) *D*

9____

98 Diagram *A* represents the wave form of an electron current entering a semiconductor device, and diagram *B* represents the wave from as the current leaves the device.

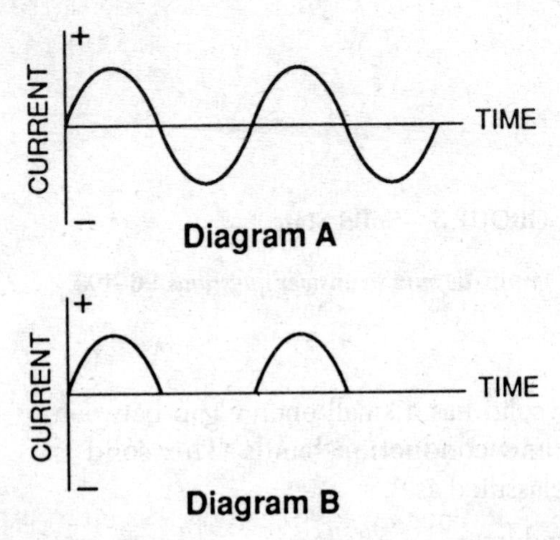

What is the device?

1 resistor
2 anode
3 cathode
4 diode

98____

99 Compared to an insulator, a conductor of electric current has
1 more free electrons per unit volume
2 fewer free electrons per unit volume
3 more free atoms per unit volume
4 fewer free atoms per unit volume
99_____

Base your answers to questions 100 through 102 on the diagram below which represents a diode.

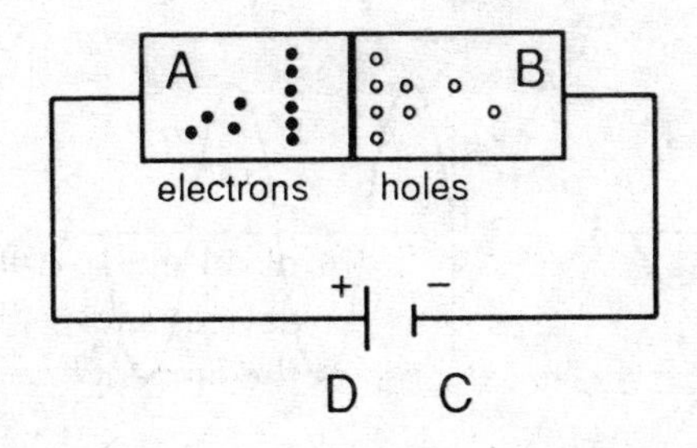

100 The *P-N* junction in the diagram is biased
(1) reverse (3) *B* to *C*
(2) forward (4) *A* to *D*
100_____

101 In the diagram, *B* represents the
(1) *N*-type silicon (3) cathode
(2) *P*-type silicon (4) diode
101_____

Note that question 102 has only three choices.

102 If the positive and negative wires of the circuit in the diagram were reversed, the current would
1 decrease
2 increase
3 remain the same
102_____

103 The graph at the right represents the alternating current signal input to a transistor amplifier. Which graph below best represents the amplified output signal from this transistor?

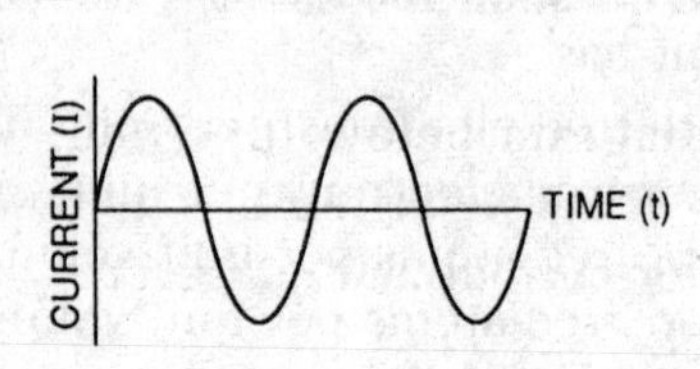

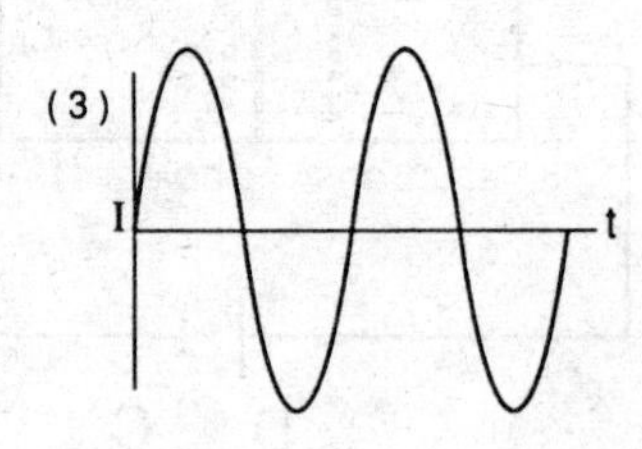

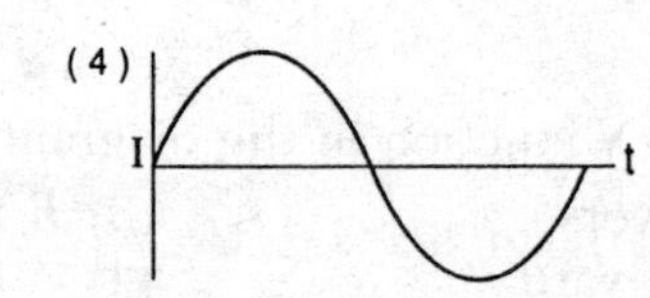

103_____

104 Which device contains a large number of transistors on a single block of silicon?

(1) junction diode
(2) conductor
(3) integrated circuit
(4) *N*-type semiconductor

104_____

Note that question 105 has only three choices.

105 The diagram below represents an operating *N-P-N* transistor circuit. Ammeter A_c reads the collector current and ammeter A_b reads the base current.

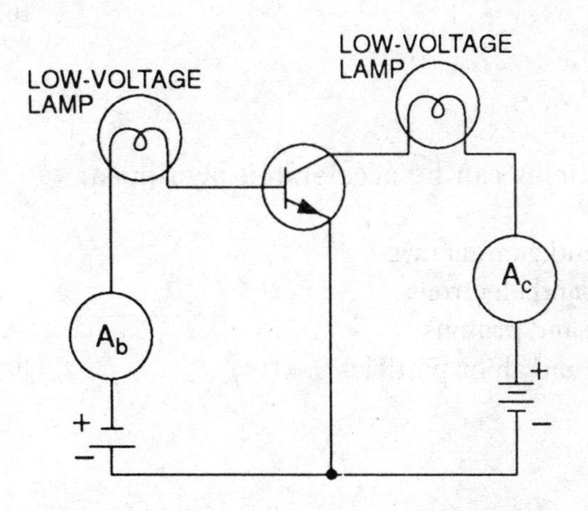

Compared to the reading of ammeter A_c, the reading of ammeter A_b is
1 less
2 greater
3 the same

105_____

GROUP 6—**Nuclear Energy**

If you choose this group, be sure to answer questions **106–115**.

106 An element has an atomic number of 63 and a mass number of 155. How many protons are in the nucleus of the element?
(1) 63
(2) 92
(3) 155
(4) 218

106_____

107 Which particle would generate the greatest amount of energy if its entire mass were converted into energy?
1 electron
2 proton
3 alpha particle
4 neutron
107_____

108 Which particles can be accelerated by a linear accelerator?
1 protons and gamma rays
2 neutrons and electrons
3 electrons and protons
4 neutrons and alpha particles
108_____

109 The equation below represents an unstable radioactive nucleus that is transmuted into another isotope (*X*) by the emission of a beta particle.

$$^{234}_{90}\mathrm{Th} \rightarrow X + {}^{0}_{-1}\mathrm{e}$$

Which new isotope is formed?
(1) $^{234}_{91}\mathrm{Pa}$
(2) $^{234}_{91}\mathrm{Th}$
(3) $^{235}_{90}\mathrm{Pa}$
(4) $^{235}_{90}\mathrm{Th}$
109_____

110 In 4.0 years, 40.0 kilograms of element *A* decays to 5.0 kilograms. The half-life of element *A* is
(1) 1.3 years
(2) 2.0 years
(3) 0.7 year
(4) 4.0 years
110_____

111 The subatomic particles that make up both protons and neutrons are known as

1 electrons
2 nuclides
3 positrons
4 quarks

111_____

112 Which equation is an example of positron emission?

(1) $^{226}_{88}Ra \rightarrow ^{222}_{86}Rn + ^{4}_{2}He$

(2) $^{210}_{82}Pb \rightarrow ^{210}_{83}Bi + ^{0}_{-1}e$

(3) $^{64}_{29}Cu \rightarrow ^{64}_{28}Ni + ^{0}_{+1}e$

(4) $^{14}_{7}N + ^{4}_{2}He \rightarrow ^{17}_{8}O + ^{1}_{1}H$

112_____

113 Which process occurs during nuclear fission?

1 Light nuclei are forced together to form a heavier nucleus.
2 A heavy nucleus splits into lighter nuclei.
3 An atom is converted to a different isotope of the same element.
4 Transmutation is produced by the emission of alpha particles.

113_____

114 In order to increase the likelihood that a neutron emitted from a nucleus will be captured by another nucleus, the neutron should be

1 accelerated through a potential difference
2 heated to a higher temperature
3 slowed down to decrease its kinetic energy
4 absorbed by a control rod

114_____

115 The energy emitted by the Sun originates from the process of

1 fission
2 fusion
3 alpha decay
4 beta decay

115_____

PART III

You must answer *all* questions in this part. [10]

116 Base your answers to parts *a* through *c* on the information below.

A newspaper carrier on her delivery route travels 200. meters due north and then turns and walks 300. meters due east.

a *On a separate paper*, draw a vector diagram following the directions below.

(1) Using a ruler and protractor and starting at point *P*, construct the sequence of two displacement vectors for the newspaper carrier's route. Use a scale of 1.0 centimeters = 50. meters. Label the vectors. [3]

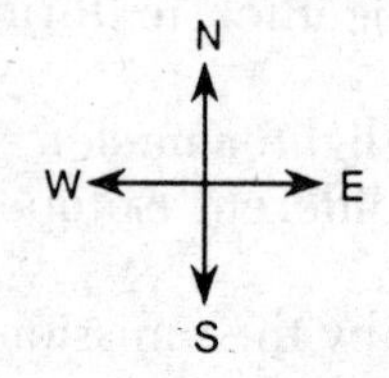

•P

(2) Construct and label the vector that represents the carrier's resultant displacement from point *P*. [1]

b What is the magnitude of the carrier's resultant displacement? [1]

c What is the angle (in degrees) between north and the carrier's resultant displacement? [1]

117 The diagram below shows a spring compressed by a force of 6.0 newtons from its rest position to its compressed position.

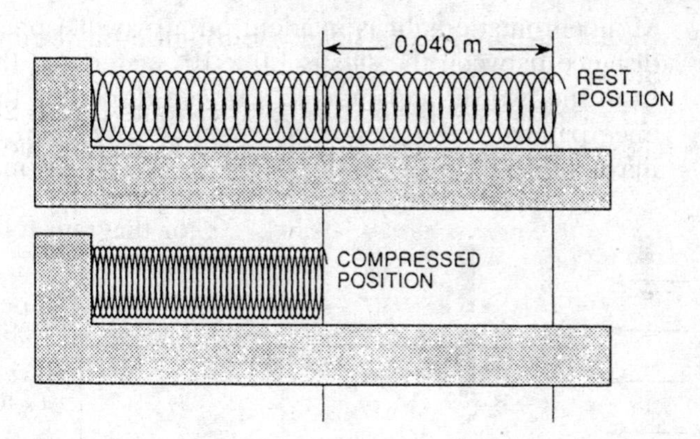

Calculate the spring constant for this spring. [Show all calculations, including equations and substitutions with units.] [2]

118 Base your answers to parts *a* and *c* on the diagram and information below.

Monochromatic light is incident on a two-slit apparatus. The distance between the slits is 1.0×10^{-3} meter, and the distance from the two-slit apparatus to a screen displaying the interference pattern is 4.0 meters. The distance between the central maximum and the first-order maximum is 2.4×10^{-3} meter.

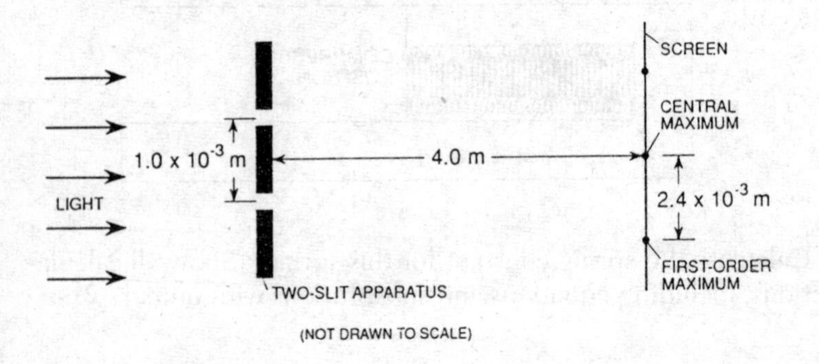

a What is the wavelength of the monochromatic light? [Show all calculations, including equations and substitutions with units.] [2]

b What is the color of the monochromatic light? [1]

c List *two* ways the variables could be changed that would cause the distance between the central maximum and the first-order maximum to increase. [2]

119 Infrared electromagnetic radiation incident on a material produces no photoelectrons. When red light of equal intensity is shone on the same material, photoelectrons are emitted from the surface.

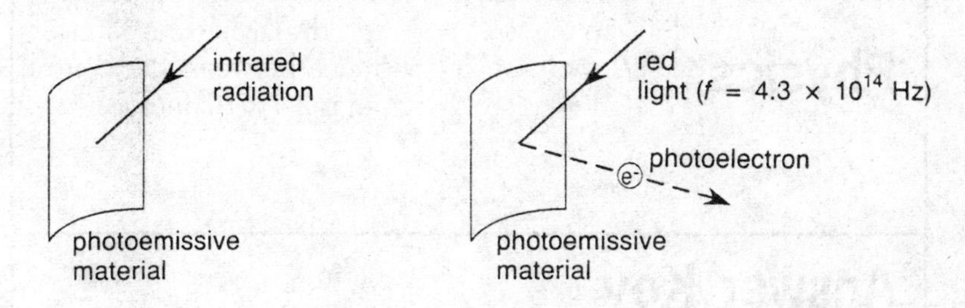

Using one or more complete sentences, explain why the visible red light causes photoelectric emission, but the infrared radiation does not. [2]

Answers June 1993

Physics

Answer Key

PART I

1. 2	11. 1	20. 4	29. 1	38. 1	47. 3
2. 2	12. 2	21. 2	30. 3	39. 2	48. 4
3. 3	13. 3	22. 3	31. 3	40. 3	49. 2
4. 1	14. 1	23. 1	32. 1	41. 4	50. 3
5. 3	15. 2	24. 4	33. 4	42. 2	51. 4
6. 3	16. 4	25. 2	34. 3	43. 2	52. 3
7. 2	17. 3	26. 2	35. 1	44. 3	53. 1
8. 4	18. 2	27. 3	36. 3	45. 3	54. 3
9. 3	19. 3	28. 4	37. 3	46. 1	55. 2
10. 3					

PART II

Group 1	Group 2	Group 3	Group 4	Group 5	Group 6
56. 2	66. 2	76. 2	86. 2	96. 2	106. 1
57. 1	67. 4	77. 1	87. 2	97. 4	107. 3
58. 2	68. 1	78. 4	88. 4	98. 4	108. 3
59. 1	69. 4	79. 3	89. 3	99. 1	109. 1
60. 4	70. 2	80. 2	90. 1	100. 1	110. 1
61. 3	71. 4	81. 4	91. 1	101. 2	111. 4
62. 3	72. 2	82. 3	92. 2	102. 2	112. 3
63. 2	73. 3	83. 3	93. 4	103. 3	113. 2
64. 2	74. 1	84. 2	94. 2	104. 3	114. 3
65. 3	75. 2	85. 1	95. 3	105. 1	115. 2

PART III — See answers explained

Answers Explained

PART I

1. **2** The *average speed* ($\overline{v}$) at which an object travels is defined as the total distance traveled (s) divided by the total time (t) required to travel that distance.

Refer to the *Mechanics* equations in the *Reference Tables*:

$$\overline{v} = \frac{\Delta s}{\Delta t}$$

Substitute the given values into the equation:

$$\overline{v} = \frac{98\text{ m}}{10.\text{ s}}$$
$$= 9.8\text{ m/s}$$

NOTE: The *Reference Table* labels Δs as *displacement*, a vector quantity. *Distance* is defined as the magnitude of the *displacement*. Since direction was not mentioned in this problem, the terms *distance* and *displacement* can be used interchangeably.

Also note that the decimal point after the zero in 10. s indicates that the zero is a significant figure. There are two significant figures in each of the given measurements; therefore the answer must also have two significant figures.

2. **2** The *meter* (m) is the SI unit of length; 1 meter is equal to 39.37 inches, a little more than 3 feet. The average classroom door is approximately 3 feet wide, 7 feet high, and 2 inches thick. The measurement closest to 1 meter is therefore the *width*.

Note that choice 4 can be eliminated because surface area is measured in square meters, not meters.

3. **3** In tackling word problems, it is always advisable to begin by organizing the stated information:

Given: $v_i = 10.\text{ m/s}$ Find: Δs
$a = 5.0\text{ m/s}^2$
$\Delta t = 10.\text{ s}$

We choose from the *Mechanics* equations in the *Reference Tables* the only equation that relates distance (Δs), time (Δt), and acceleration (a):

$$\Delta s = v_i \Delta t + \frac{1}{2}a(\Delta t)^2$$
$$= (10.\text{ m/s})10.\text{ s} + \frac{1}{2}(5.0\text{ m/s}^2)(10.\text{ s})^2$$
$$= 350\text{ m}$$

4. **1** The slope at any point on a distance-versus-time graph equals the instantaneous speed at that point. At $t = 5.0$ seconds, the slope of the given graph is zero; hence the instantaneous speed is 0 m/s, and the object is at rest.

It is instructive to analyze the graph from the start. Since the slope is decreasing, the object is slowing down or decelerating. At $t = 4.0$ seconds, when the slope becomes zero, the object is 4.0 meters from its origin and no longer moving.

5. **3** Begin by organizing the stated information:

Given: $v_i = 0$ m/s Find: $\overline{v}$

$v_f = 50.$ m/s

$\Delta t = 5.0.$ s

We choose from the *Mechanics* equations in the *Reference Tables* the equation that relates average velocity ($\overline{v}$) to initial (v_i) and final (v_f) velocities:

$$\overline{v} = \frac{v_f + v_i}{2}$$

$$= \frac{50.\text{ m/s} + 0\text{ m/s}}{2}$$

$$= 25\text{ m/s}$$

NOTE: The time, $\Delta t = 5.0$ s, is *not* needed to find the solution. A common error that some students make in this type of problem is to divide 50. by 5.0 to get 10., but 10. what? Checking units, we see that 50. m/s divided by 5.0 s yields 10. m/s^2; this is the acceleration, *not* the average speed that was asked for!

6. **3** With air resistance neglected, all objects falling near the surface of the Earth experience the same acceleration due to gravity, regardless of their mass or weight. Since the two balls in this problem, although of different weights, start together from rest and fall with the same acceleration, they will always have the same velocity and height above the ground. However, their *momenta* will differ.

Momentum is a measure of the difficulty in stopping an object and is equal to the product of the object's mass and velocity. Although both balls have the same velocity, the 10-newton ball has a greater mass (due to its greater weight) and will therefore, at the end of 3 seconds of free fall, *have a greater momentum than the 5-newton ball*.

7. **2** By the process of *resolution*, any force can be divided into two or more component forces that, when added together, yield the original force as their resultant. For example, the weight (W) of a box on an incline, which is really the force of gravity on the box, can be considered to be the resultant of a component parallel to the incline ($W_{\parallel}$) and a component perpendicular

($W_\perp$) to the incline. The component of the weight parallel to the incline tends to pull the box down the incline. See the diagram below:

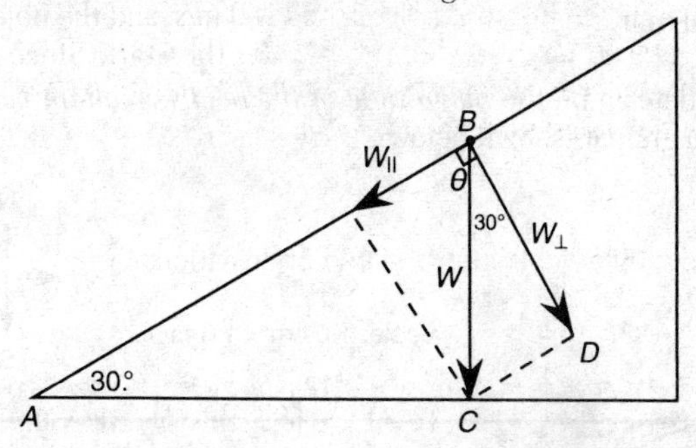

Since W points toward the center of the earth, angle ACB is a right angle. Angle θ is therefore complementary to angle A (30.°). Angle ABD is also a right angle. Therefore angle CBD is complementary to angle θ.

Since angle A and angle CBD are both complementary to the same angle, they must be equal. Hence we conclude that angle CBD equals 30.°.

Using simple trigonometry in triangle CBD, and the fact that sides CD and $W_{||}$ are of equal length, we obtain

$$\sin 30.° = \frac{W_{||}}{W}$$

Rearranging terms gives

$$W_{||} = W \sin 30.°$$
$$= 0.50W$$

As an intuitive check on your work, suppose that angle A were 0°. Since in this case the surface would be horizontal, you would expect the parallel component pulling the box down the incline to be zero. Does $W_{||} = W \sin 0° = 0$? Yes.

8. **4** *Equilibrium* occurs when the sum of the forces acting on an object equals zero. We therefore look for the pair of forces whose resultant is equal and opposite to the given force.

One method of finding the resultant of two forces is to complete the parallelogram that uses the two forces as its sides. The resultant is the diagonal drawn from the common starting point, P (see diagram below).

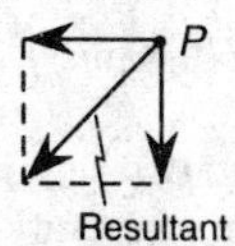

The resultant determined in this fashion for the pair of forces in choice *4* is equal and opposite to the given force, and therefore would produce equilibrium when added to it.

9. **3** To determine the *magnitude of the boat's resultant velocity*, draw a neat vector diagram as shown below.

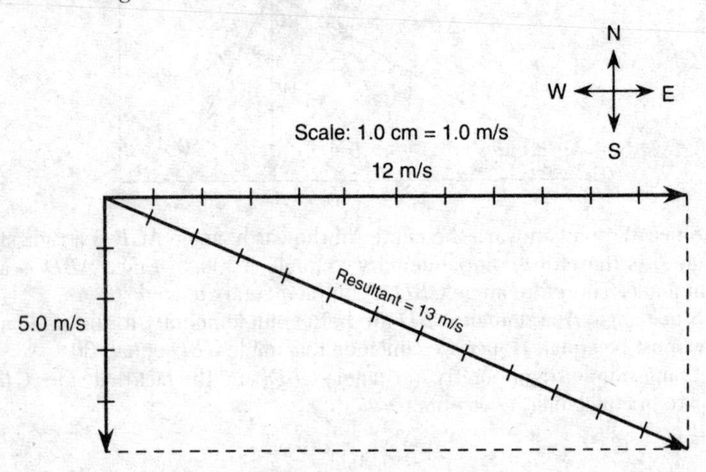

Complete the parallelogram defined by the two given vectors. The resultant is the diagonal drawn from their common starting point. When a scale of, for example, 1.0 cm = 1.0 m/s is used, the length of the diagonal is 13 centimeters, indicating that the resultant velocity is 13 m/s.

Another way of finding the resultant (R) is to use the Pythagorean theorem. You may recognize this problem as an application of the 5-12-13 right triangle. In a right triangle, the square of the length of the hypotenuse equals the sum of the squares of the lengths of the other two sides:

$$R^2 = (5.0 \text{ m/s})^2 + (12 \text{ m/s})^2$$
$$R = 13 \text{ m/s}$$

10. **3** The top wire supports the entire weight of the feeder and birds. Since the system is in equilibrium, the force in the top wire will equal the weight of the feeder and birds. The force in this wire will therefore equal the weight equivalent of a mass of 2.0 kilograms.

To convert from mass (m) to weight (w), we refer to the *Mechanics* equations in the *Reference Tables*:

$$\begin{aligned} w &= mg \\ &= (2.0 \text{ kg})(9.8 \text{ m/s}^2) \\ &= 19.6 \text{ N} \\ &= 20. \text{ N (to two significant figures)} \end{aligned}$$

NOTE: The value for g, the acceleration due to gravity, is obtained from the *List of Physical Constants* in the *Reference Tables*.

11. **1** When an unbalanced force acts on an object, there is a change in the object's momentum ($m\,\Delta v$). The product of the force (F) and the length of time that it acts (Δt) is equal to the product of the object's mass (m) and its change in velocity (Δv).

Since the car in the question is brought to a stop, the final velocity is zero, and therefore:

$$\begin{aligned}\Delta v &= v_{\text{final}} - v_{\text{initial}}\\ &= 0 \quad -(+10.\text{ m/s})\\ &= -10.\text{ m/s}\end{aligned}$$

Refer to the *Mechanics* equations in the *Reference Tables*:

$$\begin{aligned}F\Delta t &= m\Delta v\\ F(0.50\text{ s}) &= (50.\text{ kg})(-10.\text{ m/s})\\ F &= -1.0\times10^{3}\text{ N}\end{aligned}$$

12. **2** The *mechanical energy* of an object is the sum of its kinetic and potential energies. Because of nonconservative forces such as friction, some of a bouncing ball's mechanical energy is converted into heat so that on each bounce the ball rises to a lesser height.

It is assumed that the ball in this problem is dropped *from rest*, and so its initial mechanical energy is all potential. Likewise, the instant it reaches the rebound height of 0.80 meter, its energy is again all potential. Therefore the mechanical energy lost by the ball is equal to the change in its gravitational potential energy. The change in gravitational potential energy (ΔPE) is equal to the product of an object's weight (mg) and the change in its vertical height (Δh).

Refer to the *Energy* equations in the *Reference Tables*:

$$\begin{aligned}\Delta\text{PE} &= mg\,\Delta h\\ &= (0.10\text{ kg})\,(9.8\text{ m/s}^2)(0.20\text{ m})\\ &= 0.\,20\text{ J}\end{aligned}$$

NOTE: The value for g, the acceleration due to gravity, is obtained from the *List of Physical Constants* in the *Reference Tables*.

13. **3** *Kinetic energy* (KE) is the energy that an object has because of its motion. It equals one-half the product of the object's mass and the square of its velocity (or speed).

Refer to the *Energy* equations in the *Reference Tables*:

$$\begin{aligned}\text{KE} &= \frac{1}{2}mv^2\\ &= \frac{1}{2}(70.0\text{ kg})(6.00\text{ m/s})^2\\ &= 1{,}260\text{ J}\end{aligned}$$

NOTE: Kinetic energy is a *scalar*, not a vector, quantity, and therefore it is *not* affected by direction. Do not be fooled by the 30.° angle, which is extraneous information in this problem.

14. **1** According to *Newton's law of universal gravitation*, the gravitational force of attraction (F) between two masses (m_1 and m_2) is proportional to the product of the two masses, and inversely proportional to the square of the distance (r) between them.

Refer to the *Mechanics* equations in the *Reference Tables*:

$$F = \frac{Gm_1m_2}{r^2}$$

where G is the gravitational constant.

In the position shown in the diagram accompanying this problem, the gravitational force (F_1) on the spacecraft from planet P_1 is equal to the gravitational force (F_2) on the spacecraft from planet P_2. Therefore:

$$F_1 = F_2$$

and

$$\frac{Gm_{P_1}m_S}{Y^2} = \frac{Gm_{P_2}m_S}{X^2}$$

Canceling G and m_S (the mass of the spacecraft) from both sides gives

$$\frac{m_{P_1}}{Y^2} = \frac{m_{P_2}}{X^2}$$

We are given that $X > Y$. In order for the two expressions to be equal, we conclude that $m_{P_2} > m_{P_1}$. Therefore the mass of P_1 must be less than the mass of P_2.

We can also reason without equations. The gravitational force decreases with increasing distance. Since P_2 is further from spacecraft S than P_1, its gravitational force on the spacecraft should be diminished. In order to account for the fact that the forces from the two planets are equal, we conclude that the mass of P_2 must be greater that the mass of P_1.

15. **2** As explained in the answer to question 14, the gravitational force of attraction between two masses is *inversely* proportional to the square of the distance between them. When two quantities vary inversely with respect to each other, one quantity increases when the other quantity decreases.

As spacecraft S moves toward planet P_2, the distance separating them decreases; therefore the gravitational force of attraction of P_2 *increases*. At the same time the distance separating the spacecraft from planet P_1 increases; therefore the gravitational force of attraction of P_1 *decreases*. The ratio

$$\frac{\text{gravitational force of } P_2 \text{ on spacecraft}}{\text{gravitational force of } P_1 \text{ on spacecraft}}$$

therefore *increases*.

16. **4** The weight (w) of an object is equal to the product of its mass (m) and the acceleration due to gravity (g) at the object's location.

Refer to the *Mechanics* equations in the *Reference Tables*:

$$w = mg$$

The slope of a weight-versus-mass graph is equal to $\Delta w/\Delta m$. Substituting for w, we obtain

$$\text{Slope} = \frac{\Delta(mg)}{\Delta m}$$
$$= g$$

We conclude that the slope of the given graph represents the *acceleration due to gravity*.

In other words, looking at the graph of weight versus mass, if we take any two points and divide the difference between their weight values (Δw) by the difference between their mass values (Δm), we obtain the value for g.

17. **3** Since each block is moving with constant velocity, we conclude that each is in a state of *equilibrium*. *Newton's first law of motion* describes the state of equilibrium: When no unbalanced force acts on an object, an object at rest will remain at rest, and an object in motion will keep moving with constant velocity. An object is in a state of equilibrium when the vector sum of all forces acting on it is zero.

To calculate the force of friction on each block, we must realize that friction opposes motion and is therefore directed opposite to the direction of each block's motion. Since each block is in a state of equilibrium, and has no net force acting on it, we conclude that the force of friction on each block must exactly equal the applied force on that block.

Since the blocks shown in diagrams *A* and *D* have the same applied force (2 N), the forces of friction are the same (2 N) in these two diagrams.

18. **2** The work (W) done by a force (F) acting on an object is equal to the product of the force and the resultant displacement (Δs) in the direction of the force.

Refer to the *Energy* equations in the *Reference Tables*:

$$W = F\,\Delta s$$

A person must exert a force equal to his or her weight (w) in order to climb stairs. We can therefore substitute w for F in the equation for work:

$$W = w\,\Delta s$$

Since the student's weight is constant as she runs up the stairs, the work done is proportional to her changing displacement. Therefore a graph that represents the relationship between work and time will be similar in appearance to a graph of displacement versus time.

We are told that the student's speed is increasing at a constant rate, that is, she is accelerating. A graph of displacement versus time for uniform accelerated motion is a parabolic curve, as depicted in graph 2. Therefore, this graph best represents the relationship between work and time for the student.

19. **3** The work (W) done by a force (F) acting on an object is equal to the product of the force and the resultant displacement (Δs) in the direction of the force.

Refer to the *Energy* equations in the *Reference Tables*:

$$\begin{aligned} W &= F\,\Delta s \\ &= (5.0\ \text{N})\,(3.0\ \text{m}) \\ &= 15\ \text{J} \end{aligned}$$

NOTE: Do not be fooled by the extraneous information given in this problem, namely, the mass of the object and the time during which the force acts.

20. **4** The change in length or elongation (x) of a spring whose elastic limit has not been reached is proportional to the force (F) applied to stretch the spring. This relationship is known as *Hooke's law*.

Refer to the *Energy* equations in the *Reference Tables*:

$$F = kx$$

where k is the *spring constant*.

On a force-versus-elongation graph this relationship is represented by a straight line sloping obliquely upward to the right, as in graph *4*.

21. **2** An electroscope consists of a vertical metal rod having a metal knob at its upper end, and a pair of light metallic leaves suspended from its lower end. In solid bodies, especially in metals, electrons are much more easily detached than protons, which are bound up in the atomic nucleus. Hence, only the electrons can move easily through any part of the electroscope.

A positively charged electroscope has a deficiency of electrons in its knob and in its leaves. Since like charges repel, the two positively charged leaves diverge. When a positively charged rod is brought near the knob of an already positively charged electroscope, mobile electrons in the leaves will be attracted up toward the knob.

With an even greater deficiency of electrons than before, the positive leaves will repel each other with a stronger force, and therefore *diverge* further.

22. **3** The underlying principle in this problem is the law of *conservation of charge*: The net charge in a closed system is constant; that is, even though charges may be transferred from one object to another, the algebraic sum of all negative and positive charges in the system remains constant.

When the four spheres shown in the diagram for this question touch, *electrons* move from the negatively charged spheres to the positively charged spheres. If the spheres are identical in size, the net charge will be divided equally. (Note that *protons* are bound within the nucleus, and are *not* free to move).

The net charge in this problem is equal to +5 coulombs plus −3 coulombs plus +2 coulombs plus −1 coulomb, or +3 C.

23. **1** An *electric field* is said to exist in any region of space in which an electric force acts on an electric charge. An electric field exists around charged objects. This field can be represented by *lines of force* that show the direction of the electric force on an imaginary *positive* test charge at each point in the field.

A positive test charge would be repelled by other positive objects, and attracted by negative objects. Hence, lines of force always point *away* from a positive charge and *toward* a negative charge.

Looking at the diagram in the question, we conclude that *charge A is negative, and charge B is positive*.

Wrong Choices Explained:

(2) If *A* were positive and *B* were negative, the lines of force would point away from *A* and toward *B*, the exact opposite of the field in the given diagram.

(3) If both *A* and *B* were positive, *each* sphere would have lines of force pointing away from it.

(4) If both *A* and *B* were negative, *each* sphere would have lines of force pointing toward it.

24. **4** The underlying principle here is the law of *conservation of energy*: In any transfer or change of energy that occurs in a closed system, the total energy of the system remains constant. If nonconservative forces such as friction and air resistance are absent, then the total mechanical energy (kinetic + potential) of the system remains constant.

Therefore, as the block in this problem slides down the track, its loss in gravitational potential energy (ΔPE) will be accompanied by an equal gain in its kinetic energy (ΔKE):

$$\Delta PE_{lost} = \Delta KE_{gained}$$

The *gravitational potential energy* (ΔPE) of an object above the ground is equal to the product of its mass (m), the value of the acceleration due to gravity (g), and the vertical height (Δh) to which it was raised.

Refer to the *Energy* equations in the *Reference Tables*:

$$\begin{aligned}\Delta PE &= mg\,\Delta h \\ &= (10\text{ kg})(9.8\text{ m/s}^2)(5.0\text{ m}) \\ &= 490\text{ J} \\ &= 500\text{ J (to one significant figure)}\end{aligned}$$

We conclude that the total amount of gravitational potential energy changed to kinetic energy is approximately 500 joules.

NOTE: The value for g, the acceleration due to gravity, is obtained from the *List of Physical Constants* in the *Reference Tables*.

25. **2** The kinetic energy of the block at point *B* is equal to the kinetic energy that the block started with at point *A* (0 joules) plus the kinetic energy the block gained as it moved down the track from point *A* to point *B* (500 joules). The kinetic energy (KE) of an object is equal to one-half the product of its mass (m) and the speed (v) squared.

Refer to the *Energy* equations in the *Reference Tables*:

$$\text{KE} = \frac{1}{2}mv^2$$

$$500\ \text{J} = \frac{1}{2}(10\ \text{kg})v^2$$

$$v = 10\ \text{m/s}$$

26. **2** At point *C*, the block has both kinetic and potential energy. We can determine the amount of potential energy that the block has at point *C* in the same manner as we did for point *A* in the answer to question 24.

The *gravitational potential energy* (ΔPE) of an object above the ground is equal to the product of its mass (m), the value of the acceleration due to gravity (g), and the vertical height (Δh) to which it was raised.

Refer to the *Energy* equations in the *Reference Tables*:

$$\begin{aligned}\Delta\text{PE} &= mg\,\Delta h\\ &= (10\ \text{kg})\ (9.8\ \text{m/s}^2)\ (2.0\ \text{m})\\ &= 196\ \text{J}\\ &= 200\ \text{J (to one significant figure)}\end{aligned}$$

NOTE: The value for g, the acceleration due to gravity, is obtained from the *List of Physical Constants* in the *Reference Tables*.

27. **3** The *electric field intensity* (E) between two oppositely charged parallel metal plates is equal to the ratio of the potential difference (V) between the two plates to the distance (d) between the plates.

Refer to the *Electricity and Magnetism* equations in the *Reference Tables*:

$$E = \frac{V}{d}$$

If it is assumed that d remains constant, E is directly proportional to V. If V is doubled, then E is also *doubled*.

28. **4** The *potential difference* (V) between two points in an electric field is equal to the work per unit charge required to move the charge between the

two points. For any given charge, we divide the work (W) required to move that charge by the magnitude of the charge (q).

Refer to the *Electricity and Magnetism* equations in the *Reference Tables*:

$$V = \frac{W}{q}$$
$$= \frac{4.8 \times 10^{-19}\ \text{J}}{3.2 \times 10^{-19}\ \text{C}}$$
$$= 1.5\ \text{J/C}$$
$$= 1.5\ \text{V}$$

29. **1** *Resistance* (R) is the opposition to the flow of current. It is defined as the ratio of the potential difference (V) across a conductor to the current (I) flowing through it.

Refer to the *Electricity and Magnetism* equations in the *Reference Tables*:

$$R = \frac{V}{I}$$

When the temperature is held constant, the ratio V/I remains constant for metallic conductors, a relationship known as *Ohm's law*. When the potential difference applied to the ends of a resistor is plotted against the current, the graph is a straight line sloping upward.

The slope of this graph, $\Delta V/\Delta I$, represents the *resistance of the material*.

30. **3** The resistance (R) of a metallic conductor is directly proportional to its length (ℓ) and inversely proportional to its cross-sectional area (A):

$$R = \frac{\rho \ell}{A}$$

where ρ is a constant that depends on the particular material.

The wire with the smallest resistance will have the *shortest length* and the *largest cross-sectional area*. The wire shown in diagram 3 has this combination.

31. **3** An ammeter is a device used to measure current. It must be connected *in series* with the circuit element being measured; in this way the same current that flows through the circuit element flows through the ammeter.

If placed at position 3 in the diagram, an ammeter would measure the current passing *only* through resistor R_1.

Wrong Choices Explained:

(1) An ammeter placed at position 1 is connected *in parallel* to resistor R_2. An ammeter must be connected *in series*, not in parallel, with a circuit element, lest the ammeter suffer damage.

(2) An ammeter placed at position 2 would measure the current passing only through resistor R_2.

(4) An ammeter placed at position 4 would measure the *total* current in the circuit.

32. **1** The *electric energy* (W) needed to operate a device such as a toaster is equal to the product of the power (P) developed and the length of time (t) during which charge flows.

The joule is the unit for electric energy. The unit for electric power is the watt, equal to 1 J/s. Be careful not to confuse the abbreviation W, for watt, with the symbol W, which stands for energy or work.

Refer to the *Electricity and Magnetism* equations in the *Reference Tables*:

$$\begin{aligned} W \text{ (energy)} &= Pt \\ &= (1{,}500 \text{ W})(90.\text{ s}) \\ &= (1{,}500 \text{ J/s})(90.\text{ s}) \\ &= 135{,}000 \text{ J} \\ &= 1.4 \times 10^5 \text{ J (to two significant figures)} \end{aligned}$$

33. **4** When an unmagnetized steel paper clip is placed near a magnet, the steel becomes magnetized. The end of the clip that is nearer to the north pole of the magnet becomes a south pole, and is therefore attracted to the magnet. The clip has become magnetized by the process of *induction*.

As the magnet is lifted, the steel paper clip retains its magnetism and continues to be attracted to the magnet above. However, the magnetic field strength of the horseshoe magnet, which is concentrated near its poles, diminishes as the distance from the poles increases.

Therefore, as the magnet is lifted, the paper clip begins to fall because of *a decrease in the magnetic field strength near the clip*.

34. **3** When a piece of iron is placed between unlike magnetic poles, it becomes magnetized by induction, so that one end becomes a north pole and the other end a south pole. The magnetic field of the piece of iron adds to the external magnetic field, producing a high concentration of lines of force inside the iron. The magnetic field strength is therefore greatest as point C, within the iron.

Note in the accompanying diagram that magnetic lines of force go from north to south *outside* a magnet, but from south to north *inside* a magnet.

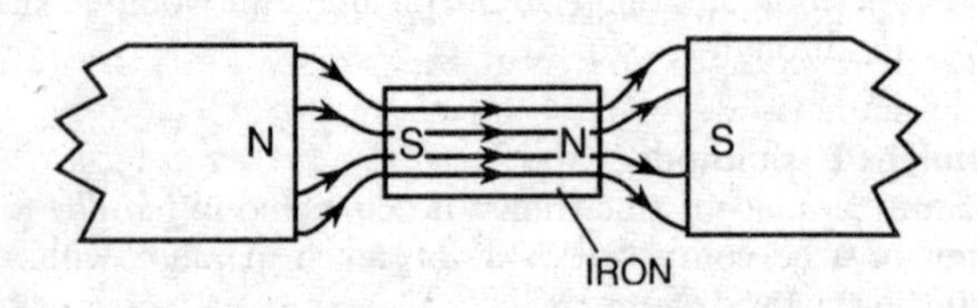

35. **1** There are various hand rules than can be used to determine the direction of the force on a current or on a current-carrying conductor in a magnetic field. Recall that the force experienced by each electron is perpendicular to both the magnetic field and the direction of the current flow, and you can eliminate choices 2 and 4.

Use the following *left-hand rule* for negatively charged particles such as electrons: Hold your outstretched left hand so that your thumb points in the direction of electron flow, and your extended fingers point in the direction of the magnetic field (north to south). Your palm will now face in the direction of the force exerted on the current.

When you apply this rule to the given situation, your palm points up, in the direction of arrow *A*.

36. **3** The magnetic field around a current-carrying coil of wire resembles that of a permanent bar magnet. To determine its polarity, use the *left-hand rule* for a coil (solenoid): Grasp the coil with your left hand so that the fingers encircling the coil point in the direction of electron flow. Your extended thumb will then point in the direction of the north pole. When you apply this rule to the given situation, your thumb points to *C*, as shown in the diagram below.

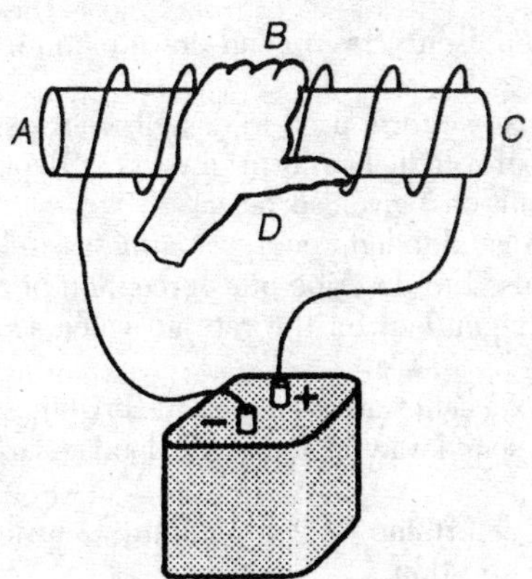

We conclude that the *N*-pole of this coil is the closest to *C*.

37. **3** This question involves the *Doppler effect*, that is, the variation in the observed wavelength and frequency when there is a relative motion between the source of waves and the receiver (observer).

The moving car in this problem reflects the radar waves back to the police officer, and therefore acts like a moving source of radar waves. The police officer is the receiver or observer of these reflected rays. The given diagram shows a *decrease* in the observed wavelength (and therefore an increase in the observed frequency) of the reflected waves since the source is moving toward the receiver.

The opposite effect would be observed if the source were moving away from the police car.

Wrong Choices Explained:

(1) *Interference* is the effect produced by two or more waves that pass through the same region at the same time. By the process of *superposition*, the resultant displacement of each particle in the medium is the algebraic sum of the displacements due to all of the individual waves. *Constructive interference* occurs when the waves from all sources arrive in phase and reinforce each other.

(2) *Refraction* is the bending that a wavefront experiences when it enters a second medium in which its speed changes.

(4) *Total internal reflection* is the phenomenon that a light beam experiences when it travels from a more optically dense medium toward a less dense medium at an angle that exceeds the critical angle.

38. **1** A *transverse pulse* is a pulse in which the particles of the medium vibrate at *right angles* to the direction in which the wave moves.

For example, in the diagram given in the question, a vertical disturbance is shown moving along a string horizontally. Each segment of the string, including point *P*, will therefore vibrate in an up-and-down direction when the disturbance passes.

Diagram *1* best represents this up-and-down motion.

39. **2** *Brightness* is a term used to describe our perception of the intensity of illumination of a light beam; brightness is dependent on the rate at which light energy falls on a given surface.

As brightness is related to light energy, *loudness* is related to sound energy. *Loudness* is a term used to describe our perception of the intensity of sound waves; loudness is dependent on the rate at which sound energy falls on a given surface.

As the amplitude of a light wave increases, the brightness of the light increases. As the amplitude of a sound wave increases, the loudness of the sound increases.

40. **3** First we determine the wavelength (λ) of the wave shown in the diagram given in the question.

The wavelength of a periodic wave is the distance between two consecutive points in phase, such as two troughs. Thus, in the diagram below, distances *AB* and *BC* are each equal to one wavelength. Since two wavelengths measure 3.0 meters, we conclude that one wavelength is equal to 1.5 meters.

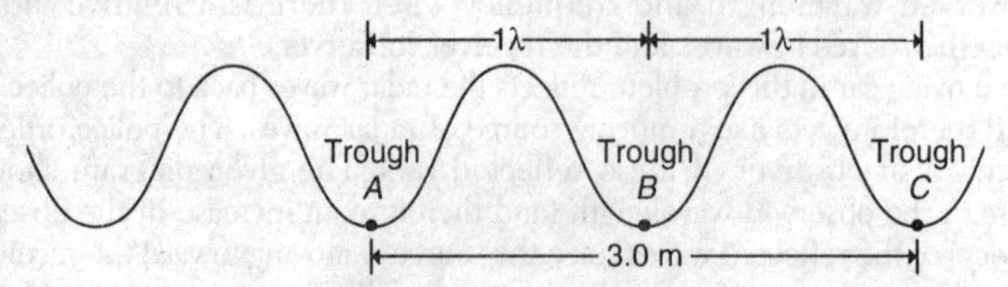

Now we can calculate the *speed* of the wave. The speed (v) of any wave is equal to the product of its frequency (f) and its wavelength (λ).

Refer to the *Wave Phenomena* equations in the *Reference Tables*:

$$\begin{aligned} v &= f\lambda \\ &= (40.\text{ Hz})(1.5\text{ m}) \\ &= 60.\text{ m/s} \end{aligned}$$

41. **4** The *frequency* (f) of a periodic wave is the number of vibrations occurring per second. The *period* (T) of a wave is the time required for the completion of one vibration. The period is the reciprocal of the frequency; that is, $T = 1/f$.

Since the frequency and the period are interrelated, if two waves have the same frequency, they must also have the same period.

Wrong Choices Explained:

(1) The fact that two waves have the same frequency does *not* mean that their *phase* relationship is determined. For example, two waves with the same frequency might even arrive at a given point with *opposite* phases, in which case destructive interference would occur.

(2) The *amplitude* of a wave is the maximum displacement of each particle in the medium from its equilibrium position. It depends on the energy used to create the wave at its source, *not* on the frequency of the wave.

(3) The *intensity* of a wave is a measure of the rate at which the wave's energy falls on a given surface. Two waves with the same frequency may have different intensities.

42. **2** The speed (v) of any wave is equal to the product of its frequency (f) and its wavelength (λ).

Refer to the *Wave Phenomena* equations in the *Reference Tables*:

$$v = f\lambda$$

$$3.0\times10^{8}\text{ m/s} = (1.0\times10^{10}\text{ Hz})(\lambda)$$

$$\lambda = 3.0\times10^{-2}\text{ m}$$

NOTE: Microwaves are part of the *electromagnetic spectrum*, which also contains radio waves, infrared waves, visible light, ultraviolet waves, X-ray waves, and gamma-ray waves. These differ in frequency and wavelength, but all travel with the *same* speed, the speed of light. We find the speed of light in a vacuum (c) in the *List of Physical Constants* in the *Reference Tables*.

43. **2** When two or more waves pass through the same region at the same time, as shown in the diagram given in the question, the phenomenon of *interference* will occur. By a process called *superposition*, the resultant displacement of each particle in the medium will equal the algebraic sum of the

displacements due to all the individual waves. Depending on a point's location in the medium, the combined effect of the two wavefronts reaching that point may be such that the resultant amplitude is either reinforced or diminished.

Maximum destructive interference occurs at the points where waves from the two sources arrive *out of phase*—for example, where the crest of one wave overlaps the trough of another. At these points the resultant amplitude is diminished.

This is the situation at point *B*, where we see a solid line (indicating a crest) overlap a dashed line (indicating a trough). Here maximum destructive interference is occurring.

Wrong Choices Explained:

(1) *Constructive interference* occurs at each point where waves from the two sources arrive *in phase*—for example, at point *A*, where two troughs arrive at the same time. At such a point, the result is a wave of increased amplitude.

(3) Point *C* lies approximately half-way between the crest and the trough of each wave. Therefore the two waves reach point *C* with approximately the *same* phase. In order for maximum destructive interference to occur, the waves must have *opposite* phases.

(4) At point *D* two crests arrive at the same time; therefore *constructive* interference will occur.

44. **3** *Standing waves* are produced when *two* waves of the same frequency and amplitude travel in opposite directions through a medium, as may occur when waves traveling down a stretched string interfere with the waves that are reflected from the far end.

Nodes, also called nodal points, represent areas of maximum *destructive* interference, as occur, for example, when a crest of the originating wave meets the trough of the reflected wave. At these points in the medium, there will be zero displacement from the equilibrium position.

Antinodes, on the other hand, represent areas of maximum *constructive* interference, as may occur, for example, when two crests or two troughs meet. In the diagram below, the antinodes are labeled *A*.

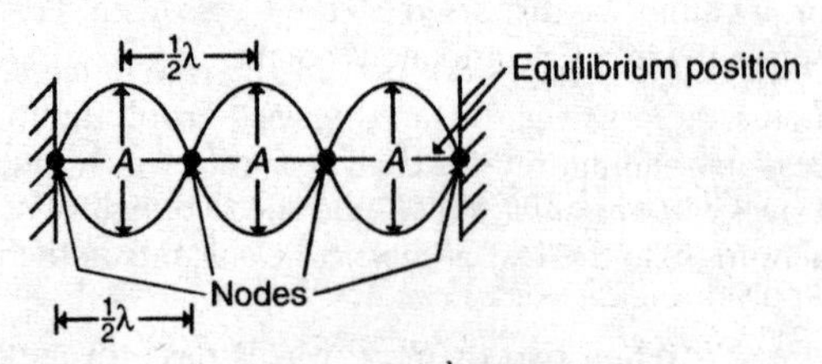

Each segment of a standing wave is equal to $\frac{1}{2}$ the wavelength; therefore the distance between two successive nodes, or between two successive antinodes, is $\frac{1}{2}$ *wavelength*.

45. **3** Use *Snell's law* to solve this problem.

Refer to the *Wave Phenomena* equations in the *Reference Tables*:

$$n_1 \sin \theta_1 = n_2 \sin \theta_2$$

where subscript 1 refers to the incident medium (medium *X*), subscript 2 refers to the second or refractive medium (air), θ_1 is the angle of incidence (30.°), and θ_2 is the angle of refraction (60.°).

The index of refraction of air (n_2) is given as 1.00 in the table entitled *Absolute Indices of Refraction* in the *Reference Tables*. Refer to the diagram below:

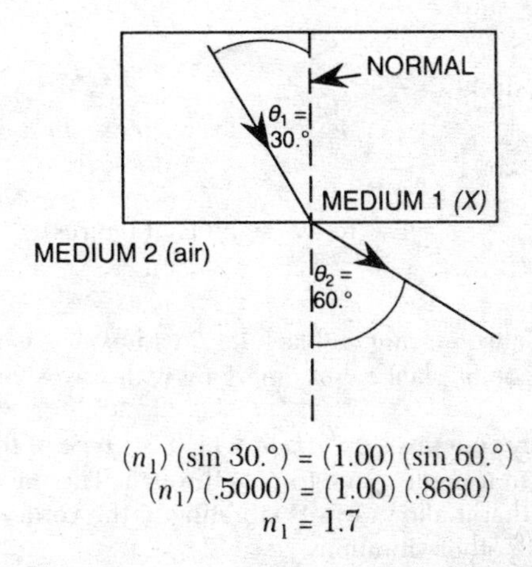

$$(n_1)(\sin 30.°) = (1.00)(\sin 60.°)$$
$$(n_1)(.5000) = (1.00)(.8660)$$
$$n_1 = 1.7$$

46. **1** *Polychromatic* light contains waves of different frequencies, and is therefore a mixture of many colors. Since each frequency of visible light travels at a different speed in glass, each color will be refracted by a different amount.

A glass prism can therefore separate polychromatic light, such as ray *B* in the given diagram, into its component colors. This phenomenon is called *dispersion*.

Ray *A*, however, consists of *monochromatic* light; that is, it contains waves of only a single frequency and a single color. Since ray *A* has light of only one color, it *cannot* be dispersed by the glass prism. Only *polychromatic* light can be separated into its component colors.

Wrong Choices Explained:

(2) *Reflection* is the bouncing back that occurs to a wave when it reaches a barrier or the interface between two different media. *Both* monochromatic and polychromatic waves can undergo reflection.

(3) *Polarization* is the separation of a beam of light so that the vibrations are all in the same plane. *Both* monochromatic and polychromatic waves can be polarized.

(4) *Refraction* is the bending of a wave from its original direction as it obliquely enters a second medium in which its speed changes. *Both* monochromatic and polychromatic waves can experience refraction.

47. **3** The sine of the *critical angle* of incidence (θ_c) relative to air is equal to the reciprocal of the *index of refraction* (n) of the optical medium.

Refer to the *Wave Phenomena* equations in the *Reference Tables*:

$$\sin\theta_c = \frac{1}{n}$$

$$\sin 44° = \frac{1}{n}$$

$$.6947 = \frac{1}{n}$$

$$n = 1.4 \text{ (to two significant figures)}$$

48. **4** We see light entering a "black box" with waves vibrating in both the vertical and horizontal planes, but emerging with waves vibrating only in the vertical plane.

The box most likely contains a *polarizer*, which is a type of filter that allows only waves vibrating in a single plane to pass through. The polarizer is oriented inside the box so that it allows waves vibrating in the vertical plane to pass through, but blocks all other vibrations.

Wrong Choices Explained:

(1) A *prism* can separate polychromatic, or white light, into its component colors.

(2) A *converging lens* can focus parallel rays to a single point.

(3) A *double slit* can create an interference pattern.

49. **2** This problem involves the *photoelectric effect*. According to quantum theory, electromagnetic radiation such as light is composed of energy-carrying particles called *photons*. The *intensity* of the light is a measure of the number of photons in the beam.

If the intensity of radiation incident on a photoemissive surface is increased, more photons will reach the surface each second. For frequencies above the threshold frequency, each photon absorbed at the surface will cause an electron to be released. Therefore the number of photoelectrons emitted by the surface is proportional to the intensity of light that falls on the surface.

This relationship is best represented by graph 2, with its straight line sloping obliquely upward from the origin.

50. **3** When light shines on the surface of a photoemissive material, the *maximum kinetic energy* (KE_{max}) of each ejected photoelectron is equal to the difference between the energy of the absorbed photon (hf) and the work function (W_0) of the material.

Refer to the *Modern Physics* equations in the *Reference Tables*:

$$\begin{aligned} KE_{max} &= hf - W_0 \\ &= 5.0\,\text{eV} - 2.0\,\text{eV} \\ &= 3.0\,\text{eV} \end{aligned}$$

NOTE: In this problem we did not have to calculate the energy of each photon because it was given. If the frequency (f) of the incident photon had been given instead of its energy, we could calculate the energy by using the equation $E_{photon} = hf$, where h is Planck's constant.

51. **4** This problem deals with Rutherford's famous alpha-particle-scattering experiment. An alpha particle is a helium nucleus; it contains two protons and two neutrons and has a charge of +2. The nucleus of each metallic atom in the foil at which the alpha particles are fired also contains protons and neutrons and is positive.

Alpha particles are scattered, that is, deflected, in hyperbolic paths as they approach the metallic nuclei because of the *electric forces of repulsion between the positive nuclei and the positive alpha particles*.

52. **3** The *momentum* (p) of a photon is inversely proportional to its wavelength (λ).

Refer to the *Modern Physics* equations in the *Reference Tables*:

$$p = \frac{h}{\lambda}$$

NOTE: The value for h, Planck's constant, is found in the *List of Physical Constants* in the *Reference Tables*.

$$\begin{aligned} &= \frac{6.6 \times 10^{-34}\ \text{J} \bullet \text{s}}{5.9 \times 10^{-7}\ \text{m}} \\ &= 1.1 \times 10^{-27}\ \text{kg} \bullet \text{m/s} \end{aligned}$$

53. **1** Electric power (P) is defined as the rate at which electric energy is supplied or used in a circuit.

Refer to the *Electricity and Magnetism* equations in the *Reference Tables*:

$$P = VI = I^2R = \frac{V^2}{R}$$

We see three different equations that relate electric power to the electric potential difference (V), current (I), and resistance (R). In this problem voltage and resistance are involved. Therefore we choose the equation:

$$P = \frac{V^2}{R}$$

We see that, with a constant voltage, the power dissipated is inversely proportional to the resistance. When two quantities vary inversely with respect to each other, one quantity decreases when the other quantity increases.

Therefore, if R increases, P *decreases*.

54. **3** The resistors in circuit A are connected in parallel. Calculate the total resistance (R_t) of circuit A.

Refer to the *Electricity and Magnetism* equations in the *Reference Tables*. For *parallel circuits*:

$$\frac{1}{R_t} = \frac{1}{R_1} + \frac{1}{R_2} + \frac{1}{R_3}$$

$$= \frac{1}{12\Omega} + \frac{1}{12\Omega} + \frac{1}{12\Omega}$$

$$R_t = 4\Omega$$

The resistors in circuit B are connected in series. Calculate the total resistance (R_t) of circuit B.

Refer to the *Electricity and Magnetism* equations in the *Reference Tables*. For *series circuits*:

$$R_t = R_1 + R_2 + R_3$$

$$= 1\Omega + 2\Omega + 1\Omega$$

$$= 4\Omega$$

We conclude that, compared to the total resistance of circuit A, the total resistance of circuit B is *the same*.

55. **2** The diagram below shows the normal drawn perpendicular to the boundary separating medium *A* from medium *B*. The arrow that represents the path of the waves in medium *A* has been extended with dashed lines to show what its direction would be without bending.

We see that the path of the waves has bent *away* from the normal as the waves entered medium *B*. When waves obliquely (at an angle) enter a medium in which they travel more quickly, they will bend away from the normal, as depicted in the diagram that accompanies this question.

We conclude that, as the waves enter medium *B*, their speed *increases*.

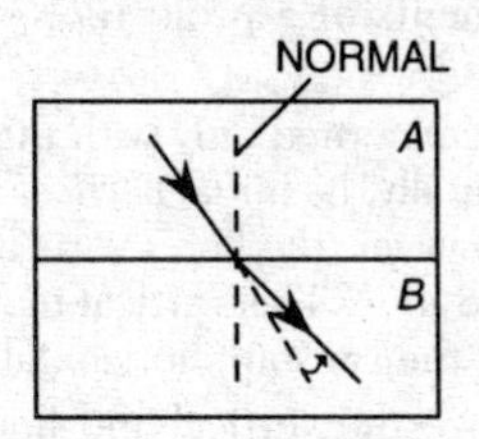

PART II

GROUP 1—**Motion in a Plane**

56. **2** An object that is thrown or shot into the air and that experiences no other forces than the pull of gravity is called a *projectile*. Its motion can be described by separating it into vertical and horizontal components. The vertical and horizontal components of a projectile's motion are *independent* of each other.

In this problem we are concerned only with the vertical component. Since the ball was thrown horizontally, its initial *vertical* velocity (v_{iy}) is zero. Under the influence of gravity, however, the ball's vertical velocity increases with an acceleration (a) equal to 9.8 m/s². This vertical motion occurs even as the ball moves horizontally along at the constant horizontal speed of 20. m/s.

It is advisable to organize the vertical and horizontal data separately. In this problem, only the vertical motion is examined.

Given vertical information: $v_{iy} = 0$
$\Delta s = 19.6$ m
$a = 9.8$ m/s²   Find: Δt (time)

We choose the equation having only these variables from the *Mechanics* equations in the *Reference Tables*:

$$\Delta s = v_i \Delta t + \frac{1}{2} a(\Delta t)^2$$

$$19.6\text{ m} = (0)\Delta t + \frac{1}{2}(9.8\text{ m/s}^2)(\Delta t)^2$$

$$\Delta t = 2.0\text{ s}$$

NOTE: The value for the acceleration due to gravity is found in the *List of Physical Constants* in the *Reference Tables*.

57. **1** As explained in the answer to question 56, the vertical and horizontal components of a projectile's motion are *independent* of each other. Since the book in this problem is pushed horizontally, with no component of velocity in the vertical direction, its initial vertical velocity is 0 m/s.

Once the book is off the desk, however, its vertical velocity will begin to increase because of the influence of gravity.

58. **2** As we can see from the sketch on the next page, the longest horizontal distance (also called the *range*) that a projectile can attain occurs when the launching angle θ is equal to 45°.

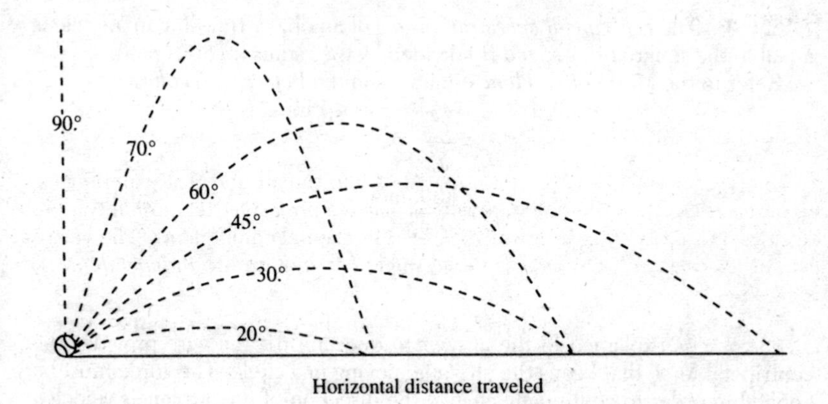

59. **1** As soon as the cyclist releases the ball, the ball becomes a projectile. As explained in the answer to question 56, the motion of a projectile can be described by separating it into vertical and horizontal components. The vertical and horizontal components of a projectile's motion are *independent* of each other.

In this problem we are concerned with only the horizontal component of the ball's motion. When released, the ball's initial horizontal speed is equal to the speed of the bicycle. Since the vertical motion of the ball does not affect its horizontal motion, and there are no horizontal forces acting on the ball (neglecting air resistance), we conclude that the horizontal motion of the ball remains constant and equal to the speed of the bicycle throughout the ball's trajectory.

On a distance-versus-time graph, an object traveling with constant speed is represented by a straight line sloping upward, as shown in the first graph in *set 1*. On a speed-versus-time graph, an object traveling with constant speed is represented by a horizontal line, as shown in the second graph in *set 1*.

60. **4** According to *Kepler's first law*, the path of each planet traveling around the Sun is an *ellipse* with the Sun at one focus. An elliptical orbit is oval in shape.

On average, Pluto has a larger orbital radius than Neptune. However, since Pluto's orbit is more elliptical than Neptune's, the two orbits *overlap* so that Pluto is sometimes on the inside track and therefore closer to the Sun.

61. **3** To maintain uniform circular motion, there must be a *centripetal force* that pulls an object toward the center of curvature of its path.

The centripetal force acting on the model airplane in this question is provided by the tension in the wire and is directed toward point *C*, the center of the horizontal circle.

62. **3** The *centripetal acceleration* (a_c) of an object traveling in a circle is equal to the square of its speed (v) divided by the radius (r) of its path.

Refer to the *Motion in a Plane* equations in the *Reference Tables*:

$$a_c = \frac{v^2}{r}$$

$$= \frac{(30.\ \text{m/s})^2}{20.\ \text{m}}$$

$$= 45\ \text{m/s}^2$$

63. **2** As explained in the answer to question 61, the wire provides the centripetal force that keeps the airplane moving in a circle. This force must be applied in order to continually change the direction of the airplane's velocity. For an object undergoing uniform circular motion, such as the airplane in this question, the velocity vector is always tangent to the circle and perpendicular to the centripetal force and acceleration.

If the wire breaks, *Newton's law of inertia* explains the motion that follows: When no net force acts on a body in motion, it will continue to move in the same direction in a straight line with a constant speed. In the absence of the centripetal force that changes the direction of its velocity, the airplane will fly off *tangentially* in a straight line, toward point *B*, as shown in the diagram below.

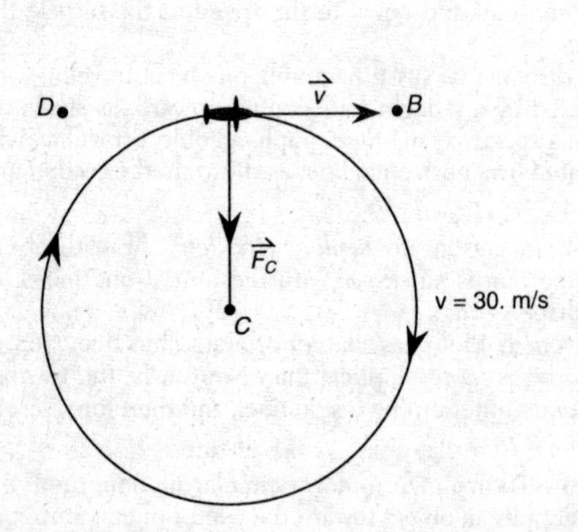

64. **2** The force required to keep the motorcycle in a circular path is the *centripetal force*, supplied by the friction between the tires and the ground.

The centripetal force (F_c) acting on an object moving at constant speed in a circular path is equal to the product of the object's mass (m) and the square of its speed (v), divided by the radius (r) of the path.

Refer to the *Motion in a Plane* equations in the *Reference Tables*:

$$F_c = \frac{mv^2}{r}$$

From this relationship we see that, if the speed of the motorcycle is increased, the force required to keep it in the same circular path also *increases*.

65. **3** *Kepler's second law* states that an imaginary line drawn from the Sun to a planet sweeps out equal areas in equal periods of time.

Since the two shaded areas in the diagram accompanying the question are equal, we conclude that time interval t_1 is *the same as* t_2.

GROUP 2—**Internal Energy**

66. **2** Checking the *Heat Constants* chart in the *Reference Tables*, we find that the melting point of ethyl alcohol is –117°C and its boiling point is 79°C. The difference is 79°C – (–117°C) = 196 Celsius degrees.

The kelvin (K) is the SI unit of temperature. A kelvin measures the *same* change in temperature as a Celsius degree. Therefore a change of 196 C° is the same as a change of *196* K.

67. **4** As a substance changes phase from a gas to a liquid, it releases energy. The amount of heat energy released when 1 kilogram of a substance changes from the gaseous to the liquid phase at its condensation point is called the *heat of vaporization*.

Checking the *Heat Constants* chart in the *Reference Tables*, we find the heats of vaporization, in kilojoules per kilogram, for the substances in the question: alcohol, 855; mercury, 295; water, 2,260; silver, 2,370.

We conclude that the substance that releases the most energy when it condenses is *silver*.

68. **1** According to the *second law of thermodynamics*, heat energy flows spontaneously from a hotter to a colder region; the reverse is *never* true.

Of the four samples of metal listed, only *platinum* at 60°C is colder than the block of lead at 100°C. Therefore heat energy will flow from the hotter lead to the colder platinum. As the platinum absorbs heat energy, its internal energy will increase.

69. **4** The *ideal gas law* states that the product of the pressure (P) and the volume (V) of an ideal gas is *directly proportional* to the product of the number of molecules (n) in the sample and their absolute temperature (T).

This relationship can be expressed as follows:

$$PV = nRT$$

where R is a constant that is the same for all gases.

The graph that best represents a direct proportion is a straight line sloping upward, that is, graph *4*.

The answer to this question could also be found by an examination of the *combined gas laws* used so often in chemistry:

$$\frac{P_1V_1}{T_1} = \frac{P_2V_2}{T_2}$$

From this relationship we can conclude that the product of the pressure and the volume ($P \bullet V$) for a given mass of ideal gas is directly proportional to the absolute temperature (T).

70. **2** The amount of heat energy (Q) that must be added to a substance to raise its temperature (without changing its phase) is equal to the product of its mass (m), the specific heat (c), and the change in temperature (ΔT_c). The change in temperature can be expressed as:

$$\Delta T_c = T_{\text{final}} - T_{\text{initial}}$$

Refer to the *Internal Energy* equations in the *Reference Tables*:

$$\begin{aligned} Q &= mc\,\Delta T_c \\ &= mc\,(T_{\text{final}} - T_{\text{initial}}) \\ &= (10\text{ kg})(4.19\text{ kJ/kg}\bullet\text{C°})(100\text{°C} - 20\text{°C}) \\ &= 3{,}352\text{ kJ} \\ &= 3{,}400\text{ kJ (approximately)} \end{aligned}$$

The value for the specific heat (c) of water can be obtained from the *Heat Constants* chart in the *Reference Tables*.

71. **4** The amount of heat energy (Q_v) required to change a liquid to a gas at *its boiling point* is equal to the product of the liquid's mass (m) and its heat of vaporization (H_v).

Refer to the *Internal Energy* equations in the *Reference Tables*:

$$\begin{aligned} Q_v &= mH_v \\ &= (5.0\text{ kg})\,(2{,}260\text{ kJ/kg}) \\ &= 11{,}300\text{ kJ} \\ &= 11{,}000\text{ kJ (approximately)} \end{aligned}$$

NOTE: The value for the heat of vaporization (H_v) of water can be obtained from the *Heat Constants* chart in the *Reference Tables*.

72. **2** A dissolved salt lowers the freezing point and *increases* the boiling point of water.

73. **3** The graph given in the question shows how the temperature of 1.0 kilogram of a substance varies as heat is added at a constant rate.

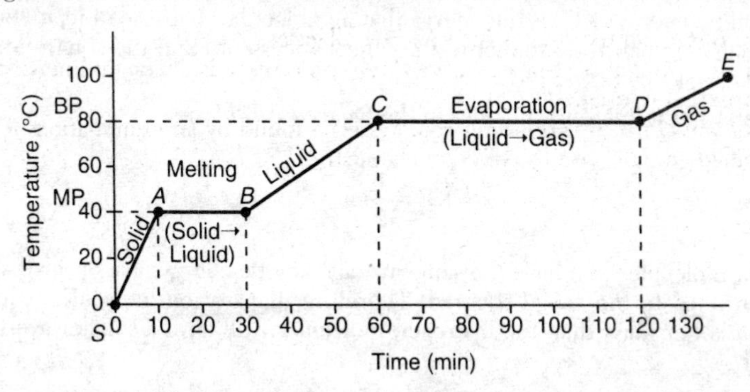

The annotated graph shown above is typical of the heating curve of any crystalline solid. The process starts at a temperature below the melting point of the substance. Interval *SA*, which slants up, represents the *solid phase* of the substance.

After 10 minutes, the graph levels off to the horizontal segment *AB*. The fact that there is no temperature change along this plateau indicates that a *change of phase* is occurring; along *AB* the solid is *melting* into a liquid. At *B*, all of the solid has been melted, and the temperature of the liquid begins to rise. Oblique segment *BC* thus represents the *liquid phase*.

The second horizontal plateau, segment *CD*, represents another phase change; during this stretch the liquid *evaporates* into a gas by boiling. At *D*, all of the liquid has been evaporated, and the temperature of the gas begins to rise above the boiling point. Segment *DE* represents the *gaseous phase* of the substance.

We are asked to determine the amount of heat that was added to change the substance from a liquid at its melting point (point *B* in the diagram above) to a vapor at its boiling point (point *D* in the diagram). Reading the graph, we see that point *B* is reached after 30 minutes, and point *D* after 120 minutes. Thus heat was added for (120–30) minutes = 90 minutes.

Since heat is added at the constant rate of 100 kJ/min, the total heat added is equal to 100 kJ/min × 90 min = 9,000 kJ.

74. **1** Increasing the pressure *raises* the boiling point of water and lowers the melting point of ice.

Therefore, as pressure is applied to a snowball, the melting point of the snow *decreases*.

75. **2** The fact that both samples are at the *same temperature* is important. The temperature of any object is a measure of the average kinetic energy of the molecules or atoms in the object. Since the samples of oxygen and

hydrogen in the problem are at the same temperature, we conclude that their molecules have the *same average kinetic energy*.

Kinetic energy (KE) is the energy that an object has because of its motion. It equals one-half the product of the object's mass (m) and the square of its speed or velocity (v).

Refer to the *Energy* equations in the *Reference Tables*:

$$KE = \frac{1}{2}mv^2$$

For molecules that have the same average kinetic energy, mass is inversely proportional to the speed squared. Therefore, the oxygen molecules, which have a larger mass than the hydrogen molecules, will have a smaller average speed.

Compared to the average speed of the oxygen molecules, the average speed of the hydrogen molecules will be *greater*.

GROUP 3—**Electromagnetic Applications**

76. **2** A *transformer* is a device that changes the voltage of an alternating current into a smaller or larger voltage of alternating current. It consists of a *primary coil* and a *secondary coil* that are wound on a common soft-iron ring called a *core*. The percent efficiency of a transformer is equal to the ratio of the power output from the secondary (s) coil to the power input supplied to the primary (p) coil, multiplied by 100.

Refer to the *Electromagnetic Applications* equations in the *Reference Tables*:

$$\% \text{ Efficiency} = \frac{V_s I_s \text{ (power output)}}{V_p I_p \text{ (power input)}} \times 100$$

Substituting the students' data for trial 1 gives

$$\% \text{ Efficiency} = \frac{(16.0 \text{ V})(2.0 \text{ A})}{(3.0 \text{ V})(12.0 \text{ A})} \times 100$$
$$= 89\% \text{ (to two significant figures)}$$

77. **1** The ratio of the number of turns of wire in the primary coil (N_p) to the number of turns in the secondary coil (N_s) is equal to the ratio of the voltage across the primary (V_p) to the voltage induced in the secondary (V_s).

Refer to the *Electromagnetic Applications* equations in the *Reference Tables*:

$$\frac{N_p}{N_s} = \frac{V_p}{V_s}$$
$$= \frac{9.0 \text{ V}}{54.0 \text{ V}}$$
$$= \frac{1}{6}$$

For trial 3 the ratio $\frac{N_p}{N_s}$ is 1 : 6.

78. **4** We know that every transformer must have an efficiency of less than 100% because of the heating that occurs in the coils and core. Therefore the power output from the secondary coil is always smaller than the power input to the primary coil, so that the ratio of output to input is less than 1.

Using the equation for transformer efficiency that was explained in the answer to question 76,

$$\% \text{ Efficiency} = \frac{V_s I_s \text{(power output)}}{V_p I_p \text{ (power input)}} \times 100$$

we analyze the data from each trial:

Trial 1. We determined that the efficiency of the first set of coils is 89% (see question 76).

Trial 2. $$\% \text{ Efficiency} = \frac{(8.0 \text{ V})(2.2 \text{ A})}{(6.0 \text{ V})(3.0 \text{ A})} \times 100$$
$$= 98\%$$

Trial 3. $$\% \text{ Efficiency} = \frac{(54.0 \text{ V})(0.7 \text{ A})}{(9.0 \text{ V})(4.3 \text{ A})} \times 100$$
$$= 98\%$$

Trial 4. $$\% \text{ Efficiency} = \frac{(5.0 \text{ V})(9.0 \text{ A})}{(12.0 \text{ V})(2.5 \text{ A})} \times 100$$
$$= 150\%$$

Since the efficiency of a real transformer cannot equal or exceed 100%, and the data for trial 4 show an efficiency of 150%, we conclude that an error was most likely made in recording the data in trial 4.

79. **3** If a straight conductor is moved through a uniform magnetic field, an electric potential difference (electromotive force) may be induced across the conductor, depending on the direction of motion. The induced potential difference is a maximum when the conductor moves in a direction that is

perpendicular to itself and the field. The magnitude of the potential difference (V) induced between the ends of the wire is equal to the product of the magnetic flux density (B), the length (l) of the conductor in the field, and the speed (v) with which the conductor moves.

Refer to the *Electromagnetic Applications* equations in the *Reference Tables*:

$$V = Blv$$
$$= (20.\ \text{T})\ (0.50\ \text{m})\ (4.0\ \text{m/s})$$
$$= 40.\ \text{V}$$

80. **2** A voltmeter is used to measure potential difference across any two points in a circuit. It is connected in parallel across the component whose potential difference is being measured, thereby creating another path for electrons.

The internal resistance of a voltmeter must be very high in comparison to the resistance of the circuit being measured so that the device draws very *little current from the circuit* and therefore has a negligible effect on the circuit.

81. **4** Refer to the *Electricity and Magnetism* equations in the *Reference Tables*.

When a particle having a charge q is placed in an electric field of strength E, it will experience a force F, given by the equation

$$E = \frac{F}{q}$$

or

$$F = qE$$

Since the charges on a proton and an electron are of equal magnitude, the particles will experience an equal force when they enter the same electric field.

The relationship between force (F) and acceleration (a) is expressed by *Newton's second law of motion*, which is found in the *Mechanics* equations in the *Reference Tables*:

$$F = ma$$

where m is the mass of the object. If the force is held constant, then mass and acceleration are inversely proportional to one another.

In the *List of Physical Constants* in the *Reference Tables*, we find that the mass of the electron (m_e) is less than the mass of the proton (m_p). Since the electron has a smaller mass, the acceleration of the electron is *greater* than the acceleration of the proton.

However, since the charge on an electron is negative, but the charge on a proton is positive, we conclude that the force and acceleration experienced by an electron will be *in the opposite direction* to the force and acceleration experienced by a proton.

82. **3** The electron current that is induced in a conductor arises from the forces on the individual electrons in the conductor. Therefore the easiest way to find the direction of the electric current in the question is to concentrate on one electron in the wire as the wire moves through position P. The direction of the induced current will be the same as the direction of the magnetic force on this electron.

There are various hand rules that can be used to determine the direction of the force on an electron moving in a magnetic field. Recall that the force experienced by the moving electron is perpendicular to both its own velocity and the magnetic field, and you can eliminate choices 1 and 2.

Use the following *left-hand rule*: Hold your outstretched left hand so that your thumb points in the direction of the electron's velocity (toward the left), and your extended fingers point in the direction of the magnetic field. Your palm will now face in the direction of force exerted on the electron.

When you apply this rule to the given situation, your palm points *into the page*. This is the direction of the induced electron current.

NOTE: This question can also be solved using Lenz's law.

83. **3** The *force* (F) acting on a charged particle moving in a direction perpendicular to a magnetic field is equal to the product of the particle's charge (q), its velocity (v), and the strength of the magnetic field, also called the magnetic flux density (B).

Refer to the *Electromagnetic Applications* equations in the *Reference Tables*:

$$\begin{aligned} F &= qvB \\ &= (1.6 \times 10^{-19}\text{ C})(3.0 \times 10^{7}\text{ m/s})(2.0\text{ T}) \\ &= 9.6 \times 10^{-12}\text{ N} \end{aligned}$$

NOTE: The charge of the electron is obtained from the *List of Physical Constants* in the *Reference Tables*.

84. **2** In diagrams 1 and 2, we see a negative electron (e^-) move in the region between what appear to be two oppositely charged parallel plates. In each case, the electron would be repelled by the negatively charged plate and attracted by the positively charge plate.

The electron would therefore experience an upward acceleration and deflection in diagram 1, but a *downward* acceleration and deflection in diagram 2.

Wrong Choices Explained:

(1) See the explanation above.

(3), (4) There are various hand rules that can be used to determine the direction of the force on an electron moving in a magnetic field. Use the following *left-hand rule*: Hold your outstretched left hand so that your thumb points in the direction of the electron's velocity (toward the right), and your extended fingers point in the direction of the magnetic field (north to south).

Your palm will now face in the direction of force exerted on the electron. When you apply this rule to diagram 3, your palm points *out of the page*. When you apply this rule to diagram 4, your palm points *into the page*.

85. **1** A *laser* (light amplification by stimulated emission of radiation) is a device that produces electromagnetic radiation by stimulating the emission of photons from atoms. The light produced by a laser *consists of coherent waves*. This means that all the waves are exactly *in phase* (in step) with each other. Ordinary light is incoherent and disorganized, consisting of waves of many different frequencies with many different phase relationships.

There are other characteristics of lasers that you should know. The light is monochromatic (single frequency); travels as a parallel beam with little divergence, or spreading; and is of very high intensity.

Wrong Choices Explained:

(2) Only white light, which is *polychromatic* and consists of many different fequencies, can be dispersed (separated) into a complete continuous spectrum of colors. Since a light beam produced by a monochromatic laser has only a single frequency, it *cannot be dispersed*.

(3) Like any other kind of wave, the waves of light in a laser beam can be *reflected* and *refracted*.

(4) Laser light *does exhibit wave properties*.

GROUP 4—**Geometric Optics**

86. **2** One of the characteristics of the image formed by a *plane mirror* is that the distance of every point on the image behind the mirror is exactly equal to the distance of the corresponding point on the object in front of the mirror.

Refer to the annotated diagram below. Only in diagram 2 in the answer choices is the distance from each vertex on the image (A', B', C') to the mirror the same as the distance from the corresponding point on the object (A, B, C) to the mirror.

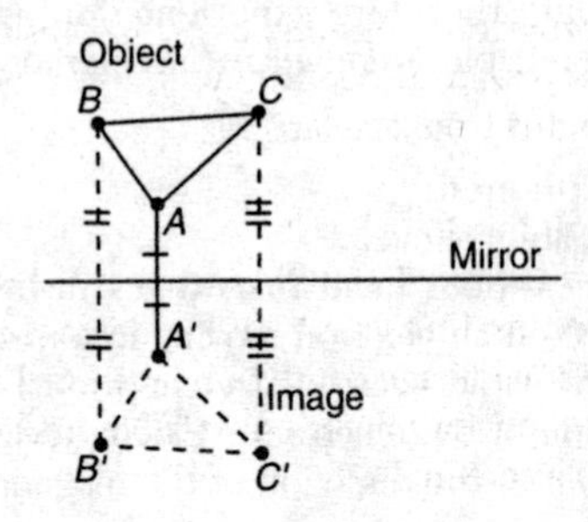

87. **2** All incident rays that are parallel to the principal axis of a *spherical concave (converging) mirror,* such as ray *R,* are reflected so that they pass through the focal point (*F*) of the mirror.

88. **4** For a large spherical mirror, the focal length (*f*) is equal to one-half the radius of curvature (*R*):

$$f = \frac{R}{2}$$

$$= \frac{16 \text{ cm}}{2}$$

$$= 8 \text{ cm}$$

For a *concave (converging) spherical mirror,* the focal length is positive. The relationship between the object distance (d_o), the image distance (d_i), and the focal length of a spherical mirror is found in the *Geometric Optics* equations in the *Reference Tables:*

$$\frac{1}{d_o} + \frac{1}{d_i} = \frac{1}{f}$$

$$\frac{1}{12 \text{ cm}} + \frac{1}{d_i} = \frac{1}{8 \text{ cm}}$$

$$d_i = 24 \text{ cm}$$

89. **3** *Every* image produced by a convex (diverging) mirror is *virtual and smaller than the object.* A convex mirror *cannot* produce images that are real or larger than the object.

Therefore the image of the shoplifter as seen in a convex mirror is *virtual and smaller than the shoplifter.*

90. **1** Only a *converging* lens can produce an image that can be projected onto a screen; such an image is called a *real* image.

In order to produce an image that is real and larger than the object, as required for movie projection, the object (film) must be placed at a *distance greater than the focal length.* (Specifically, the film must be placed between one and two focal lengths from the lens.)

Wrong Choices Explained:

(2) When an object is placed at a distance *less than one focal length* from a converging lens, such as fine print under a magnifying glass, the image formed is virtual. A virtual image *cannot* be projected onto a screen.

(3), (4) A *diverging* lens produces only virtual images. A virtual image *cannot* be projected onto a screen.

91. **1** A converging lens can produce a real image of an object because light diverging from every point on the object is made to *converge to a single point* by the lens to produce the corresponding point of the image.

For example, the diagram below shows how two light rays leaving a common point A on the arrow are refracted by the lens so that they converge to single point, A'. If a screen were placed at A', a real image of the arrowhead would be seen. In similar fashion, light rays leaving all other points on the arrow are brought to focus to form a real image of the entire arrow.

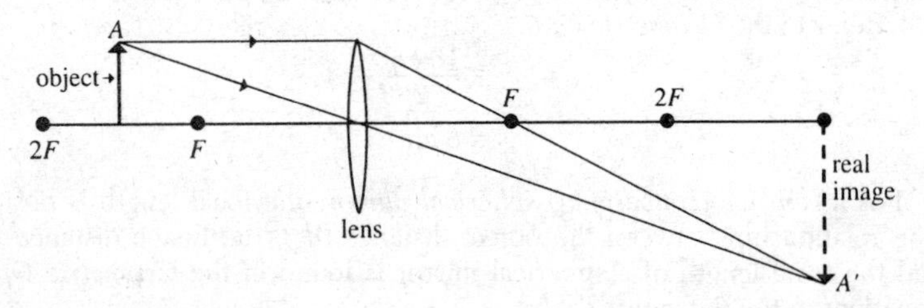

92. **2** When an object is placed at exactly two focal lengths ($2F$) from a convex (converging) lens, the image formed will be *real, inverted, and the same size as the object.*

An application of this arrangement is a copy machine. To help in analyzing problems involving convex lenses, you should associate an application with each of the possible object locations, and be able to draw the corresponding sketch.

Wrong Choices Explained:

(1) Only a *concave (diverging)* lens can produce an image that is virtual, erect, and *smaller* than the object.

(3) The application of this (convex) arrangement is the slide or movie projector, in which case the object is placed *between* F and $2F$, *not* at $2F$.

(4) An image that is virtual, erect, and the same size as the object *cannot* be attained with a single lens.

93. **4** The phenomenon illustrated in the diagram in the question is *chromatic aberration.* It results from the fact that glass is a *dispersive* medium for light, so that each frequency (color) of visible light travels at a different speed in glass. Consequently, each color is refracted by a different amount, and therefore the colors will not come to the same focus after passing through a glass lens.

Chromatic aberration is a defect that can be corrected by using a diverging lens of a different material with the converging lens.

Wrong Choices Explained:

(1) *Total internal reflection* occurs when light passes from a more optically dense to a less optically dense medium at an angle that exceeds the critical angle.

(2) There is no phenomenon known as *critical angle reflection*.

(3) *Spherical aberration* occurs because rays passing through the edges of a converging lens do not come to focus at the focal point, so that a blurred image of the object is created. This defect can be corrected by using a diaphragm in front of the converging lens to cover the edges of the lens.

94. **2** For any thin lens, the ratio of object size (S_o) to image size (S_i) is equal to the ratio of object distance (d_o) to image distance (d_i).

Refer to the *Geometric Optics* equations in the *Reference Tables*:

$$\frac{S_o}{S_i} = \frac{d_o}{d_i}$$

$$\frac{2.0 \text{ m}}{S_i} = \frac{4.0 \text{ m}}{0.050 \text{ m}}$$

$$S_i = 0.025 \text{ m}$$

95. **3** The image formed by a plane mirror is the same size as the object.

Since the true size of an object is independent of its distance from a mirror, the size of the image *remains the same* even if the distance between the man and the mirror changes.

GROUP 5—**Solid State**

96. **2** According to the *band model*, the electrons of any substance are restricted to particular energy bands, the most important of which are the *conduction band* and the *valence band*. These bands are separated by an *energy gap*.

Electrons in the outermost shell of an atom are in the valence band and are not mobile bacause they are held to the atom by attractive bonds. Electrons in the conduction band, however, are practically free of bonds and can be set in motion by a small applied voltage.

A solid that has a small energy gap between its valence and conduction bands is classified as a *semiconductor*. Electrons in the valence band need only a small amount of energy, which internal thermal energy can provide, to jump to the conduction band. Therefore, as the temperature is increased, more electrons jump the gap, and so the semiconductor's ability to conduct improves.

Wrong Choices Explained:

(1) In a good *conductor*, the valence and conduction bands overlap so that there is *no* appreciable energy gap. This overlapping places a large number of valence electrons in the conduction band, where they can move easily.

(3), (4) Solids that have *large* energy gaps are classified as *insulators*. The effect of the large gap is that very few electrons in the valence band can acquire enough energy to jump to the conduction band. With few conduction electrons, an insulator, such as *glass*, is a poor conductor of electricity.

97. **4** A *hole* is a spot in a crystalline structure that is missing an electron. This hole, or empty space, can accept a valence electron from an adjacent atom if the electron gains enough energy to break from the atomic bond.

This electron, in turn, leaves a hole behind in the atom from which it came, a hole that can be filled by an electron from yet another neighboring atom, creating yet another hole, and so on. Holes therefore seem to move in the opposite direction to the electron flow, as if they were positive charges.

Under the influence of an electric field from a battery, electrons move away from the negative end of the semiconductor toward the positive end. Simultaneously, "positive" holes move in the opposite direction.

We conclude, therefore, that the hole in the diagram accompanying the question will move toward letter *D*.

98. **4** The simplest of all semiconductor devices, the *diode* acts like a one-way valve, allowing current to flow through it in one direction only, and blocking current flow in the opposite direction.

If alternating current, which is current that reverses its direction each half cycle, is passed through a diode, the diode will allow the current to flow only during alternating half cycles, blocking the current when its direction is reversed.

The effect is to change or *rectify* the alternating current, as represented by diagram *A*, into a pulsating direct current, as represented by diagram *B*.

Wrong Choices Explained:

(1) A *resistor* is a part of a circuit that is made out of a metal conductor and obeys Ohm's law.

(2), (3) An *anode* and a *cathode* are electrodes that are found in various electrical devices, such as gas discharge tubes and electrochemical and electrolytic cells.

99. **1** *Insulators* are typically nonmetals such as glass and rubber, and have very few free electrons. *Conductors* are typically metallic solids such as copper and silver, and have many free electrons.

Compared to an insulator, a conductor of electric current has *more free electrons per unit volume*.

100. **1** Although the semiconductors that make up a diode are electrically neutral, the crystal structure of the *N*-type material has an excess of free electrons, and the *P*-type material has an excess of holes, or vacancies, that can accept electrons. Therefore we conclude that *A* in the diagram in the question represents the *N*-type semiconductor and *B* represents the *P*-type semiconductor.

As shown in the diagram below, some of the electrons from the *N*-type semiconductor naturally diffuse across the *P-N* junction to fill holes in the *P*-type semiconductor. This drift current is curtailed, however, by the buildup of negative charge on the *P*-side of the junction, which creates a potential or electric field barrier. When a potential difference is placed across the diode, as in the diagram, this barrier may be either reinforced or diminished, depending on the polarity of the voltage source.

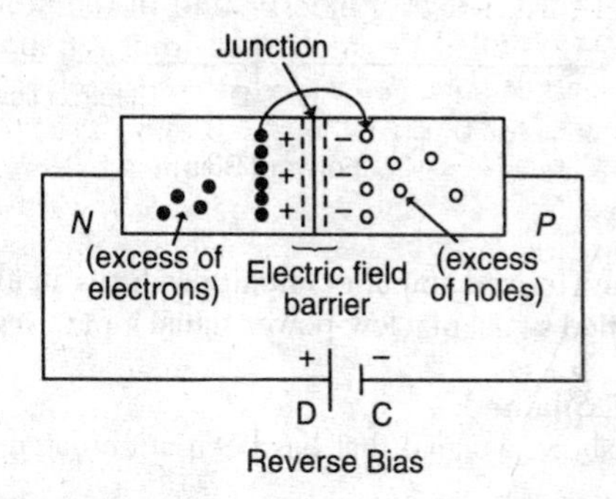

Reverse Bias

When a *P-N* junction is *reverse biased*, the applied potential difference reinforces the junction electric field barrier, therefore further retarding the flow of current. This situation occurs when the positive terminal of the battery is connected to the *N*-type material, and the negative terminal to the *P*-side of the diode, as shown in the diagram given in the question.

101. **2** As explained in the answer to question 100, the crystal structure of an *N*-type semiconductor has an excess of free electrons, while a *P*-type semiconductor has an excess of holes, or vacancies, than can accept electrons.

We conclude that in the diagram, *A* represents the *N*-type semiconductor and *B* represents the *P-type semiconductor* (such as silicon).

102. **2** If the positive and negative wires of the circuit in the diagram were reversed, the *P-N* junction would be *forward biased* and the current would *increase*. This situation occurs when the positive terminal of a battery is connected to the *P*-type material, and the negative terminal to the *N*-type material. The electric field produced by the battery would then oppose and overcome the small, naturally occurring junction electric field barrier.

Under the influence of this larger field, the small drift current would be reinforced, and conduction electrons in the *N*-type material would move toward the junction and flow across it. At the same time, holes in the *P*-type material would flow toward the junction, adding to the diode current.

Note, however, that only electrons would flow in the external circuit. See the diagram on the following page.

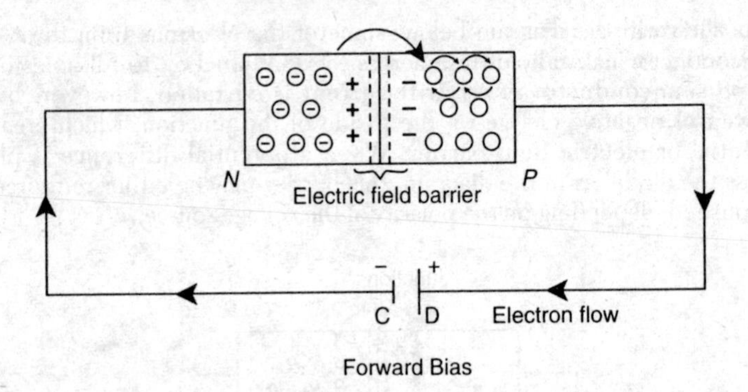

Forward Bias

103. **3** A transistorized amplifier circuit produces an identical (same frequency) but amplified signal of a low-power signal input, as shown in graph 3.

Wrong Choices Explained:

(1) This graph shows a signal that has been attentuated (diminished), not amplified.

(2) This graph shows a signal with greater frequency than that of the input signal.

(4) This graph shows a signal with a smaller frequency than that of the input signal.

104. **3** An *integrated circuit* consists of a small semiconductor block, called a *chip*, that is typically made of silicon and that contains a large number of interconnected transistors and other electronic parts. An integrated circuit can take the place of a full-scale conventional circuit, and therefore has led to the miniaturization of electronic devices and appliances.

Wrong Choices Explained:

(1) A *junction diode*, sometimes called just a *diode*, is a *two-element* device, made by joining a *P*-type semiconductor with an *N*-type semiconductor.

(2) A *conductor* is typically a metallic solid, such as copper or silver, that has great ability to carry an electric current because of the presence of many free electrons.

(4) An *N-type semiconductor* is a semiconducting material that has an excess of free electrons. The *N* stands for "negative," referring to the electrons, which are the majority charge carriers.

105. **1** The diagram in this problem shows the basic operating circuit for an *N-P-N* transistor. In the annotated diagram on the following page, note that the emitter-base junction is forward biased. In this connection, the polarity of the potential difference (V_{eb}) will support the emitter-base current,

which starts with a diffusion of electrons from the *N*-type emitter across the *P-N* junction to fill the holes in the *P*-type base. Some of these electrons will complete the circuit and return to the positive terminal of the battery in the emitter-base circuit.

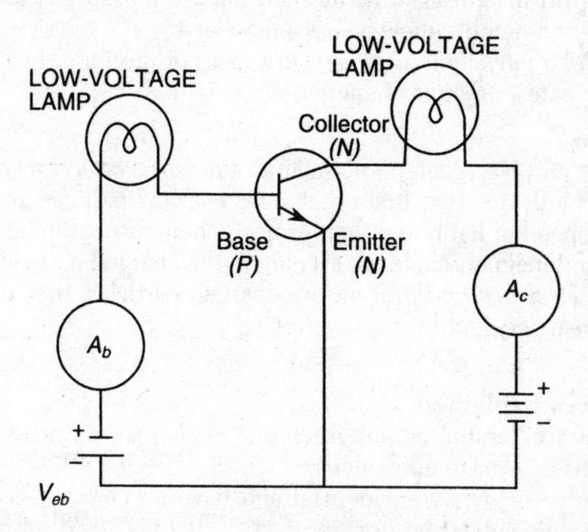

However, the emitter is generally heavily doped so that it has a large number of current carriers—in this case, electrons. Moreover, the base is extremely thin, so it has comparatively few holes. As a result, *most* of the electrons pass from the emitter, through the base, and into the collector. These electrons then continue on to the positive terminal of the battery in the collector circuit. As a result, the collector current is much larger than the base current.

We conclude, therefore, that, compared to the reading of ammeter A_c (which reads the collector current), the reading of ammeter A_b (which reads the base current), is *less*.

GROUP 6—**Nuclear Energy**

106. **1** The *atomic number* is the number of protons in the nucleus of an atom. Since the atomic number of the element in the problem is 63, we conclude that there are *63* protons in its nucleus.

NOTE: The *mass number*, also called the *atomic mass*, is the sum of the number of protons and neutrons in an atom's nucleus.

107. **3** Mass and energy are interchangeable, that is, under certain conditions mass can be converted into energy and vice versa. According to Einstein's famous *mass-energy equation*, the energy equivalent (E) of a given mass is *proportional* to the mass (m) and the speed of light (c) in a vacuum, squared.

Refer to the *Nuclear Energy* equations in the *Reference Tables*:

$$E = mc^2$$

We conclude that the particle with the greatest mass would generate the greatest amount of energy. In terms of atomic mass units, an electron has a mass of 0, a proton and a neutron each have a mass of 1, and an *alpha particle*, which is a helium nucleus, has a mass of 4.

Since an *alpha particle* has the greatest mass of the four choices, it would generate the greatest amount of energy.

108. **3** A particle accelerator, such as a *linear accelerator*, cyclotron, or synchrotron, is a device that first accelerates *charged* particles to speeds that approach the speed of light, and then projects them into samples of matter. It uses electric and magnetic fields to accelerate the charged particles.

Since both *electrons and protons* are charged particles, they can be accelerated by a linear accelerator.

Wrong Choices Explained:

(1) A linear accelerator *cannot* accelerate *uncharged* particles, such as the neutral photons that make up *gamma rays*.

(2), (4) Since *neutrons* are neutral and have no charge, they *cannot* be accelerated by a linear accelerator.

109. **1** *Atomic number*, which represents the amount of nuclear charge, is conserved in all nuclear reactions. Therefore the sums of the atomic numbers on both sides of the equation must be equal.

Recalling that atomic number is represented by the subscript for each symbol, we conclude that the atomic number of X must be 91, in order for the sum, $90 = 91 + (-1)$, to be the same on both sides of the equation.

Mass number (atomic mass) is also conserved in nuclear reactions. It is represented by the superscript for each symbol. The given reaction, the mass number of X must be 234, in order for the sum, $234 = 234 + 0$, to be the same on both sides.

Both choices 1 and 2 show an isotope that has an atomic number (nuclear charge) of 91 and a mass number of 234. Since it is the atomic number that distinguishes one element from another, and thorium (Th) has atomic number 90, not 91, we can eliminate choice 2.

Choice *1*, $^{234}_{91}Pa$, is correct.

NOTE: Lacking a periodic table of elements to verify the identity of the element with atomic number 91, we can check the *Uranium Disintegration Series* in the *Reference Tables*, where we confirm that the element in question is indeed protactinium (Pa).

110. **1** The *half-life* of a radioactive element is defined as the time required for one-half of the nuclei in any sample to disintegrate or decay. Each radioisotope has a specific half-life. In the decay of a radioactive isotope,

the final mass (m_f) of the original sample is equal to the fraction $1/2^n$ times the initial mass (m_i).

Refer to the *Nuclear Energy* equations in the *Reference Tables*:

$$m_f = \frac{m_i}{2^n}$$

where n is the number of half-lives that have elapsed.

$$5.0 \text{ kg} = \frac{40.0 \text{ kg}}{2^n}$$

$$n = 3$$

We conclude that three half-lives have elapsed during the 4.0-year period. One half-life of element A is therefore equal to 4.0 years/3 = *1.3 years*.

111. **4** Recent research in atomic physics has suggested the existence of elementary subatomic particles called *quarks*, which are smaller than and make up some nuclear particles, such as protons and neutrons.

Wrong Choices Explained:

(1) Although a neutron in the nucleus of an atom can disintegrate into a proton and an electron, the quark, *not* the *electron*, is believed to be the elemental building block of neutrons (and protons).

(2) A *nuclide* is a single nuclear species, having a single atomic number and mass number, such as $^{14}_{7}N$.

(3) A *positron* is a positive electron that is emitted from a nucleus during positive beta decay.

112. **3** A *positron* is a positive electron and is represented by the symbol $^{0}_{+1}e$. Only equation 3 shows this symbol.

A positron is emitted from a nucleus during positive beta decay. When a nucleus emits a positron, the atomic number of the nucleus decreases by 1, but the mass number remains the same.

An example of positron emission is represented by the nuclear equation in choice 3, where copper-64 emits a positron; in the process the mass number remains unchanged (64), but the atomic number is decreased by 1 (from 29 to 28), and a different element (nickel-64) is formed.

Wrong Choices Explained:

(1) This equation is an example of *alpha decay*.

(2) This equation is an example of *negative beta decay*.

(4) This equation is an example of *artificial transmutation*.

113. **2** In most nuclear fission reactions, *a heavy nucleus* captures a neutron and then *splits into lighter nuclei*, releasing two or more neutrons and great amounts of energy.

Wrong Choices Explained:

(1) This process occurs during *nuclear fusion*.

(3) This process occurs when a neutron is absorbed or captured by a nucleus.

(4) This process occurs during radioactive *alpha decay*, a type of natural transmutation.

114. **3** In a nuclear reactor, fission takes place when a heavy nucleus captures a neutron and then breaks apart into two lighter fragments of intermediate mass, releasing two or more neutrons and great amounts of energy. To maintain a self-sustaining chain reaction, each of these emitted neutrons must trigger one or more additional fissions.

In order to increase the likelihood that a neutron emitted from a nucleus will be captured by another nucleus, the neutron should be *slowed down to decrease its kinetic energy*.

Substances that consist of particles of mass comparable to that of the neutron, such as heavy water, are used to slow down the neutrons released during fission. Such substances are called *moderators*.

115. **2** The energy emitted by the Sun originates from the process of *fusion*. Nuclear fusion is the combination of two light nuclei to form a heavier nucleus, with the loss of some mass and the production of energy equivalent to the amount of mass lost. The fuel must be in plasma form and extremely high temperatures and pressure are required, as are found in the Sun. This type of reaction also occurs in the hydrogen bomb.

The Sun's energy is produced in a series of fusion reactions, during which hydrogen is converted into the heavier element helium.

PART III

116. **a.** (1) To receive full credit, each of the two vectors must include an arrowhead, have an appropriate label, and be drawn using the specified scale. Also, each vector must be drawn in the correct direction, the second displacement vector must originate at the arrowhead end of the first displacement vector, and the two vectors must be drawn at right angles (see diagrams below).

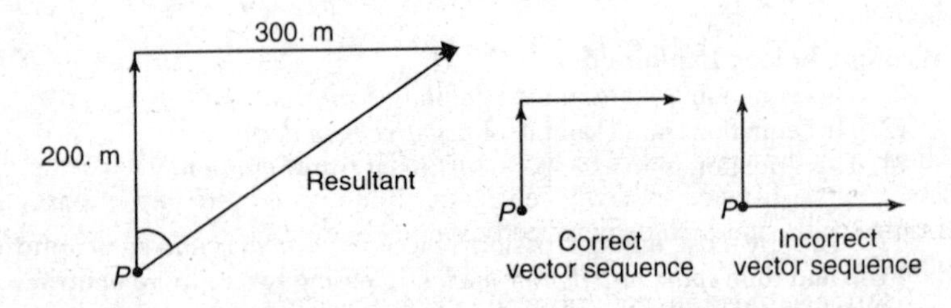

(2) Construct the resultant displacement by drawing a vector from the starting point (*P*) of the first vector (200.0 m) to the endpoint of the second vector (300. m). Label this vector "Resultant." See the diagram on the previous page.

b. Since your diagram was drawn to scale, measure the length of the resultant and convert to meters. The length turns out to be 7.21 centimeters.

The scale specified in the question is 1.0 cm = 50. m. Therefore the magnitude of the resultant is

$$7.21 \text{ cm} \times \frac{50. \text{ m}}{1.0 \text{ cm}} = 361 \text{ m}$$

(An answer that is within ±15 meters of 361 meters is also accepted.)

c. Place your protractor at point *P* and measure the angle between the first vector, which points north, and the resultant vector. This angle turns out to be 56°. (An answer that is within ±2° of 56° is also accepted.)

117. Within a certain range the amount that a spring is stretched or compressed (x) is proportional to the force (F) applied. This relationship is known as *Hooke's law.*

Refer to the *Energy* equations in the *Reference Tables:*

$$F = kx$$

where k is the spring constant. Substitute the given values:

$$6.0 \text{ N} = k\,(0.040 \text{ m})$$

Solve for k:

$$\frac{6.0 \text{ N}}{0.040 \text{ m}} = k = 150 \text{ N/m}$$

NOTE: Units must appear in the substitution and in the final answer for full credit.

118. **a.** The relationship that exists among the various factors involved in the double-slit experiment can be found in the *Wave Phenomena* equations in the *Reference Tables:*

$$\frac{\lambda}{d} = \frac{x}{L}$$

where λ is the wavelength of light used, d is the distance between the two slits, x is the distance from the central maximum to the first-order maximum (bright band), and L is the distance between the double-slit apparatus and the screen.

Rearrange the terms and substitute the given values:

$$\lambda = \frac{dx}{L}$$
$$= \frac{(1.0 \times 10^{-3}\ \text{m})\ (2.4 \times 10^{-3}\ \text{m})}{4.0\ \text{m}}$$
$$= 0.6 \times 10^{-6}\ \text{m}$$
$$= 6.0 \times 10^{-7}\ \text{m}$$

NOTE: Units must appear in the substitution and in the final answer for full credit.

b. Refer to the table entitled *Wavelengths of Light in a Vacuum* in the *Reference Tables.* The value for the wavelength found in part **a**, 6.0×10^{-7} meter, is seen to fall within the range of wavelengths listed for *orange* light.

c. Rearrange the terms of the equation written in part **a** so that x, the distance between the central maximum and the first-order maximum, stands alone:

$$\frac{\lambda L}{d} = x$$

From the equation we see that x is directly proportional to both the wavelength (λ) and the distance to the screen (L), but inversely proportional to the distance between the slits (d). Therefore, x will *increase* if we:

(1) *decrease the distance between the slits (decrease* d), *or*
(2) *move the screen further away (increase* L), *or*
(3) *use light with greater (longer) wavelength (increase* λ)
NOTE: Any two of the above responses are required for full credit.

119. For full credit, you must write one or more *complete* sentences. A complete sentence begins with a capital letter, has a subject and verb, and ends with a period. The following are two sample correct answers:

The frequency of red light equals or exceeds the threshold frequency of the material, while the frequency of infrared radiation does not.

The energy of red light equals or exceeds the work function of the material, while the energy of infrared radiation does not.

Topic	Question Numbers (total)	Wrong Answers (x)	Grade
Mechanics	1–11, 14–17: (15)		$\frac{100(15-x)}{15} = \%$
Energy	12, 13, 18–20, 24–26: (8)		$\frac{100(8-x)}{8} = \%$
Electricity and Magnetism	21–23, 27–36, 53, 54: (15)		$\frac{100(15-x)}{15} = \%$
Wave Phenomena	37–48, 55: (13)		$\frac{100(13-x)}{13} = \%$
Modern Physics	49–52: (4)		$\frac{100(4-x)}{4} = \%$
Motion in a Plane	56–65: (10)		$\frac{100(10-x)}{10} = \%$
Internal Energy	66–75: (10)		$\frac{100(10-x)}{10} = \%$
Electromagnetic Applications	76–85: (10)		$\frac{100(10-x)}{10} = \%$
Geometrical Optics	86–95: (10)		$\frac{100(10-x)}{10} = \%$
Solid State Physics	96–105: (10)		$\frac{100(10-x)}{10} = \%$
Nuclear Energy	106–115: (10)		$\frac{100(10-x)}{10} = \%$

To further pinpoint your weak areas, use the Topic Outline in the front of the book.

Examination June 1994

Physics

PART I

Answer all 55 questions in this part. [65]

Directions (1–55): For *each* statement or question, select the word or expression that, of those given, best completes the statement or answers the question. Record the answers to these questions in the spaces provided.

1 A car travels 20. meters east in 1.0 second. The displacement of the car at the end of this 1.0-second interval is

(1) 20. m
(2) 20. m/s
(3) 20. m east
(4) 20. m/s east

1_____

2 Which two graphs represent the motion of an object on which the net force is zero?

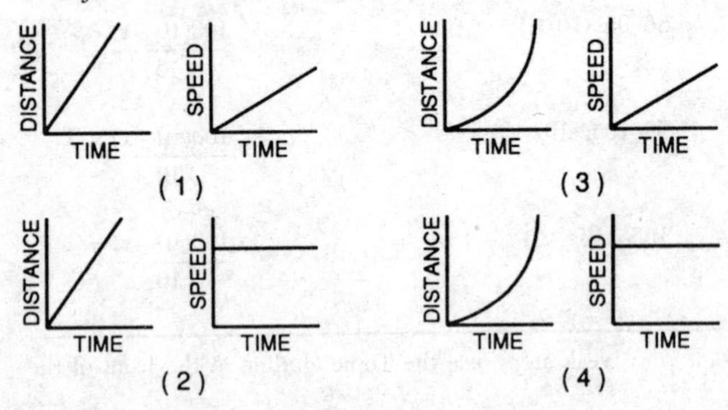

2_____

3 As shown in the diagram below, an astronaut on the Moon is holding a baseball and a balloon. The astronaut releases both objects at the same time.

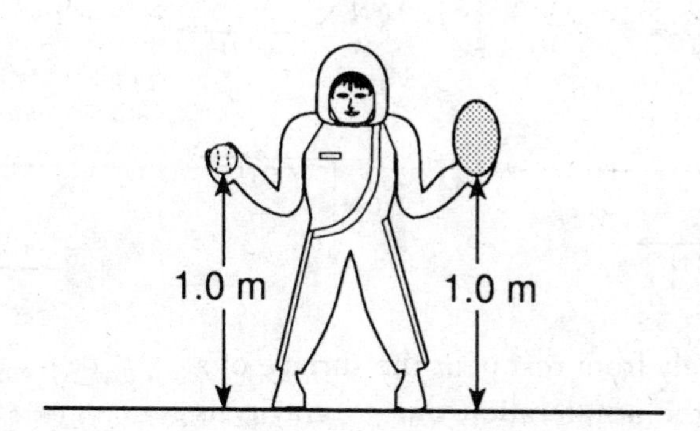

What does the astronaut observe? [Note: The Moon has no atmosphere.]

1 The baseball falls slower than the balloon.
2 The baseball falls faster than the balloon.
3 The baseball and balloon fall at the same rate.
4 The baseball and balloon remain suspended and do not fall.

3_____

4 Which is a vector quantity?

1 distance
2 time
3 speed
4 acceleration

4_____

5 A 3.0-newton force and a 4.0-newton force act concurrently on a point. In which diagram below would the orientation of these forces produce the greatest net force on the point?

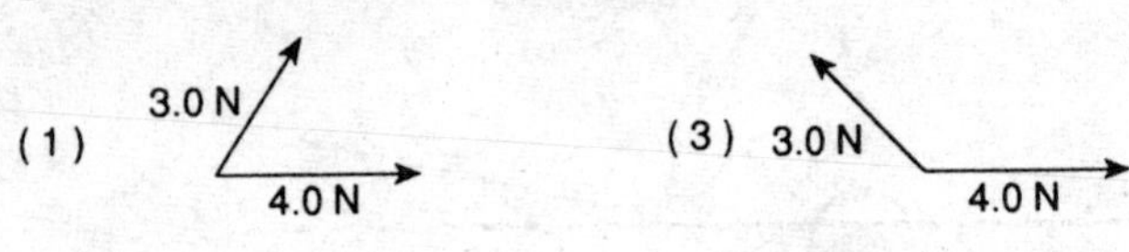

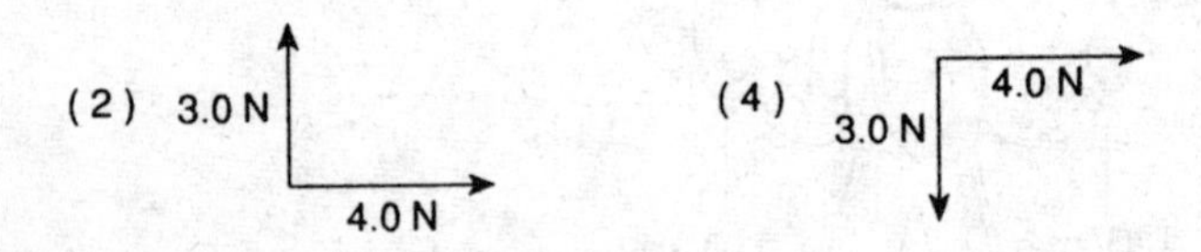

5____

6 A rock falls freely from rest near the surface of a planet where the acceleration due to gravity is 4.0 meters per second^2. What is the speed of this rock after it falls 32 meters?

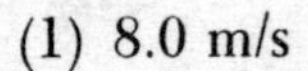

(1) 8.0 m/s
(2) 16 m/s
(3) 25 m/s
(4) 32 m/s

6____

7 The diagram below represents a 10.-newton block sliding down a 30.° incline at a constant speed.

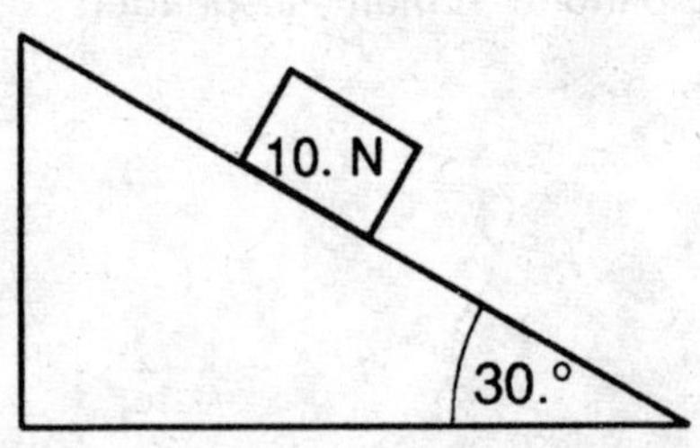

The force of friction on the block is approximately

(1) 5.0 N
(2) 10. N
(3) 49 N
(4) 98 N

7____

8 The graph below shows speed as a function of time for four cars, A, B, C, and D, in straight-line motion.

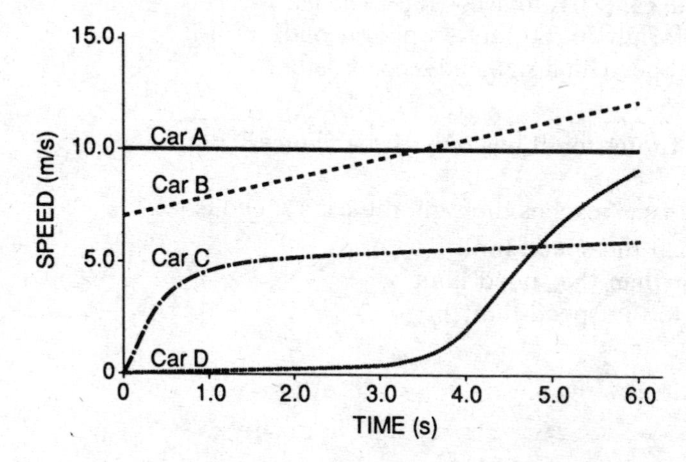

Which car experienced the greatest average acceleration during this 6.0-second interval?

1 car A
2 car B
3 car C
4 car D

8____

Base your answers to questions 9 and 10 on the information and diagram below.

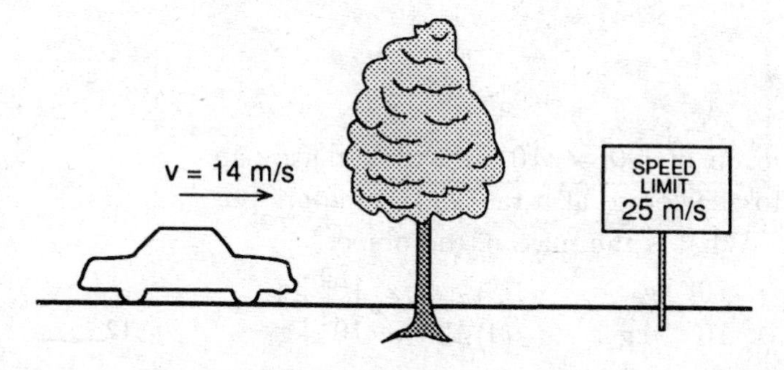

A car is traveling at a constant speed of 14 meters per second along a straight highway. A tree and a speed limit sign are beside the highway. As it passes the tree, the car starts to accelerate. The car is accelerated uniformly at 2.0 meters per second^2 until it reaches the speed limit sign, 5.0 seconds later.

Note that question 9 has only three choices.

9 When the car reaches the sign, the car's speed is
1 less than the speed limit
2 greater than the speed limit
3 equal to the speed limit

9_____

10 What is the distance between the tree and the sign?
(1) 10. m
(2) 25 m
(3) 70. m
(4) 95 m

10_____

11 The approximate mass of a nickel is
(1) 0.0005 kg
(2) 0.005 kg
(3) 0.5 kg
(4) 5 kg

11_____

12 A net force of 5.0×10^2 newtons causes an object to accelerate at a rate of 5.0 meters per second^2. What is the mass of the object?
(1) 1.0×10^2 kg
(2) 2.0×10^{-1} kg
(3) 6.0×10^2 kg
(4) 2.5×10^3 kg

12_____

13 The magnitude of the gravitational force between two objects is 20. newtons. If the mass of each object were doubled, the magnitude of the gravitational force between the objects would be

(1) 5.0 N
(2) 10. N
(3) 20. N
(4) 80. N

13____

14 The mass of a space shuttle is approximately 2.0×10^6 kilograms. During lift-off, the net force on the shuttle is 1.0×10^7 newtons directed upward. What is the speed of the shuttle 10. seconds after lift-off? [Neglect air resistance and the mass change of the shuttle.]

(1) 5.0×10^0 m/s
(2) 5.0×10^1 m/s
(3) 5.0×10^2 m/s
(4) 5.0×10^3 m/s

14____

15 A 2.0-kilogram toy cannon is at rest on a frictionless surface. A remote triggering device causes a 0.005-kilogram projectile to be fired from the cannon. Which equation describes this system after the cannon is fired?

1 mass of cannon + mass of projectile = 0
2 speed of cannon + speed of projectile = 0
3 momentum of cannon + momentum of projectile = 0
4 velocity of cannon + velocity of projectile = 0

15____

16 Which statement explains why a book resting on a table is in equilibrium?

1 There is a net force acting downward on the book.
2 The weight of the book equals the weight of the table.
3 The acceleration due to gravity is 9.8 m/s^2 for both the book and the table.
4 The weight of the book and the table's upward force on the book are equal in magnitude, but opposite in direction.

16_____

17 A student pulls a block 3.0 meters along a horizontal surface at constant velocity. The diagram below shows the components of the force exerted on the block by the student.

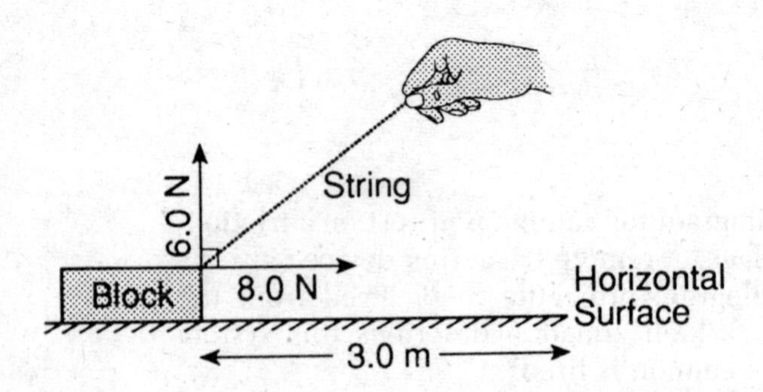

How much work is done against friction?

(1) 18 J
(2) 24 J
(3) 30. J
(4) 42 J

17_____

18 The diagram below shows a 1.0×10^3-newton crate to be lifted at constant speed from the ground to a loading dock 1.5 meters high in 5.0 seconds.

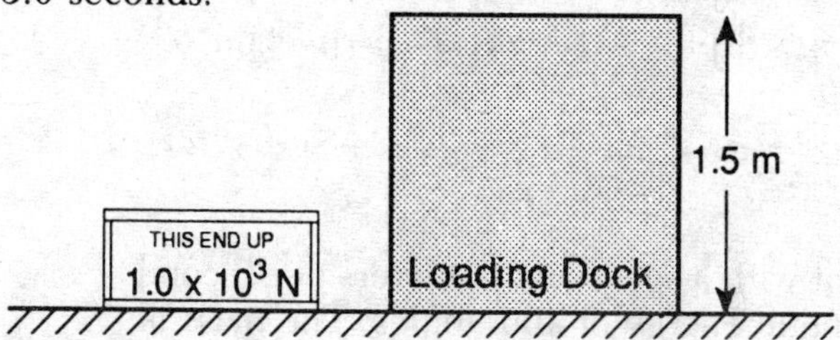

What power is required to lift the crate?

(1) 1.5×10^3 W
(2) 2.0×10^2 W
(3) 3.0×10^2 W
(4) 7.5×10^3 W

18_____

19 Graphs *A* and *B* below represent the results of applying an increasing force to stretch a spring which did not exceed its elastic limit.

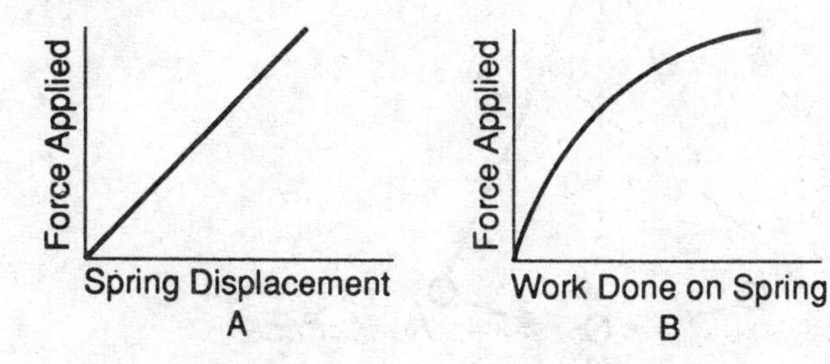

The spring constant can be represented by the

1 slope of graph *A*
2 slope of graph *B*
3 reciprocal of the slope of graph *A*
4 reciprocal of the slope of graph *B*

19_____

20 A force of 0.2 newton is needed to compress a spring a distance of 0.02 meter. The potential energy stored in this compressed spring is

(1) 8×10^{-5} J
(2) 2×10^{-3} J
(3) 2×10^{-5} J
(4) 4×10^{-5} J

20_____

21 An object with a speed of 20. meters per second has a kinetic energy of 400. joules. The mass of the object is

(1) 1.0 kg
(2) 2.0 kg
(3) 0.50 kg
(4) 40. kg

21_____

22 In the diagram below, an ideal pendulum released from point *A* swings freely through point *B*.

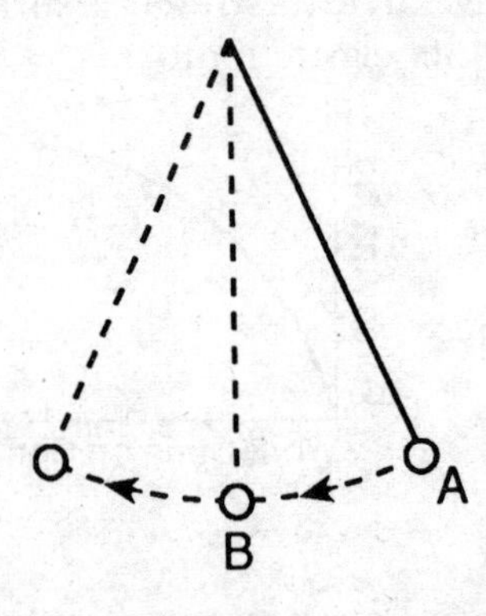

Compared to the pendulum's kinetic energy at *A*, its potential energy at *B* is

1 half as great
2 twice as great
3 the same
4 four times as great

22_____

23 As shown in the diagram below, a neutral pith ball suspended on a string is attracted to a positively charged rod.

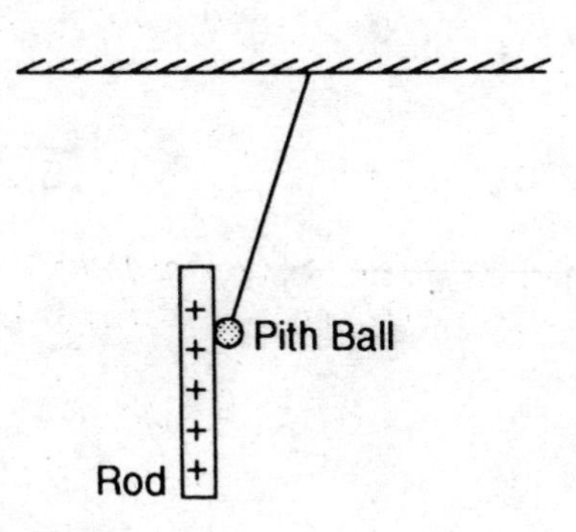

During contact with the rod, the pith ball

1 loses electrons
2 gains electrons
3 loses protons
4 gains protons

23____

24 The electrostatic force between two positive point charges is F when the charges are 0.1 meter apart. When these point charges are placed 0.05 meter apart, the electrostatic force between them is

(1) $4F$, and attracting
(2) $\frac{F}{4}$, and attracting
(3) $4F$, and repelling
(4) $\frac{F}{4}$, and repelling

24____

25 An electron is located 1.0 meter from a +2.0-coulomb charge, as shown in the diagram below.

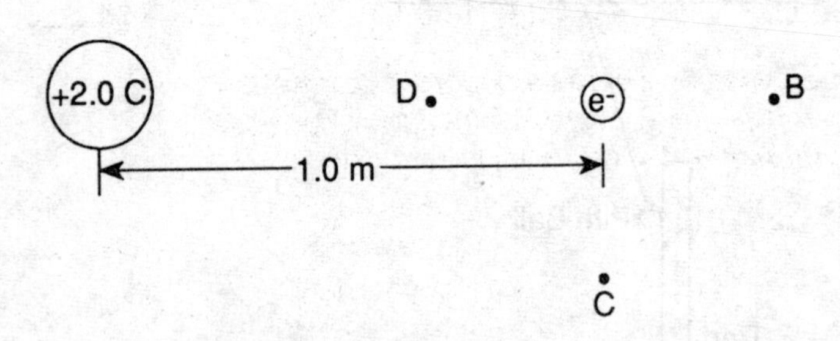

The electrostatic force acting on the electron is directed toward point

(1) *A* (3) *C*
(2) *B* (4) *D* 25____

26 What is the magnitude of the electric force acting on an electron located in an electric field with an intensity of 5.0×10^3 newtons per coulomb?

(1) 3.2×10^{-23} N (3) 5.0×10^3 N
(2) 8.0×10^{-16} N (4) 3.2×10^{22} N 26____

27 The unit "volts per meter" measures the same quantity as

1 joules per volt
2 newtons per ampere-meter
3 newton-meters^2 per coulomb^2
4 newtons per coulomb 27____

28 A series circuit has a total resistance of 1.00×10^2 ohms and an applied potential difference of 2.00×10^2 volts. The amount of charge passing any point in the circuit in 2.00 seconds is

(1) 1.26×10^{19} C
(2) 2.00 C
(3) 2.52×10^{19} C
(4) 4.00 C

28_____

29 A copper wire is connected across a constant voltage source. The current flowing in the wire can be increased by increasing the wire's

1 cross-sectional area
2 length
3 resistance
4 temperature

29_____

30 Which two of the resistor arrangements shown below have equivalent resistance?

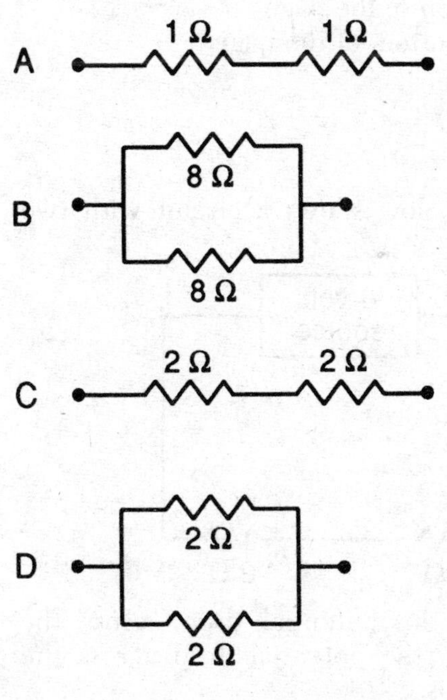

(1) *A* and *B*
(2) *B* and *C*
(3) *C* and *D*
(4) *D* and *A*

30_____

31 A clothes dryer connected to a 240-volt line draws 30. amperes of current for 20. minutes (1,200 seconds). Approximately how much electrical energy is consumed by the dryer?

(1) 4.8×10^3 J
(2) 7.2×10^3 J
(3) 1.4×10^5 J
(4) 8.6×10^6 J

31____

32 Electrons are flowing in a conductor as shown in the diagram at the right. What is the direction of the magnetic field at point *P*?

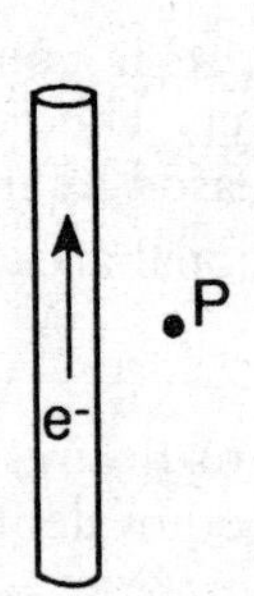

1 toward the top of the page
2 toward the bottom of the page
3 into the page
4 out of the page

32____

33 The diagram below shows a circuit with two resistors.

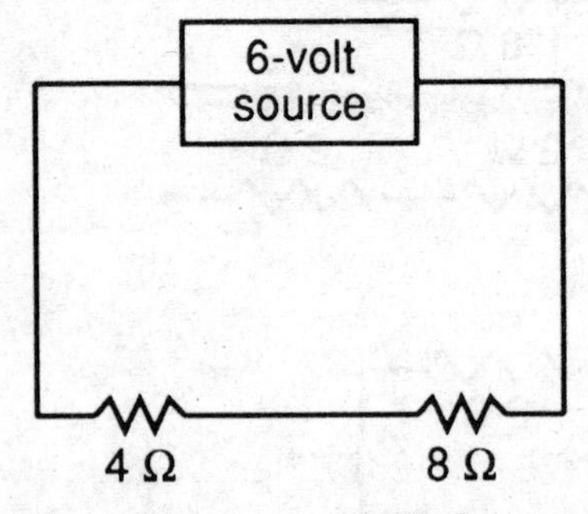

Compared to the potential drop across the 8-ohm resistor, the potential drop across the 4-ohm resistor is

1 the same
2 twice as great
3 one-half as great
4 four times as great

33____

34 An electromagnet would have the greatest strength if its wire were wrapped around a core made of

1 wood
2 iron
3 aluminum
4 copper

34____

35 An electron moving in a uniform magnetic field experiences the maximum magnetic force when the angle between the direction of the electron's motion and the direction of the magnetic field is

(1) 0°
(2) 45°
(3) 90°
(4) 180°

35____

36 An accelerating particle that does *not* generate electromagnetic waves could be

1 a neutron
2 a proton
3 an electron
4 an alpha particle

36____

37 A characteristic common to sound waves and light waves is that they

1 are longitudinal
2 are transverse
3 transfer energy
4 travel in a vacuum

37____

38 As a longitudinal wave passes through a medium, the particles of the medium move

1 in circles
2 in ellipses
3 parallel to the direction of wave travel
4 perpendicular to the direction of wave travel

38____

Base your answers to questions 39 through 41 on the diagram below which shows a parked police car with a siren on top. The siren is producing a sound with a frequency of 680 hertz, which travels first through point *A* and then through point *B*, as shown. The speed of the sound is 340 meters per second.

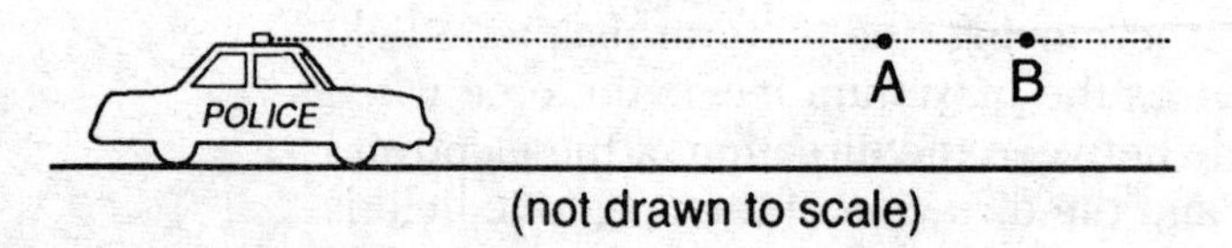

(not drawn to scale)

39 If the sound waves are in phase at points *A* and *B*, the distance between the points could be

(1) 1λ
(3) $\frac{3}{2}\lambda$
(2) $\frac{1}{2}\lambda$
(2) $\frac{1}{4}\lambda$

39____

40 What is the wavelength of the sound produced by the car's siren?

(1) 0.50 m
(3) 2.3×10^{5} m
(2) 2.0 m
(4) 2.3×10^{-6} m

40____

Note that question 41 has only three choices.

41 If the car were to accelerate toward point *A*, the frequency of the sound heard by an observer at point *A* would

1 decrease
2 increase
3 remain the same

41____

42 The diagram below shows two pulses, each of length λ, traveling toward each other at equal speed in a rope.

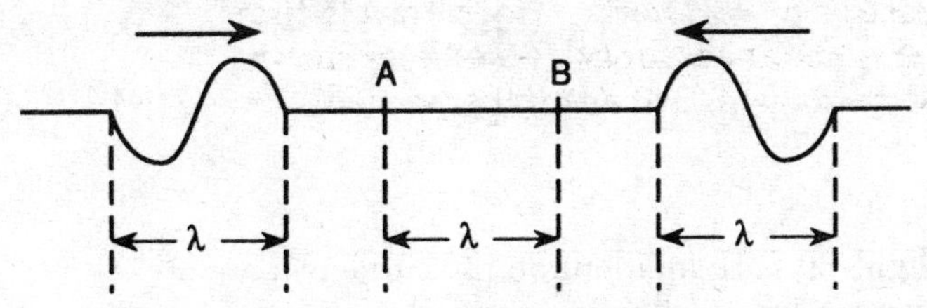

Which diagram best represents the shape of the rope when both pulses are in region *AB*?

(1) A 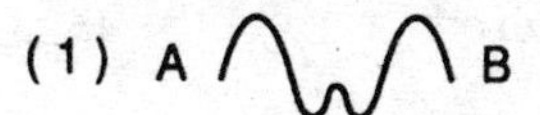 B

(2) A B

(3) A  B

(4) A  B

42_____

43 The diagram below shows two waves traveling in the same medium for the same length of time.

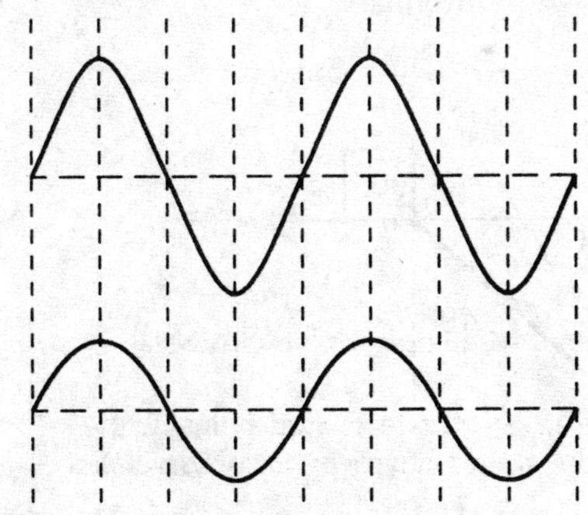

The two waves have different

1 amplitudes
2 frequencies
3 speeds
4 wavelengths

43_____

44 The distance from the Moon to Earth is 3.9×10^8 meters. What is the time required for a light ray to travel from the Moon to Earth?

(1) 0.65 s (3) 2.6 s
(2) 1.3 s (4) 3.9 s 44_____

45 Parallel light rays are incident on the surface of a plane mirror. Upon reflection from the mirror, the light rays will

1 converge 3 be parallel
2 diverge 4 be scattered 45_____

46 In the diagram below, a ray of monochromatic light ($\lambda = 5.9 \times 10^{-7}$ meter) reaches the boundary between medium *X* and air and follows the path shown.

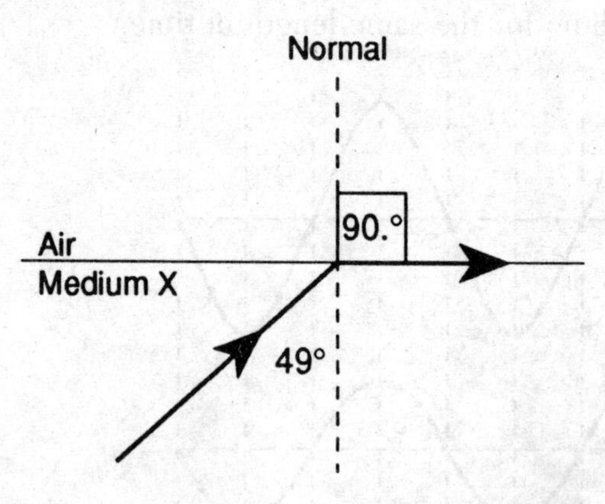

Which medium is most likely medium *X*?

1 diamond 3 Lucite
2 flint glass 4 water 46_____

47 Which phenomenon can *not* be exhibited by longitudinal waves?

1 reflection
2 refraction
3 diffraction
4 polarization

47____

48 A metal surface emits photoelectrons when illuminated by green light. This surface must also emit photoelectrons when illuminated by

1 blue light
2 yellow light
3 orange light
4 red light

48____

49 A ray of light ($\lambda = 5.9 \times 10^{-7}$ meter) traveling in crown glass is incident on a diamond interface at an angle of 30.°, as shown in the diagram below.

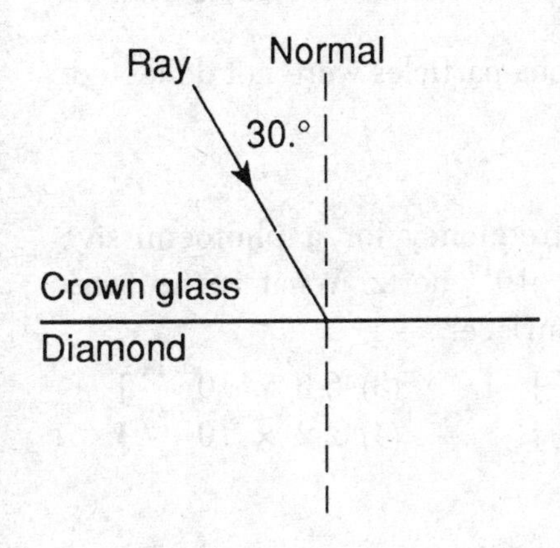

The angle of refraction for the light ray is closest to

(1) 12°
(2) 18°
(3) 30.°
(4) 53°

49____

50 An x-ray photon collides with an electron in an atom, ejecting the electron and emitting another photon. During the collision, there is conservation of

1 momentum, only
2 energy, only
3 both momentum and energy
4 neither momentum nor energy

50____

51 After Rutherford bombarded gold foil with alpha particles, he concluded that the volume of an atom is mostly empty space. Which observation led Rutherford to this conclusion?

1 Some of the alpha particles were deflected 180°.
2 The paths of deflected alpha particles were hyperbolic.
3 Many alpha particles were absorbed by gold nuclei.
4 Most of the alpha particles were not deflected.

51____

52 The threshold frequency for a photoemissive surface is 1.0×10^{14} hertz. What is the work function of the surface?

(1) 1.0×10^{-14} J
(2) 6.6×10^{-20} J
(3) 6.6×10^{-48} J
(4) 2.2×10^{-28} J

52____

53 What is the *minimum* amount of energy needed to ionize a mercury electron in the *c* energy level?

(1) 0.57 eV
(2) 4.86 eV
(3) 5.52 eV
(4) 10.38 eV

53____

Note that questions 54 and 55 have only three choices.

54 As the color of light changes from red to yellow, the frequency of the light

1 decreases
2 increases
3 remains the same

54_____

55 A car is driven from Buffalo to Albany and on to New York City, as shown in the diagram below.

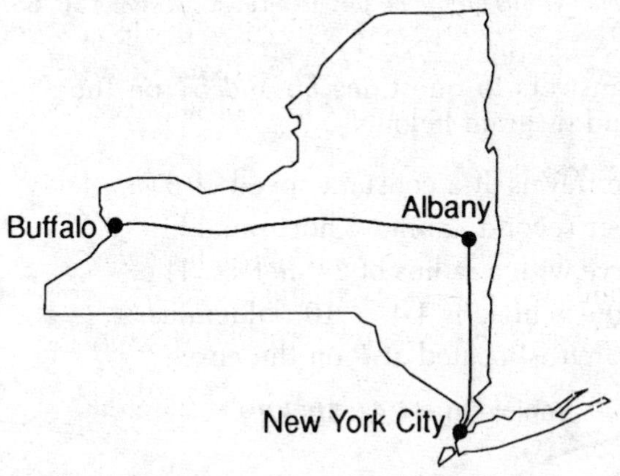

Compared to the magnitude of the car's total displacement, the distance driven is

1 shorter
2 longer
3 the same

55_____

PART II

This part consists of six groups, each containing ten questions. Each group tests an optional area of the course. Choose two of these six groups. Be sure that you answer all ten questions in each group chosen. Record the answers to these questions in the spaces provided. [20]

GROUP 1—Motion in a Plane

If you choose this group, be sure to answer questions **56–65.**

Base your answers to questions 56 and 57 on the information and diagram below.

A vehicle travels at a constant speed of 6.0 meters per second around a horizontal circular curve with a radius of 24 meters. The mass of the vehicle is 4.4×10^3 kilograms. An icy patch is located at P on the curve.

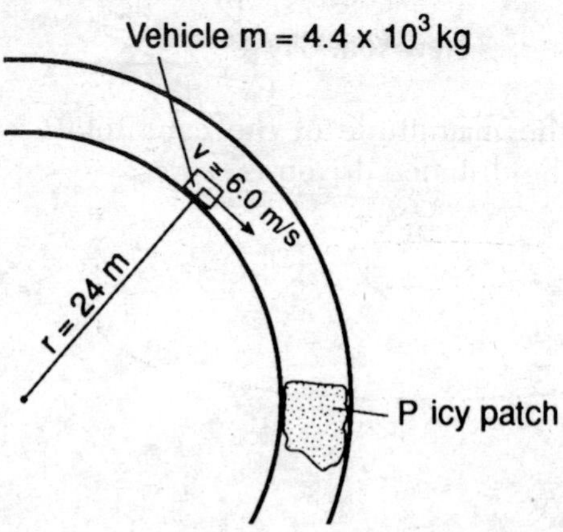

56 What is the magnitude of the frictional force that keeps the vehicle on its circular path?

(1) 1.1×10^3 N
(2) 6.6×10^3 N
(3) 4.3×10^4 N
(4) 6.5×10^4 N

56_____

57 On the icy patch of pavement, the frictional force on the vehicle is zero. Which arrow best represents the direction of the vehicle's velocity when it reaches icy patch *P*?

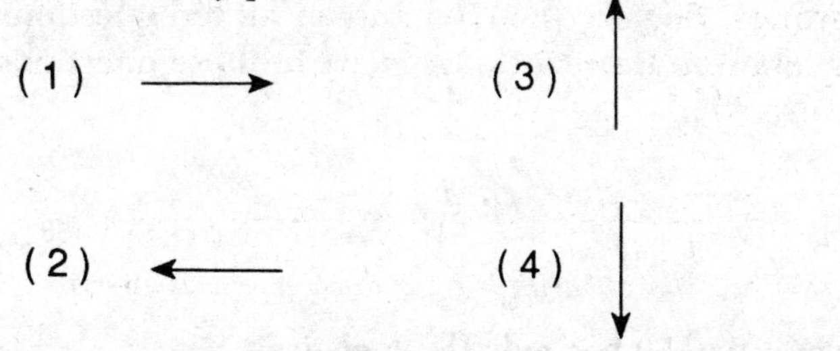

57____

Base your answers to questions 58 and 59 on the information and diagram below.

A 60.-kilogram adult and a 30.-kilogram child are passengers on a rotor ride at an amusement park. When the rotating hollow cylinder reaches a certain constant speed, v, the floor moves downward. Both passengers stay "pinned" against the wall of the rotor, as shown in the diagram below.

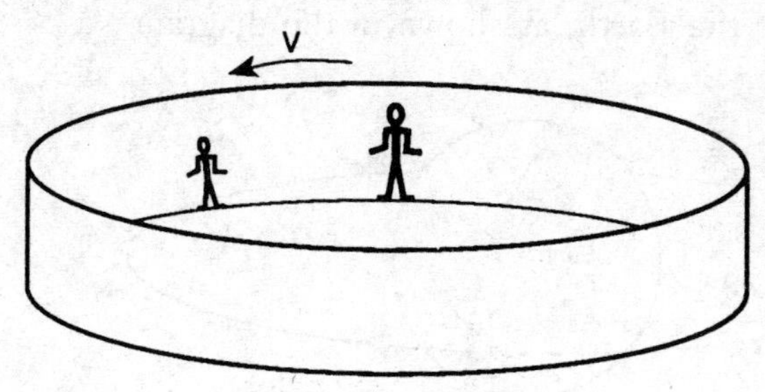

58 The magnitude of the frictional force between the adult and the wall of the spinning rotor is F. What is the magnitude of the frictional force between the child and the wall of the spinning rotor?

(1) F
(2) $2F$
(3) $\frac{F}{2}$
(4) $\frac{F}{4}$

58_____

Note that question 59 has only three choices.

59 Compared to the magnitude of the acceleration of the adult, the magnitude of the acceleration of the child is

1 less
2 greater
3 the same

59_____

60 A satellite is moving at constant speed in a circular orbit about the Earth, as shown in the diagram below.

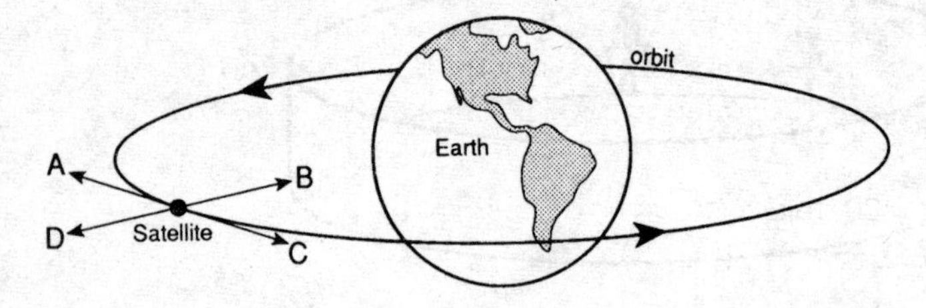

The net force acting on the satellite is directed toward point

(1) A
(2) B
(3) C
(4) D

60_____

61 Which diagram best represents the orbit of the planet Pluto around the Sun?

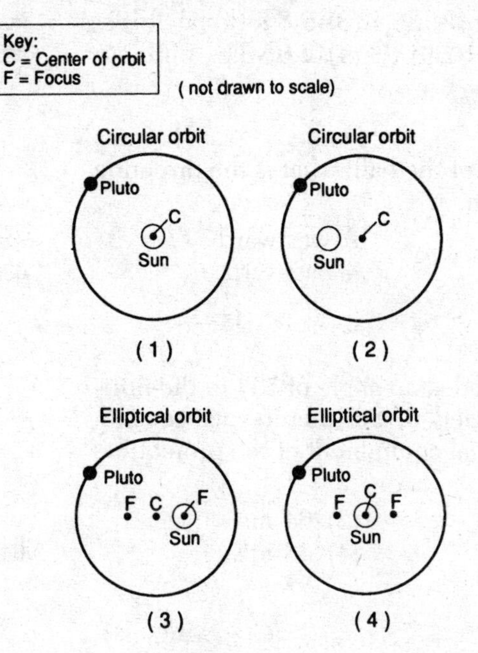

61_____

Base your answers to questions 62 and 63 on the diagram below which shows a ball projected horizontally with an initial velocity of 20. meters per second east, off a cliff 100. meters high. [Neglect air resistance.]

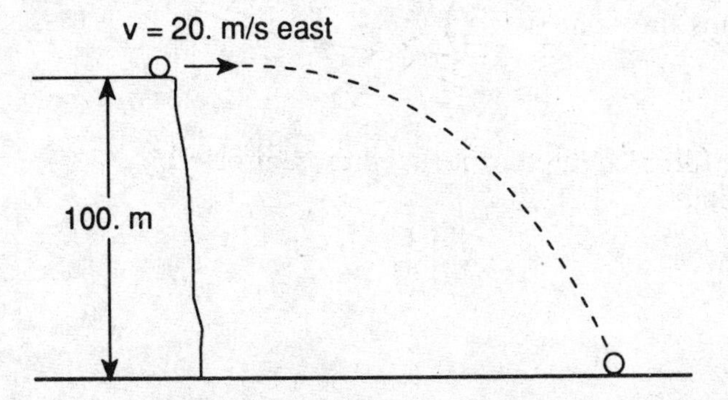

62 How many seconds does the ball take to reach the ground?

(1) 4.5 s
(2) 20. s
(3) 9.8 s
(4) 2.0 s

62____

63 During the flight of the ball, what is the direction of its acceleration?

1 downward
2 upward
3 westward
4 eastward

63____

64 A projectile is fired at an angle of 53° to the horizontal with a speed of 80. meters per second. What is the vertical component of the projectile's initial velocity?

(1) 130 m/s
(2) 100 m/s
(3) 64 m/s
(4) 48 m/s

64____

Note that question 65 has only three choices.

65 As the distance between the Moon and Earth increases, the Moon's orbital speed

1 decreases
2 increases
3 remains the same

65____

GROUP 2—**Internal Energy**

If you choose this group, be sure to answer questions **66–75.**

66 Which line on the graph below best represents the relationship between the average kinetic energy of the molecules of an ideal gas and absolute temperature?

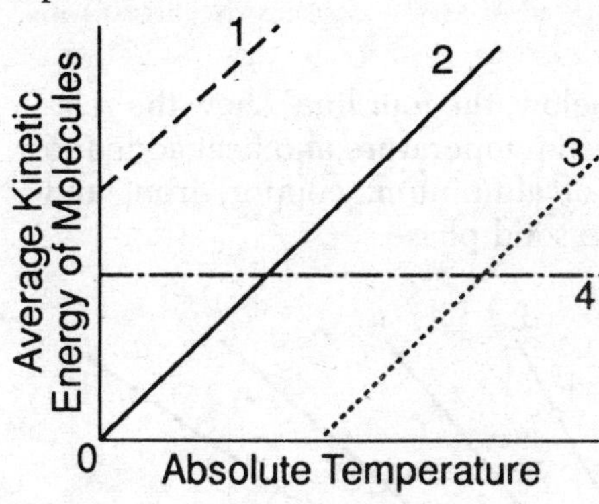

(1) 1
(2) 2
(3) 3
(4) 4

66_____

67 The difference between the boiling point of lead and the freezing point of lead is

(1) 328 K
(2) 1412 K
(3) 1658 K
(4) 1740 K

67_____

68 Equal masses of aluminum and copper, both at 0°C, are placed in the same insulated can of hot water. Which statement describes this system at equilibrium (the net exchange of internal energy is zero)?

1 The water has a higher temperature than the aluminum and copper.
2 The aluminum has a higher temperature than the copper and water.
3 The copper has a higher temperature than the aluminum and water.
4 The aluminum, copper, and water have the same temperature.

68_____

69 An unknown liquid with a mass of 0.010 kilogram absorbs 0.032 kilojoule of heat. Its temperature rises 8.0 C°, with no change in phase. What is the specific heat of the unknown liquid?

(1) 0.0040 kJ/kg•C°
(2) 0.40 kJ/kg•C°
(3) 26 kJ/kg•C°
(4) 260 kJ/kg•C°

69____

70 On the graph below, the four lines show the relationship between temperature and heat added to equal masses of aluminum, copper, iron, and platinum in the solid phase.

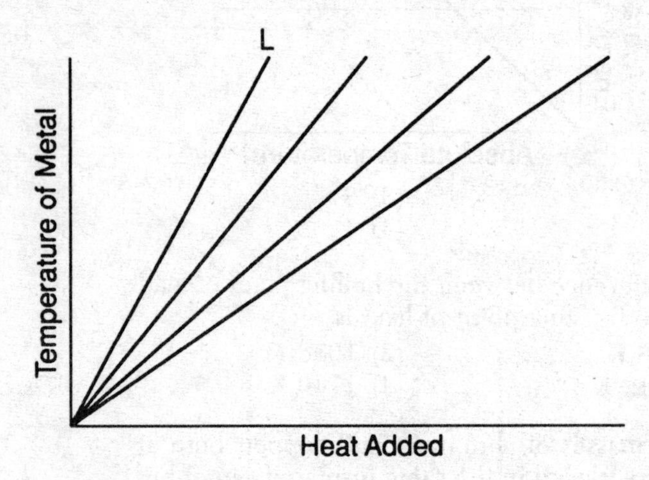

Which metal is represented by line *L*?

1 aluminum
2 copper
3 iron
4 platinum

70____

71 How much heat is required to change 3.0 kilograms of ice at 0.0°C to water at 0.0°C?

(1) 1.0×10^3 kJ
(2) 3.3×10^2 kJ
(3) 2.3×10^3 kJ
(4) 6.8×10^3 kJ

71____

72 According to the second law of thermodynamics, as time passes, the total entropy in the universe

1 decreases, only
2 increases, only
3 remains the same
4 cyclically increases and decreases

72_____

Note that question 73 has only three choices.

73 As the number of gas molecules in a rigid container at constant temperature is increased, the pressure on the walls of the container

1 decreases
2 increases
3 remains the same

73_____

74 In which diagram below does the water have the highest boiling point?

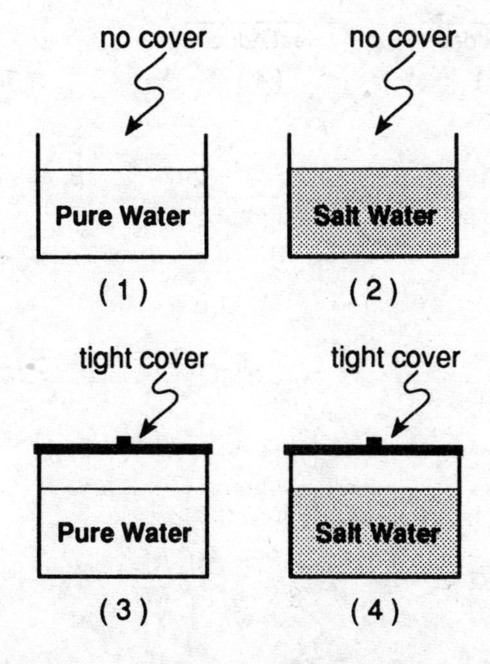

74_____

75 A crystalline solid at a temperature below its melting point is heated at a constant rate to a temperature above its melting point. Which graph best represents the average internal kinetic energy ($\overline{KE}$) of the substance as a function of heat added?

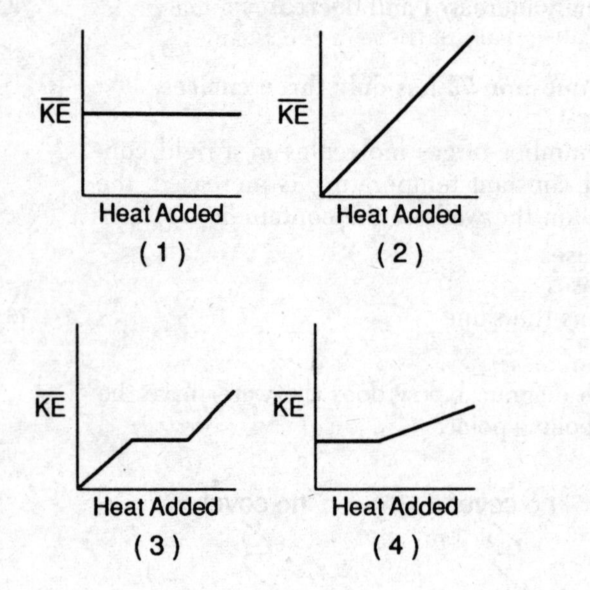

75 ______

GROUP 3—**Electromagnetic Applications**

If you choose this group, be sure to answer questions **76–85.**

76 A beam of particles is produced in a cathode-ray tube. The beam may be deflected by a magnetic field because each particle in the beam

1 possesses a charge
2 is at rest
3 has a rest mass greater than 9.1×10^{-31} kilogram
4 has a speed of 3.0×10^{8} meters per second

76_____

77 Which device does *not* operate by means of torque exerted on a current-carrying loop of wire in a magnetic field?

1 ammeter
2 electric motor
3 transformer
4 voltmeter

77_____

Base your answers to questions 78 and 79 on the diagram below which represents an electron about to enter uniform magnetic field *B*. The velocity of the electron (v) is 6.0×10^{7} meters per second to the right. The flux density of the magnetic field is 4.0×10^{-2} tesla, directed into the page.

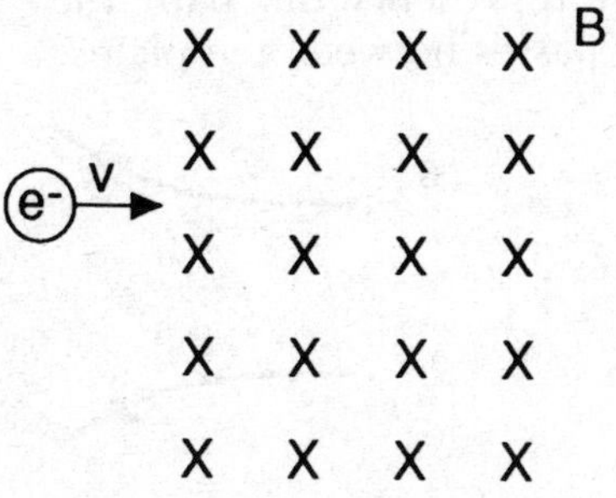

X X X X B
X X X X
e⁻ V → X X X X
X X X X
X X X X

78 When the electron first enters the magnetic field, the electron experiences a magnetic force directed toward the

1 top of the page
2 bottom of the page
3 left of the page
4 right of the page

78____

79 The magnitude of the magnetic force acting on the electron in the field is approximately

(1) 2.4×10^{-11} N
(2) 3.8×10^{-13} N
(3) 1.6×10^{-18} N
(4) 2.2×10^{-24} N

79____

80 In the diagram below, an electron moving with speed v enters the space between two oppositely charged parallel plates.

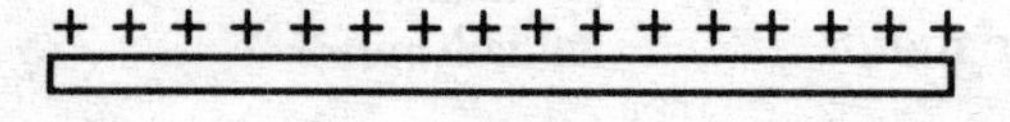

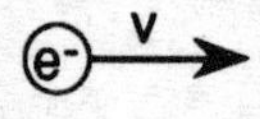

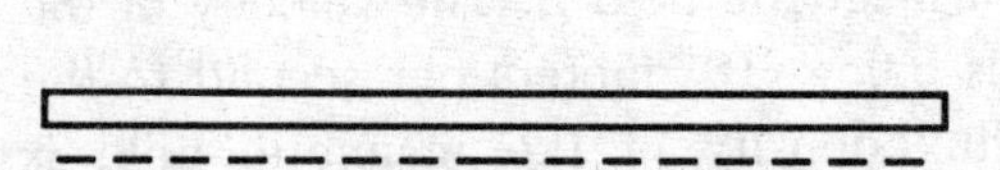

Which diagram best represents the path the electron follows as it passes between the plates?

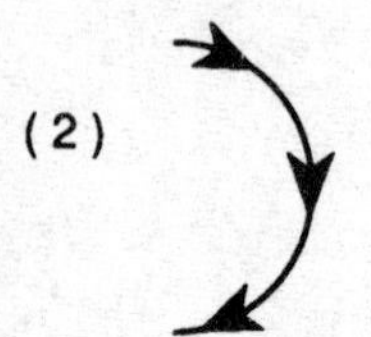

80____

81 After Millikan performed his oil drop experiment, he concluded that

1 there is a minimum amount of charge that particles can acquire
2 oil drops exhibit gravitational attraction for other oil drops
3 oil drops are largely empty space
4 there is a minimum amount of mass that particles can acquire

81_____

82 When a 12-volt potential difference is applied to the primary coil of a transformer, an 8.0-volt potential difference is induced in the secondary coil. If the primary coil has 24 turns, how many turns does the secondary coil have? [Assume 100% efficiency.]

(1) 36 (3) 3
(2) 16 (4) 4

82_____

Base your answers to questions 83 through 85 on the diagram below which shows a loop of wire being rotated at a constant rate about an axis in a uniform magnetic field.

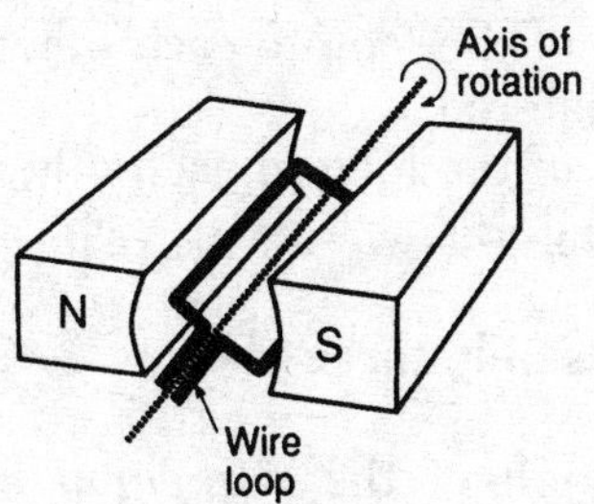

83 Which graph best represents the relationship between induced potential difference across the ends of the loop and time, for one complete rotation?

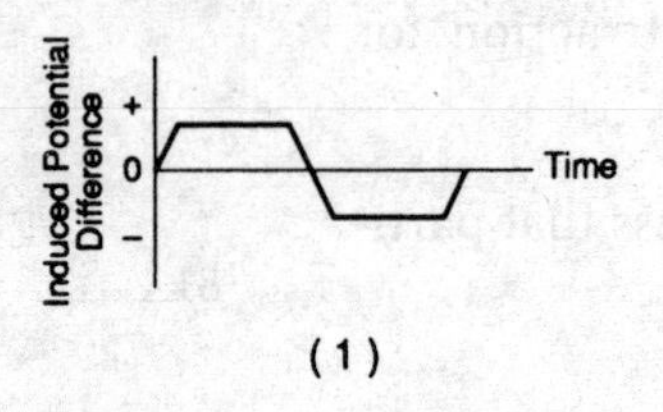

(1)

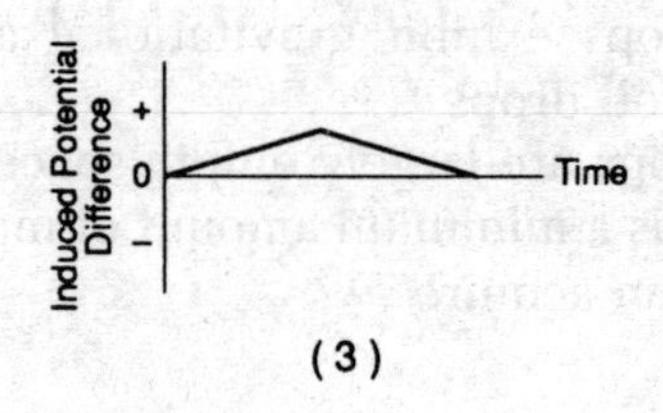

(3)

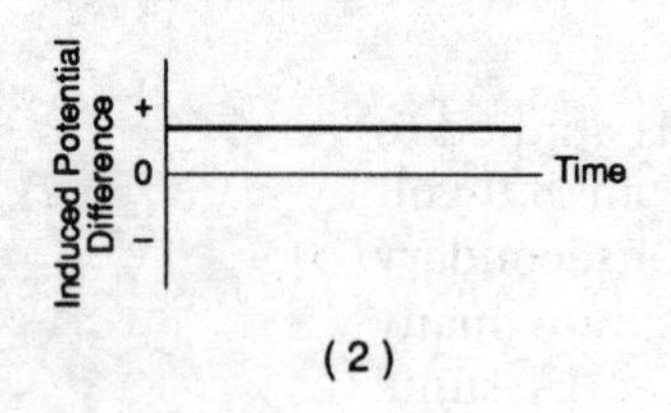

(2)

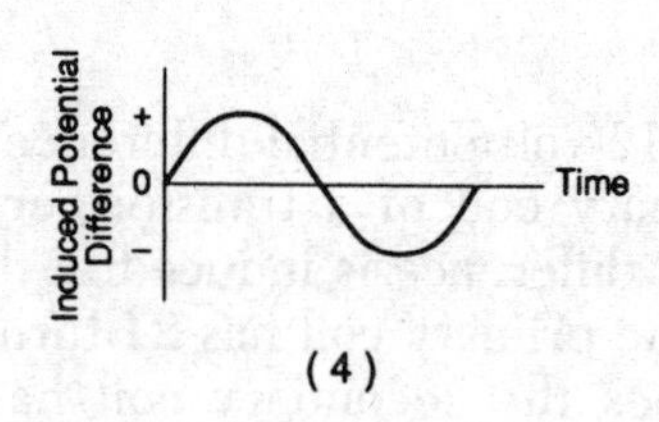

(4)

83_____

84 Which procedure would enable a current to flow in the loop, due to the induced potential difference?

1 turning the loop in the opposite direction at the same rate of rotation
2 increasing the distance between the ends of the loop
3 connecting the ends of the loop to each other with an insulating material
4 connecting the ends of the loop to each other with a conducting material

84_____

Note that question 85 has only three choices.

85 As the speed of rotation of the wire loop is increased, the maximum electromotive force induced in the loop

1 decreases
2 increases
3 remains the same

85_____

GROUP 4—**Geometric Optics**

If you choose this group, be sure to answer questions **86–95.**

86 A plane mirror produces an image of an object. Compared to the object, the image appears
1 inverted and the same size
2 reversed and the same size
3 inverted and larger
4 reversed and larger
86_____

87 A candle is located beyond the center of curvature, *C*, of a concave spherical mirror having principal focus *F*, as shown in the diagram below.

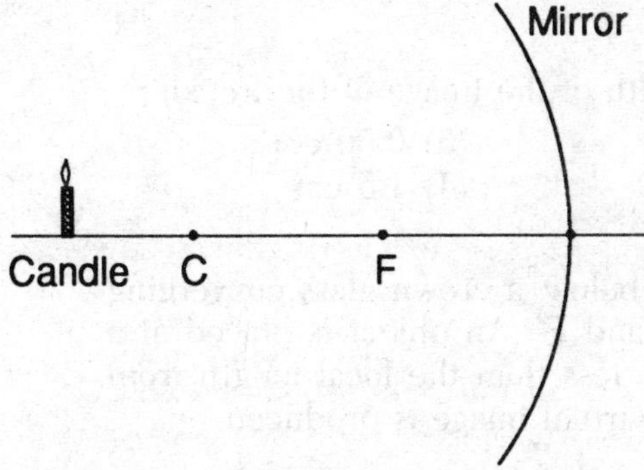

Where is the candle's image located?
1 beyond *C*
2 between *C* and *F*
3 between *F* and the mirror
4 behind the mirror
87_____

88 The convex spherical mirror found on the passenger side of many cars contains the warning: "Objects are closer than they appear." Which phrase best describes the image of an object viewed in this mirror?
1 real and smaller than the object
2 real and larger than the object
3 virtual and smaller than the object
4 virtual and larger than the object
88_____

Base your answers to questions 89 and 90 on the information below.

> A concave mirror with a focal length of 20. centimeters is used to examine a 0.50-centimeter-wide freckle on a person's face. The person's face is located 10. centimeters from the mirror.

89 The image of the freckle produced by the mirror is

1 real and inverted
2 real and erect
3 virtual and inverted
4 virtual and erect

89_____

90 What is the width of the image of the freckle?

(1) 1.0 cm
(2) 2.0 cm
(3) 0.50 cm
(4) 1.5 cm

90_____

91 In the diagram below, a crown glass converging lens has foci *F* and *F'*. An object is placed at a distance slightly less than the focal length from the lens, and a virtual image is produced.

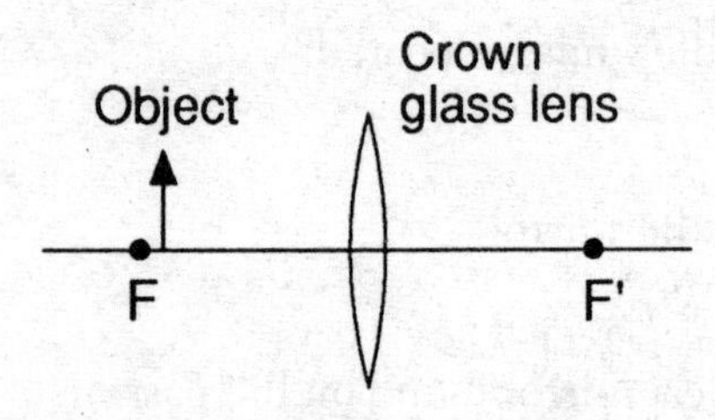

The object remains in the same position and the crown glass lens is replaced with another lens of identical shape. If the new lens produces a real image of the object, the new lens is most likely made of

1 water
2 Lucite
3 fused quartz
4 flint glass

91_____

92 Which diagram below shows the path of light rays as they pass from an object at $2F$ through a converging lens to the image formed at $2F'$?

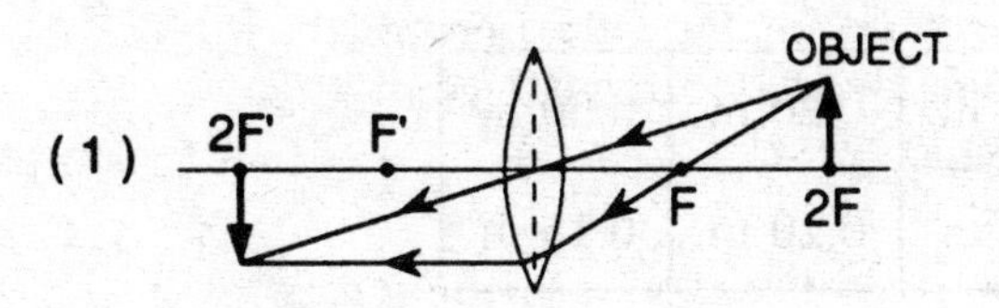

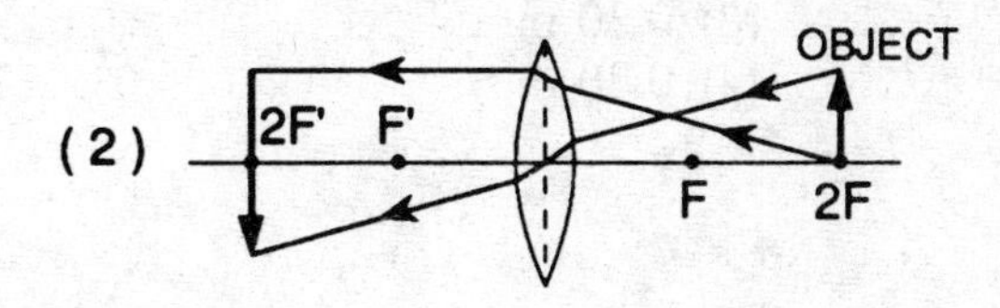

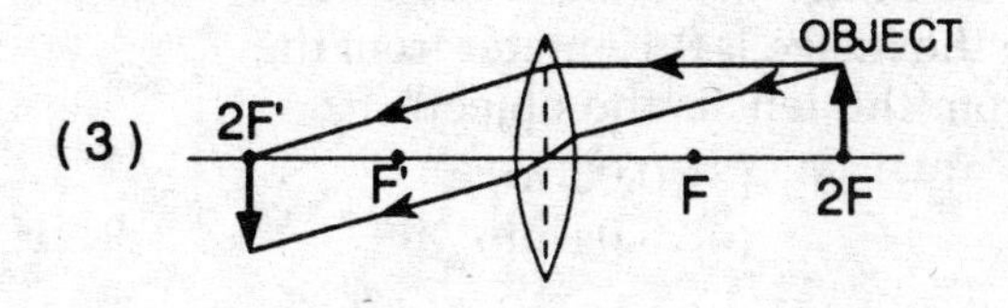

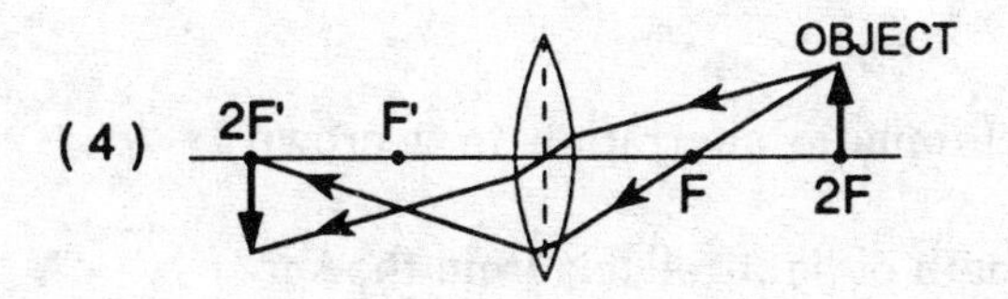

92_____

93 A student placed an object at various distances (d_o) from a converging lens. The corresponding image distance (d_i) was measured and recorded in the data table below.

d_o	0.15 m	0.20 m	0.30 m
d_i	0.30 m	0.20 m	0.15 m

What is the focal length of the lens?
(1) 0.10 m (3) 0.20 m
(2) 0.15 m (4) 0.30 m 93____

94 A lens forms a real image three times the size of the object when the image is 0.12 meter from the lens. How far from the lens is the object?
(1) 0.36 m (3) 0.03 m
(2) 0.09 m (4) 0.04 m 94____

95 What causes chromatic aberration in a crown glass lens?
1 Each wavelength of light reflects from the surface of the lens.
2 Each wavelength of light is refracted a different amount by the lens.
3 White light waves interfere inside the lens.
4 White light waves diffract around the edge of the lens. 95____

GROUP 5—**Solid State**

If you choose this group, be sure to answer questions **96–105.**

96 Which quantity is equal to the reciprocal of resistivity?

1 conductivity
2 current
3 resistance
4 voltage

96____

97 A solid is determined to be a conductor, an insulator, or a semiconductor on the basis of

1 its melting point and boiling point
2 its atomic number and mass number
3 the number of electrons in the conduction band and the energy gap between bands
4 the number of electrons per square meter of cross-sectional area of the material

97____

98 *P*-type semiconducting material functions primarily as

1 an acceptor of protons
2 an acceptor of electrons
3 a donor of protons
4 a donor of electrons

98____

99 What permits current to flow through a semiconductor when it is connected to a battery, as shown in the diagram at the right?

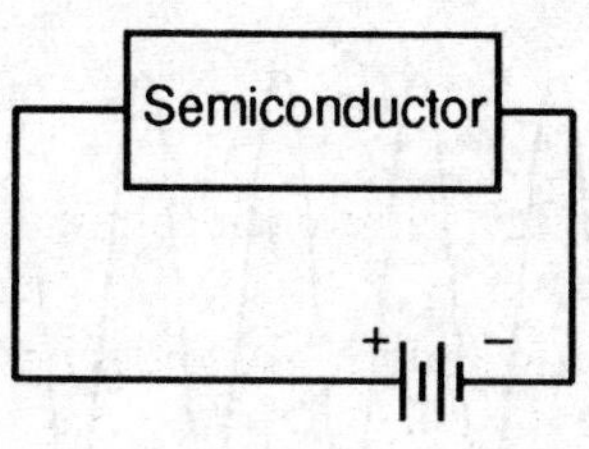

1 holes moving toward the right, only
2 electrons moving toward the left, only
3 both electrons and holes moving toward the left
4 electrons moving left and holes moving right

99____

100 The diagram below shows the alternating current input signal to a diode.

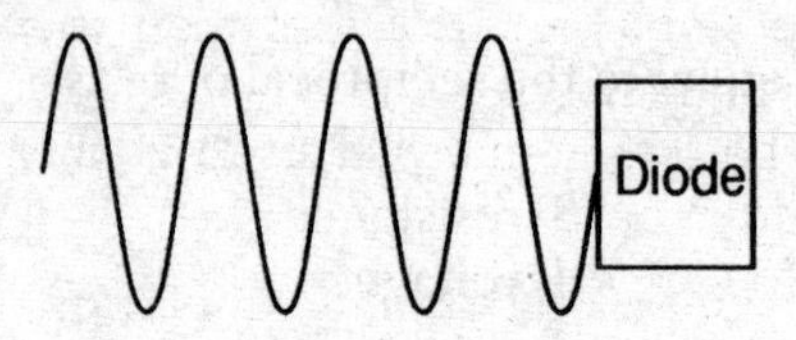

Which diagram below best represents the output signal?

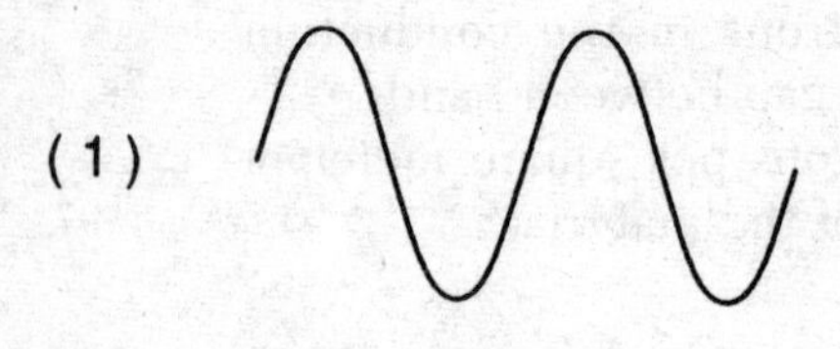

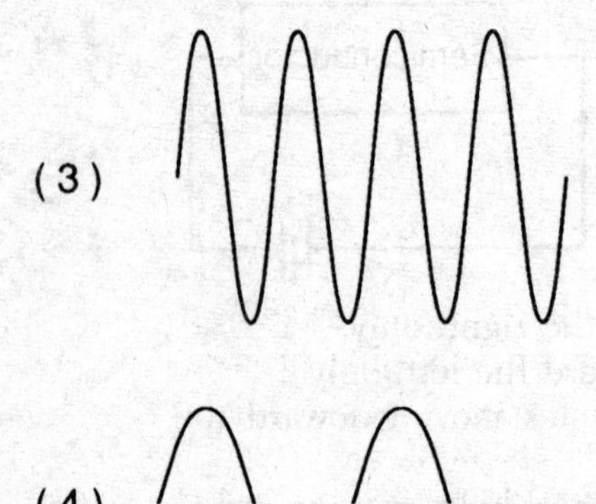

100_____

101 What is a basic difference between a transistor and a diode?

1 the number of junctions between *P*-type and *N*-type material
2 the size of the single junction between the *P*-type and *N*-type material
3 the nature of the donor material used in the *N*-type semiconductor
4 the amount of current applied to the semiconductor material

101_____

102 In the diagram below, which part of the operating *N-P-N* transistor is the collector?

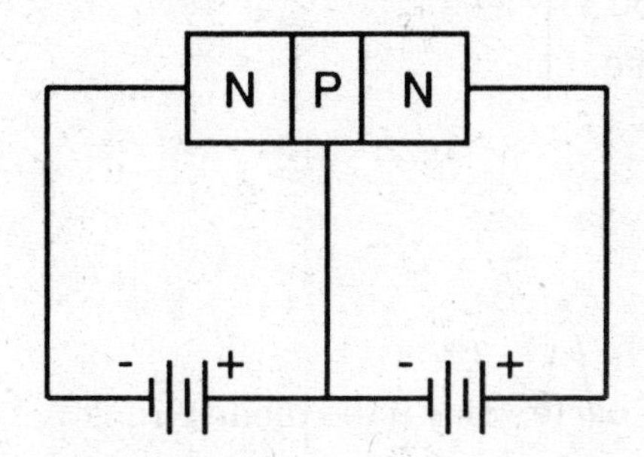

1 the left *N*-type section
2 the right *N*-type section
3 the left *P-N* junction
4 the right *P-N* junction

102_____

103 A small change in the emitter-base current in a transistor brings about a large change in the collector current. This current-increasing property of a transistor is called

1 amplification
2 biasing
3 Ohm's law
4 rectification

103_____

Base your answers to questions 104 and 105 on the graph below of current versus potential difference for a diode.

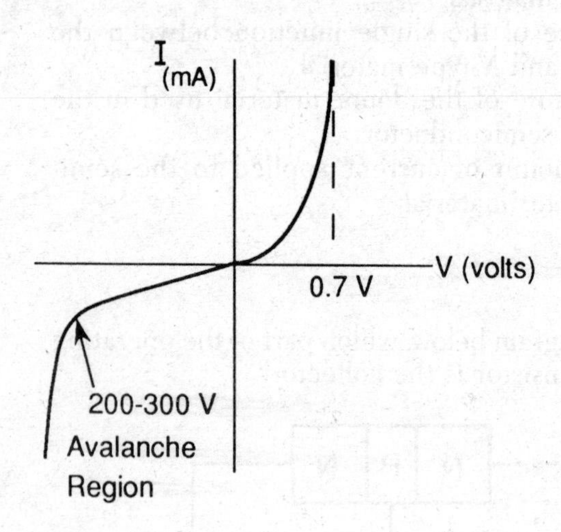

104 The "avalanche" occurs only if the diode circuit is

1 open
2 closed
3 reverse biased
4 forward biased

104_____

Note that question 105 has only three choices.

105 As the amount of doping material in the semiconductor increases, the magnitude of the voltage required for the "avalanche" to occur

1 decreases
2 increases
3 remains the same

105_____

GROUP 6—**Nuclear Energy**

If you choose this group, be sure to answer questions **106–115.**

106 How many neutrons are in an atom of $^{222}_{86}Rn$?

(1) 84 (3) 136
(2) 86 (4) 222 106_____

107 The chart below shows the masses of selected particles.

Particle	Mass
$^{235}_{92}U$	235.0 u
$^{138}_{56}Ba$	137.9 u
$^{95}_{36}Kr$	94.9 u
$^{1}_{0}n$	1.0 u

In the equation

$$^{235}_{92}U + ^{1}_{0}n \rightarrow ^{138}_{56}Ba + ^{95}_{36}Kr + 3^{1}_{0}n + E,$$

the energy E is equivalent to a mass of

(1) 0.2 u (3) 2.2 u
(2) 2.0 u (4) 0.0 u 107_____

108 Isotopes of the same element have nuclei with identical

1 mass numbers
2 binding energies
3 numbers of neutrons
4 numbers of protons 108_____

109 Which subatomic particle can *not* be accelerated by an electromagnetic field?

1 alpha 3 electron
2 neutron 4 positron 109_____

110 According to the Uranium Disintegration Series, the immediate decay product of $^{234}_{90}Th$ is

(1) $^{230}_{92}U$
(2) $^{230}_{89}Ac$
(3) $^{238}_{92}U$
(4) $^{234}_{91}Pa$

110____

111 In the reaction $^{27}_{13}Al + ^{4}_{2}He \rightarrow ^{30}_{15}P + ^{1}_{0}n + X$, what could X represent?

1 proton
2 gamma radiation
3 alpha particle
4 beta particle

111____

112 A radioactive isotope has a half-life of 3 minutes. If 10 kilograms of this isotope remains after 15 minutes, the original mass of the isotope must have been

(1) 50 kg
(2) 160 kg
(3) 250 kg
(4) 320 kg

112____

113 When an atomic nucleus captures an electron, the atomic number of that nucleus

1 decreases by 1
2 decreases by 2
3 increases by 1
4 increases by 2

113____

114 The equation $^{3}_{1}H + ^{1}_{1}H \rightarrow ^{4}_{2}He$ + energy is an example of

1 alpha decay
2 positron capture
3 fusion
4 fission

114____

115 Which equation represents nuclear fission?

(1) $^{214}_{82}Pb \rightarrow ^{214}_{83}Bi + ^{0}_{-1}e$

(2) $4^{1}_{1}H \rightarrow ^{4}_{2}He + 2^{0}_{+1}e$

(3) $^{235}_{92}U + ^{1}_{0}n \rightarrow ^{138}_{56}Ba + ^{95}_{36}Kr + 3^{1}_{0}n$

(4) $^{238}_{92}U \rightarrow ^{234}_{90}Th + ^{4}_{2}He$

115____

PART III

You must answer *all* questions in this part. [15]

116 Base your answers to parts *a* through *c* on the information and data table below.

A resistor was held at constant temperature in an operating electric circuit. A student measured the current through the resistor and the potential difference across it. The measurements are shown in the data table below.

Data Table

Current (A)	Potential Difference (V)
0.010	2.3
0.020	5.2
0.030	7.4
0.040	9.9
0.050	12.7

a Using the information in the data table, construct a graph on the grid provided, following the directions below.

(1) Mark an appropriate scale on the axis labeled "Current (A)." [1]

(2) Plot the data points for potential difference versus current. [1]

(3) Draw the best-fit line. [1]

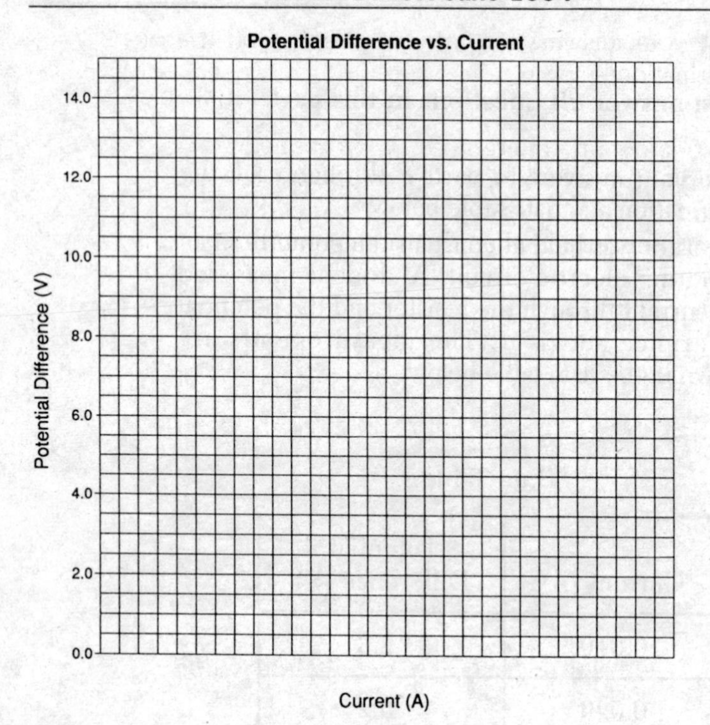

b Using your graph, find, the slope of the best-fit line. [Show all calculations, including the equation and substitution with units.] [2]

c What physical quantity does the slope of the graph represent? [1]

117 Base your answers to parts *a* through *d* on the information below.

A 6.0-kilogram concrete block is dropped from the top of a tall building. The block has fallen a distance of 55 meters and has a speed of 30. meters per second when it hits the ground.

a At the instant the block was released, what was its gravitational potential energy with respect to the ground? [Show all calculations, including the equation and substitution with units] [2]

b Calculate the kinetic energy of the block at the point of impact. [Show all calculations, including the equation and substitution with units.] [2]

c How much mechanical energy was "lost" by the block as it fell? [1]

d Using one or more complete sentences, explain what happened to the mechanical energy that was "lost" by the block. [1]

118 Base your answers to parts *a* through *c* on the information and diagram below.

A ray of light *AO* is incident on a plane mirror as shown.

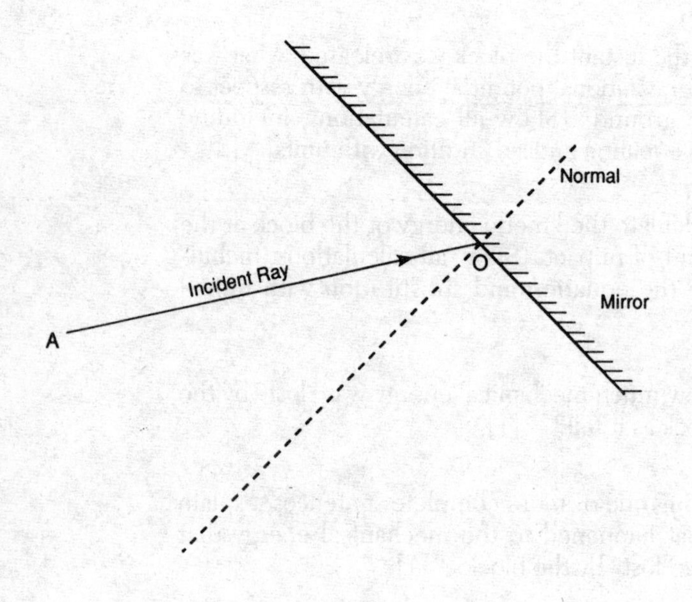

a Using a protractor, measure the angle of incidence for light ray *AO* and record the value on your answer sheet. [1]

b What is the angle of reflection of the light ray? [1]

c Using a protractor and straightedge, construct the reflected ray on the diagram. [1]

Answers June 1994

Physics

Answer Key

PART I

1. 3	11. 2	20. 2	29. 1	38. 3	47. 4
2. 2	12. 1	21. 2	30. 2	39. 1	48. 1
3. 3	13. 4	22. 3	31. 4	40. 1	49. 2
4. 4	14. 2	23. 1	32. 4	41. 2	50. 3
5. 1	15. 3	24. 3	33. 3	42. 4	51. 4
6. 2	16. 4	25. 4	34. 2	43. 1	52. 2
7. 1	17. 2	26. 2	35. 3	44. 2	53. 3
8. 4	18. 3	27. 4	36. 1	45. 3	54. 2
9. 1	19. 1	28. 4	37. 3	46. 4	55. 2
10. 4					

PART II

Group 1	Group 2	Group 3	Group 4	Group 5	Group 6
56. 2	66. 2	76. 1	86. 2	96. 1	106. 3
57. 4	67. 2	77. 3	87. 2	97. 3	107. 1
58. 3	68. 4	78. 2	88. 3	98. 2	108. 4
59. 3	69. 2	79. 2	89. 4	99. 4	109. 2
60. 2	70. 4	80. 3	90. 1	100. 2	110. 4
61. 3	71. 1	81. 1	91. 4	101. 1	111. 2
62. 1	72. 2	82. 2	92. 1	102. 2	112. 4
63. 1	73. 2	83. 4	93. 1	103. 1	113. 1
64. 3	74. 4	84. 4	94. 4	104. 3	114. 3
65. 1	75. 3	85. 2	95. 2	105. 1	115. 3

PART III — See answers explained

Answers Explained

PART I

1. **3** *Displacement* is defined as the change in an object's position. To fully describe this change we must state how far the object has moved, and in which direction. Thus displacement is a vector quantity.

In this problem a car travels 20. m east. The displacement of the car from its starting position is therefore *20. m east.*

Wrong Choices Explained:

(1) 20. m is the *distance* traveled. An indication of direction is required to describe the displacement.

(2) 20. m/s is the *speed.*

(4) 20. m/s east is the *velocity.*

2. **2** According to Newton's second law of motion, when an unbalanced force acts on an object, the object undergoes accelerated motion; on the other hand, if the net force is zero, the object will *not* be accelerated. Such an object is in a state of equilibrium, and may be at rest or may be moving at constant speed in a straight line.

On a distance-versus-time graph, an object traveling at constant speed is represented by a straight line sloping upward, as shown in the first graph in set 2. On a speed-versus-time graph, an object traveling at constant speed is represented by a horizontal line, as shown in the second graph in set 2.

3. **3** The acceleration due to gravity for objects near the surface of the moon is a constant, equal to approximately 1/6 the value of g on earth, that is, $1/6 \times 9.8$ m/s^2. This value is *independent of mass;* all masses fall *at the same rate* no matter how small or large they may be, provided that we can neglect air resistance. Since the moon has no atmosphere, there is no air resistance and the astronaut observes that *the baseball and balloon fall at the same rate.*

4. **4** A *vector quantity* has both magnitude and direction. *Acceleration,* the time rate of change of velocity, is a vector quantity. To fully describe acceleration, we must stipulate both its magnitude and its direction, for example, 2.0 m/s^2 east.

Wrong Choices Explained:

(1) *Distance* is a *scalar quantity.* A scalar quantity has magnitude only. Distance represents the magnitude of the displacement vector.

(2) *Time* is a *scalar quantity.* It has magnitude but no direction.

(3) *Speed* is a *scalar quantity.* It represents the magnitude of the velocity vector.

5. **1** *Concurrent forces* are forces that act on the same point in an object at the same time. When two concurrent forces act in the same direction, their resultant is a maximum; when the two forces act in opposite directions, their resultant is a minimum. For angles in between these two extremes, the smaller the angle separating the two forces, the larger the resultant. The pair of forces separated by the smallest angle is shown in *diagram 1.* This orientation would produce the greatest resultant and hence the greatest net force on the point.

This idea can be shown graphically by finding the resultant of the two forces in each diagram. One method of finding the resultant of two forces is to complete the parallelogram that uses the two forces as its sides. The resultant ($\vec{R}$) is the diagonal drawn from the common starting point. See the diagrams of the four choices below.

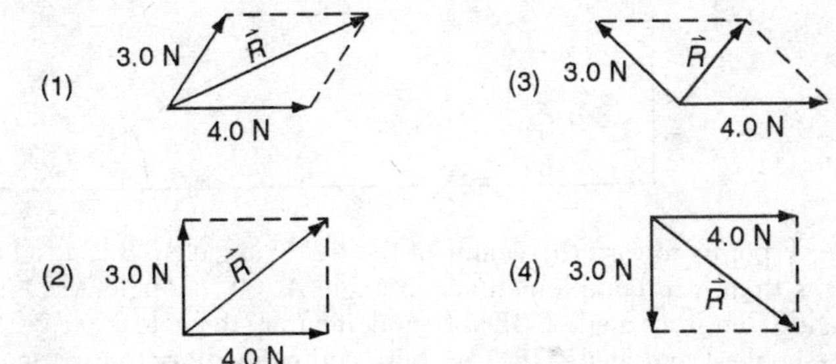

The resultant determined in this fashion for the pair of forces in *diagram 1* is the largest, and would therefore produce the greatest net force on the point.

6. **2** It is always advisable to tackle word problems by organizing the stated information:

Given: $v_i = 0$
$a = 4.0\ \text{m/s}^2$
$\Delta s = 32\ \text{m}$

Find: v_f

Note that *time* is not involved.

Choose the appropriate equation from the *Mechanics* equations in the *Reference Tables.* The only equation that involves each of the variables in this problem is

$$\begin{aligned} v_f^2 &= v_i^2 + 2a\Delta s \\ &= 0 + (2)(4.0\ \text{m/s}^2)(32\ \text{m}) \\ &= 256\ \text{m}^2/\text{s}^2 \\ v_f &= 16\ \text{m/s} \end{aligned}$$

7. **1** The diagram below shows a free-body diagram of the block as it slides down the incline. By the process of *resolution,* the weight (W) of the block, which is really the force of gravity on the block, can be divided into a component perpendicular ($W_{\perp}$) to the incline, and a component parallel to the incline ($W_{\parallel}$). The component parallel to the incline tends to pull the block down the incline. Since the block slides at a constant speed and does not accelerate, we conclude that the retarding force of friction on the block is equal in magnitude to the component of the block's weight that pulls it down the incline.

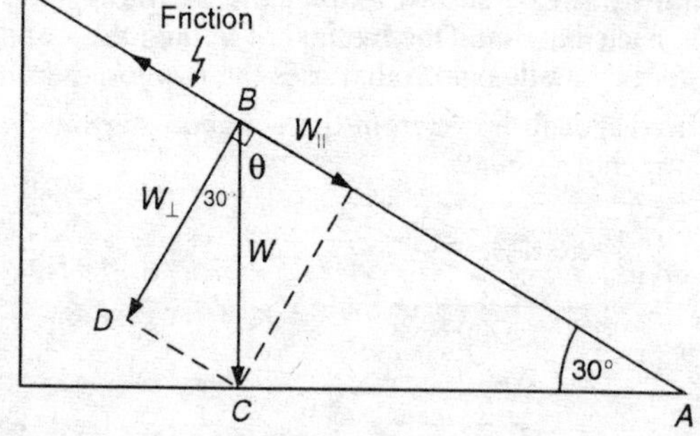

Since W points toward the center of the earth, angle ACB is a right angle. Angle θ is therefore complementary to angle A (30.°). Angle ABD is also a right angle. Therefore angle CBD is complementary to angle θ.

Since angle A and angle CBD are both complementary to the same angle, they must be equal. Hence we conclude that angle CBD equals 30.°.

Using simple trigonometry in triangle CBD, and the fact that sides CD and $W_{\parallel}$ are of equal length, we obtain:

$$\sin 30.° = \frac{W_{\parallel}}{W}$$

Rearranging terms gives

$$W_{\parallel} = W \sin 30.°$$
$$= 0.50W$$

Since $W = 10.$ N,

$$W_{\parallel} = (0.50)\,(10.\text{ N}) = 5.0\text{ N},$$

and hence the force of friction also equals *5.0 N*.

8. **4** The *average acceleration* ($\overline{a}$) of a body over a given time interval equals the change in its velocity (Δv) divided by the time (Δt) during which the change takes place. Refer to the *Mechanics* equations in the *Reference Tables:*

$$\overline{a} = \frac{\Delta v}{\Delta t}$$

However, $\Delta v/\Delta t$ also represents the slope of the line connecting two points on a speed-versus-time graph. Therefore the average acceleration of each car can be determined by calculating the slope of the line connecting the starting point at $t = 0$ and the endpoint at $t = 6.0$ s.

Since *car D* experienced the greatest Δv during the 6.0-s interval, it has the greatest slope and the *greatest average acceleration*.

9. **1** It is always advisable to tackle word problems by organizing the stated information:

Given: $v_i = 14$ m/s   Find: v_f

$\bar{a} = 2.0$ m/s^2

$\Delta t = 5.0$ s

Note that *distance* is not involved.

Now choose the appropriate equation from the *Mechanics* equations in the *Reference Tables.* If you realize that Δv is equal to $v_f - v_i$, you will see that the only equation that involves each of the variables in this problem is

$$\bar{a} = \frac{\Delta v}{\Delta t}$$

$$= \frac{v_f - v_i}{\Delta t}$$

In this problem, the *final speed* is the unknown. Therefore rearrange the terms so that the unknown, v_f, stands alone:

$$v_f = v_i + \bar{a}\,\Delta t$$

$$= 14 \text{ m/s} + (2.0 \text{ m/s}^2)(5.0 \text{ s})$$

$$= 24 \text{ m/s}$$

You can also analyze this problem without the use of equations. Accelerating at the rate of 2.0 m/s^2 means that the car's speed gains 2.0 m/s *each* second. After 5.0 s, the speed will have increased by 2.0 m/s^2 × 5.0 s = 10 m/s. Since the car started with a speed of 14 m/s, its final speed will be 14 + 10 = 24 m/s.

The traffic sign indicates that the speed limit is 25 m/s. Therefore, when the car reaches the sign, the car's speed (24 m/s) will be *less than the speed limit.*

10. **4** Organize the given data as you did in problem 9, noting that *distance* is now the unknown:

Given: $v_i = 14$ m/s   Find: Δs

$a = 2.0$ m/s^2

$\Delta t = 5.0$ s

Now choose the appropriate equation from the *Mechanics* equations in the *Reference Tables.* The only equation that involves each of the variables in this problem is

$$\Delta s = v_i \Delta t + \frac{1}{2} a(\Delta t)^2$$
$$= (14 \text{ m/s})(5.0 \text{ s}) + \frac{1}{2}(2.0 \text{ m/s}^2)(5.0 \text{ s})^2$$
$$= 95 \text{ m}$$

11. **2** The kilogram (kg) is the SI unit of mass; a 1-kg object weighs about 2.2 lb. Think in terms of the standard brass kilogram masses used in the physics lab.

Convert 0.005 kg to pounds, and then ounces:

$$0.005 \text{ kg} \times 2.2 \text{ lb/kg} \times 16 \text{ oz/lb} = 0.2 \text{ oz.}$$

This seems about right for the mass of a nickel, especially when you consider the other choices.

Wrong Choices Explained:

(1) 0.0005 kg is 10 times smaller, equivalent to 0.02 oz.
(3) 0.5 kg is equivalent to more than 1 lb.
(4) 5 kg is equivalent to 11 lb.

12. **1** The net force *(F)* acting on an object is equal to the product of the object's mass *(m)* and the resulting acceleration *(a)*. This relationship is known as *Newton's second law of motion*:

$$F = ma$$

Solving for the unknown, *m*, in this question gives:

$$m = \frac{F}{a}$$
$$= \frac{5.0 \times 10^2 \text{N}}{5.0 \text{ m/s}^2}$$
$$= 1.0 \times 10^2 \text{ kg}$$

13. **4** According to *Newton's law of universal gravitation*, the gravitational force of attraction *(F)* between two masses (m_1 and m_2) is proportional to the product of the two masses, and inversely proportional to the square of the distance *(r)* between them. Refer to the *Mechanics* equations in the *Reference Tables:*

$$F = \frac{Gm_1m_2}{r^2}$$

where G is the universal gravitational constant. Since distance is not involved in this problem, the relationship can be simplified:

$$F \propto m_1m_2$$

If the mass of each object were doubled, the product of their masses would quadruple, and so the magnitude of the gravitational force between the objects (originally 20. N) would be four times larger, that is, 4 × 20. N = *80. N*.

14. **2** When an unbalanced force acts on an object, there is a change in the object's momentum ($m\Delta v$). The product of the force (F) and the length of time that it acts (Δt) is equal to the product of the object's mass (m) and its change in velocity (Δv). Refer to the *Mechanics* equations in the *Reference Tables*:

$$\begin{aligned} F\Delta t &= m\Delta v \\ (1.0 \times 10^7\text{N})(10.0\text{s}) &= (2.0 \times 10^6\text{ kg})(\Delta v) \\ \Delta v &= 5.0 \times 10^1\text{ m/s} \end{aligned}$$

Since the space shuttle starts from rest, the change in its speed (Δv) is the final speed attained 10.0 s after lift-off.

15. **3** The *momentum* of an object is equal to the product of its mass and its velocity. Refer to the *Mechanics* equations in the *Reference Tables*:

$$p = mv$$

During an "explosion" such as occurs when a cannon fires a projectile, the total momentum of the system is *conserved*. Therefore the total momentum before the explosion is equal to the total momentum after the explosion. Since both the cannon and its projectile are at rest ($v = 0$) before the explosion, neither object has momentum, and the total momentum before the explosion equals zero.

Therefore, by the *law of conservation of momentum* described above, the total momentum *after* the explosion must also equal zero, that is, the *momentum of cannon + momentum of projectile = 0*.

16. **4** *Equilibrium* occurs when the sum of the forces acting on an object equals zero. A book resting on a table is acted on by two forces (ignoring the effects of the air): the earth's gravitational pull (which is the weight of the book) and the table's upward force on the book.

The book is in equilibrium because *the weight of the book and the table's upward force on the book are equal in magnitude, but opposite in direction,* and therefore add up to zero.

17. **2** The work (W) done by a force (F) acting on an object is equal to the product of the force and the resultant displacement (Δs) *in the direction of the force.* Only the horizontal component (8.0 N) of the force exerted on the block by the student does work, moving the block 3.0 m along the horizontal surface. The vertical force does no work and can be ignored.

Since the block moves with a constant velocity, and does *not* accelerate, we conclude that the sum of the forces acting on it equals zero. Hence, the horizontal component of the force exerted on the block by the student is equal in magnitude to the retarding force of friction. All the work done by the horizontal component is work done against friction. No work is converted to kinetic energy since the block does not gain speed.

Refer to the *Energy* equations in the *Reference Tables:*

$$\begin{aligned} W &= F\Delta s \\ &= (8.0 \text{ N})(3.0 \text{ m}) \\ &= 24 \text{ J} \end{aligned}$$

18. **3** *Power* is defined as the rate at which work is done, that is, power = work/time. Refer to the *Energy* equations in the *Reference Tables:*

$$P = \frac{W}{\Delta t} = \frac{F\Delta s}{\Delta t}$$

where W is the work done by a force (F) acting on an object and is equal to the product of the force and the resultant displacement (Δs) in the direction of the force.

When an object is to be lifted vertically at constant speed, a force equal to the weight of the object is required. We can therefore substitute the weight of the crate in the equation:

$$\begin{aligned} P &= \frac{F\Delta s}{\Delta t} \\ &= \frac{(1.0 \times 10^3 \text{ N})(1.5 \text{ m})}{5.0 \text{ s}} \\ &= 3.0 \times 10^2 W \end{aligned}$$

19. **1** Within a certain range the amount a spring is stretched or displaced (x) is proportional to the force (F) applied to stretch it.

Refer to the *Energy* equations in the *Reference Tables:*

$$F = kx$$

where k is the *spring constant.* This relationship is known as *Hooke's law.*

The graph that shows the relationship between the applied force and the resulting spring displacement is a straight line from the origin, typical for a directly proportional relationship. The slope of a force-displacement graph is equal to the change in the ordinate divided by the change in the abscissa, that is,

$$\text{Slope} = \frac{\Delta F}{\Delta x}.$$

If we solve the Hooke's law equation for the spring constant, we also get

$$k = \frac{F}{x}$$

We conclude that the spring constant can be represented by the *slope of graph A.*

20. **2** Within a certain range the amount a spring is stretched or compressed (x) is proportional to the force (F) applied to displace it. This relationship is known as *Hooke's law.* Refer to the *Energy* equations in the *Reference Tables*:

$$F = kx$$

where k is the *spring constant.*

The *potential energy* stored in a stretched or compressed spring (PE_s) is equal to one-half the product of its spring constant (k) and the square of the displacement (x). Refer to the *Energy* equations in the *Reference Tables*:

$$PE_s = \frac{1}{2}kx^2$$

Before we can solve for the potential energy we must determine the spring constant. Solving the Hooke's law equation for the spring constant, k, we get

$$\begin{aligned} k &= \frac{F}{x} \\ &= \frac{0.2 \text{ N}}{0.02 \text{ m}} \\ &= 10 \text{ N/m} \end{aligned}$$

Substituting this value into the equation for elastic potential energy gives

$$\begin{aligned} PE_s &= \frac{1}{2}kx^2 \\ &= \frac{1}{2}(10 \text{ N/m})(0.02 \text{ m})^2 \\ &= 2 \times 10^{-3} \text{ J} \end{aligned}$$

21. **2** *Kinetic energy* (KE) is the energy that an object has because of its motion. It equals one-half the product of the object's mass and the square of its velocity (or speed). Refer to the *Energy* equations in the *Reference Tables:*

$$KE = \frac{1}{2}mv^2$$

$$400.\ \text{J} = \frac{1}{2}m(20.\ \text{m/s})^2$$

$$m = 2.0\ \text{kg}$$

22. **3** *Kinetic energy* is the energy that an object has because of its motion. It equals one-half the product of the object's mass and the square of its velocity (or speed). At point *A* in the diagram, the pendulum is at rest. Its kinetic energy therefore equals zero at point *A*.

The *gravitational potential energy* of an object above the ground level is equal to the product of its mass, the value of the acceleration due to gravity, and the vertical height to which the object was raised. Point *B* in the diagram is the lowest point in the pendulum's swing. This point is generally considered to be the ground level. The potential energy of the pendulum at point *B* is therefore zero.

We conclude that, compared to the pendulum's kinetic energy at *A*, its potential energy at *B* is *the same.*

23. **1** Within an atom, electrons are lighter and more loosely bound than protons, and are therefore more easily detached than protons, which are bound up in the atomic nucleus. Therefore electrons, rather than protons, are transferred during a charging process.

A pith ball is made of soft plant tissue that has been coated with a metallic paint. There are many mobile electrons on the metallic surface of a pith ball. When a neutral pith ball makes contact with a positively charged rod, the negatively charged electrons on the pith ball will be attracted to the positively charged rod, and some of these electrons will be transferred to the rod.

We conclude that, during contact with the rod, the pith ball *loses electrons.*

24. **3** According to *Coulomb's law,* the electrostatic force *(F)* between two point charges (q_1 and q_2) is directly proportional to the product of the two charges and *inversely* proportional to the square of the distance (*r*) between them. Refer to the *Electricity and Magnetism* equations in the *Reference Tables:*

$$F = \frac{kq_1q_2}{r^2}$$

In this question the distance between the two charges is *halved,* from 0.1 m to 0.05 m. Remember that electric force is *inversely* proportional to the square of the distance. When two quantities vary inversely with respect to each other, one quantity increases when the other quantity decreases. If r is decreased by a factor of 1/2, then the force will *increase* by a factor of $1/(1/2)^2$, or by a factor of 4.

Since positive charges repel, we conclude that, when these point charges are placed 0.05 meter apart, the electrostatic force between them is *4F, and repelling.*

25. **4** Like charges repel, and unlike charges attract. Since the electron in the question is negatively charged, it will be attracted to a positively charged object, as shown in the diagram. Moreover, the electrostatic or Coulomb force of attraction is directed along the line connecting the two charges.

We conclude that the electrostatic force acting on the electron in the diagram is directed toward point *D.*

26. **2** The electric field intensity (E) at any point in an electric field is defined as the electric force (F) exerted on a unit charge (q) placed in the field at that point. Refer to the *Electricity and Magnetism* equations in the *Reference Tables:*

$$E = \frac{F}{q}$$

Solving for the unknown, F, in this question gives:

$$\begin{aligned} F &= qE \\ &= (1.6 \times 10^{-19}\ \text{C})\ (5.0 \times 10^{3}\ \text{N/C}) \\ &= 8.0 \times 10^{-16}\ \text{N} \end{aligned}$$

NOTE: The charge of the electron, 1.6×10^{-19}C, is found in the *List of Physical Constants* in the *Reference Tables.*

27. **4** The unit "volts per meter" measures the same quantity as *newtons per coulomb.* The equivalence of these units can be shown as follows:

$$\begin{aligned} 1\ \text{volt} &= 1\ \text{joule/coulomb} \\ &= 1\ \text{newton} \cdot \text{meter/coulomb} \\ 1\ \text{volt/meter} &= 1\ \text{newton/coulomb} \end{aligned}$$

You may also have noticed that two different equations are given for electric field intensity (E) in the *Electricity and Magnetism* equations in the *Reference Tables:*

$$E = \frac{F}{q}$$

and

$$E = \frac{V}{d}$$

Substituting units into these two equations gives:

$$E = \frac{F}{q} = \text{newtons/coulomb}$$

and

$$E = \frac{V}{d} = \text{volts/meter}$$

We conclude that the unit "volts per meter" measures the same quantity as *newtons per coulomb.*

28. **4** Apply Ohm's law to determine how much total current (I_t) flows in the circuit. Refer to the *Electricity and Magnetism* equations in the *Reference Tables:*

$$R_t = \frac{V_t}{I_t}$$

where V_t stands for the electric potential difference applied to the circuit, and R_t for the total circuit resistance. Solving for I_t, we obtain:

$$\begin{aligned} I_t &= \frac{V_t}{R_t} \\ &= \frac{2.00 \times 10^2 \text{V}}{1.00 \times 10^2 \, \Omega} \\ &= 2.00 \text{ A} \end{aligned}$$

Electric current is the time-rate at which electric charge flows past a given point in a circuit. Thus, the size of the current (I) is equal to the quantity of charge (Δq) that passes, divided by the time (Δt) required for the charge to pass. A current of 1 *ampere* (A) transfers charge at the rate of 1 coulomb (C) per second; thus, 1 A = 1 C/s.

Refer to the *Electricity and Magnetism* equations in the *Reference Tables:*

$$I = \frac{\Delta q}{\Delta t}$$

Solving for the unknown, Δq, in this problem gives:

$$\begin{aligned} \Delta q &= I\Delta t \\ &= (2.00 \text{ A})(2.00 \text{ s}) \\ &= (2.00 \text{ C/s})(2.00 \text{ s}) \\ &= 4.00 \text{ C} \end{aligned}$$

29. **1** The resistance (R) of a metallic conductor is directly proportional to its length (ℓ) and inversely proportional to its cross-sectional area (A):

$$R = \frac{\rho \ell}{A}$$

where ρ is a constant that depends on the particular material. If the cross-sectional area A were increased, the wire's resistance would decrease.

According to Ohm's law, resistance (R) and current (I) are inversely proportional to each other. Refer to the *Electricity and Magnetism* equations in the *Reference Tables:*

$$R = \frac{V}{I}$$

or

$$I = \frac{V}{R}$$

where V stands for the electric potential difference. Therefore, if R were decreased, I would increase.

We conclude that the current flowing in the wire can be increased by increasing the wire's *cross-sectional area.*

Wrong Choices Explained:

(2) If the *length* were increased, the resistance would also increase, and the current would *decrease.*

(3) If the *resistance* were increased, then by Ohm's law the current would *decrease.*

(4) When the *temperature* of a metal is increased, the resistance increases, and the current will therefore decrease.

30. **2** Calculate the total resistance for each resistor arrangement.

The resistors in arrangements A and C are connected *in series.* In a series connection, the total resistance (R_t) is equal to the sum of the resistances of all the components. Refer to the *Electricity and Magnetism* equations in the *Reference Tables:*

$$A:\quad R_t = R_1 + R_2 + R_3 + \ldots$$
$$= 1\ \Omega + 1\ \Omega$$
$$= 2\ \Omega$$
$$C:\quad R_t = R_1 + R_2 + R_3 + \ldots$$
$$= 2\ \Omega + 2\ \Omega$$
$$= 4\ \Omega$$

The resistors in arrangements B and D are connected *in parallel.* We know this because there is more than one path in which the current can flow. Refer to the *Electricity and Magnetism* equations in the *Reference Tables:*

$$B: \quad \frac{1}{R_t} = \frac{1}{R_1} + \frac{1}{R_2} + \frac{1}{R_3} + \ldots$$

$$= \frac{1}{8}\ \Omega + \frac{1}{8}\ \Omega$$

$$R_t = 4\ \Omega$$

$$D: \quad \frac{1}{R_t} = \frac{1}{R_1} + \frac{1}{R_2} + \frac{1}{R_3} + \ldots$$

$$= \frac{1}{2}\ \Omega + \frac{1}{2}\ \Omega$$

$$R_t = 1\ \Omega$$

We conclude that the two resistor arrangements that have equivalent resistance are *B and C.*

31. **4** The *electric energy* (W) used by a device is equal to the product of the power (P) developed and the length of time (t) during which charge flows. The joule is the unit for electric energy. Refer to the *Electricity and Magnetism* equations in the *Reference Tables:*

$$W = Pt = VIt = I^2Rt$$

In this problem the voltage (V) and current (I) are given, but not the resistance (R). Therefore we choose the equation

$$\begin{aligned} W &= VIt \\ &= (240\ \text{V})(30.\ \text{A})(1200\ \text{s}) \\ &= 8.6 \times 10^6\ \text{J} \end{aligned}$$

32. **4** An electric current flowing through a straight conductor produces a magnetic field that can be represented by magnetic flux lines that form concentric circles in a plane perpendicular to the conductor and that have the wire as common center.

Use the following *left-hand rule* to determine the direction of these lines of force: Hold the wire in your left hand so that your extended thumb points in the direction of electron flow, and your fingers curl around the wire. Your curled fingers will now point in the direction of the magnetic lines of force. At point P, your curled fingers point *out of the page.* This is the direction of the magnetic field at point P, as shown in the diagram below.

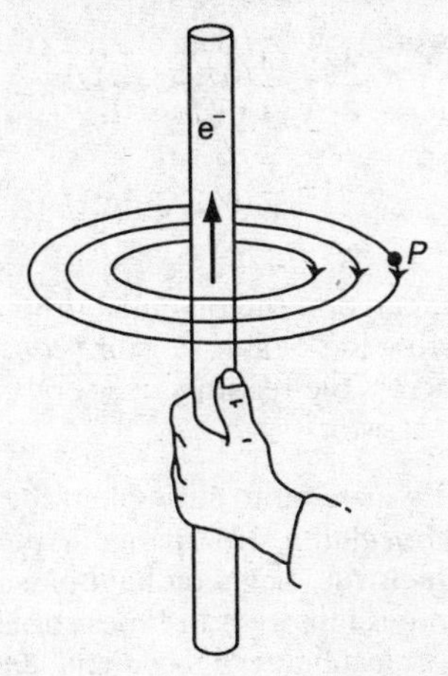

33. **3** Since there is only one path in which current can flow in the circuit shown in the diagram in this question, we conclude that the circuit is a *series connection.* In a series connection, the total resistance (R_t) is equal to the sum of the resistances of all the components. Refer to the *Electricity and Magnetism* equations in the *Reference Tables:*

$$R_t = R_1 + R_2 + R_3 + \dots$$
$$= 4\ \Omega + 8\ \Omega$$
$$= 12\ \Omega$$

Apply Ohm's law to the entire circuit to determine how much total current (I_t) flows. Refer to the *Electricity and Magnetism* equations in the *Reference Tables:*

$$R_t = \frac{V_t}{I_t}$$

where V_t stands for the electric potential difference applied to the circuit. Solving for I_t, we obtain

$$I_t = \frac{V_t}{R_t}$$
$$= \frac{6\text{V}}{12\ \Omega}$$
$$= 0.5\ \text{A}$$

Since this is a series connection, the same current flows through each component in the circuit. Now apply Ohm's law to determine the potential drop across each resistor:

Across the 4-ohm resistor: $V = IR$
$= (0.5 \text{ A})(4 \ \Omega)$
$= 2 \text{ V}$

Across the 8-ohm resistor: $V = IR$
$= (0.5 \text{ A})(8 \ \Omega)$
$= 4 \text{ V}$

We conclude that, compared to the potential drop across the 8-ohm resistor, the potential drop across the 4-ohm resistor is *one-half as great.* (In general, the potential drops across the resistors in a series circuit are proportional to the resistances of these resistors.)

34. **2** The ability of a material to strengthen the magnetic field passing through it is called its *permeability.* Iron is highly permeable, and therefore concentrates and strengthens the magnetic field passing through it. An electromagnet with an iron core is hundreds of times stronger than an empty coil. In contrast to iron, the permeabilities of *wood* and *aluminum* are very low. A *copper* core actually reduces the strength of an electromagnet.

We conclude that an electromagnet would have the greatest strength if its wires were wrapped around a core made of *iron.*

35. **3** The magnetic force acting on a charged particle such as an electron moving through a uniform magnetic field is a maximum when the angle between the direction of the electron's motion and the direction of the magnetic field is *90°.* There is no magnetic force when the angle between the direction of motion of the electron and the direction of the field is 0°.

36. **1** Electromagnetic waves are generated by accelerating electric charges, such as the oscillating electrons on the antenna of a broadcasting television or radio station.

Of the particles listed, only an accelerating *neutron* does *not* generate electromagnetic waves, because a neutron has no charge.

37. **3** Both sound waves and light waves *transfer energy.*

Wrong Choices Explained:

(1) Sound waves *are longitudinal*, but light waves are transverse.

(2) Light waves *are transverse*, but sound waves are longitudinal.

(4) Light waves *travel in a vacuum*, as through space, but sound waves need a medium within which to travel.

38. **3** As a longitudinal wave, such as a sound wave, travels through a medium, the particles of the medium vibrate back and forth in a direction that is *parallel to the direction of wave travel.*

Wrong Choices Explained:

(1), (2) *Water waves* are considered to be elliptical, and particles of the medium move *in ellipses* as the waves pass. However, when water waves pass through very deep ocean water, the particles on the surface of the water move in ellipses that are nearly *circles*.

(4) When a *transverse* wave passes through a medium, the particles of the medium move *perpendicular to the direction of wave travel*.

39. **1** The wavelength of a periodic wave is the distance between two consecutive points on the wave that are in phase. If the sound waves from the police car's siren are in phase at points *A* and *B*, the distance between the points could be one wavelength (if *A* and *B* are *consecutive* points in phase) or some other whole number of wavelengths (if *A* and *B* are not consecutive points in phase). However, the distance between *A* and *B* could *not* be a fraction of a wavelength.

The distance between the points could be **1λ.**

40. **1** The speed (v) of any wave is equal to the product of its frequency (f) and its wavelength (λ). Refer to the *Wave Phenomena* equations in the *Reference Tables:*

$$v = f\lambda$$
$$340 \text{ m/s} = (680 \text{ Hz})(\lambda)$$
$$\lambda = 0.50 \text{ m}$$

41. **2** This question illustrates the *Doppler effect,* that is, the variation in the observed frequency when there is a relative motion between the source of the waves and the receiver. There is a *decrease* in the observed frequency when the distance between the source and the receiver is increasing, for example, when the source is moving away from the receiver. However, there is an *increase* in the observed frequency if the source is moving toward the receiver, as in this question.

Even before the police car accelerates, an observer at point *A* would hear a higher frequency than the true frequency of the siren because the siren is moving toward the receiver. If the car were to accelerate toward point *A*, the frequency of the sound heard by an observer at point *A* would *increase*, and keep increasing until the accelerating car reached the observer. See the diagram below.

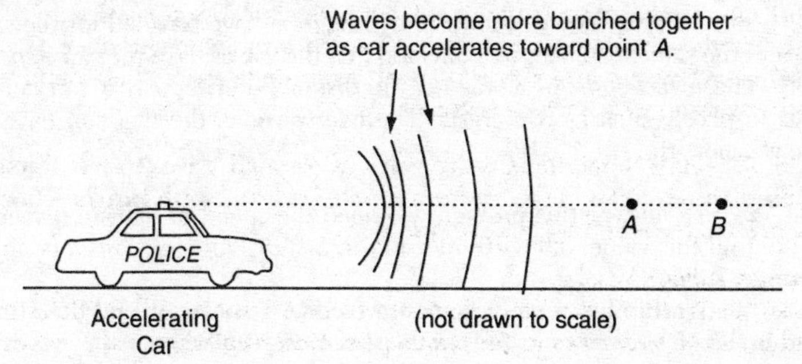

42. **4** When two or more waves pass through the same region at the same time, as shown in the diagram given in this question, the phenomenon of *intereference* occurs. By a process called *superposition*, the resultant displacement of each particle in the medium will equal the algebraic sum of the displacements due to all of the individual waves.

Maximum destructive interference occurs at the points where waves from the two sources arrive *out of phase*, for example, when the crest of one wave overlaps the trough of another. At these points the resultant amplitude is diminished. This is the situation when both pulses in the given diagram reach region *AB*.

In the diagram below, which shows the two pulses when they have both reached region *AB*, we see that the crest of one wave overlaps the trough of the other. Since both pulses appear to have the same amplitude, the resultant displacement of the medium in region *AB* will be zero, as shown in diagram 4.

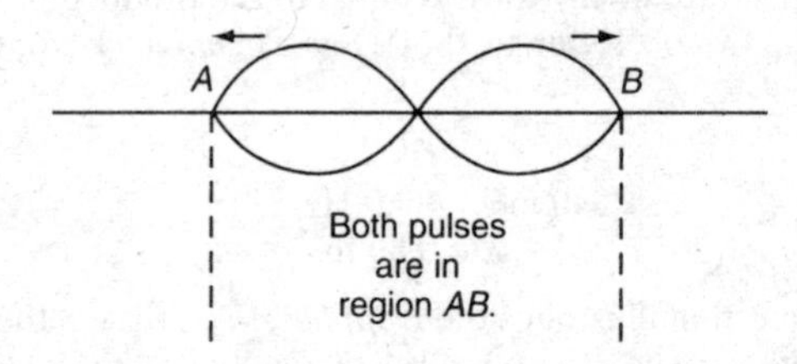

43. **1** The *amplitude* of a wave is the maximum displacement of each particle in the medium through which the wave travels from its rest position. In transverse waves, as shown in the diagram in this question, the amplitude is simply the height of a crest or the depth of a trough, that is, the distance of the wave's peak from the x-axis.

In the diagram we note that the two waves have different *amplitudes*.

Wrong Choices Explained:

(2) The *frequency* of a periodic wave is the number of vibrations occurring per second. The two waves shown in the diagram have the same frequency.

(3) Since the two waves shown in the diagram have traveled for the same time and the same distance, we conclude that the waves have the same *speed*.

(4) The *wavelength* of a wave is the distance between two consecutive points in phase, such as two crests. The two waves in the diagram have the same wavelength.

44. **2** To answer this problem we need the speed of light in a vacuum (c). We find this value, 3.0×10^8 m/s, in the *List of Physical Constants* in the *Reference Tables*.

The relationship between distance traveled (Δs), speed (v) and time (Δt) is found in the *Mechanics* equations in the *Reference Tables:*

$$\bar{v} = \frac{\Delta s}{\Delta t}$$

Solving for the unknown, Δt, in this question, gives

$$\Delta t = \frac{\Delta s}{\bar{v}}$$

$$= \frac{3.9 \times 10^8 \text{m}}{3.0 \times 10^8 \text{m/s}}$$

$$= 1.3 \text{ s}$$

45. **3** The diagram below shows several parallel light rays incident on the surface of a plane mirror. Note that each of the rays is reflected so that its angle of reflection is equal to its angle of incidence.

Since the incident rays are parallel, they each have the same angle of incidence. Therefore the reflected rays will have the same angle of reflection. We conclude that, upon reflection from the mirror, the light rays will *be parallel.*

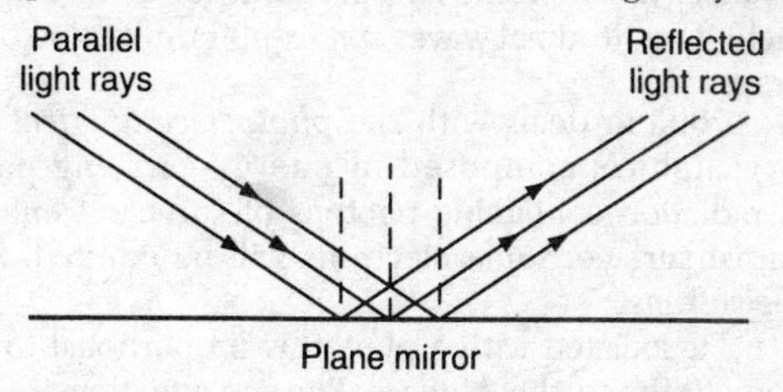

46. **4** The light ray in the diagram shown in this question reaches the boundary at its *critical angle.* The critical angle is the angle of incidence such that the angle of refraction is 90°, and it occurs only for a light ray that travels into a medium in which its speed increases, as from water to air.

The sine of the critical angle of incidence (θ_c) relative to air is equal to the reciprocal of the index of refraction (n) of the incident medium. Refer to the *Wave Phenomena* equations in the *Reference Tables:*

$$\sin \theta_c = \frac{1}{n}$$

Substitute 49° for the critical angle, and solve for n:

$$\sin 49° = \frac{1}{n}$$

$$0.755 = \frac{1}{n}$$

$$n = 1.3 \text{ (to two significant figures)}$$

Checking the *Absolute Indices of Refraction table* in the *Reference Tables,* we see that water has an index of refraction equal to 1.33.

We conclude that medium X is most likely *water.*

47. **4** *Polarization* is the separation of a beam of light so that the vibrations are all in the same plane. Only tranverse waves, such as light, can be polarized. Polarization can *not* be exhibited by longitudinal waves, such as sound.

Wrong Choices Explained:

(1) *Reflection* is the bouncing back that occurs to a wave when it reaches a barrier or the interface between two different media. Longitudinal waves *can* exhibit reflection.

(2) *Refraction* is the bending of a wave from its original direction as it obliquely enters a second medium in which its speed changes. Longitudinal waves *can* exhibit refraction.

(3) *Diffraction* is the bending or spreading of a wave into the region behind an obstacle. Longitudinal waves *can* exhibit diffraction.

48. **1** This problem deals with the *photoelectric effect.* According to quantum theory, light is composed of energy-carrying particles called *photons.* When radiation containing photons of sufficient energy is incident on a photosensitive surface, some electrons will be emitted. Such electrons are called *photoelectrons.*

The energy (E_f) associated with a photon is proportional to the frequency (f) of the photon. Refer to the *Modern Physics* equations in the *Reference Tables:*

$$E_{\text{photon}} = hf$$

where h is Planck's constant.

For each photosensitive surface there is a minimum frequency below which no photoelectrons will be emitted. This frequency is known as the *threshold frequency* (f_0). If the photons in green light have sufficient energy to cause a metal surface to emit photoelectrons, then the photons in light of any color with a *higher* frequency than green will also cause the surface to emit photoelectrons. Of the colors in the four answer choices—blue, yellow, orange, and red—only blue light has a higher frequency and therefore greater energy than green light.

We conclude that the metal surface must also emit photoelectrons when illuminated by *blue light.*

NOTE: If you didn't remember that colors toward the violet end of the visible spectrum, such as blue, have a higher frequency than colors toward the red end of the spectrum, you could deduce this fact from the *Wavelengths of Light in a Vacuum* chart in the *Reference Tables.* Violet light is seen to have the shortest wavelength. From the wave equation, $v = f\lambda$, we see that the wavelength and frequency of a wave are inversely proportional to each other.

Since violet light has the shortest wavelength, it must have the highest frequency of the visible colors.

Since the energy of a photon is proportional to its frequency, we conclude that violet photons are the most energetic of the photons in the visible range, and red photons the least energetic.

49. **2** Use *Snell's law* to solve this problem. Refer to the *Wave Phenomena* equations in the *Reference Tables:*

$$n_1 \sin \theta_1 = n_2 \sin \theta_2$$

where the subscript 1 refers to the incident medium (crown glass), the subscript 2 refers to the refractive medium (diamond), θ_1 is the angle of incidence, and θ_2 is the angle of refraction. The indices of refraction of crown glass (n_1) and diamond (n_2) are listed as 1.52 and 2.42, respectively, in the table entitled *Absolute Indices of Refraction* in the *Reference Tables*. Refer to the diagram below:

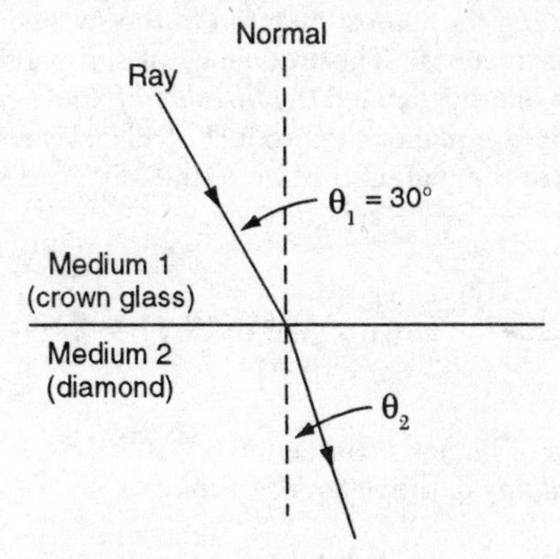

Substitute the given values into the equation above:

$$(1.52)(\sin 30.^\circ) = (2.42)(\sin \theta_2)$$

$$(1.52)(0.500) = (2.42)(\sin \theta_2)$$

$$\theta_2 = 18^\circ$$

50. **3** The American physicist Arthur H. Compton first noticed that, when a high-energy x-ray photon collides with an electron, the electron travels off in one direction and a photon of a lower frequency moves off in another direction. He explained the so-called *Compton effect* in terms of the *conservation of both momentum and energy.* Thus the energy and momentum lost by the x-ray photon equal the energy and momentum gained by the targeted electron.

One reason why this experiment is so important is that it demonstrates that photons have momentum and obey the law of conservation of momentum just as do particles of mass.

51. **4** In his famous scattering experiments, the British physicist Ernest Rutherford observed that most of the alpha particles (positive helium nuclei) that he directed at gold foil passed right through, unhindered in their path. Because *most of the alpha particles were not deflected*, Rutherford concluded that the volume of an atom is mostly empty space, with a small, dense core called the *nucleus*. All of the positive charge of the atom, and most of its mass, are concentrated in this core. The relatively light electrons were assumed to be widely separated from the nucleus.

52. **2** This problem involves the *photoelectric effect*. As explained for question 48, when photons of sufficient energy are directed at a photoemissive surface, some electrons will be emitted. The energy required to just overcome the forces holding the loosest electron to the surface is called the *work function* (W_0) of the material. The frequency of the photons that have this minimum amount of energy is called the *threshold frequency* (f_0).

The work function is equal to the product of Planck's constant (h) and the threshold frequency of the material. Refer to the *Modern Physics* equations in the *Reference Tables:*

$$\begin{aligned} W_0 &= hf_0 \\ &= (6.6 \times 10^{-34}\ \mathrm{J \cdot s})\,(1.0 \times 10^{14}\ \mathrm{Hz}) \\ &= 6.6 \times 10^{-20}\ \mathrm{J} \end{aligned}$$

NOTE: The value of Planck's constant, 6.6×10^{-34} J • s, is obtained from the *List of Physical Constants* in the *Reference Tables*.

53. **3** In the diagram labeled *A few energy levels for the mercury atom* in the *Reference Tables*, we note that in the *c* energy level the energy of the atom is –5.52 eV. In order to be ionized, an electron in this level would have to absorb enough energy to be completely removed from the atom, thereby forming an ion. The electron would have to jump to level *j*, which is designated as the *ionization* level with an energy of 0.00 eV.

The *minimum* amount of energy needed to ionize a mercury electron in the *c* energy level is the difference between the energies of these two levels, that is, 5.52 eV.

54. **2** In the *Wavelengths of Light in a Vacuum* chart in the *Reference Tables*, yellow light is seen to have a shorter wavelength than red light.

For the relationship between wavelength (λ) and frequency (f), we refer to the *Wave Phenomena* equations in the *Reference Tables:*

$$v = f\lambda$$

where v stands for the speed of the wave.

Since all colors of light travel at the same speed in air or in a vacuum, we conclude that the wavelength and the frequency of light are inversely proportional to each other. When two quantities vary inversely with respect to each other, one quantity increases when the other quantity decreases.

As the color of light changes from red to yellow, the wavelength decreases, and therefore the frequency *increases.*

55. **2** *Displacement* is defined as the change in position of a body. To fully describe displacement, we must state how far the body moved, and in which direction. Thus displacement is a vector quantity. We draw a sketch to find the total or resultant displacement:

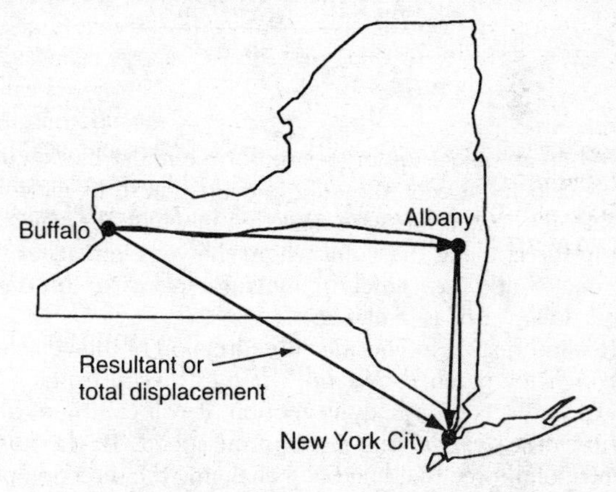

First we draw a vector from the starting point (Buffalo) to Albany to represent the displacement of the car along the first leg of its trip. Following the *head-to-tail* method of vector addition, we draw a second vector that starts at Albany and ends at New York City. This vector represents the displacement of the car along the second leg of its trip.

To find the resultant displacement, we draw a vector from the starting point (Buffalo) to the head of the second displacement vector (New York City). The length of this vector represents the magnitude of the car's total displacement, or change in position. The *distance driven*, however, is represented by the sum of the lengths of the other two vectors.

In any triangle, the sum of the lengths of any two sides is always greater than the length of the third side. This is called the *triangle inequality,* derived from the fact that the shortest distance between two points is a straight line.

We conclude that, compared to the magnitude of the car's total displacement, the distance driven is *longer.*

PART II

GROUP 1—**Motion in a Plane**

56. **2** In this question friction provides a centripetal, or "center-seeking," force that keeps the vehicle on its circular path. The centripetal force (F_c) acting on an object moving at constant speed in a circular path is directly proportional to the product of the object's mass (m) and the square of its speed (v), divided by the radius (r) of the path. Refer to the *Motion in a Plane* equations in the *Reference Tables*:

$$F_c = \frac{mv^2}{r}$$

$$= \frac{(4.4 \times 10^3 \text{ kg})(6.0 \text{ m/s})^2}{24 \text{ m}}$$

$$= 6.6 \times 10^3 \text{ N}$$

57. **4** When an object undergoes uniform circular motion, the direction of the object's velocity at each moment is tangential to its path, and is in the same clockwise or counterclockwise direction in which the object is traveling. The tangent to the circle at the point where the vehicle reaches the icy patch is a vertical line. Since the vehicle is moving clockwise, the direction of its velocity points downward, as in choice *4*.

NOTE: It is interesting to consider the direction of the vehicle's motion as it *passes through* icy patch *P*. *Newton's law of inertia* explains the motion: When no net force acts on a body in motion, it will continue to move in the same direction in a straight line at constant speed. In the absence of the frictional centripetal force that had been changing the direction of its velocity, the vehicle will *continue* to travel *tangentially* in a straight-line path downward.

58. **3** The following diagram shows the forces acting on a passenger on a rotor ride:

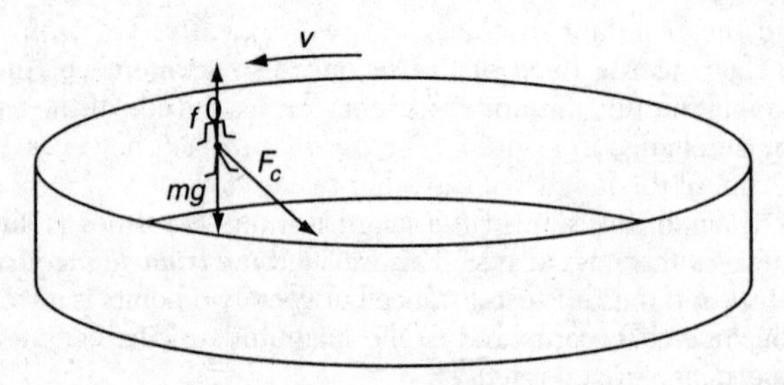

We see that three forces are acting on the passenger: the centripetal force *(F_c)*, which gives the passenger a centripetal acceleration directed toward the center of the rotor; the passenger's weight *(mg)*, which acts in a downward direction; and the frictional force *(f)*, which acts upward.

When passengers stay "pinned" against the wall of the rotor, as in this question, there is no acceleration in the vertical plane because the passengers do not rise or fall once they're "pinned." Therefore the net force in the vertical plane must equal zero. Since only two forces are acting on a passenger in the vertical plane, these forces must be equal and opposite. We conclude that the frictional force on each passenger is equal in magnitude to his or her weight.

Weight *(mg)* is proportional to mass. A 30.-kg child has one-half the mass of a 60.-kg adult, and therefore one-half the weight. Since the frictional force in this question is equal to the passenger's weight, we conclude that the child experiences one-half the frictional force that the adult experiences, or *F*/2.

59. **3** The *centripetal acceleration* (a_c) of an object traveling in a circle is equal to the square of the object's speed *(v)* divided by the radius *(r)* of its path. Refer to the *Motion in a Plane* equations in the *Reference Tables*:

$$a_c = \frac{v^2}{r}$$

Since the adult and the child are rotating in a circular path with the same speed and the same radius, the magnitude of the acceleration of the child is *the same* as that of the adult.

60. **2** To keep an object moving at constant speed in a circular path requires a centripetal, or "center-seeking," force that pulls the object toward the center of curvature of its path.

The centripetal force acting on the satellite in this question is provided by the gravitational force of attraction between the satellite and the Earth, and is directed toward point *B*, which is the direction toward the center of the circular path.

61. **3** According to *Kepler's first law*, the path of each planet (Pluto in this question) traveling around the Sun is an *ellipse* with the Sun at one focus, as shown in diagram 3.

62. **1** An object that is thrown or shot into the air and that experiences no other forces than the pull of gravity is called a *projectile*. Its motion can be described by separating it into vertical and horizontal components. The vertical and horizontal components of a projectile's motion are *independent* of each other.

In this question we are concerned only with the vertical component. Since the ball was thrown horizontally, its initial *vertical* velocity (v_{iy}) is zero. Under the influence of gravity, however, the ball's vertical velocity increases with an acceleration (a) equal to 9.8 m/s². This vertical motion occurs even as the ball moves horizontally along at a constant horizontal speed of 20. m/s.

It is advisable to organize the vertical and horizontal data separately. In this question, only the vertical motion is examined.

Given vertical information: $v_{iy} = 0$ Find: Δt (time)
$\Delta s = 100.$ m
$a = 9.8$ m/s² ("down" is considered positive)

We choose the equation having only these variables from the *Mechanics* equations in the *Reference Tables:*

$$\Delta s = v_i \Delta t + \frac{1}{2} a (\Delta t)^2$$

$$100.\ \text{m} = (0)\,\Delta t + \frac{1}{2}(9.8\ \text{m/s}^2)(\Delta t)^2$$

$$\Delta t = 4.5\ \text{s}$$

NOTE: The value for the acceleration due to gravity, 9.8 m/s², is found in the *List of Physical Constants* in the *Reference Tables.*

63. **1** If air resistance is neglected, the only accelerating force acting on an object, such as a ball, that has been projected into the air is gravity. The effect of gravity is to cause an acceleration in the vertical component of the object's motion. During the flight of the object, at *each* point in its trajectory, the direction of this acceleration is *downward.*

64. **3** Make a vector diagram to represent the initial velocity of the projectile: 80. m/s at an angle of 53° to the horizontal. This vector must be resolved into its vertical and horizontal components. Using the initial velocity vector (v_i) as the diagonal, draw a parallelogram around it as shown below:

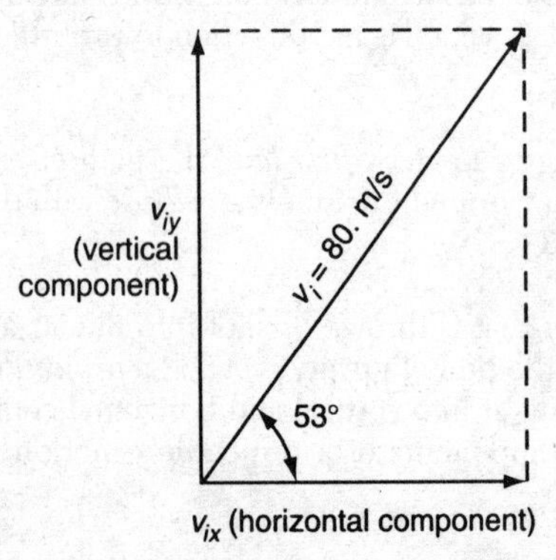

The vertical line drawn to complete the parallelogram's side represents the vertical component (v_{iy}) of the initial velocity. Refer to the *Motion in a Plane* equations in the *Reference Tables*, or use simple trigonometry, to obtain

$$\sin 53° = \frac{v_{iy}}{v_i}$$

Solve for v_{iy}:

$$\begin{aligned} v_{iy} &= v_i \sin 53° \\ &= (80.\ \text{m/s})\ (0.799) \\ &= 64\ \text{m/s} \end{aligned}$$

65. **1** Related to *Kepler's second law* is the fact that, as a planet moves further from the Sun, the planet's potential energy increases and its kinetic energy decreases. The underlying principle is the *law of conservation of energy:* In any transfer of energy among objects in a closed system, the total energy of the system remains constant. If it is assumed that there are no frictional forces, any increase in the potential energy will be accompanied by an equivalent decrease in the kinetic energy. As an object that is thrown straight up rises, for example, the kinetic energy that it loses is converted into an equal quantity of potential energy gained.

The same concept holds for planets orbiting the Sun and for satellites, such as the Moon, orbiting the Earth. As the distance between the Moon and the Earth increases, the Moon gains potential energy at the expense of kinetic energy lost. Since the Moon loses kinetic energy, which is related to speed, the Moon's orbital speed *decreases.*

GROUP 2—**Internal Energy**

66. **2** The average kinetic energy of the molecules of an ideal gas is directly proportional to the absolute temperature. The graph for such a relationship is a straight line pointing toward the origin as shown by line 2 on the graph.

NOTE: Since absolute zero is unreachable, the line generally approaches but does *not* touch the origin.

67. **2** Checking the *Heat Constants* chart in the *Reference Tables*, we find that the boiling point of lead is 1740°C and its freezing point (which is always equivalent to the melting point) is 328°C. The difference is 1740°C – 328°C = 1412 Celsius degrees.

The kelvin (K) is the SI unit of temperature. A kelvin measures the *same* change in temperature as does a Celsius degree. Therefore a change of 1412 C° is the same as a change of *1412 K.*

68. **4** According to the *second law of thermodynamics*, heat energy flows spontaneously from a hotter to a colder region. Thus, when cold masses

of aluminum and copper are placed in contact with hot water, heat energy will flow from the hot water into the cold metals. In general, heat energy will continue to pass from a warmer body to a colder body until both are at the same temperature. In this question, when equilibrium is reached, *the aluminum, copper, and water have the same temperature.*

Also, the principle of conservation of energy demands that no heat energy be created or destroyed during this process.

69. **2** The *specific heat (c)* of a substance is the amount of heat absorbed or released when the temperature of a unit mass of the substance is changed by 1 Celsius degree (1 C°). According to the *Internal Energy* equations in the *Reference Tables:*

$$Q = mc\Delta T_c$$

where Q is the amount of heat absorbed or released when a substance of mass m has a change in temperature, ΔT_c. Solving for the specific heat gives

$$c = \frac{Q}{m\Delta T_c}$$

$$= \frac{0.032 \text{ kJ}}{(0.010 \text{ kg})(8.0 \text{ C}^\circ)}$$

$$= 0.40 \text{ kJ/kg} \cdot \text{C}^\circ$$

70. **4** The slope of each oblique line segment on the graph is $\Delta T_c/\Delta Q$ (temperature change of metal/heat added). Except for mass *(m)*, which is not a factor in this question because each metal has the same mass, this is the *reciprocal* of the expression that we used in question 69 to determine the specific heat, that is,

$$c = \frac{Q}{\text{m}\Delta T_c}$$

We conclude that the slope of each line on the graph in this question is inversely related to the specific heat. Since line L has the steepest slope, it must therefore represent the metal with the *smallest* specific heat. Checking the *Heat Constants* chart in the *Reference Tables*, we find that, of the four metals listed in the answer choices, platinum has the smallest specific heat. We conclude that *platinum* is represented by line L.

71. **1** The *Heat Constants* chart in the *Reference Tables* gives the melting point of ice as 0°C. At the melting point, the heat (Q_f) required to change a solid to a liquid is equal to the product of the solid's mass *(m)* and its *heat of fusion* (H_f). Refer to the *Internal Energy* equations in the *Reference Tables:*

$$Q_f = mH_f$$
$$= (3.0 \text{ kg}) (334 \text{ kJ/kg})$$
$$= 1.0 \times 10^3 \text{ kJ}$$

NOTE: The value for the heat of fusion of ice, 334 kJ/kg, can be obtained from the *Heat Constants* chart in the *Reference Tables.*

72. **2** *Entropy* is a quantitative measure of the disorder or randomness of a system. According to the *second law of thermodynamics,* there is a natural tendency for a system to move toward a state of greater disorder. The second law also states that, as time passes, the total entropy in the universe *increases, only.*

73. **2** According to the *kinetic theory of gases,* the pressure exerted by a gas is caused by the repeated bombardment of gas molecules on the walls of the container. The pressure is proportional to the number and strength of the collisions. As the number of gas molecules is increased at constant temperature in a rigid container, so that volume is also constant, there are more collisions and therefore the pressure on the walls of the container *increases.*

The answer to this question could also be determined by an examination of the *ideal gas law,* which states that the product of the pressure (P) and the volume (V) of an ideal gas is *directly proportional* to the product of the number of moles (n) in the sample and the absolute temperature (T). This relationship can be expressed as follows:

$$PV = nRT$$

where R is a constant that is the same for all gases. As the number of gas molecules is increased at constant temperature and volume, n increases and therefore the pressure (P) on the walls of the container *increases.*

74. **4** The factors that affect the boiling and freezing points of water are the presence of a dissolved salt and the pressure. A dissolved salt lowers the freezing point and *raises the boiling point* of water. Increased pressure lowers the melting point of ice and *raises the boiling point* of water.

In diagram *4,* the sample of water has a dissolved salt and is covered, so that pressure builds up when it is heated. Both the presence of a dissolved salt and the increased pressure raise the boiling point of water, giving the water in diagram 4 the highest boiling point.

75. **3** Before answering this question, let us study the following annotated graph.

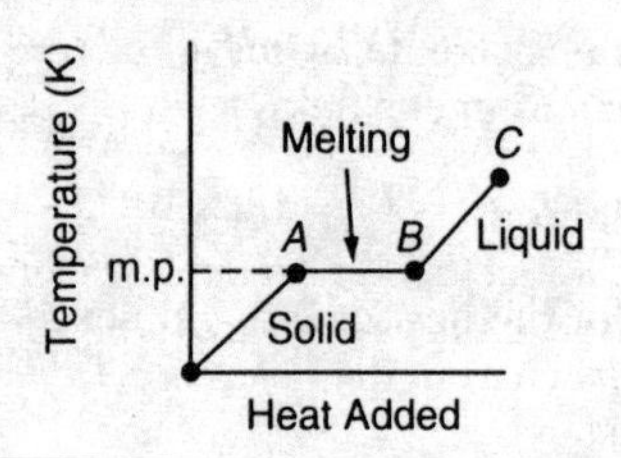

This graph shows how the *temperature* of a crystalline solid varies as heat is added at a contant rate. The process starts at a temperature below the melting point (M.P.) of the solid. Interval *SA*, which slants up, represents the *solid phase* of the substance. At point *A*, the graph levels off to horizontal segment *AB*. We notice that there is no temperature change along this plateau. This fact indicates that a *change of phase* is occurring; along *AB* the solid is *melting*, at its melting points, into a liquid. At *B*, all of the solid has been melted, and the temperature of the liquid begins to rise. Oblique segment *BC* thus represents the *liquid phase*. (Note that this diagram does not show the gaseous state.) Now we are ready to answer the question.

The *average internal kinetic energy* ($\overline{\text{KE}}$) of a substance is proportional to the substance's absolute temperature. Therefore, in the graph above, as the temperature rises along *SA*, $\overline{\text{KE}}$ will also rise. Along segment *AB*, which represents the phase change, the temperature is constant and so is $\overline{\text{KE}}$. After all the solid has melted at point *B*, the temperature and $\overline{\text{KE}}$ begin to rise again.

Since the graphs of $\overline{\text{KE}}$-versus-heat added and temperature-versus-heat added should be similar in appearance, we choose graph *3*.

GROUP 3—**Electromagnetic Applications**

76. **1** A cathode-ray tube is a device found in television sets and computer monitors. It consists of an evacuated glass tube with a source of electrons at one end, a fluorescent screen at the other end, and deflecting plates in between.

The beam of electrons is controlled by deflecting plates that produce either electric or magnetic fields. The beam may be deflected by a magnetic field because each particle in the beam *possesses a charge*. A charged particle moving through a magnetic field experiences a force that is perpendicular to both the path of the particle and the magnetic field.

77. **3** A *transformer* is a device that changes the voltage of an alternating current into a smaller or larger voltage of alternating current. It consists of a primary coil and a secondary coil that are wound on a common soft-iron ring called a *core*. A continually changing current in the primary produces a continually changing magnetic field that induces an alternating voltage in the secondary coil.

A transformer does *not* operate by means of torque exerted on a current-carrying loop of wire in a magnetic field.

Wrong Choices Explained:

(1),(2), and (4) The ammeter, electric motor, and voltmeter *do* operate by means of torque exerted on a current-carrying loop of wire in a magnetic field.

78. **2** There are various hand rules that can be used to determine the direction of the force on an electron moving in a magnetic field.

Use the following *left-hand rule:* Hold your outstretched left hand so that your thumb points in the direction of the electron's velocity (toward the right), and your extended fingers point in the direction of the magnetic field (north to south, into the page). Your palm will now face in the direction of the magnetic force exerted on the electron.

When you apply this rule in this question, your palm points toward *the bottom of the page.*

79. **2** The force (F) acting on a charged particle moving in a direction perpendicular to a magnetic field is equal to the product of the particle's charge (q), its velocity (v), and the strength of the magnetic field, also called the *magnetic flux density* (B).

Refer to the *Electromagnetic Applications* equations in the *Reference Tables:*

$$\begin{aligned} F &= qvB \\ &= (1.6 \times 10^{-19}\,\text{C})(6.0 \times 10^{7}\ \text{m/s})(4.0 \times 10^{-2}\,\text{T}) \\ &= 3.8 \times 10^{-13}\ \text{N} \end{aligned}$$

NOTE: The charge of the electron, 1.6×10^{-19}C, is obtained from the *List of Physical Constants* in the *Reference Tables.*

80. **3** The negative electron will be repelled by the lower, negative plate and attracted by the upper, positive plate. Therefore it will be deflected upward, toward the top of the page, as shown in diagram 3.

81. **1** In his famous oil drop experiment, the American physicist Robert Millikan observed that the electric charge on each oil drop is always an integral multiple of a certain indivisible unit of charge, called the *elementary charge* and equal to 1.6×10^{-19} C. This is the charge on a single electron. Millikan found no oil drops having less charge.

Millikan concluded that *there is a minimum amount of charge that particles can acquire.*

82. **2** A transformer is a device that changes the voltage of an alternating current into a smaller or larger voltage of alternating current. It consists of

a *primary coil* and a *secondary coil* that are wound on a common soft-iron ring called a *core*. The ratio of the number of turns of wire in the primary coil (N_p) to the number of turns in the secondary (N_s) is equal to the ratio of the voltage across the primary (V_p) to the voltage induced in the secondary (V_s).

Refer to the *Electromagnetic Applications* equations in the *Reference Tables:*

$$\frac{N_p}{N_s} = \frac{V_p}{V_s}$$

$$\frac{24 \text{ turns}}{N_s} = \frac{12 \text{ V}}{8.0 \text{ V}}$$

$$N_s = 16 \text{ turns}$$

83. **4** The diagram in the question shows a simple generator—a wire loop that is rotated in a uniform magnetic field. It works on the principle that, when a conductor moves across the lines of force of a magnetic field, a potential difference is induced in the loop. As the loop turns, there is a continual change in the total magnetic linkage. This change creates an induced potential difference that alternates in direction and varies in magnitude between zero and some maxiumum value.

For example, in the diagrams below, the loop cuts across the most lines of force per unit time in position 1, creating a maximum induced potential difference. When the loop passes through its vertical position (position 2), however, it moves parallel to the field, temporarily cutting across no lines of force. At this time, the induced potential difference is zero. Another quarter-turn later, the loop is again in the horizontal position (position 3), experiencing the maximum induced potential difference, but in the opposite direction.

During one complete rotation, the rotating loop produces an *alternating* voltage, as in graph *4*.

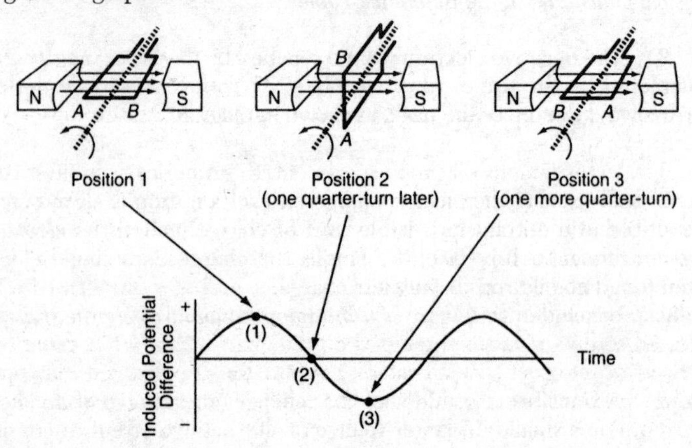

84. **4** In order for current to flow in the loop, there must be a closed path through which the electrons can travel. This closed path, called a *circuit,* could be achieved by *connecting the ends of the loop to each other with a conducting material.*

Generally, the loop in a generator is connected to an external circuit to form a closed path. The alternating induced potential difference in the loop then causes an alternating current to flow in the external circuit.

85. **2** According to *Faraday's law of induction,* moving a conductor across magnetic lines of force induces an electric potential difference in the conductor. This induced potential difference is often called an *electromotive force* or, simply, *EMF,* even though it is not a force. The induced EMF is a maximum when the conductor moves in a direction that is *perpendicular* to itself and the field. The magnitude of the EMF or electric potential difference (V) induced between the ends of a wire that is moving across the lines of force of a magnetic field at right angles to the field is equal to the product of the magnetic flux density (B), the length (ℓ) of the conductor in the field, and the speed (v) with which the conductor moves.

Refer to the *Electromagnetic Applications* equations in the *Reference Tables:*

$$V = B\ell v$$

From this equation, we see that the induced EMF is proportional to the speed of the conductor across the magnetic field.

We conclude that, as the speed of rotation of the wire loop in this question is increased, the maximum electromotive force induced in the loop *increases.*

GROUP 4—**Geometric Optics**

86. **2** The image formed by a plane mirror is the *same size* as the object. It is also virtual, erect, and *reversed.*

The image is *not* inverted; if it were, you would see yourself upside down in a bathroom mirror!

87. **2** We construct a neat ray diagram, shown below, to answer this question. To locate the image we must follow the path of only two light rays, both coming from the top (B) of the candle. First we draw the ray (ray 1) that passes through the center of curvature (C); the mirror reflects this ray back upon itself. Next we draw the ray (ray 2) that is parallel to the principal axis; it is reflected so that is passes through the principal focus (F). The point where these two rays intersect (B') indicates where the top of the image will appear.

From our sketch we conclude that the candle's image is located *between C and F.*

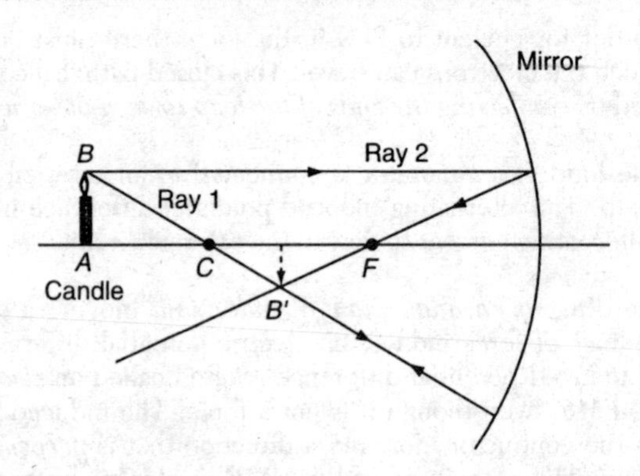

88. **3** *Every* image produced by a convex (diverging) mirror is *virtual and smaller than the object.* A convex mirror *cannot* produce images that are real or larger than the object.

89. **4** When an object, such as a person's face, is placed at a distance less than one focal length from a concave (converging) spherical mirror, as in this question, the image formed is *virtual and erect.* Since the image is also enlarged, a concave mirror is often used as a shaving or make-up mirror.

Not recalling the details of this application of a concave mirror, we could answer the question by constructing a neat ray diagram as shown below. We draw a horizontal line to represent the principal axis of the mirror. We indicate the center of curvature as *C.* For a large spherical mirror, the focal length is equal to half the radius of curvature; therefore, halfway between the mirror and *C*, we indicate the principal focus as *F.* We use arrow *AB* to represent the object (in this case, a freckle on a person's face). In this question it is given that the freckle is located at a distance equal to half the focal length from the mirror.

To locate the image we must follow the path of only two light rays, both coming from the top of the arrow. First we draw the ray (ray 1) that passes through the center of curvature (*C*); the mirror reflects this ray back upon itself. Next we draw the ray (ray 2) that is parallel to the principal axis; it is reflected so that it passes through the principal focus (*F*). Since ray 1 and ray 2 diverge after reflection, we extend them backward until they meet at point (*B′*) behind the mirror. Since the reflected rays only *appear* to originate from this point behind the mirror, a *virtual* image of *B* will appear at *B′*.

From our sketch we conclude that the image of the freckle produced by the mirror is *virtual and erect.*

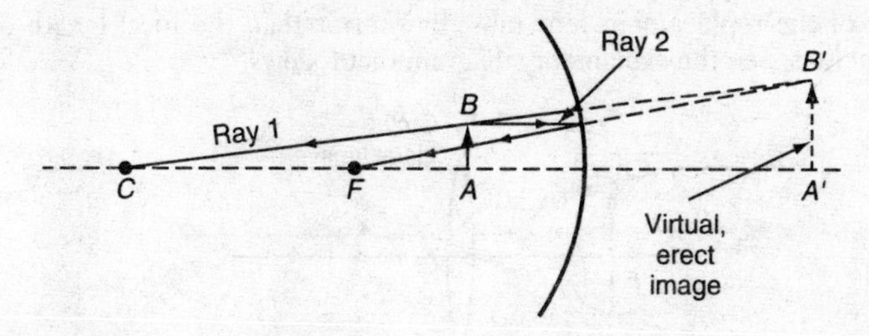

90. **1** This problem involves two steps: first we must calculate the image distance, and then we can find the image size.

For a spherical mirror, the reciprocal of the object distance (d_o) plus the reciprocal of the image distance (d_i) is equal to the reciprocal of the focal length (f) of the mirror. For a concave (converging) mirror, the focal length is positive.

Refer to the *Geometric Optics* equations in the *Reference Tables:*

$$\frac{1}{d_o} + \frac{1}{d_i} = \frac{1}{f}$$

$$\frac{1}{10.\text{ cm}} + \frac{1}{d_i} = \frac{1}{20.\text{ cm}}$$

$$d_i = -20.\text{ cm}$$

The negative sign means that the image is virtual and located 20. cm *behind* the mirror.

For any spherical mirror, the ratio of the object size (S_o) to the image size (S_i) is equal to the ratio of the object distance (d_o) to the image distance (d_i).

Again refer to the *Geometric Optics* equations in the *Reference Tables:*

$$\frac{S_o}{S_i} = \frac{d_o}{d_i}$$

$$\frac{0.50\text{ cm}}{S_i} = \frac{10.\text{ cm}}{20.\text{ cm}}$$

$$S_i = 1.0\text{ cm}$$

91. **4** When an object is placed at a distance less than one focal length from a converging lens, as indicated by the arrow in the diagram in this question, the image formed is virtual. In order for a converging lens to produce a real image of an object, the object must be placed at a distance *greater* than the focal length. Since the replacement lens in this question produces a real image, we conclude that, with the new lens in place, the arrow is situated beyond one focal length.

However, when the original lens is replaced, the object "distance" does not change. Only the focal length could have changed. We conclude that the focal length of the replacement lens must by *shorter* than the focal length of the original lens. See the explanatory diagram that follows.

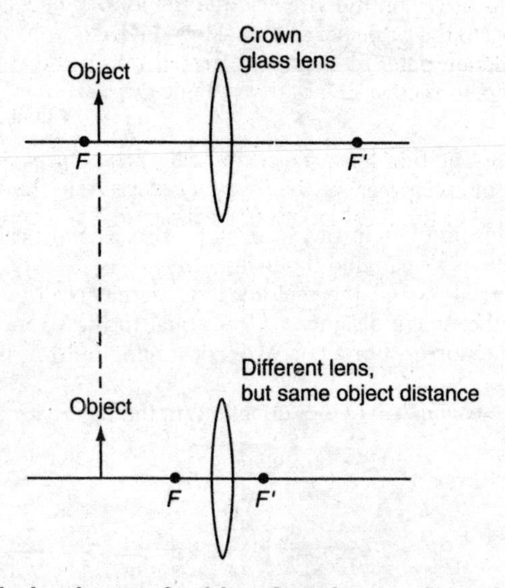

The lens with the shortest focal length is the one that refracts or bends light the most. The amount of bending depends on the speed with which light travels through the material; the lens in which light travels the most slowly will bend light the most. The speed of light (v) in a transparent medium is inversely proportional to the index of refraction (n) of the medium. Refer to the *Wave Phenomena* equations in the *Reference Tables:*

$$n = \frac{c}{v}$$

where c is the speed of light in a vacuum. Therefore a larger value of n is associated with a smaller value for v, a larger amount of bending, and a shorter focal length.

Use the table entitled *Absolute Indices of Refraction* to compare the indices of refraction for the four answer choices. Of the materials listed, only flint glass has a higher n (1.6l) than crown glass (1.52). Therefore light travels more slowly in flint glass than in crown glass, and bends more.

A replacement lens made of *flint glass* would therefore have a shorter focal length than the orginal lens made of crown glass and, since the object distance did not change, would produce a real image of the object.

92. **1** In diagram 1 we see a ray that leaves the top of the object at $2F$ and passes through the optical center of the lens. Such a ray is expected to

emerge parallel to its original direction, as shown in diagram *1.* In diagram 1 we also see another ray that leaves the top of the object and passes thorugh the principal focus (F) on the same side of the lens. Such a ray is expected to emerge parallel to the principal axis, as shown in diagram *1*.

The intersection point of the two refracted rays marks the location of the top of the image formed at $2F'$, as shown in diagram *1*.

93. **1** For any thin lens, the reciprocal of the object distance from the lens (d_o) plus the reciprocal of the image distance from the lens (d_i) is equal to the reciprocal of the focal length (f) of the lens. For a converging lens, the focal length is positive.

Refer to the *Geometric Optics* equations in the *Reference Tables:*

$$\frac{1}{d_o} + \frac{1}{d_i} = \frac{1}{f}$$

Choose the data from any one of the student's three trials, for example, $d_o = 0.15$ m and $d_i = 0.30$ m. Substituting into the above equation gives

$$\frac{1}{0.15\text{ m}} + \frac{1}{0.30\text{ m}} = \frac{1}{f}$$

$$f = 0.10\text{ m}$$

94. **4** For any thin lens, the ratio of the object size (S_o) to the image size (S_i) is equal to the ratio of the object distance (d_o) to the image distance (d_i). Since the image is given as three times the size of the object in this question, let $3S_o$ represent the image size.

Refer to the *Geometric Optics* equations in the *Reference Tables:*

$$\frac{S_o}{S_i} = \frac{d_o}{d_i}$$

$$\frac{S_o}{3S_o} = \frac{d_o}{0.12\text{ m}}$$

$$d_o = 0.04\text{ m}$$

95. **2** *Chromatic aberration* results from the fact that glass is a *dispersive* medium for light, so that each wavelength (color) of visible light travels at a different speed in glass. Consequently, *each wavelength of light is refracted a different amount by the lens,* and therefore the colors do not come to the same focus after passing through the lens.

Chromatic aberration is a defect that can be corrected by using a diverging lens of a different material with the converging lens.

GROUP 5—**Solid State**

96. **1** Do not confuse resistivity with resistance. *Resistance* is opposition to current flow. The resistance (R) of a material is directly proportional to its length (ℓ) and inversely proportional to its cross-sectional area (A):

$$R = \frac{\rho \ell}{A}$$

where ρ is a proportionality constant called resistivity. *Resistivity* is an inherent property of a material that is a quantitative indicator of the material's electrical resistance at a specific temperature. The higher the resistivity, the greater is the resistance. Conductors such as copper and silver have small resistivities. Insulators such as rubber have large resistivities.

The reciprocal of resistivity is *conductivity*. Electrical conductivity is a measure of a material's ability to carry an electric current. Metallic conductors have greater electrical conductivity than insulators.

97. **3** According to the *band model*, the electrons of any substance are restricted to particular energy bands, the most important of which are the *conduction band* and the *valence band*. These bands are separated by an *energy gap*.

Electrons in the outermost shell of an atom are in the valence band and are immobile because they are held to the atom by attractive bonds. Electrons in the conduction band, however, are practically free of bonds and can be set in motion by a small applied voltage.

A solid is determined to be a conductor, an insulator, or a semiconductor on the basis of *the number of electrons in the conduction band and the energy gap between bands.*

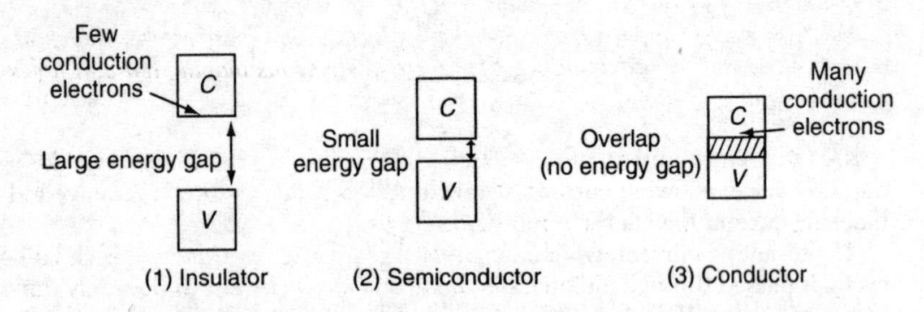

In the explanatory diagram above, Figure 1 shows a large energy gap with few electrons in the conduction (C) band. Solids that have large energy gaps are classified as *insulators.* The effect of the large gap is that very few electrons in the valence (V) band can acquire enough energy to jump to the conduction band. With few conduction electrons, an insulator is a poor conductor of electricity.

Figure 2 shows a small energy gap. A solid that has a small energy gap between its valence and conduction bands is classified as a *semiconductor*. Only a small amount of energy is needed by electrons in the valence band to jump to the conduction band. Internal themal energy can provide this energy. As the temperature is increased, more electrons jump the gap, and so the semiconductor's ability to conduct improves.

Figure 3 shows an overlapping of the two energy bands. In a *conductor*, the valence and conduction bands *overlap* so that there is no appreciable energy gap. This overlapping places a large number of valence electrons in the conduction band, where they can move easily.

98. **2** As a result of doping (adding small amounts of another element to a semiconducting substance), a semiconductor may be produced that has either an excess of free electrons or a deficiency of electrons.

A semiconductor with an excess of free electrons is called an *N-type semiconductor* and serves as a donor of electrons. A semiconductor with a deficiency of free electrons is called a *P-type semiconductor* and functions primarily as *an acceptor of electrons*.

99. **4** A *hole* is a spot in a crystalline structure that is missing an electron. This hole, or empty space, can accept a valence electron from an adjacent atom if the electron gains enough energy to break from the atomic bond.

This electron, in turn, leaves a hole behind in the atom from which it came, a hole that can be filled by an electron from yet another neighboring atom, creating yet another hole, and so on. Holes therefore seem to move in the *opposite* direction to the electron flow, *as if they were positive charges*.

Under the influence of an electric field from a battery, electrons move away from the negative end of the semiconductor toward the positive end. Simultaneously, "positive" holes move in the opposite direction.

We conclude, therefore, that in the diagram accompanying the question, current flows in the semiconductor because of *electrons moving left and holes moving right*.

100. **2** The simplest of all semiconductor devices, the *diode* acts like a one-way valve, allowing current to flow through it in one direction only, and blocking current flow in the opposite direction.

If alternating current, which is current that reverses its direction each half-cycle, is passed through a diode, the diode will allow current to flow only during alternating half-cycles, blocking current when its direction is reversed.

The effect is to change or *rectify* the alternating current into a pulsating direct current (with the same frequency), as represented by diagram 2.

101. **1** A *diode* is a two-element device, made by joining a *P*-type semiconductor with an *N*-type semiconductor. There is one *P-N* junction in a diode.

A *transistor*, on the other hand, is a three-element semiconductor device, consisting of two junction diodes that share a common "base." If *P*-type material is shared, the transistor is called *N-P-N*. If *N*-type material is shared, the transistor is called *P-N-P*. There are two *P-N* junctions in a transistor.

We conclude that the basic difference between a transistor and a diode is *the number of junctions between P-type and N-type material.*

102. **2** A transistor is a three-element semiconductor device, consisting of two junction diodes that share a common "*base.*" The semiconductors to the sides of the base are called the *emitter* and the *collector*. To determine which side of the transistor functions as the collector, you must examine how the transistor is connected in the circuit.

The diagram given in this question shows the basic operating circuit for a transistor. You must remember that in such a circuit the emitter-base junction is forward biased, and the *base-collector junction is reverse biased.*

Since a transistor has two *P-N* junctions, two sources of potential difference are needed to provide the necessary biases. A *P-N* junction is said to be *forward biased* when the positive terminal of a battery is connected to the *P*-type material, and the negative terminal to the *N*-type material. In the diagram that accompanies the question we see that the left side of the transistor is forward biased.

A *P-N* junction is said to be *reverse biased* when the positive terminal of a battery is connected to the *N*-type material, and the negative terminal to the *P*-type material. In the diagram in the question we see that the right side of the transistor is reverse biased.

Since in an operating *N-P-N* transistor circuit the *base-collector junction is reverse biased*, we conclude that, in the diagram, the collector is *the right N-type section*. See the diagram below.

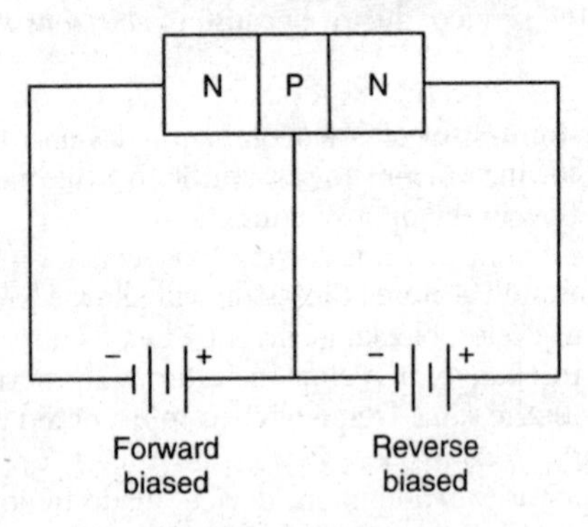

An *N-P-N* Transistor Circuit

103. **1** The current-increasing property of a transistor is called *amplification.* This property is used in a microphone-speaker circuit, for example, when a small input signal from a microphone is connected to the emitter-base circuit, and an amplified reproduction of the input signal is produced in the collector circuit.

Wrong Choices Explained:

(2) *Biasing* refers to the phenomenon that occurs when a potential difference is placed across a *P-N* junction. Depending on the polarity of the applied voltage, the electric field barrier at the junction may be either strengthened or diminished.

(3) *Ohm's law* states that the ratio of voltage to current remains constant for a metallic conductor when the temperature is held constant.

(4) *Rectification* is the process in which an alternating current is changed into a pulsating direct current, as occurs when an alternating current is passed through a diode. See question 100.

104. **3** When a diode is *reverse biased*, practically no current flows until the applied potential difference becomes very large (200–300 V). When this "breakdown voltage" is reached, the semiconductor crystal begins to break down or "avalanche," allowing a large reverse current to flow. The "avalanche" occurs only if the diode circuit is *reverse biased.*

105. **1** As the amount of doping material in the semiconductor increases, the magnitude of the voltage required for the "avalanche" to occur *decreases.*

GROUP 6—Nuclear Energy

106. **3** In the notation system used, the *superscript* represents the atomic mass (mass number), which is the sum of the number of protons and neutrons in the atom's nucleus. The *subscript* represents the atomic number, which is the number of protons in the nucleus.

To determine the number of neutrons in an atom of ${}^{222}_{86}\text{Rn}$, subtract the atomic number (86) from the atomic mass (222): 222 – 86 = 136.

107. **1** Substitute the values listed in the chart in this question into the given equation:

$${}^{235}_{92}\text{U} + {}^{1}_{0}\text{n} \rightarrow {}^{138}_{56}\text{Ba} + {}^{95}_{36}\text{Kr} + 3{}^{1}_{0}\text{n} + E$$

$$235.0\ \mu + 1.0\ \mu \rightarrow 137.9\ \mu + 94.9\ \mu + 3(1.0\ \mu) + E$$

$$E = 0.2\ \mu$$

The reaction in this question is an example of *nuclear fission,* in which a heavy nucleus captures a neutron and then splits into lighter nuclei, releasing two or more neutrons and great amounts of energy. In nuclear physics mass and energy are equivalent. The energy released in a fission reaction is equal to the mass lost.

108. **4** *Isotopes* are different forms of the same element whose nuclei have identical *numbers of protons,* but different numbers of neutrons. Isotopes of the same element have the same chemical symbol and the same atomic number but different atomic masses (mass numbers).

109. **2** Only *charged* particles can be accelerated by an electromagnetic field. Since neutrons are neutral and have no charge, *neutrons cannot* be accelerated by an electromagnetic field.

Wrong Choices Explained:

(1) An *alpha* particle is a helium nucleus. It has a charge of +2 and therefore *can* be accelerated by an electromagnetic field.

(3) An *electron* has a negative charge and therefore *can* be accelerated by an electromagnetic field.

(4) A *positron* is a positive electron that is emitted from a nuclues during positive beta decay. Since it is a charged particle, it *can* be accelerated by an electromagnetic field.

110. **4** In the notation system used, the *subscript* represents the atomic number and the *superscript* represents the mass number (atomic mass). Let us examine the *Uranium Disintegration Series* in the *Reference Tables.* The mass numbers are on the right of the grid, and the atomic numbers and chemical symbols of certain elements are at the top.

The first dot of the graph in the upper right-hand corner locates uranium-238 and the beginning of its disintegration series. Following the first arrow, we see that $^{238}_{92}U$ decays into $^{234}_{90}Th$. (In this reaction, a decrease of 2 in the atomic number and a decrease of 4 in the mass number indicate the emission of one alpha particle, $^{4}_{2}He$. Each diagonal arrow represents the emission of an alpha particle.)

The second arrow represents the second step in the disintegration series. This arrow points horizontally to the right, indicating that $^{234}_{90}Th$ decays into $^{234}_{91}Pa$ (In this reaction, an increase of 1 in the atomic number and no change in the mass number indicate the emission of one electron, or *beta* particle, $^{0}_{-1}e$. Each short horizontal arrow to the right represents the emission of a beta particle.)

We conclude that the immediate decay product of ${}^{234}_{90}\text{Th}$ is ${}^{234}_{91}\text{Pa}$.

111. **2** *Atomic number*, which represents the amount of nuclear charge, is conserved in all nuclear reactions. Therefore the sums of the atomic numbers on both sides of the equation must be equal. In this question the given reaction is

$${}^{27}_{13}\text{Al} + {}^{4}_{2}\text{He} \rightarrow {}^{30}_{15}\text{P} + {}^{1}_{0}\text{n} + X$$

Recalling that atomic number is represented by the subscript for each symbol, we conclude that the atomic number of X must be 0 in order for the sum, 13 + 2 = 15 + 0 + 0, to be the same on both sides of the equation.

Mass number (atomic mass) is also conserved in nuclear reactions. It is represented by the superscript for each symbol. In the given reaction, the mass number of X must be 0 in order for the sum, 27 + 4 = 30 + 1 + 0, to be the same on both sides.

Since X has an atomic number of 0 and a mass number of 0, X could represent *gamma radiation*. Gamma radiation consists of high energy photons that are emitted during a nuclear reaction. Since gamma radiation is a type of electromagnetic radiation, its emission has no effect on atomic number or mass number.

Wrong Choices Explained:

(1) A *proton* has an atomic number of 1 and a mass number of 1.

(3) An *alpha particle* is a helium nucleus and has an atomic number of 2 and a mass number of 4.

(4) A *beta particle* is an electron and has an atomic number of –1 and a mass number of 0.

112. **4** The *half-life* of a radioactive isotope is defined as the time required for one-half of the nuclei in any sample of the isotope to disintegrate or decay. Each radioisotope has a specific half-life. The half-life of the isotope in this question is 3 mins. Therefore, after 15 mins, 15/3 = 5 half-lives will have elapsed.

In the decay of a radioactive isotope, the final mass (m_f) that remains from the original sample is equal to the fraction $1/2^n$ times the initial mass (m_i), where n is the number of half-lives that have elapsed.

Refer to the *Nuclear Energy* equations in the *Reference Tables:*

$$m_f = \frac{m_i}{2^n}$$

$$10 \text{ kg} = \frac{m_i}{2^5}$$

$$m_i = 320 \text{ kg}$$

113. **1** *Electron capture* (*K*-capture) occurs when an atomic nucleus absorbs an orbital electron, generally from the innermost electron shell, the *K*-shell. When this phenomenon occurs, the mass number of the nucleus is unchanged but its atomic number *decreases by 1.*

The following nuclear equation is an example of electron capture:

$$^{40}_{19}K + {}^{0}_{-1}e \rightarrow {}^{40}_{18}Ar$$

Here potassium-40 has captured an electron; in the process the mass number (40) remains unchanged, but the atomic number is decreased by 1 (from 19 to 18). Since a different element (argon-40) is formed, this is a type of transmutation.

114. **3** Nuclear *fusion* is the combination of two light nuclei to form a heavier nucleus, with the loss of some mass and the production of energy equivalent to the mass lost. This was the process used in making the hydrogen bomb. Fusion occurs naturally on the Sun.

The nuclear equation in this problem shows the fusion of two light hydrogen isotopes $\left({}^{3}_{1}H \text{ and } {}^{1}_{1}H\right)$ to produce a heavier helium nucleus $\left({}^{4}_{2}He\right)$ and energy.

Wrong Choices Explained:

(1) *Alpha decay* is the spontaneous emission of an alpha particle from a nucleus. Great amounts of energy are not released in alpha decay.

(2) A *positron* is a positive electron that is emitted from a nucleus during positive beta decay. The symbol for a positron is ${}^{0}_{+1}e$.

(4) In most nuclear *fission* reactions, a heavy nucleus captures a neutron and then breaks apart into two fragments of intermediate mass, releasing two or more neutrons and great amounts of energy.

115. **3** In most *nuclear fission* reactions, a heavy nucleus captures a neutron and then breaks apart into two fragments of intermediate mass, releasing two or more neutrons and great amounts of energy. Equation *3* fits this description. It represents the fission reaction of uranium-235, triggered by the absorption of a neutron. Two fragments of intermediate mass (barium-138 and krypton-95) have been produced, along with the release of three neutrons.

Wrong Choices Explained:

(1) This equation is an example of *beta decay.*

(2) This equation is an example of the nuclear *fusion* (or "proton-proton chain") that occurs in the interior of the Sun and in other stars that are composed primarily of hydrogen.

(4) This equation is an example of *alpha decay.*

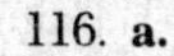

PART III

116. **a.**

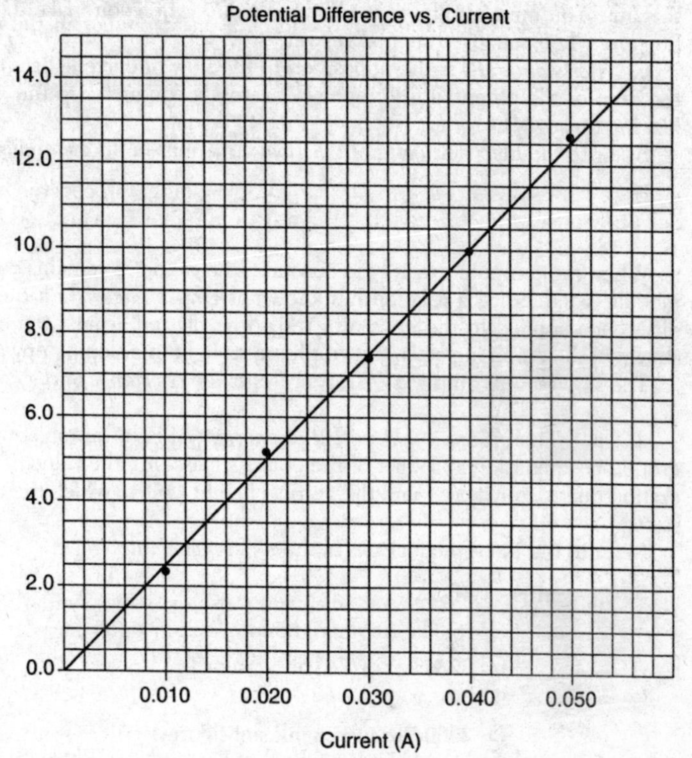

b. The *slope* is the change in the ordinate or y value, divided by the change in the abscissa or x value, for any two points *on the best-fit line,* such as the points (0.040 A, 10.0 V) and (0.010 A, 2.5V):

$$\text{Slope} = \frac{\Delta y}{\Delta x}$$

or

$$\begin{aligned}\text{Slope} &= \frac{\Delta V}{\Delta I} \\ &= \frac{10.0\text{V} - 2.5\text{V}}{0.040\text{A} - 0.010\text{A}} \\ &= 250\Omega\end{aligned}$$

NOTE: The values substituted must come from points that lie on the best-fit straight line. The values from the given data table do *not* necessarily fall on this line. You must also be sure to include *units* when you make the substitution *and* when you give the final answer.

c. *Resistance (R)* is the opposition to the flow of current. It is defined as the ratio of the potential difference (V) across a conductor to the current (I) flowing through it.

Refer to the *Electricity and Magnetism* equations in the *Reference Tables:*

$$R = \frac{V}{I}$$

When the temperature is held constant, the ratio V/I remains constant for metallic conductors, a relationship known as *Ohm's law*. When the potential difference applied to the ends of a resistor is plotted against the current, as shown in the graph in part a, the graph is a straight line sloping upward.

The slope of this graph, $\Delta V/\Delta I$, represents the *resistance of the conductor.*

117. **a.** The *gravitational potential energy* (ΔPE) of an object above the ground is equal to the product of the object's mass *(m)*, the value of the acceleration due to gravity *(g)*, and the vertical height *(Δh)* to which the object was raised.

Refer to the *Energy* equations in the *Reference Tables*:

$$\begin{aligned} \Delta \text{PE} &= mg\Delta h \\ &= (6.0\text{ kg})(9.8\text{m/s}^2)(55\text{ m}) \\ &= 3234\text{ kg}\cdot\text{m}^2/\text{s}^2 \text{ or N}\cdot\text{m or J} \end{aligned}$$

or

$$= 3200\text{ J to two significant figures}$$

NOTE: The value for g, the acceleration due to gravity, 9.8 m/s^2, is obtained from the *List of Physical Constants* in the *Reference Tables.*

b. The *kinetic energy* (KE) of an object is equal to one-half the product of its mass *(m)* and the speed *(v)* squared.

Refer to the *Energy* equations in the *Reference Tables:*

$$\begin{aligned} \text{KE} &= \frac{1}{2}mv^2 \\ &= \frac{1}{2}(6.0)\,(30.\text{ m/s})^2 \\ &= 2700\text{ kg}\cdot\text{m}^2/\text{s}^2 \text{ or N}\cdot\text{m or J} \end{aligned}$$

c. *Mechanical energy* is the sum of the kinetic and potential energies of an object. At the instant the block was released from the top of the 55-m building, it had 0 kinetic energy (because it was at rest) plus 3200 J of gravitational potential energy. Therefore the block started with a total mechanical energy of 3200 J.

When the block hit the ground, it had 0 gravitational potential energy (because it was on the ground level and $\Delta h = 0$) plus 2700 J of kinetic energy. Therefore, when the block hit the ground it had a total mechanical energy of 2700 J.

We conclude that the block lost 3200 J – 2700 J = *500 J* as it fell.

d. For full credit, you must write one or more *complete* sentences that correctly explain what happens to the "lost" mechanical energy. A complete sentence begins with a capital letter, has a subject and verb, and ends with a period. The following are four sample correct answers:

The energy was lost due to air friction.
The energy was converted to heat energy.
The energy was lost due to work done against friction.
Work was done on the air by the block.

This is a sample of an *unacceptable* answer:
It was friction.

118. **a.** The *angle of incidence* is the angle between the incident ray and the normal, or imaginary line drawn perpendicular to the surface. In this question, the angle of incidence for light ray *AO* measures *33°*. (An answer that is within plus or minus of 2° from 33° is also acceptable.)

b. According to the *law of reflection,* the angle of incidence is equal to the *angle of reflection*. Therefore the angle of reflection of the light ray also equals *33°*.

c.

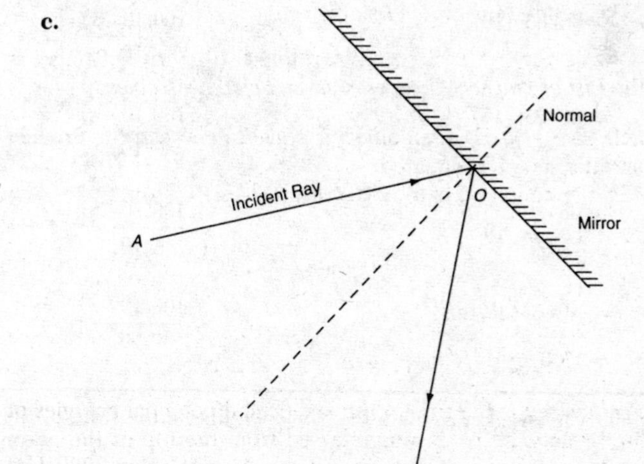

Topic	Question Numbers (total)	Wrong Answers (x)	Grade
Mechanics	1–16, 55, 58: (18)		$\frac{100(18-x)}{18} = \%$
Energy	17–22: (6)		$\frac{100(6-x)}{6} = \%$
Electricity and Magnetism	23–36: (14)		$\frac{100(14-x)}{14} = \%$
Wave Phenomena	37–47, 49, 54: (13)		$\frac{100(13-x)}{13} = \%$
Modern Physics	48, 50–53: (5)		$\frac{100(5-x)}{5} = \%$
Motion in a Plane	56, 57, 59–65: (9)		$\frac{100(9-x)}{9} = \%$
Internal Energy	66–75: (10)		$\frac{100(10-x)}{10} = \%$
Electromagnetic Applications	76–85: (10)		$\frac{100(10-x)}{10} = \%$
Geometrical Optics	86–95: (10)		$\frac{100(10-x)}{10} = \%$
Solid State Physics	96–105: (10)		$\frac{100(10-x)}{10} = \%$
Nuclear Energy	106–115: (10)		$\frac{100(10-x)}{10} = \%$

To further pinpoint your weak areas, use the Topic Outline in the front of the book.

Examination June 1995

Physics

PART I

Answer all 55 questions in this part. [65]

Directions (1–55): For *each* statement or question, select the word or expression that, of those given, best completes the statement or answers the question. Record the answers to these questions in the spaces provided.

1 The thickness of a dollar bill is closest to

(1) 10^{-4} m
(2) 10^{-2} m
(3) 10^{-1} m
(4) 10^{1} m

1_____

2 A jogger accelerates at a constant rate as she travels 5.0 meters along a straight track from point *A* to point *B*, as shown in the diagram below.

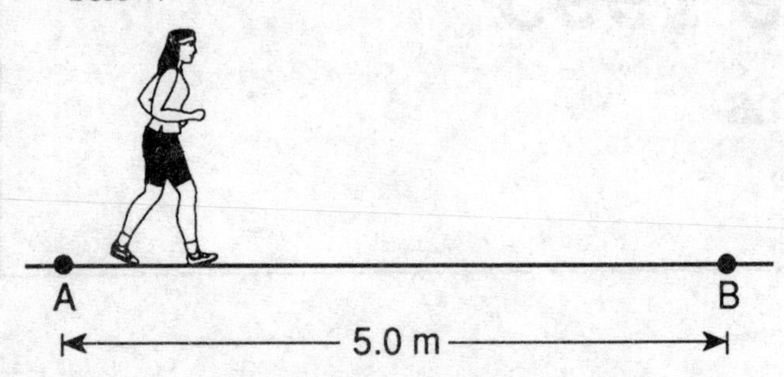

If her speed was 2.0 meters per second at point *A* and will be 3.0 meters per second at point *B*, how long will it take her to go from *A* to *B*?

(1) 1.0 s
(2) 2.0 s
(3) 3.3 s
(4) 4.2 s

2 ____

3 Which graph best represents the motion of an object falling from rest near the Earth's surface? [Neglect friction.]

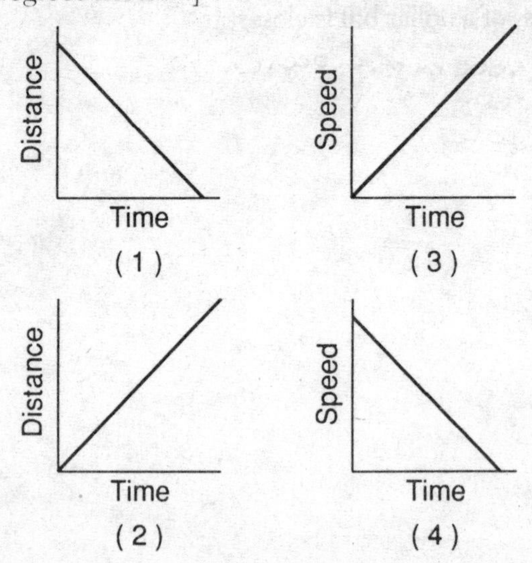

3 ____

4 An object falls freely from rest near the surface of the Earth. What is the speed of the object when it has fallen 4.9 meters from its rest position?

(1) 4.9 m/s (3) 24 m/s
(2) 9.8 m/s (4) 96 m/s

4 2

5 Which term represents a vector quantity?

1 work 3 force
2 power 4 distance

5 3

6 A river flows due east at 1.5 meters per second. A motorboat leaves the north shore of the river and heads due south at 2.0 meters per second, as shown in the diagram below.

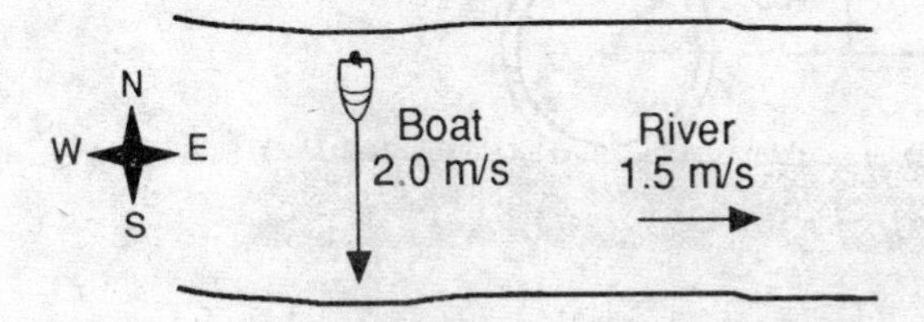

Which vector best represents the resultant velocity of the boat relative to the riverbank?

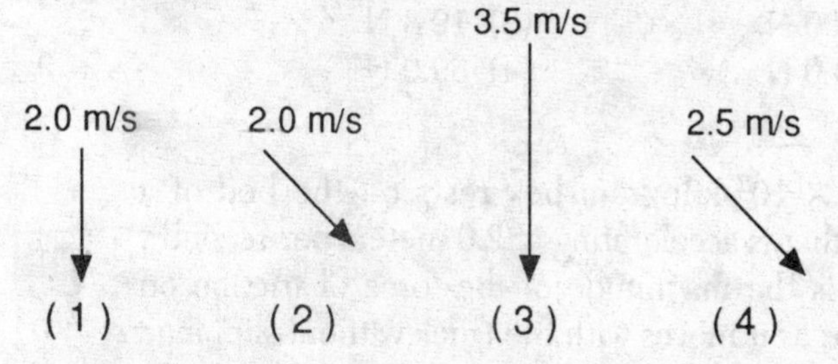

6 4

7 Which combination of concurrent forces could *not* produce equilibrium?

(1) 10. N, 20. N, and 50. N
(2) 20. N, 30. N, and 50. N
(3) 30. N, 40. N, and 50. N
(4) 40. N, 40. N, and 50. N

7 4

8 A 60.-kilogram astronaut weighs 96 newtons on the surface of the Moon. The acceleration due to gravity on the Moon is

(1) 0.0 m/s^2 (3) 4.9 m/s^2
(2) 1.6 m/s^2 (4) 9.8 m/s^2

8 2

9 The handle of a lawn roller is held at 45° from the horizontal. A force, *F*, of 28.0 newtons is applied to the handle as the roller is pushed across a level lawn, as shown in the diagram below.

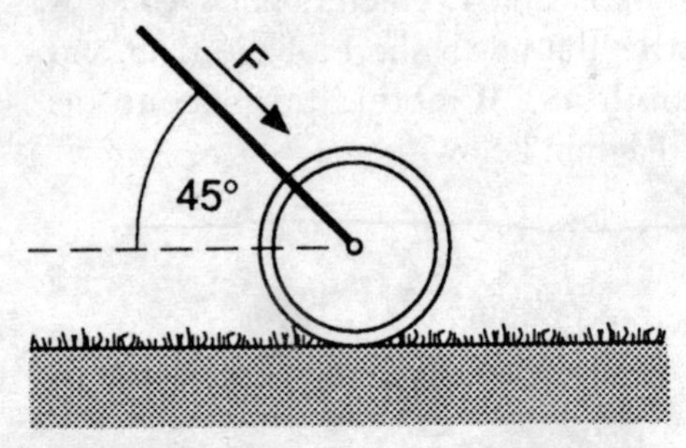

What is the magnitude of the force moving the roller forward?

(1) 7.00 N (3) 19.8 N
(2) 14.0 N (4) 39.0 N

9 4

10 A 1.0×10^2-kilogram box rests on the bed of a truck that is accelerating at 2.0 meters per second^2. What is the magnitude of the force of friction on the box as it moves with the truck without slipping?

(1) 1.0×10^3 N (3) 5.0×10^2 N
(2) 2.0×10^2 N (4) 0.0 N

10 2

11 A student weighing 500. newtons stands on a spring scale in an elevator. If the scale reads 520. newtons, the elevator must be

1 accelerating upward
2 accelerating downward
3 moving upward at constant speed
4 moving downward at constant speed

11 1

12 A box decelerates as it moves to the right along a horizontal surface, as shown in the diagram at the right. Which vector best represents the force of friction on the box?

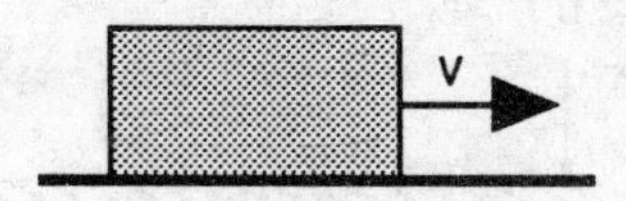

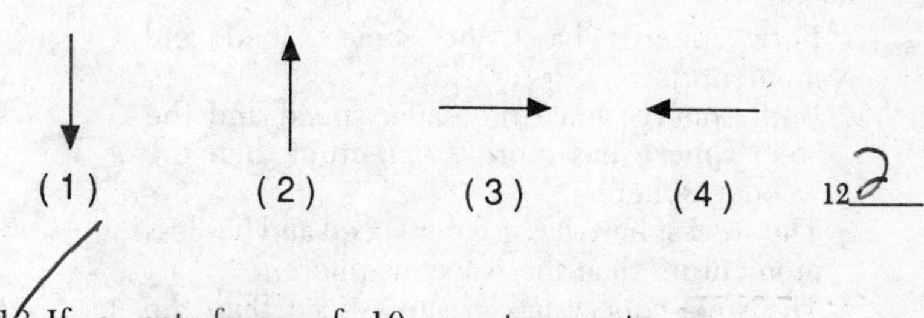

12 2

13 If a net force of 10. newtons acts on a 6.0-kilogram mass for 8.0 seconds, the total change of momentum of the mass is

(1) 48 kg•m/s
(2) 60. kg•m/s
(3) 80. kg•m/s
(4) 480 kg•m/s

13 3

14 In the diagram below, a 0.4-kilogram steel sphere and a 0.1-kilogram wooden sphere are located 2.0 meters above the ground. Both spheres are allowed to fall from rest.

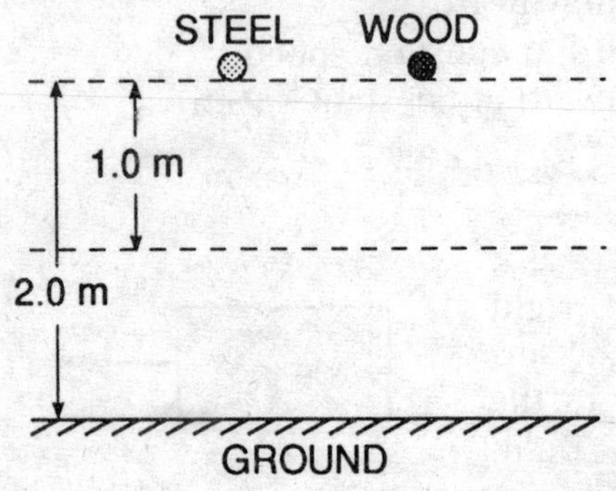

Which statement best describes the spheres after they have fallen 1.0 meter? [Neglect air resistance.]

1 Both spheres have the same speed and momentum.
2 Both spheres have the same speed and the steel sphere has more momentum than the wooden sphere.
3 The steel sphere has greater speed and has less momentum than the wooden sphere.
4 The steel sphere has greater speed than the wooden sphere and both spheres have the same momentum.

14 ____

15 A constant force of 2.0 newtons is used to push a 3.0-kilogram mass 4.0 meters across the floor. How much work is done on the mass?

(1) 6.0 J
(2) 8.0 J
(3) 12 J
(4) 24 J

15 ____

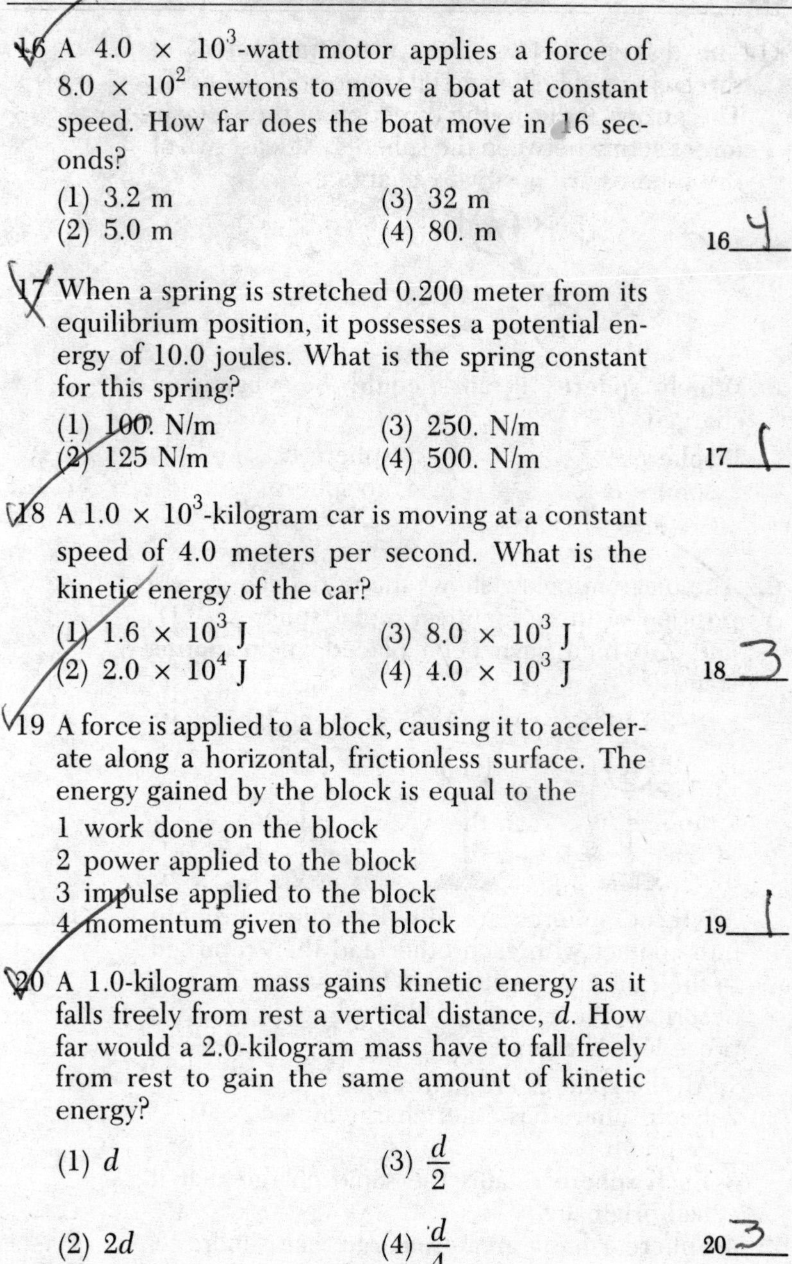

16 A 4.0×10^3-watt motor applies a force of 8.0×10^2 newtons to move a boat at constant speed. How far does the boat move in 16 seconds?

(1) 3.2 m
(2) 5.0 m
(3) 32 m
(4) 80. m

16_____

17 When a spring is stretched 0.200 meter from its equilibrium position, it possesses a potential energy of 10.0 joules. What is the spring constant for this spring?

(1) 100. N/m
(2) 125 N/m
(3) 250. N/m
(4) 500. N/m

17_____

18 A 1.0×10^3-kilogram car is moving at a constant speed of 4.0 meters per second. What is the kinetic energy of the car?

(1) 1.6×10^3 J
(2) 2.0×10^4 J
(3) 8.0×10^3 J
(4) 4.0×10^3 J

18_____

19 A force is applied to a block, causing it to accelerate along a horizontal, frictionless surface. The energy gained by the block is equal to the

1 work done on the block
2 power applied to the block
3 impulse applied to the block
4 momentum given to the block

19_____

20 A 1.0-kilogram mass gains kinetic energy as it falls freely from rest a vertical distance, *d*. How far would a 2.0-kilogram mass have to fall freely from rest to gain the same amount of kinetic energy?

(1) d
(2) $2d$
(3) $\frac{d}{2}$
(4) $\frac{d}{4}$

20_____

21 The diagram below shows the arrangement of three charged hollow metal spheres, *A*, *B*, and *C*. The arrows indicate the direction of the electric forces acting between the spheres. At least two of the spheres are positively charged.

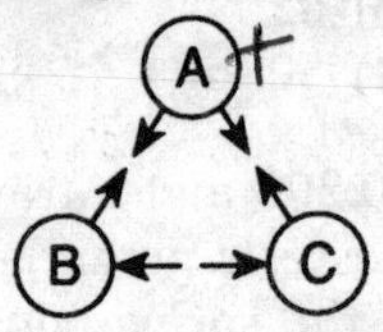

Which sphere, if any, could be negatively charged?

1 sphere *A*
2 sphere *B*
3 sphere *C*
4 no sphere

21____

22 The diagram below shows the initial charge and position of three identical metal spheres, *X*, *Y*, and *Z*, which have been placed on insulating stands.

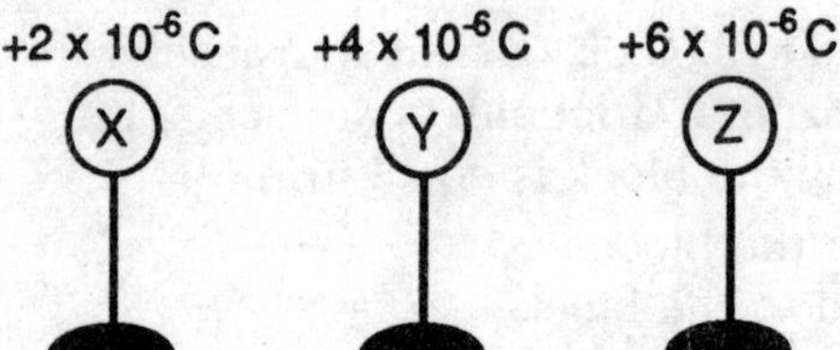

All three spheres are simultaneously brought into contact with each other and then returned to their original positions. Which statement best describes the charge of the spheres after this procedure is completed?

1 All the spheres are neutral.
2 Each sphere has a net charge of $+4 \times 10^{-6}$ coulomb.
3 Each sphere retains the same charge that it had originally.
4 Sphere *Y* has a greater charge than spheres *X* or *Z*.

22____

23 Which diagram best represents the electric field of a point charge?

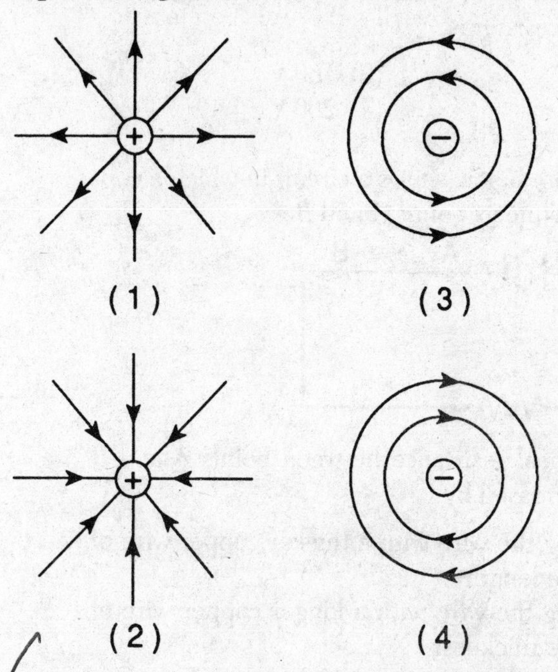

23 1

24 If 20. joules of work is done in transferring 5.0 coulombs of charge between two points, the potential difference between these two points is

(1) 100 V
(2) 50. V
(3) 0.25 V
(4) 4.0 V

24 4

25 A neutral atom must contain equal numbers of

1 protons and neutrons, only
2 protons and electrons, only
3 electrons and neutrons, only
4 protons, neutrons, and electrons

25 4

26 A 20.-ohm resistor has 4.0 coulombs passing through it in 5.0 seconds. The potential difference across the resistor is

(1) 8.0 V
(2) 100 V
(3) 160 V
(4) 200 V

26____

27 The diagram below shows a circuit in which a copper wire connects points *A* and *B*.

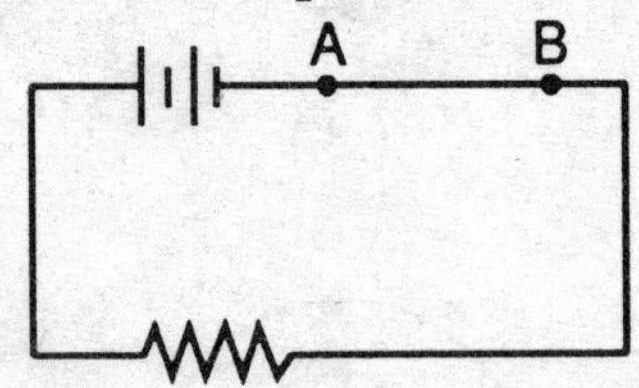

The electrical resistance between points *A* and *B* can be decreased by

1 replacing the wire with a thicker copper wire of the same length
2 replacing the wire with a longer copper wire of the same thickness
3 increasing the temperature of the copper wire
4 increasing the potential difference supplied by the battery

27____

28 In the circuit diagram below, ammeter *A* measures the current supplied by the 10.-volt battery.

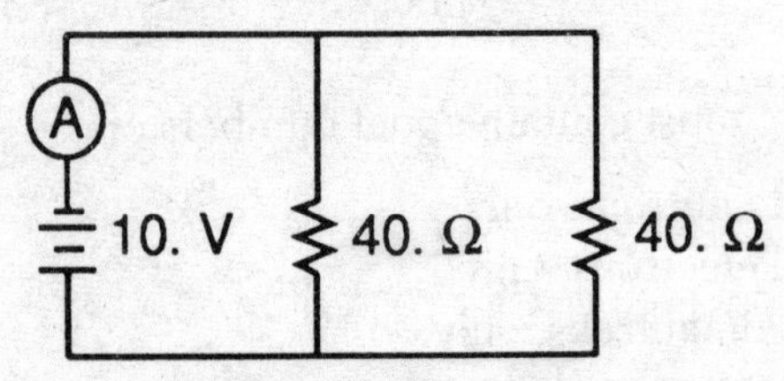

The current measured by ammeter *A* is

(1) 0.13 A
(2) 2.0 A
(3) 0.50 A
(4) 4.0 A

28____

29 Which unit is equivalent to a watt, the SI unit of power?

1 joule/second
2 joule/volt
3 joule/ohm
4 joule/coulomb

29_____

30 An electric fan draw 1.7 amperes of current when operated at a potential difference of 120 volts. How much electrical energy is needed to run this fan for 1 hour [1 hour = 3600 seconds]

(1) 7.1×10^1 J
(2) 2.0×10^2 J
(3) 2.5×10^5 J
(4) 7.3×10^5 J

30_____

31 Two solenoids are wound on soft iron cores and connected to batteries, as shown in the diagram below.

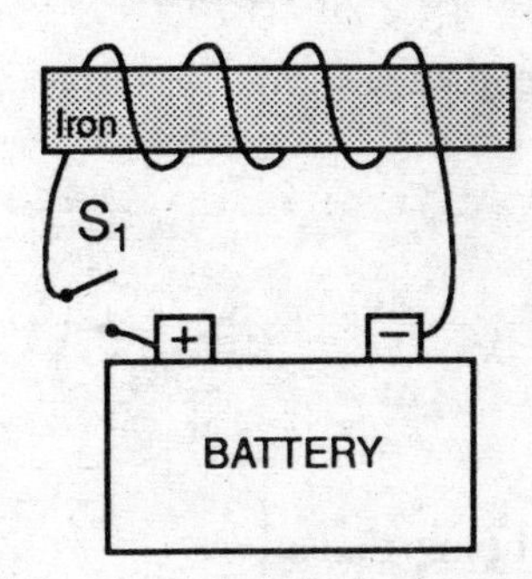

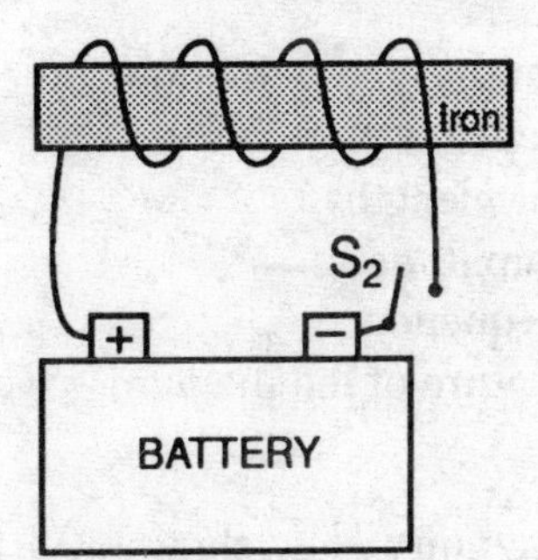

When switches S_1 and S_2 are closed, the solenoids

1 repel because of adjacent north poles
2 repel because of adjacent south poles
3 attract because of adjacent north and south poles
4 neither attract nor repel

31_____

32 Which diagram below best represents the magnetic field near a bar magnet?

32____

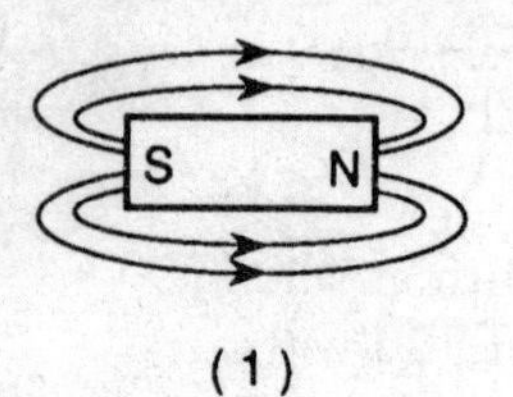

(1)

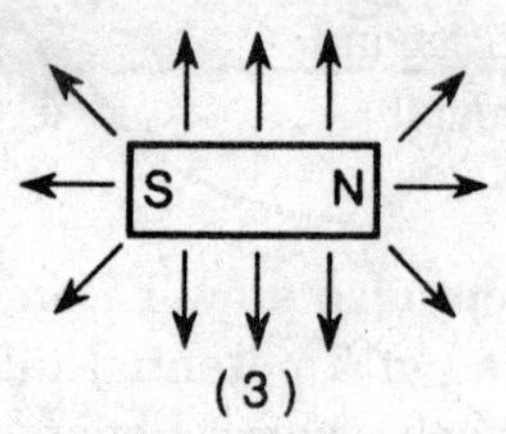

(3)

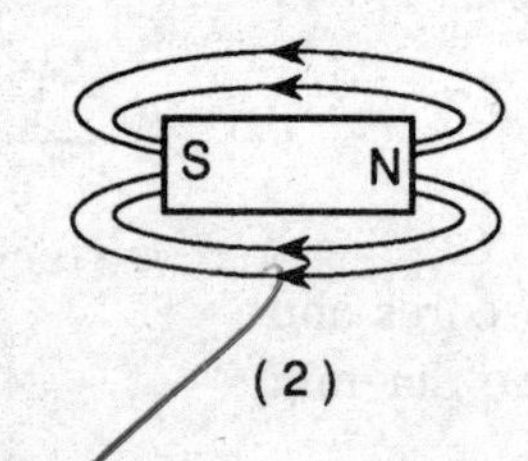

(2)

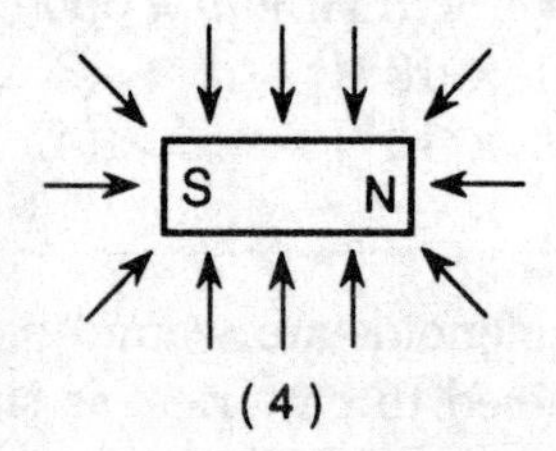

(4)

32____

33 In a nondispersive medium, the speed of a light wave depends on
1 its wavelength
2 its amplitude
3 its frequency
4 the nature of the medium

33____

34 The diagram below shows a piston being moved back and forth to generate a wave. The piston produces a compression, *C*, every 0.50 second.

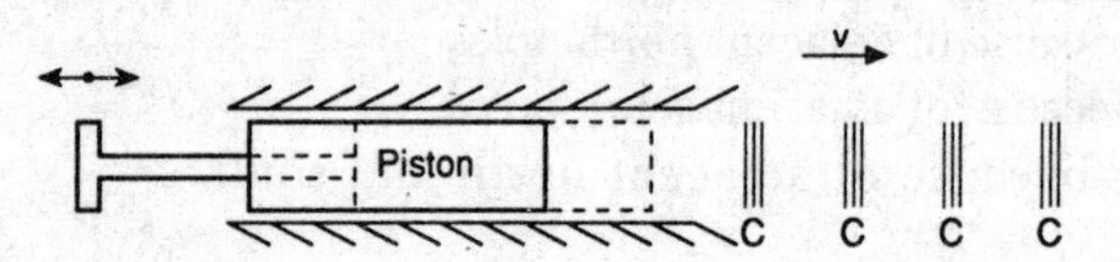

The frequency of this wave is
(1) 1.0 Hz
(2) 2.0 Hz
(3) 5.0×10^{-1} Hz
(4) 3.3×10^{2} Hz

34____

35 The diagram below represents lines of magnetic flux within a region of space.

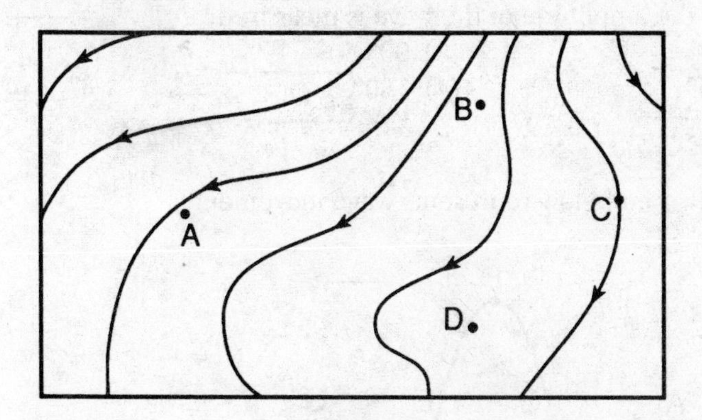

The magnetic field strength is greatest at point

(1) *A* (3) *C*
(2) *B* (4) *D*

35 3

36 The diagram below shows a current-carrying wire located in a magnetic field which is directed toward the top of the page. The electromagnetic force on the wire is directed out of the page.

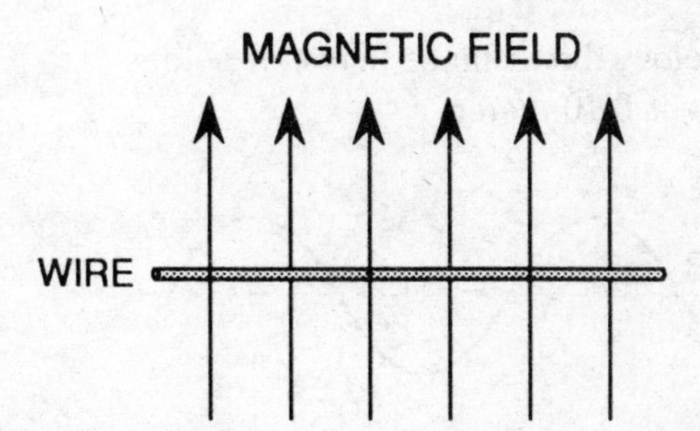

In the wire, the electron flow is directed toward the

1 left
2 right
3 top of the page
4 bottom of the page

36 2

37 What is the angle between the direction of propagation of a transverse wave and the direction in which the amplitude of the wave is measured?

(1) 0° (3) 90°
(2) 45° (4) 180°

37 3

38 The diagram below represents wave movement.

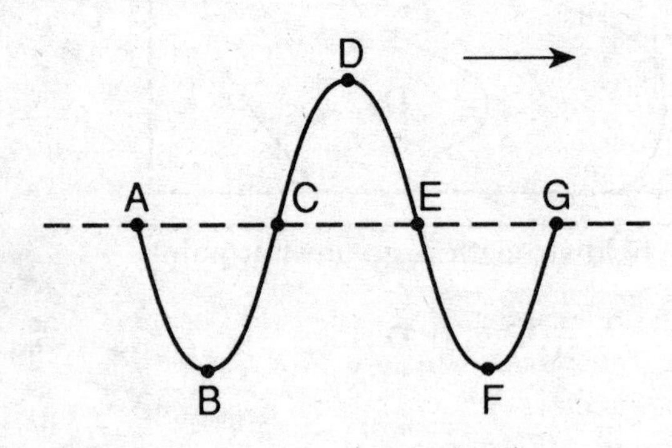

Which two points are in phase?

(1) *A* and *G* (3) *C* and *E*
(2) *B* and *F* (4) *D* and *F*

38 2

39 In the diagram below, the distance between points *A* and *B* on a wave is 0.10 meter.

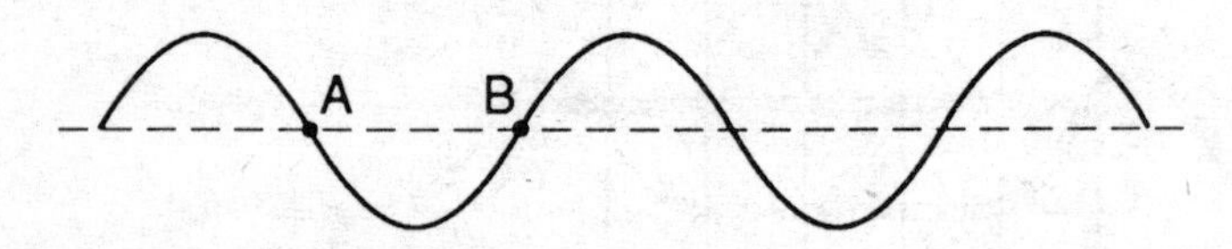

This wave must have

1 an amplitude of 0.10 m
2 an amplitude of 0.20 m
3 a wavelength of 0.10 m
4 a wavelength of 0.20 m

39 4

40 Which diagram best represents the reflection of light from an irregular surface?

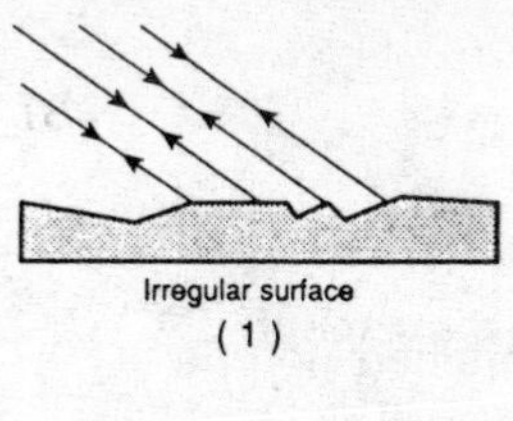

(1)

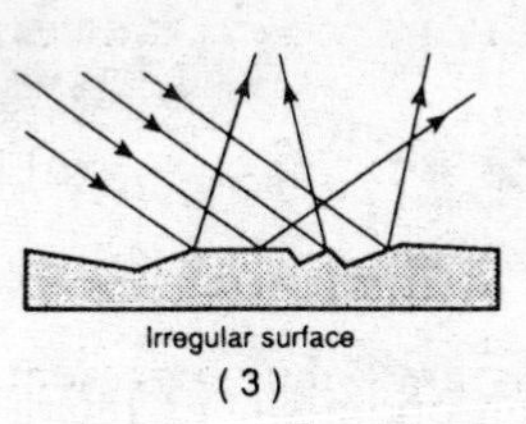

(3)

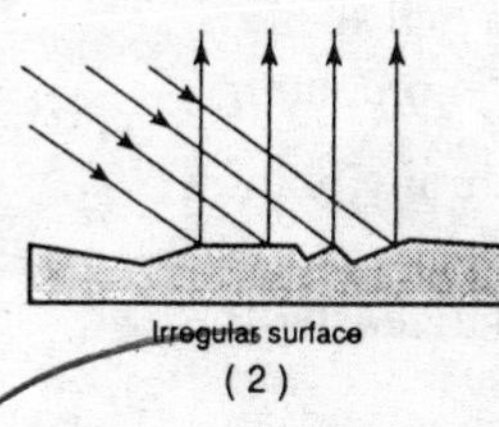

(2)

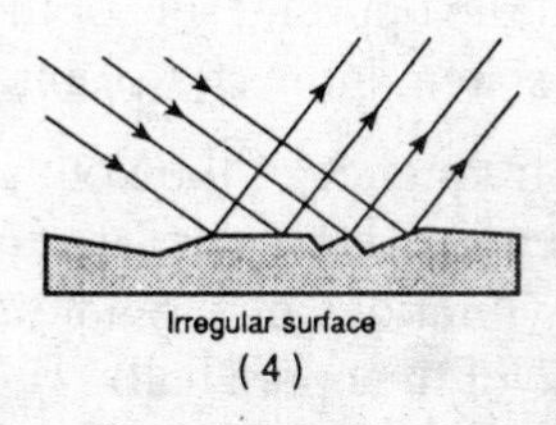

(4)

40 3

41 A stationary radar gun can determine the speed of a pitched baseball by measuring the difference in frequency between incident and reflected radar waves. This process illustrates

1 the Doppler effect
2 standing waves
3 the critical angle
4 diffraction

41 1

42 The diagram below represents a rope along which two pulses of equal amplitude, *A*, approach point *P*.

When the two pulses meet at *P*, the vertical displacement of the rope at point *P* will be

(1) A
(2) $2A$
(3) 0
(4) $\frac{A}{2}$

42 2

43 In a vacuum, a monochromatic beam of light has a frequency of 6.3×10^{14} hertz. What color is the light?

1 red
2 yellow
3 green
4 blue

43 4

44 When an opera singer hits a high-pitched note, a glass on the opposite side of the opera hall shatters. Which statement best explains this phenomenon?

1 The frequency of the note and natural vibration frequency of the glass are equal.
2 The vibrations of the note are polarized by the shape of the opera hall.
3 The amplitude of the note increases before it reaches the glass.
4 The singer and glass are separated by an integral number of wavelengths.

44 2

45 A beam of light crosses a boundary between two different media. Refraction can occur if

1 the angle of incidence is 0°
2 there is no change in the speed of the wave
3 the media have different indices of refraction
4 all of the light is reflected

45 4

46 What is the energy of a photon with a frequency of 5.0×10^{14} hertz?

(1) 3.3 eV
(2) 3.2×10^{-6} eV
(3) 3.0×10^{48} J
(4) 3.3×10^{-19} J

46 4

47 The diagram below shows white light being dispersed as it passes from air into a glass prism.

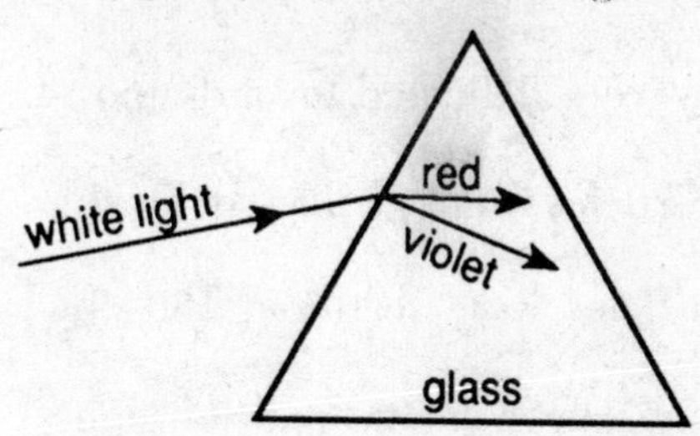

This phenomenon occurs because, in glass, each frequency of light has a different

1 intensity
2 amplitude
3 angle of incidence
4 absolute index of refraction

47 2

48 Which graph below best represents the relationship between the frequency of a light source causing photoemission and the maximum kinetic energy of the photoelectrons produced?

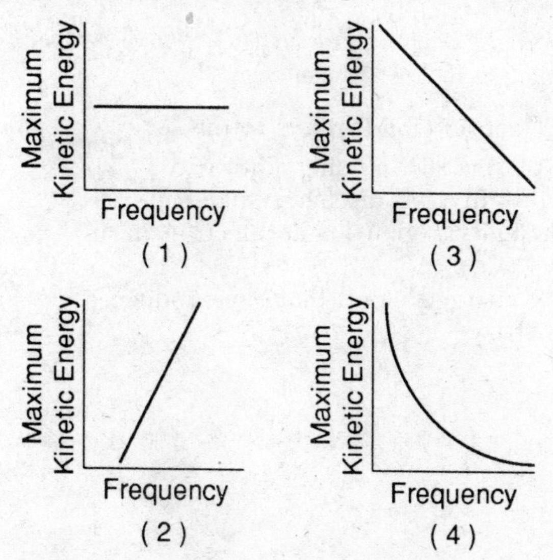

48 1

49 Which observation was made by Rutherford when he bombarded gold foil with alpha particles?

1 Alpha particles were deflected toward a positive electrode.
2 Some alpha particles were deflected by the gold foil.
3 Most alpha particles were scattered 180° by the gold foil.
4 Gold foil had no effect on the path of alpha particles.

49 1

50 Which electron transition in the hydrogen atom results in the emission of a photon of greatest energy?

(1) $n = 2$ to $n = 1$
(2) $n = 3$ to $n = 2$
(3) $n = 4$ to $n = 2$
(4) $n = 5$ to $n = 3$

50 1

51 The term "electron cloud" refers to the

1 electron plasma surrounding a hot wire
2 cathode rays in a gas discharge tube
3 high-probability region for an electron in an atom
4 negatively charged cloud that can produce a lightning strike

51 3

Note that questions 52 through 55 have only three choices.

36/55 , 49

52 As the angle between a force and level ground decreases from 60° to 30°, the vertical component of the force

1 decreases
2 increases
3 remains the same

52 1

53 In a baseball game, a batter hits a ball for a home run. Compared to the magnitude of the impulse imparted to the ball, the magnitude of the impulse imparted to the bat is

1 less
2 greater
3 the same

53 2

54 As the mass of a body increases, its gravitational force of attraction on the Earth

1 decreases
2 increases
3 remains the same

54 2

55 An interference pattern is observed as light passes through two closely spaced slits. As the distance between the two slits is decreased, the distance between adjacent bright bands in the interference pattern

1 decreases
2 increases
3 remains the same

55 1

PART II

This part consists of six groups, each containing ten questions. Each group tests an optional area of the course. Choose two of these six groups. Be sure that you answer all ten questions in each group chosen. Record the answers to these questions in the spaces provided. [20]

GROUP 1—**Motion in a Plane**

If you choose this group, be sure to answer questions **56–65**.

Base your answers to questions 56 through 58 on the information and diagram below.

In the diagram below, a 10.-kilogram sphere, *A*, is projected horizontally with a velocity of 30. meters per second due east from a height of 20. meters above level ground. At the same instant, a 20.-kilogram sphere, *B*, is projected horizontally with a velocity of 10. meters per second due west from a height of 80. meters above level ground. [Neglect air friction.]

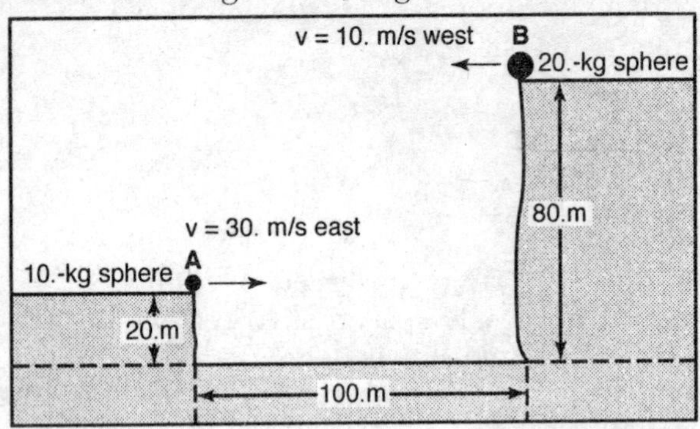

56 Initially, the spheres are separated by a horizontal distance of 100. meters. What is the horizontal separation of the spheres at the end of 1.5 seconds?

(1) 15 m
(2) 30. m
(3) 40. m
(4) 45 m

56 

57 The magnitude of the horizontal acceleration of sphere *A* is

(1) 0.0 m/s²   (3) 9.8 m/s²
(2) 2.0 m/s²   (4) 15 m/s²

57_____

58 Compared to the vertical acceleration of sphere *A*, the vertical acceleration of sphere *B* is

1 the same   3 one-half as great
2 twice as great   4 four times as great

58_____

Base your answers to questions 59 through 62 on the information and diagram below.

The diagram shows a 5.0-kilogram cart traveling clockwise in a horizontal circle of radius 2.0 meters at a constant speed of 4.0 meters per second.

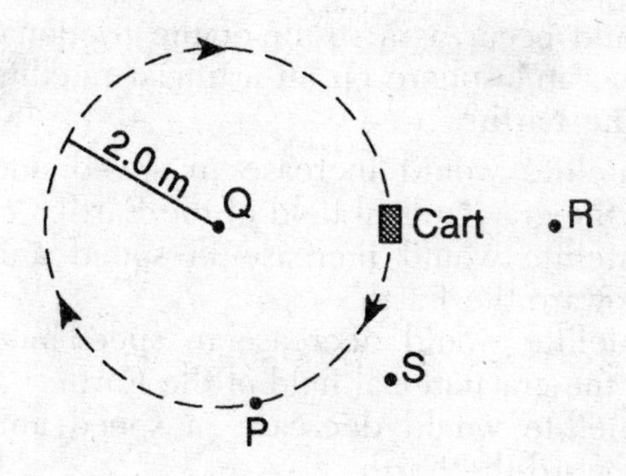

59 At the position shown, the velocity of the cart is directed toward point

(1) *P*   (3) *R*
(2) *Q*   (4) *S*

59_____

60 At the position shown, the centripetal acceleration of the cart is directed toward point

(1) *P*
(2) *Q*
(3) *R*
(4) *S*

60____

61 If the mass of the cart was doubled, the magnitude of the centripetal acceleration of the cart would be

1 unchanged
2 doubled
3 halved
4 quadrupled

61____

62 What is the magnitude of the centripetal force acting on the cart?

(1) 8.0 N
(2) 20. N
(3) 40. N
(4) 50. N

62____

---

63 What would occur as a result of the frictional drag of the atmosphere on an artificial satellite orbiting the Earth?

1 The satellite would increase in speed and escape the gravitational field of the Earth.
2 The satellite would increase in speed and spiral toward the Earth.
3 The satellite would decrease in speed and escape the gravitational field of the Earth.
4 The satellite would decrease in speed and spiral toward the Earth.

63____

64 A cannon with a muzzle velocity of 500. meters per second fires a cannonball at an angle of 30.° above the horizontal. What is the vertical component of the cannonball's velocity as it leaves the cannon?

(1) 0.0 m/s
(2) 250. m/s
(3) 433 m/s
(4) 500. m/s

64____

Note that question 65 has only three choices.

65 Satellites A and B are orbiting the Earth in circular orbits as shown below. The mass of satellite A is twice as great as the mass of satellite B. Earth has radius R.

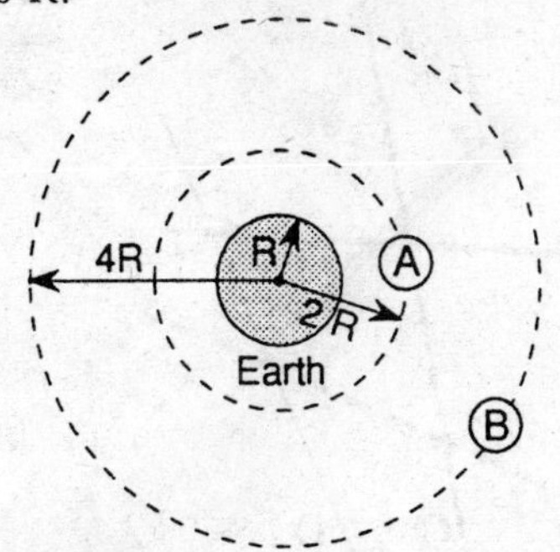

Compared to the orbital period of satellite A, the orbital period of satellite B is

1 shorter
2 longer
3 the same

65 2

GROUP 2—**Internal Energy**

If you choose this group, be sure to answer questions **66–75**.

66 Absolute zero represents a substance's minimum

1 internal molecular energy
2 gravitational potential energy
3 specific heat
4 heat of fusion

66_____

67 At 1.0 atmosphere of pressure, which substance is a solid at $-51°C$ and a liquid at $-33°C$?

1 ammonia
2 ethyl alcohol
3 mercury
4 lead

67_____

68 Equal masses of four different solids, *A*, *B*, *C*, and *D*, are heated at a constant rate. The graph below represents the temperature of each solid as a function of the heat added to the solid.

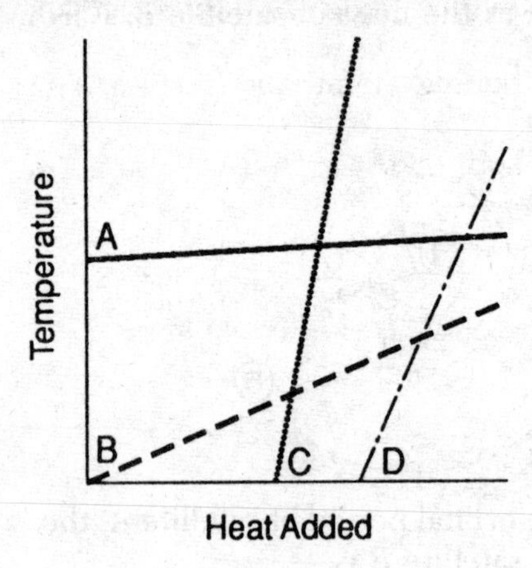

Which solid has the greatest specific heat?

(1) *A*
(2) *B*
(3) *C*
(4) *D*

68_____

69 A change of 10. Celsius degrees is produced by adding 2.4 kilojoules of heat to 1.0 kilogram of a substance. The substance could be

1 silver
2 lead
3 platinum
4 aluminum

69_____

70 The amount of heat required to melt 0.50 kilogram of iron at its melting point is approximately

(1) 0.23 kJ
(2) 0.90 kJ
(3) 13 kJ
(4) 130 kJ

70_____

71 Increasing the external pressure on a sample of water will

1 increase its boiling point and increase its freezing point
2 increase its boiling point and decrease its freezing point
3 decrease its boiling point and increase its freezing point
4 decrease its boiling point and decrease its freezing point

71____

72 According to the kinetic theory of gases, an ideal gas of low density has relatively large

1 molecules
2 energy loss in molecular collisions
3 forces between molecules
4 distances between molecules

72____

73 If the pressure on a fixed mass of an ideal gas is doubled at a constant temperature, the volume of this gas sample will be

1 the same
2 doubled
3 halved
4 quartered

73____

74 According to the second law of thermodynamics, which phenomenon will most likely occur?

1 The entropy of the universe will steadily decrease.
2 The universe will steadily become more disordered.
3 The universe will eventually reach equilibrium at absolute zero.
4 Within the universe, more heat will flow from colder to warmer regions than from warmer to colder regions.

74____

Note that question 75 has only three choices.

75 A sample of liquid ethyl alcohol is boiling. As more heat is added, the temperature of the liquid alcohol will
1 decrease
2 increase
3 remain the same
75____

GROUP 3—**Electromagnetic Applications**

If you choose this group, be sure to answer questions **76–85.**

76 Which graph best represents the relationship between the degree of deflection of a galvanometer needle and the current passing through its coil?

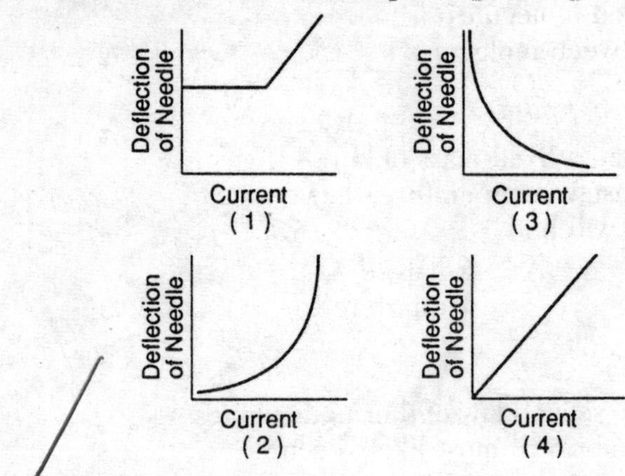

76 2

77 The torque on the armature of an operating electric motor may be increased by
1 decreasing the current in the armature
2 decreasing the magnetic field strength of the field poles
3 increasing the potential difference applied to the armature
4 increasing the distance between the armature and the field poles
77 3

78 In an operating practical motor, the magnetic field produced by the current-carrying coil is strengthened and concentrated by the

1 split-ring commutator
2 back emf
3 field pole
4 iron core

78 ____

79 The diagram below represents an electron beam entering the region between two oppositely charged parallel plates.

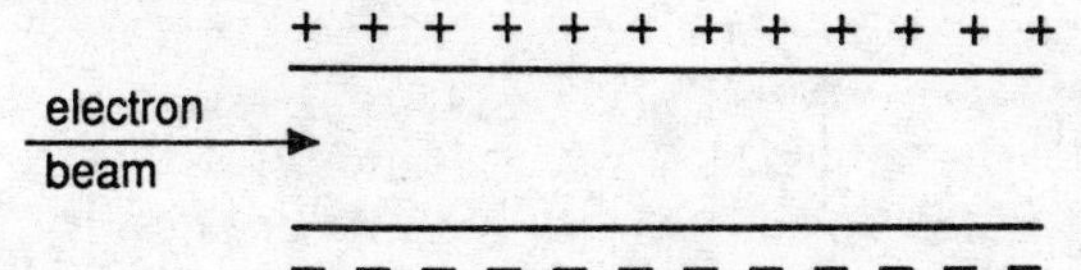

In which direction will the beam of electrons be deflected?

1 out of the page
2 into the page
3 toward the top of the page
4 toward the bottom of the page

79 ____

80 A proton having a velocity of 1.5×10^6 meters per second to the right is projected into a magnetic field having a flux density of 3.0 teslas directed out of the page, as shown in the diagram below.

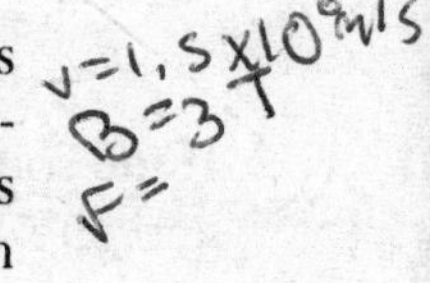

What is the magnitude of the magnetic force on the proton?

(1) 4.1×10^{-24} N
(2) 7.2×10^{-13} N
(3) 4.5×10^{6} N
(4) 7.2×10^{6} N

80 ____

Base your answers to questions 81 and 82 on the information and diagram below.

Four electron beams, *A*, *B*, *C*, and *D*, are projected into a magnetic field directed out of the page.

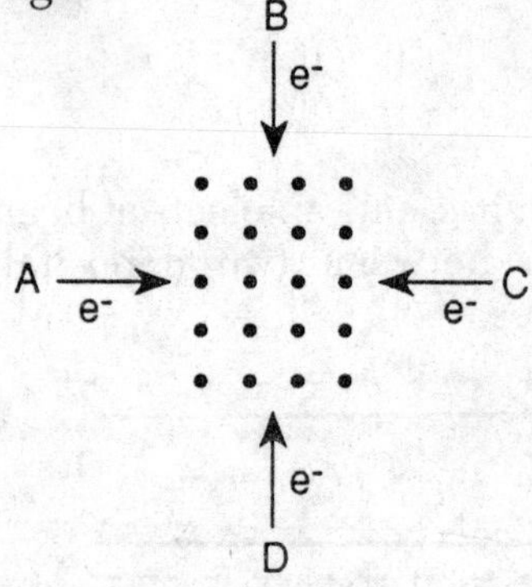

81 Which beam of electrons will initially be deflected toward the top of the page by the magnetic field?
(1) *A* (3) *C*
(2) *B* (4) *D* 81 3

82 If the speed of the electrons in beam *B* is doubled and the magnetic field strength is halved, the magnitude of the deflecting force on the electrons will be
1 unchanged 3 halved
2 doubled 4 quadrupled 82 1

83 The charge-to-mass ratio of an electron is
(1) 9.1×10^{-31} C/kg (3) 5.7×10^{-12} C/kg
(2) 1.6×10^{-19} C/kg (4) 1.8×10^{11} C/kg 83 4

84 A potential difference of 10. volts is induced in a wire as it is moved at a constant speed of 5.0 meters per second perpendicular to a magnetic field having a flux density of 4.0 newtons per ampere-meter. What is the length of the wire in the field?

(1) 0.50 m (3) 8.0 m
(2) 2.0 m (4) 200 m

84____

85 The primary coil of an operating transformer has 200 turns and the secondary coil has 40 turns. This transformer is being used to

1 decrease voltage and decrease current
2 decrease voltage and increase current
3 increase voltage and decrease current
4 increase voltage and increase current

85____

GROUP 4—**Geometric Optics**

If you choose this group, be sure to answer questions **86–95**.

86 A truck has the letters OWOW painted on the front of its hood. A person in a car driving ahead of the truck views these letters in the rear-view mirror. How do the letters appear?

(1) WOWO (3) OMOM
(2) OWOW (4) MOMO

86____

87 A concave mirror has a radius of curvature of 0.60 meter. When an object is placed 0.40 meter from the reflecting surface, the image distance will be

(1) 0.10 m (3) 0.83 m
(2) 0.20 m (4) 1.2 m

87____

88 The diagram below shows parallel monochromatic incident light rays being reflected from a concave mirror.

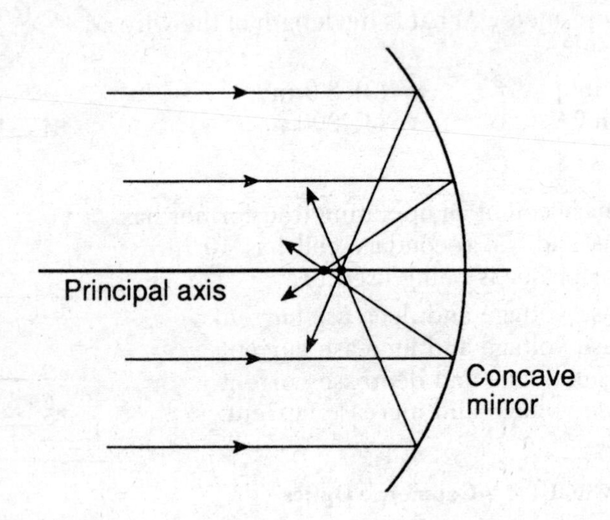

Which phenomenon does the diagram illustrate?

1 chromatic aberration
2 spherical aberration
3 refraction
4 dispersion

88 ______

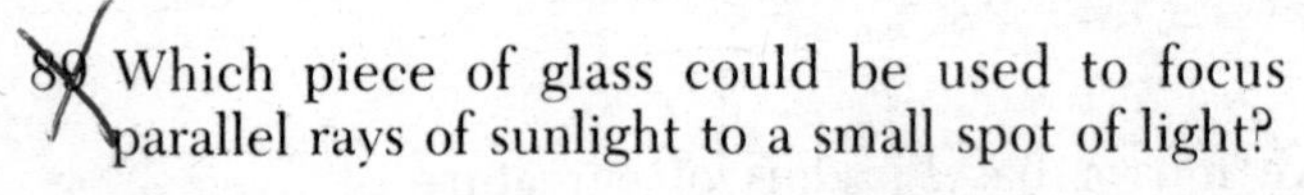

89 Which piece of glass could be used to focus parallel rays of sunlight to a small spot of light?

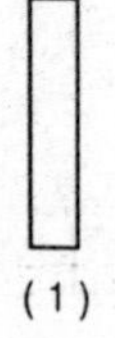

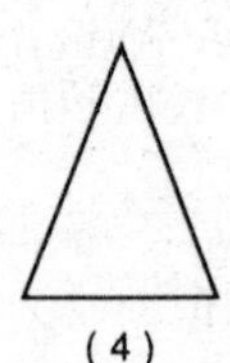

(1) (2) (3) (4)

89 ______

90 In the diagram below, a lamp 0.4 meter tall is placed 0.6 meter in front of a convex mirror.

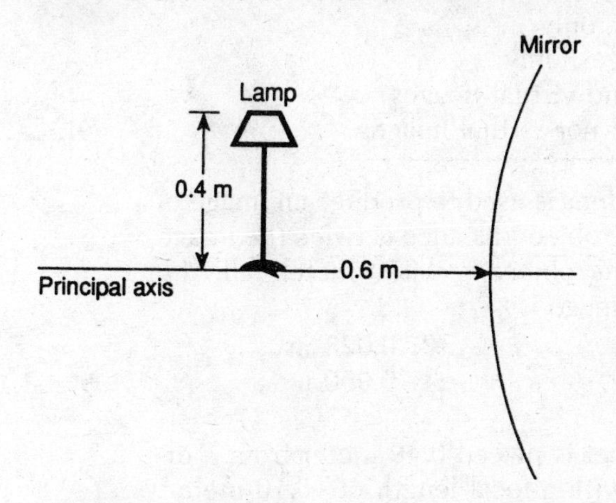

Which diagram best represents an image of the lamp that could be formed by this mirror?

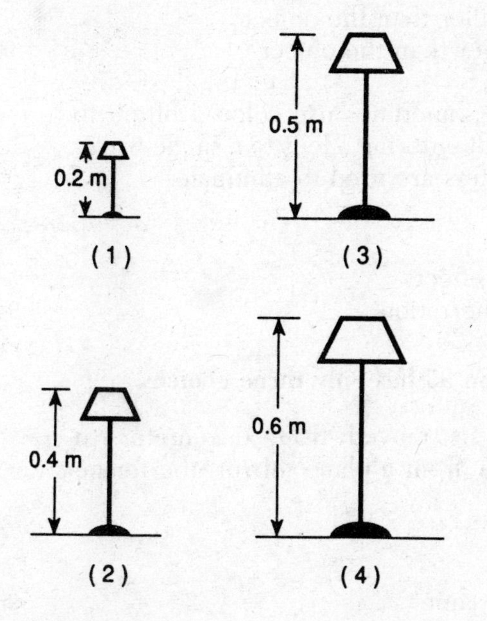

90 ____

91 Which type of images can be formed by a converging lens?

1 real images, only
2 virtual images, only
3 both real and virtual images
4 neither real nor virtual images

91 ___

92 A converging lens is used to produce an image of an object. The object distance is twice the image distance. If the object is 0.050 meter tall, the height of its image is

(1) 0.010 m
(2) 0.020 m
(3) 0.025 m
(4) 0.050 m

92 ___

93 When an object is placed 0.40 meter from a diverging lens with a focal length of –0.10 meter, the image produced will be

1 virtual and smaller than the object
2 virtual and larger than the object
3 real and smaller than the object
4 real and larger than the object

93 ___

94 Photographers sometimes use colored filters to restrict the light entering a lens to a single wavelength. The filters are used to eliminate

1 diffusion
2 diffraction
3 polarization effects
4 chromatic aberration

94 ___

Note that question 95 has only three choices.

95 As an object is moved from 0.2 meter to 0.3 meter away from a plane mirror, the image distance

1 decreases
2 increases
3 remains the same

95 ___

GROUP 5—**Solid State**

If you choose this group, be sure to answer questions **96–105**.

96 The circuit diagram below shows a *P*-type semiconductor in series with a lamp, a resistor, and a battery.

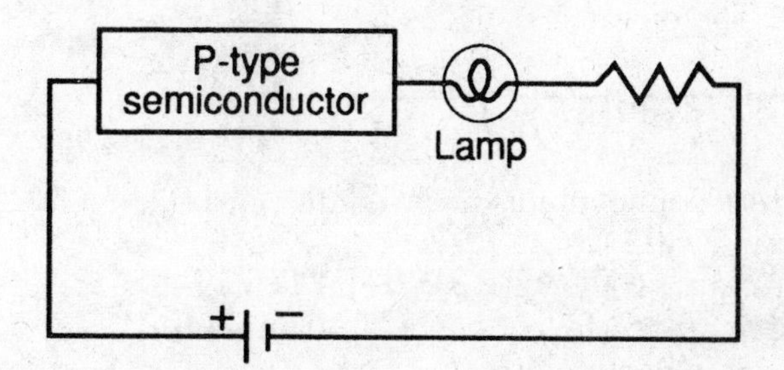

What would increase the current in the circuit?

1 increasing the resistance of the resistor
2 reversing the battery polarity
3 reversing the connections to the semiconductor
4 increasing the temperature of the semiconductor

96_____

97 According to the energy band model, in which material is the energy gap between the conduction band and the valence band greatest?

1 a conductor
2 an insulator
3 an extrinsic semiconductor
4 an intrinsic semiconductor

97_____

98 The diagram below shows a circuit connecting a battery to semiconductor *A*.

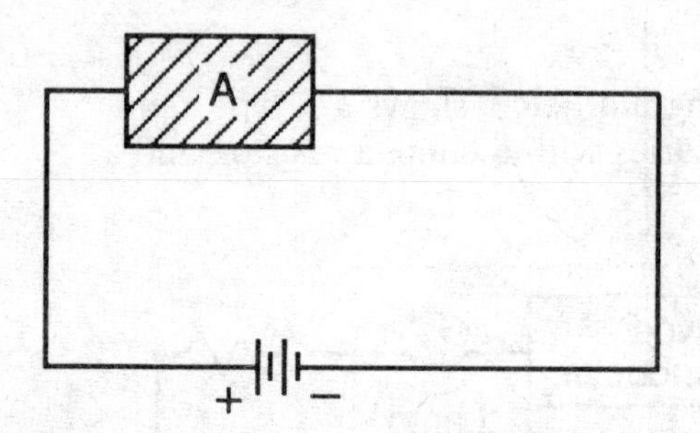

If the battery connection is reversed, the current in the circuit will

1 decrease only if *A* is *N*-type
2 decrease only if *A* is *P*-type
3 increase if *A* is either *N*-type or *P*-type
4 remain the same if *A* is either *N*-type or *P*-type 98_____

99 The majority charge carriers in a *P*-type semiconductor are holes. When a section of *P*-type semiconductor is connected across the terminals of a battery, where do the holes flow?

1 toward the negative terminal, only
2 toward the positive terminal, only
3 equally toward both the negative and positive terminals
4 toward neither terminal 99_____

100 The simplest circuit element that will allow electric current to pass through a circuit in only one direction is

1 a transistor
2 a diode
3 an *N*-type semiconductor
4 a *P*-type semiconductor 100_____

101 The graph below shows the relationship between current and applied potential difference for an electrical device.

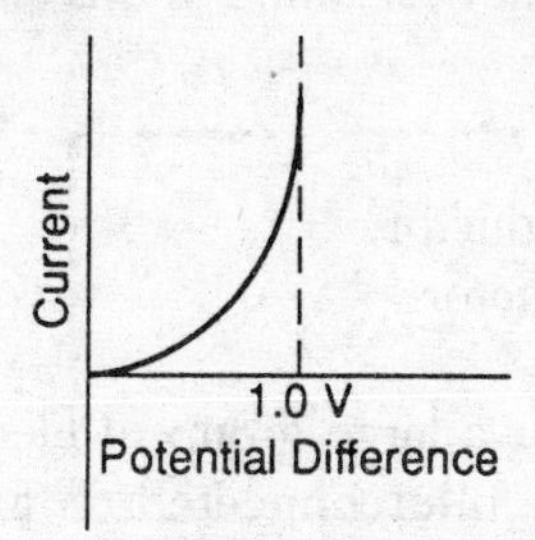

This device is most likely a

1 forward-biased *P*-type semiconductor
2 reverse-biased *P*-type semiconductor
3 forward-biased *P-N* junction
4 reverse-biased *P-N* junction 101____

102 The symbol  is to a *P-N* junction

as the symbol 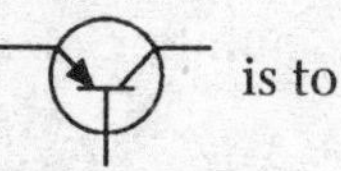 is to

1 an *N-P-N* transistor
2 a *P-N-P* transistor
3 a zener diode
4 an integrated circuit 102____

103 The diagram below shows an *N-P-N* semiconductor device.

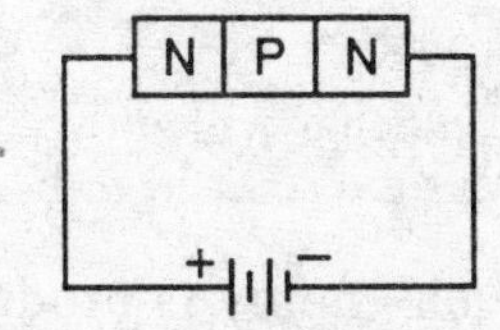

The base in the semiconductor shown is the

(1) right *N* (3) *P*
(2) left *N* (4) *P-N* junction 103____

104 A student is designing a circuit to amplify a small voltage change into a larger voltage change. The electric circuit element best suited to this task is

1 a transistor
2 a diode
3 an *N*-type semiconductor
4 a *P*-type semiconductor

104____

105 What is the name for a large group of electronic components that are interconnected on a single block of silicon inside a computer?

1 a thermistor
2 an integrated circuit
3 a transistor
4 a collector

105____

GROUP 6—**Nuclear Energy**

If you choose this group, be sure to answer questions **106-115.**

106 Which atom has the same number of neutrons as $^{16}_{8}O$?

(1) $^{16}_{7}N$
(2) $^{17}_{8}O$
(3) $^{15}_{7}N$
(4) $^{15}_{8}O$

106____

107 The force that holds the nucleons of an atom together is

1 weak and short-ranged
2 weak and long-ranged
3 strong and short-ranged
4 strong and long-ranged

107____

108 Approximately how much energy is produced when 0.50 atomic mass unit of matter is completely converted into energy?

(1) 9.3 MeV
(2) 9.3×10^2 MeV
(3) 4.7 MeV
(4) 4.7×10^2 MeV

108____

109 Atoms of different isotopes of the same element contain the same number of
1 neutrons, but a different number of protons
2 neutrons, but a different number of electrons
3 electrons, but a different number of protons
4 protons, but a different number of neutrons 109____

110 The disintegration of the nucleus of an atom of a naturally occurring radioactive element may produce more
1 neutrons in the nucleus
2 electrons in the nucleus
3 protons in the nucleus
4 atomic mass 110____

111 In the nuclear equation ${}^{14}_{6}C \rightarrow {}^{14}_{7}N + X$, the X represents a
1 beta particle
2 gamma ray
3 neutron
4 positron 111____

112 The half-life of a radium isotope is 1,600 years. After 4,800 years, approximately how much of an original 10.0-kilogram sample of this isotope will remain?
(1) 0.125 kg
(2) 1.25 kg
(3) 1.67 kg
(4) 3.33 kg 112____

113 In nuclear reactors, neutrons are slowed down by
1 moderators
2 control rods
3 fuel rods
4 accelerators 113____

114 For nuclear fusion to occur, the reacting nuclei must
1 absorb thermal neutrons
2 have large kinetic energies
3 be fissionable
4 have a critical mass 114____

Note that question 115 has only three choices.

115 If the mass defect for nucleus X is larger than the mass defect for nucleus Y, then nucleus X has

1 a smaller binding energy than nucleus Y
2 a larger binding energy than nucleus Y
3 the same binding energy as nucleus Y 115____

PART III

You must answer *all* questions in this part. [15]

Base your answers to questions 116 through 118 on the speed-time graph below, which represents the linear motion of a cart.

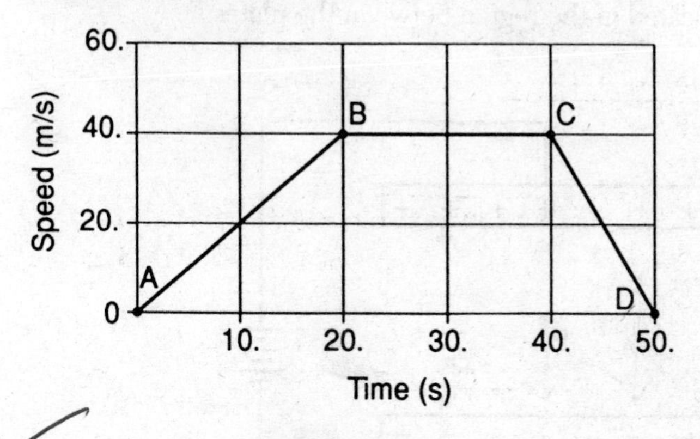

116 Determine the magnitude of the acceleration of the cart during interval *AB*. [Show all calculations, including the equation and substitution with units.] [2]

117 Calculate the distance traveled by the cart during interval *BC*. [Show all calculations, including the equation and substitution with units.] [2]

118 What is the average speed of the cart during interval *CD*? [1]

Base your answers to questions 119 and 120 on the information and diagram below.

Two parallel plates separated by a distance of 2.0×10^{-2} meter are charged to a potential difference of 1.0×10^{2} volts. Points *A*, *B*, and *C* are located in the region between the plates.

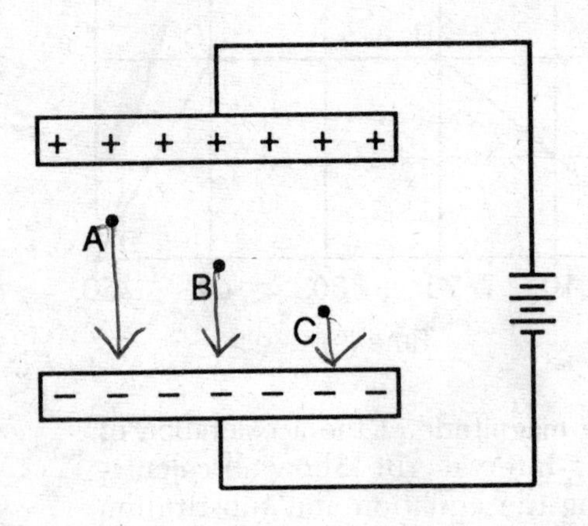

119 *On the diagram,* sketch the electric field lines between the oppositely charged parallel plates through points *A*, *B*, and *C*. [Draw lines with arrowheads in the proper direction.] [2]

120 Calculate the magnitude of the electric field strength between the plates. [Show all calculations, including the equation and substitutions with units.] [2]

Base your answers to questions 121 through 123 on the diagram below, which shows light ray AO in Lucite. The light ray strikes the boundary between Lucite and air at point O with an angle of incidence of 30.°. The dotted line represents the normal to the boundary at point O.

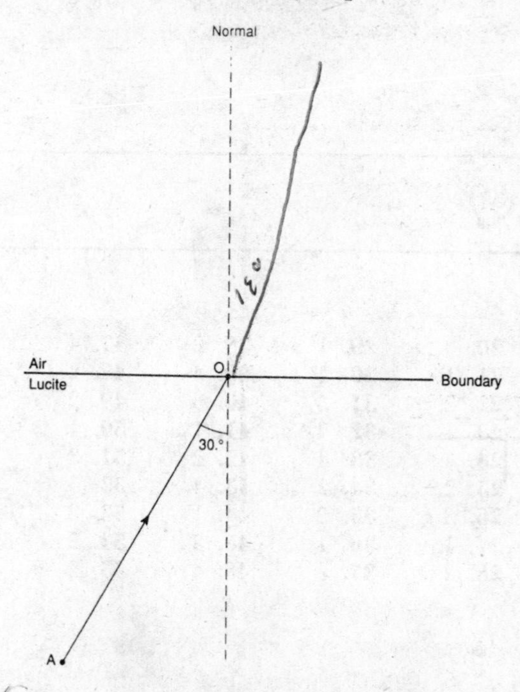

121 Calculate the angle of refraction for incident ray AO. [Show all calculations, including the equation and substitution with units.] [2]

122 *On the diagram,* using your answer from question 121, construct an arrow with a protractor and straightedge, to represent the refracted ray. [2]

123 Calculate the critical angle for a Lucite-air boundary. [Show all calculations, including the equation and substitution with units.] [2]

Answers June 1995

Physics

Answer Key

PART I

1. 1	11. 1	20. 3	29. 1	38. 2	47. 4
2. 2	12. 4	21. 1	30. 4	39. 4	48. 2
3. 3	13. 3	22. 2	31. 3	40. 3	49. 2
4. 2	14. 2	23. 1	32. 2	41. 1	50. 1
5. 3	15. 2	24. 4	33. 4	42. 2	51. 3
6. 4	16. 4	25. 2	34. 2	43. 4	52. 1
7. 1	17. 4	26. 3	35. 2	44. 1	53. 3
8. 2	18. 3	27. 1	36. 1	45. 3	54. 2
9. 3	19. 1	28. 3	37. 3	46. 4	55. 2
10. 2					

PART II

Group 1	Group 2	Group 3	Group 4	Group 5	Group 6
56. 3	66. 1	76. 4	86. 1	96. 4	106. 3
57. 1	67. 3	77. 3	87. 4	97. 2	107. 3
58. 1	68. 1	78. 4	88. 2	98. 4	108. 4
59. 4	69. 1	79. 3	89. 2	99. 1	109. 4
60. 2	70. 4	80. 2	90. 1	100. 2	110. 3
61. 1	71. 2	81. 1	91. 3	101. 3	111. 1
62. 3	72. 4	82. 1	92. 3	102. 2	112. 2
63. 4	73. 3	83. 4	93. 1	103. 3	113. 1
64. 2	74. 2	84. 1	94. 4	104. 1	114. 2
65. 2	75. 3	85. 2	95. 2	105. 2	115. 2

PART III — See answers explained

Answers Explained

PART I

1. **1** The *meter* (m) is the SI unit of length. A meter is approximately equal to 3.3 feet. Of the four choices, 10^{-4} m, one tenth of a millimeter, is the only reasonable one for the thickness of a dollar bill.

Wrong Choices Explained:

(2) 10^{-2} m = 1 cm, or a little less than 0.5 in.
(3) 10^{-1} m = 1 dm = 10 cm, or approximately 4 ins.
(4) 10^{1} m is approximately 33 ft.

2. **2** The best way to approach any word problem is to correctly identify the quantities given and the quantity to be found.

Given: Acceleration is constant. Find: $\Delta t = ?$

$\Delta s = 5.0$ m
$v_i = 2.0$ m/s
$v_f = 3.0$ m/s

Refer to the *Mechanics* section in the *Reference Tables* to find an equation(s) that relates time with displacement and velocity.

Solution:
$$\bar{v} = \frac{v_i + v_f}{2} = \frac{2.0 \text{ m/s} + 3.0 \text{ m/s}}{2} = 2.5 \text{ m/s}$$

$$\bar{v} = \frac{\Delta s}{\Delta t}$$

$$\Delta t = \frac{\Delta s}{\bar{v}} = \frac{5.0 \not{\text{m}}}{2.5 \not{\text{m}}/\text{s}} = 2.0 \text{ s}$$

3. **3** A freely falling object near the Earth's surface accelerates at a constant rate. Since $a = \Delta v/\Delta t$, acceleration is equal to the slope on a speed versus time graph, and a constant acceleration will have a constant slope.

Wrong Choices Explained:

(1) On a distance versus time graph, the slope represents average velocity since $\bar{v} = \Delta s/\Delta t$. A constant negative slope represents a constant negative velocity (zero acceleration).

(2) A constant positive slope on a distance versus time graph represents constant positive velocity (zero acceleration).

(3) A constant negative slope on a velocity versus time graph indicates negative acceleration or deceleration.

4. **2** Write out the problem.

Given: $v_i = 0$ Find: $v_f = ?$

$\Delta s = 4.9$ m

$a = 9.8$ m/s^2

Refer to the *Mechanics* section in the *Reference Tables* to find an equation(s) that relates velocity with displacement and acceleration.

Solution: $v_f^2 = v_i^2 + 2a\,\Delta s$

$= (0 \text{ m/s})^2 + 2(9.8 \text{ m/s}^2)(4.9 \text{ m})$

$= 96.04 \text{ m}^2/\text{s}^2$

$v_f = 9.8$ m/s

5. **3** A *vector* quantity has both magnitude and direction. A *scalar* quantity has only magnitude. *Force* has both magnitude and direction (e.g., 20 N at 30° north of east) and is therefore a *vector* quantity.

Wrong Choices Explained:

(1) *Work* is equal to force times displacement, using only the magnitude of the force applied in the direction of displacement; therefore, work is a *scalar* quantity.

(2) *Power* is equal to work done per unit time. Since neither work nor time is a vector quantity, power is a *scalar* quantity.

(4) *Distance* is a *scalar* quantity. *Displacement* is a *vector* quantity.

6. **4** To find a *resultant vector*, place the vectors head to tail, and draw the resultant vector from the tail of the first vector to the head of the last vector.

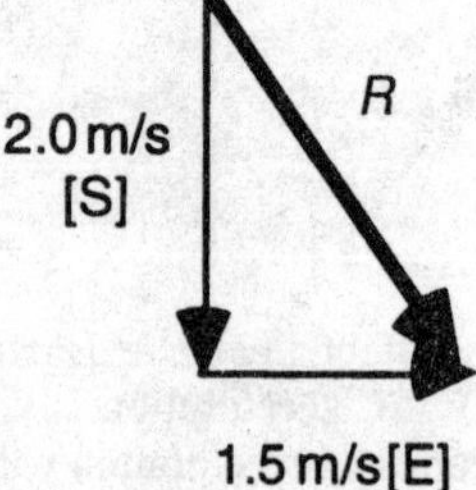

Since this is a right triangle, use the Pythagorean theorem:

$$a^2 + b^2 = c^2$$

to determine the magnitude of the resultant.

$R^2 = (2.0 \text{ m/s})^2 + (1.5 \text{ m/s})^2$

$R = 2.5$ m/s

Wrong Choices Explained:

(1) This choice ignores the effect of the river current (incorrect magnitude and direction).

(2) This choice uses the effect of the direction of the current, but the magnitude is ignored (correct direction, incorrect magnitude).

(3) This choice simply adds the magnitudes of the vectors and ignores their directions (incorrect magnitude and direction).

7. **1** For an object to be in *equilibrium,* the vector sum of all the forces must be zero. In other words, if the vectors are placed head to tail, the starting point of the first vector is the ending point of the last vector. Since there are three vectors in this problem, we could get either a straight line or a triangle.

From geometry: If we have two sides of a triangle, the length of the third side must be less than the sum of the lengths of the other two sides but greater than the magnitude of the difference between the lengths. Only the combination in choice (1) cannot produce a triangle or a straight line.

Wrong Choices Explained:

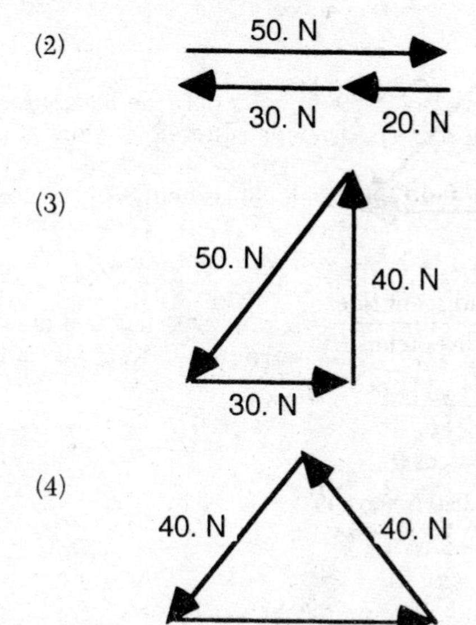

8. **2** Write out the problem.

Given: $m = 60.$ kg Find: $a = ?$

$w = 96$ N

Refer to the *Mechanics* section in the *Reference Tables* to find an equation(s) that relates acceleration with mass and weight.

Solution:
$$w = mg$$
$$g = \frac{w}{m}$$
$$= \frac{96 \text{ N}}{60. \text{ kg}}$$
$$= 1.6 \text{ m/s}^2$$

9. **3** Write out the problem.

Given: $F = 28.0$ N at 45° to horizontal Find: $F_{horizontal} = ?$

To solve this problem, resolve the vector F into its horizontal and vertical components.

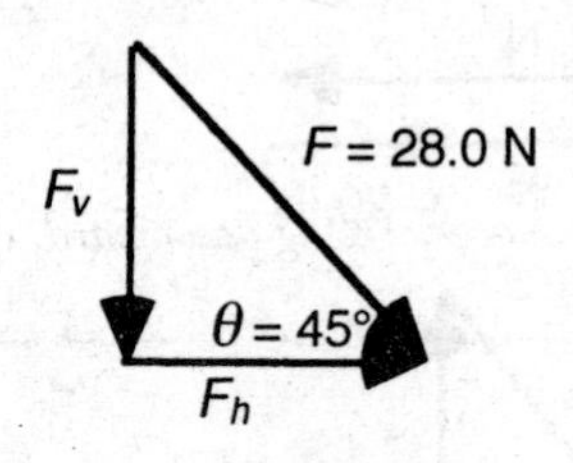

Solution:
$$\cos\theta = \frac{\text{adjacent side}}{\text{hypotenuse}}$$
$$= \frac{F_{horizontal}}{F}$$
$$F_{horizontal} = F \cdot \cos\theta$$
$$= (28.0 \text{ N})(\cos 45°)$$
$$= 19.8\ N$$

10. **2** Write out the problem.

Given: $m = 1.0 \times 10^2$ kg Find: $F_{friction} = ?$

$a = 2.0 \text{ m/s}^2$

Note that the box is accelerating at the same rate as the truck with respect to the ground.

Refer to the *Mechanics* section in the *Reference Tables* to find an equation(s) that relates force with mass and acceleration.

Solution:
$$F = ma$$
$$= (1.0 \times 10^2 \text{ kg})(2.0 \text{ m/s}^2)$$
$$= 2.0 \times 10^2 \frac{\text{kg} \cdot \text{m}}{\text{s}^2} = 2.0 \times 10^2 \text{ N}$$

This is the value of the force that produces the acceleration of the box. Since the box is not moving with respect to the truck, the force of friction is the force that is accelerating the box, and its value is equal to the accelerating force (2.0×10^2 N) calculated above.

11. **1** *Newton's third law* states that, when an object exerts a force on a second object, the second object exerts a force on the first object that is equal in magnitude but opposite in direction. For the scale to read a larger weight (force) pushing down, there must be a greater force pushing up. An elevator *accelerating upward* would have a net force acting *upward*.

Wrong Choices Explained:

(2) If the elevator were *accelerating downward*, the scale would read less than 500 N.

(3), (4) If the elevator were *moving upward or downward at constant speed*, the scale would read 500 N, no net force.

12. **4** The box in this problem is moving toward the right. The force of friction always acts to *oppose* the direction of motion; therefore, it acts toward the *left*.

13. **3** Write out the problem.

Given: $F = 10.$ N
$m = 6.0$ kg
$t = 8.0$ s

Find: $\Delta p = ?$

Refer to the *Mechanics* section in the *Reference Tables* to find an equation(s) that relates momentum with force, mass, and time.

Solution:
$$p = mv$$
$$\therefore \Delta p = m\Delta v$$
$$F\Delta t = m\Delta v$$
$$\therefore \Delta p = F\Delta t$$
$$= (10.\text{ N})(8.0\text{ s})$$
$$= 80.\text{ N}\cdot\text{s} = 80.\frac{\text{kg}\cdot\text{m}}{\text{s}^2}\cdot\text{s}$$
$$= 80.\frac{\text{kg}\cdot\text{m}}{\text{s}} \text{ or } 80.\text{ kg}\cdot\text{m/s}$$

(Note that mass is not needed in this calculation.)

14. **2** Acceleration in free fall is *constant* and *independent of mass*. Since both spheres start from the same height at the same time, their final *velocities* are the *same*. Since momentum (p) equals mass times velocity and the steel sphere has a greater mass than the wooden sphere, it will have a *greater momentum* than the wooden sphere.

15. **2** Write out the problem.

Given: $F = 2.0\text{ N}$ Find: $W = ?$
$m = 3.0\text{ kg}$
$\Delta s = 4.0\text{ m}$

Refer to the *Energy* section in the *Reference Tables* to find an equation(s) that relates work with force, mass, and displacement.

Solution:
$$W = F\Delta s$$
$$= (2.0\text{ N})(4.0\text{ m})$$
$$= 8.0\text{ N}\cdot\text{m} = 8.0\text{ J}$$

(Note that mass is not needed in this calculation.)

16. **4** Write out the problem.

Given: $P = 4.0 \times 10^3\text{ W}$ Find: $\Delta s = ?$
$F = 8.0 \times 10^2\text{ N}$
$\Delta t = 16\text{ s}$

Refer to the *Energy* section in the *Reference Tables* to find an equation(s) that relates displacement with power, force, and time.

Solution:
$$P = \frac{F\Delta s}{\Delta t}$$
$$\Delta s = \frac{P\Delta t}{F}$$
$$= \frac{(4.0 \times 10^3 \text{ W})(16 \text{ s})}{8.0 \times 10^2 \text{ N}}$$
$$= 80. \text{ m}$$

17. **4** Write out the problem.

Given: $x = 0.200$ m Find: $k = ?$

$PE_s = 10.0$ J

Refer to the *Energy* section in the *Reference Tables* to find an equation(s) that relates a spring constant with the potential energy stored in a spring, and displacement.

Solution:
$$PE_s = \frac{1}{2}kx^2$$
$$k = \frac{2PE_s}{x^2}$$
$$= \frac{2(10.0 \text{ J})}{(0.200 \text{ m})^2}$$
$$= 500. \frac{\text{N}}{\text{m}} \quad \text{or} \quad 500. \text{ N/m}$$

18. **3** Write out the problem.

Given: $m = 1.0 \times 10^3$ kg Find: $KE = ?$

$v = 4.0$ m/s

Refer to the *Energy* section in the *Reference Tables* to find an equation(s) that relates kinetic energy with mass, and velocity.

Solution:
$$KE = \frac{1}{2}mv^2$$
$$k = \frac{1}{2}(1.0 \times 10^3 \text{kg})(4.0 \text{ m/s})^2$$
$$= 8.0 \times 10^3 \frac{\text{kg} \cdot \text{m}^2}{\text{s}^2}$$
$$= 8.0 \times 10^3 \text{N} \cdot \text{m} = 8.0 \times 10^3 \text{ J}$$

19. **1** *Energy* is the ability to do work. If an object gains energy, then, by the law of conservation of energy, *work was done* on the object.

20. **3** By the *law of conservation of energy, potential energy* lost as an object falls is converted to *kinetic energy*. $PE = mgh$; therefore, an object that has twice the mass of another needs to fall only half the distance ($d/2$) to gain the same amount of energy.

21. **1** In the diagram, *A* and *B* attract, *A* and *C* attract, and *B* and *C* repel each other. *Like* charges *repel* each other, and *opposites attract* each other. Since two of the three spheres are positively charged and would repel each other, we conclude that *B* and *C* are positive. *A* would be negatively charged since it attracts both *B* and *C*.

22. **2** Since spheres *X*, *Y*, and *Z* are identical, charge will be transferred among them until the charge on all three spheres *is the same*:

$$\frac{(+2\times10^{-6}\text{C}) + (+4\times10^{-6}\text{C}) + (+6\times10^{-6}\text{C})}{3} = +4\times10^{-6}\text{C}$$

23. **1** Electric field lines are drawn in such a way that the direction of the arrow represents the direction that a small positive test charge would move when placed in the field. A *positive* test charge would be *repelled* by a positive point charge; therefore, diagram (1) is an accurate representation of the electric field around a positive point charge.

24. **4** Write out the problem.

Given: $W = 20.\text{ J}$ Find: $V = ?$

$q = 5.0\text{ C}$

Refer to the *Electricity and Magnetism* section in the *Reference Tables* to find an equation(s) that relates electric potential difference with work and charge.

Solution: $$V = \frac{W}{q}$$

$$= \frac{20.\text{ J}}{5.0\text{ C}}$$

$$= 4.0\frac{\text{J}}{\text{C}} = 4.0\text{ V}$$

25. **2** "Neutral" means that the sum of all charges equals zero. Since each proton holds a +1 elementary charge, each electron holds a −1 elementary charge, and a neutron has no charge, all that is necessary for *neutrality* is to have the same number of *protons* and *electrons*.

26. **3** Write out the problem.

Given: $R = 20.\ \Omega$ Find: $V = ?$

$q = 40.\ \text{C}$

$\Delta t = 5.0\ \text{s}$

Refer to the *Electricity and Magnetism* section in the *Reference Tables* to find an equation(s) that relates electric potential difference with resistance, charge, and time.

Solution:

$$I = \frac{\Delta q}{\Delta t}$$

$$R = \frac{V}{I}$$

$$\therefore V = IR$$

$$V = \frac{\Delta q}{\Delta t}R$$

$$= \frac{40.\ \text{C}}{5.0\ \text{s}}\ 20.\ \Omega$$

$$= 160\ \text{V}$$

27. **1** *Resistance* is directly proportional to the resistivity (ρ) of the wire for a given temperature and to the length of the wire, and inversely proportional to the cross-sectional area of the wire:

$$R = \rho\frac{L}{A}$$

A *thicker* wire *of the same length* would have a larger cross-sectional area and thus have *lower resistance.*

Wrong Choices Explained:

(2) A *longer wire* would have *increased resistance.*

(3) In conductors, resistivity increases with *increasing temperature,* resulting in *increased resistance.*

(4) For objects that obey Ohm's law, resistance is *independent* of the *applied voltage.*

28. **3** Write out the problem.

Given: $V = 10.\ \text{V}$ Find: $I_T = ?$

$R_1 = 40.\ \Omega$

$R_2 = 40.\ \Omega$

Refer to the *Electricity and Magnetism* section in the *Reference Tables* to find an equation(s) that relates current with resistance and electric potential difference.

Solution:

$$\frac{1}{R_T} = \frac{1}{R_1} + \frac{1}{R_2}$$
$$= \frac{1}{40.\ \Omega} + \frac{1}{40.\ \Omega}$$
$$= \frac{2}{40.\ \Omega} = \frac{1}{20.\ \Omega}$$
$$R_T = 20.\ \Omega$$

$$R_T = \frac{V_T}{I_T}$$
$$\therefore I_T = \frac{V_T}{R_T}$$
$$= \frac{10.\text{V}}{20.\Omega}$$
$$= 0.50\ \text{A}$$

29. **1** *Power,* the work done per unit time ($P = W/\Delta t$), is measured in watts. *Work* is measured in joules and *time* in seconds, so a watt is equivalent to a *joule/second.*

Wrong Choices Explained:

(2) *Joule/volt* $= \frac{\text{J}}{\text{V}} = \frac{\text{J}}{\text{J/C}} = \text{C}$. The coulomb is the SI unit for charge.

(3) *Joule/ohm* is not equivalent to any SI unit.

(4) *Joule/coulomb* $= \frac{\text{J}}{\text{C}} = \text{V}$. The volt is the SI unit for electric potential difference.

30. **4** Write out the problem.

Given: $I = 1.7\ \text{A}$
$V = 120\ \text{V}$
$t = 3600\ \text{s}$

Find: $W = ?$
Energy is the ability to do work, so W = energy.

Refer to the *Electricity and Magnetism* section in the *Reference Tables* to find an equation(s) that relates work, current, electric potential difference, and time.

Solution:

$$W = VIt$$
$$= (120\ \text{V})(117\ \text{A})(3600\ \text{s})$$
$$= \left(120\frac{\text{J}}{\not{\text{C}}}\right)\left(117\frac{\not{\text{C}}}{\not{\text{s}}}\right)(3600\ \not{\text{s}})$$
$$= 7.3 \times 10^5\ \text{J}$$

31. **3** To solve this problem, use the following *left-hand rule*: Wrap the fingers of the left hand in the direction of the electron flow, and the thumb will point in the direction of the north end of the solenoid.

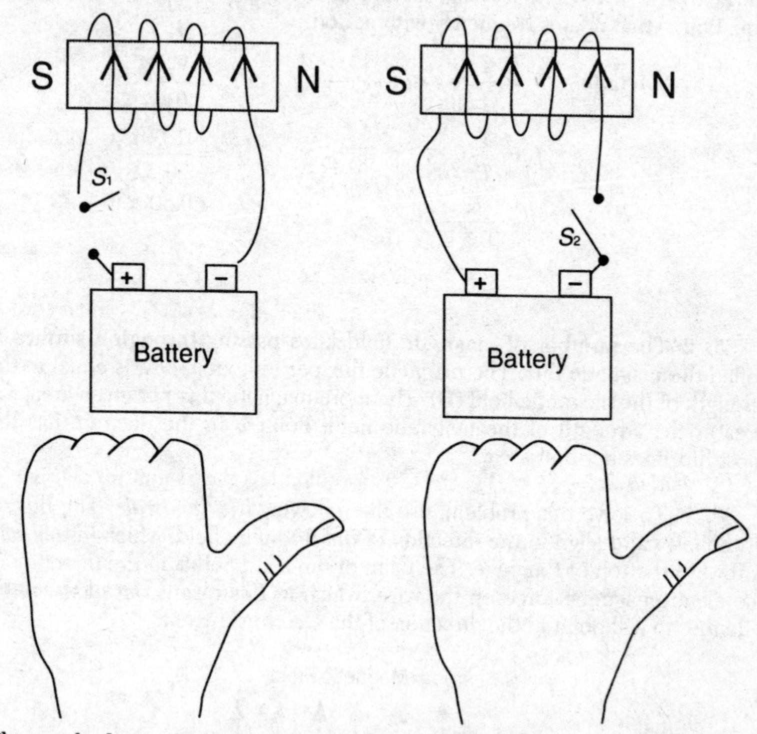

This method reveals that the north pole of the first solenoid is adjacent to the south pole of the second solenoid. *Opposite* poles *attract* and *like* poles *repel*; therefore, these two solenoids will *attract* because of their adjacent north and south poles.

32. **2** The direction of a magnetic field line is defined as the direction that the north pole of a compass needle points to when it is placed in the magnetic field. On the outside of a magnet, the field lines leave from the north pole and enter through the south pole. There are no isolated poles from which the field lines can start or stop, so field lines always form closed loops. In the diagram in choice (2) the arrows of the field lines point in the correct direction.

33. **4** In a nondispersive medium, the speed of a light wave depends on the *nature of the medium* and is independent of the wave frequency, wavelength, and amplitude.

34. **2** Write out the problem.

Given: $T = 0.5$ s Find: $f = ?$

Refer to the *Wave Phenomena* section in the *Reference Tables* to find an equation(s) that relates frequency with period.

Solution: $T = \frac{1}{f}$

$$f = \frac{1}{T}$$

$$= \frac{1}{0.5\ \text{s}}$$

$$= 2\ \text{s}^{-1} = 2\ \text{Hz}$$

35. **2** The number of magnetic field lines passing through a surface is called the magnetic flux. The magnetic flux per unit area, ø/A, is equal to the strength of the magnetic field (**B**). The more magnetic flux per given area, the greater the strength of the magnetic field. Point *B* in the diagram has the most flux lines per unit area.

36. **1** To solve this problem, use the following *left-hand rule*: The fingers of the left hand align in the direction of the magnetic field, which in this case is toward the top of the page. The palm of the hand points in the direction of the electromagnetic force on the wire, which in this case is out of the page. The thumb will point in the direction of the electron current.

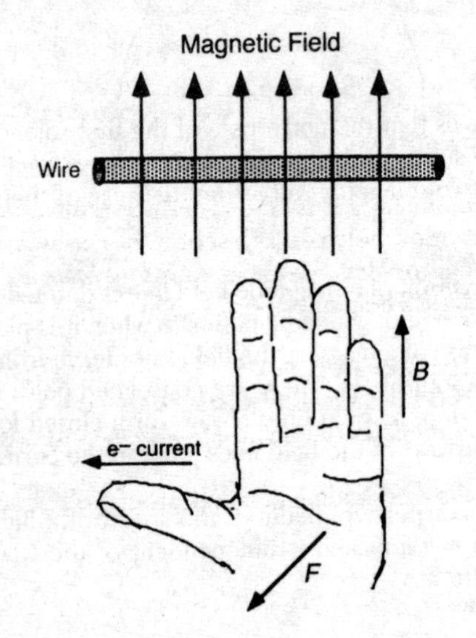

As can be seen from the diagram, the thumb in this instance points to the *left*.

37. **3** In transverse waves, the *amplitude*, or displacement from the equilibrium position, is *perpendicular* (i.e., at a 90° angle) to the direction of propagation.

38. **2** Two points are *in phase* when they are at identical displacements from equilibrium and in the next instant of time will move in the same direction. Points such as *B* and *F* that are a *whole number* of wavelengths apart are in phase. Points that are an odd number of half wavelengths apart are 180° out of phase.

Wrong Choices Explained:

(1) *A* and *G* are both at the equilibrium point, but *A* will go up and *G* will go down. They are $1\frac{1}{2}$ wavelengths apart.

(3) *C* and *E* are both at the equilibrium point, but *C* will go down and *E* will go up. They are $\frac{1}{2}$ wavelength apart.

(4) *D* and *F* are equidistant from equilibrium, but *D* will go down and *F* will go up. They are $\frac{1}{2}$ wavelength apart.

39. **4** A wavelength distance is equal to the distance between two consecutive points that are in phase. Points *A* and *B* are 180° out of phase, so the distance between them, 0.10 m, is equal to one-half the wavelength, λ. Therefore, $\lambda = 0.20$ m. Since we cannot assume that the diagram has been drawn to scale, we can make no measurements of the magnitude of the amplitude.

40. **3** The *law of reflection* states that the angle that an incident ray makes with a normal to the surface is equal to the angle that the reflected ray makes with a normal to the surface. Only in diagram (3) is this law represented.

41. **1** The *Doppler effect* is the variation in observed frequency when there is relative motion between the source of the waves and the observer. The moving baseball reflects the radar waves and therefore acts as a moving source of radar waves. The gun is the observer.

Wrong Choices Explained:

(2) *Standing waves* are waves with stationary nodes.

(3) The *critical angle* is the angle of incidence that produces a refracted angle of 90°.

(4) *Diffraction* is the bending of waves around objects in their path.

42. **2** When waves pulses interfere, their amplitudes are combined. Since the amplitudes of the two pulses are in the same direction, there is constructive interference and the vertical displacement of the rope at P will be the sum of the individual amplitudes, $A + A = 2A$.

43. **4** Colors of light and their wavelengths are listed in the *Wavelengths of Light in a Vacuum* table in the *Reference Tables*. In this problem, however, we are given the frequency of light. We need to refer to the *Wave Phenomena* section of the *Reference Tables* to find an equation(s) that relates frequency to wavelength.

Solution: $v = f\lambda$

$\lambda = \frac{v}{f}$ In this problem, v is the speed of light in a vacuum, which is a constant and can be found in the *List of Physical Constants* in the *Reference Tables*.

$$= \frac{3.0 \times 10^8 \text{ m/s}}{6.3 \times 10^{14} \text{ Hz}}$$

$$= \frac{3.0 \times 10^8 \text{ m/}\not{s}}{6.3 \times 10^{14} \text{ /}\not{s}}$$

$$= 4.8 \times 10^{-7} \text{m}$$

The wavelength calculated falls within the wavelength range for *blue* light.

44. **1** The *natural vibrational frequency* of the glass is known as its resonant frequency. When the glass is set vibrating at this frequency by a *note of the same frequency*, a standing wave is set up that builds in amplitude until the energy level is enough to break the glass.

45. **3** *Refraction* is the bending of a wave front, which occurs as a result of a change in speed of the wave when it enters a new medium at an angle. The *index of refraction* of a medium is the ratio of the speed of light in a vacuum to the speed of light in the medium: $n = c/v$. If two *media have different indices of refraction*, the speed of light is different in the two media and refraction can occur.

Wrong Choices Explained:

(1) If the *incident angle is 0°*, there is no refraction because the wave has not entered the medium obliquely.

(2) If *there is no change in the speed of the wave*, the second medium is not considered a new medium and no refraction occurs.

(4) Refraction occurs only if the wave enters a new medium. If all *of the light is reflected*, none is transmitted to a new medium and no refraction occurs.

46. **4** Write out the problem.

Given: $f = 5.0 \times 10^{14}$ Hz Find: $E = ?$

Refer to the *Modern Physics* section in the *Reference Tables* to find an equation(s) that relates the energy of a photon with frequency.

Solution: $E_{photon} = hf$

Now refer to the *List of Physical Constants* in the *Reference Tables* to find the value of (h), Planck's constant.

$$E_{photon} = (6.6 \times 10^{-34}\ \text{J} \cdot \text{s})(5.0 \times 10^{14}\ \text{Hz})$$
$$= (6.6 \times 10^{-34}\ \text{J} \cdot \not{\text{s}})(5.0 \times 10^{14}\ /\not{\text{s}})$$
$$= 3.3 \times 10^{-19}\ \text{J}$$

47. **4** The degree of bending of light as it enters a new medium is related to the refractive index of a material via *Snell's law*:

$$n_1 \sin \theta_1 = n_2 \sin \theta_2$$

Since the angle of incidence and the refractive index of the incident material are the same for all colors, the only way that the colors could refract at different angles is if they have different *refractive indices* in the material.

48. **2** The kinetic energy of a photoelectron is related to the frequency of the incident photons via the equation

$$KE_{max} = hf - W_0$$

where $W_0 = hf_0$, and f_0, is the threshold frequency. When the work function has been met by having an incident photon with a frequency greater than the threshold frequency, any extra energy from the photon gives kinetic energy to the photoelectron. Only graph (2) shows the relationship of *increasing frequency* yielding *increased kinetic energy.*

Wrong Choices Explained:

(1) This graph shows that *increasing frequency* has *no effect* on kinetic energy.

(3) This graph shows that *increasing frequency decreases* kinetic energy.

(4) This graph shows an *inverse relationship* between frequency and kinetic energy.

49. **2** Alpha particles are helium nuclei and have a +2 elementary electric charge. *Some of these particles are deflected* away from a small, dense, positive nucleus such as that in gold and are deflected toward the orbiting electrons, which are negative. This is the conclusion that Rutherford drew from his observation.

Wrong Choices Explained:

(1) Like charges repel, so alpha particles could *not* be deflected toward a positive electrode. Also, Rutherford did not use electrodes in this experiment.

(3) Only a *few* alpha particles were scattered at angles greater than 90°, leading Rutherford to conclude that atoms consist of small, dense, positive nuclei.

(4) Most of the particles passed straight through, *unaffected by the gold foil,* which led Rutherford to conclude that the atom was mostly empty space. However, some of the particles were deflected by the gold foil.

50. **1** To answer this question, look at the *Energy Level Diagrams for Mercury and Hydrogen* in the *Reference Tables*. You also need to refer to the *Modern Physics* section of the *Reference Tables* to find the equation for the energy of a photon.

Using the values for the electron transition in choice (1) gives

$$\begin{aligned} E_{photon} &= E_i - E_f \\ &= -3.4 \text{ eV} - (-13.60 \text{ eV}) \\ &= 10.2 \text{ eV} \end{aligned}$$

Wrong Choices Explained:

(2) The energy of the photon equals 1.89 eV.
(3) The energy of the photon equals 2.55 eV.
(4) The energy of the photon equals 0.97 eV.

51. **3** The quantum model of the atom predicts only the probability that an electron is at a specific location. A three-dimensional plot of the points of equal probability of finding an electron at any given radius from the nucleus can be constructed. The denser the points, the higher is the probability of finding the electron. This *region of high probability* is known as the *electron cloud.*

52. **1** From principles of trigonometry :

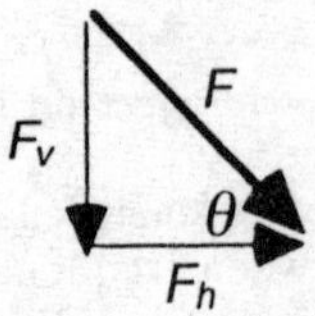

$$\sin\theta = \frac{\text{opposite}}{\text{hypotenuse}} = \frac{F_v}{F}$$

$$F_v = F \sin\theta$$

As the angle between the applied force and the ground decreases, the sine of the angle decreases; therefore, the vertical component of the force *decreases.*

53. **3** For this problem, refer to the *Mechanics* section of the *Reference Tables* to find an equation for impulse:

$$J = F\Delta t$$

The time that the bat is in contact with the ball is the same amount of time that the ball is in contact with the bat. By *Newton's third law* we know that the force exerted by the bat on the ball is the same as the force exerted by the ball on the bat. Therefore, the impulses are *equal* in magnitude.

54. **2** *Newton's law of universal gravitation* states that the force of attraction between two masses is directly proportional to the product of the masses and inversely proportional to the square of the distance between them. Refer to the *Mechanics* section of the *Reference Tables* to find an equation for force:

$$F = \frac{Gm_1m_2}{r^2}$$

If the mass of a body increases, its gravitational force of attraction on the Earth also *increases.*

55. **2** The relationship between the distance between bright bands formed by a double slit and the distance between the two slits is represented by the following equation, which can be found in the *Wave Phenomena* section of the *Reference Tables*:

$$\frac{\lambda}{d} = \frac{x}{L}$$

$$\lambda = \frac{xd}{L}$$

For a given wavelength, if d, the distance between the slits, decreases, x, the distance between the bright bands, will *increase.*

PART II

GROUP 1— Motion in a Plane

56. **3** Write out the problem.

Given: $m_A = 10.\text{ kg}$ $m_B = 20.\text{ kg}$
$v_{i_A} = 30.\text{ m/s E}$ $v_{i_B} = 10.\text{ m/s W}$
$h_A = 20.\text{ m}$ $h_B = 80.\text{ m}$
$\Delta t = 1.5\text{ s}$
Initial horizontal distance between
A and *B* = 100. m.

Find: Horizontal distance between *A* and *B* after 1.5 s = ?

Refer to the *Mechanics* section in the *Reference Tables* to find an equation(s) that relates displacement with velocity, time, mass, and height.

Solution:

$$v_{x_A} = \frac{\Delta s_{x_A}}{\Delta t} \qquad v_{x_B} = \frac{\Delta s_{x_B}}{\Delta t}$$

$$\Delta s_{x_A} = v_{x_A}\Delta t \qquad \Delta s_{x_B} = v_{x_B}\Delta t$$

$$= (30.\text{ m/s E})(1.5\text{ s}) \qquad = (10.\text{ m/s W})(1.5\text{ s})$$

$$= 45.\text{ m E} \qquad = 15.\text{ m W}$$

At the end of 1.5 s, the spheres have traveled toward each other a distance of 60. m (45. m + 15. m). The horizontal separation is 40. m (100. m – 60. m). (Note that mass and height are not required for this calculation.)

57. **1** No force is acting in the horizontal direction; therefore, there is *no horizontal acceleration* of sphere A.

58. **1** The *vertical acceleration* is a constant due to gravity (9.8 m/s^2) and is independent of mass. Therefore, the vertical acceleration is *the same* for the two spheres.

59. **4** The *direction of the velocity* of an object, such as a cart, moving in uniform circular motion is tangent to the circle at any point, and points in the direction the object would take if the centripetal force were to be suddenly removed. At the position shown, the velocity of the cart is directed toward point *S*.

60. **2** *Centripetal acceleration*, like linear acceleration, occurs as a result of an unbalanced force and has a direction equal to the direction of the unbalanced force. Since the force is directed toward the center of the circle (centripetal means "center seeking"), the acceleration is also directed toward the center of the circle (in this case, point Q).

61. **1** *Centripetal acceleration*, like linear acceleration, is independent of mass. Refer to the *Motion in a Plane* section of the *Reference Tables* to find an equation for centripetal acceleration:

$$a_c = \frac{v^2}{r}$$

A change in mass has *no effect* on centripetal acceleration.

62. **3** Write out the problem.

Given: $m = 5.0 \text{ kg}$ Find: $F_c = ?$
$r = 2.0 \text{ m}$
$v = 4.0 \text{ m/s}$

Refer to the *Motion in a Plane* section in the *Reference Tables* to find an equation(s) that relates centripetal force with mass, radius, and velocity.

Solution:
$$F_c = \frac{mv^2}{r}$$
$$= \frac{(5.0 \text{ kg})(4.0 \text{ m/s})^2}{2.0 \text{ m}}$$
$$= 40.\frac{\text{kg} \cdot \text{m}}{\text{s}^2} = 40. \text{ N}$$

63. **4** An artificial satellite orbiting the Earth is similar to an object moving in uniform circular motion and can be described by the equation

$$F_c = \frac{mv^2}{r}$$

given in the *Motion in a Plane* section of the *Reference Tables*. The centripetal force required for the uniform circular motion is provided by the force of gravitational attraction exerted on the satellite by the Earth; therefore, the centripetal acceleration is equal to g, the acceleration due to gravity. In other words,

$$F_c = mg \quad \text{or} \quad g = \frac{v^2}{r}$$

Since g is a constant, if friction (a force that acts to oppose the direction of motion), were to continually slow down the satellite, a smaller and smaller radius of curvature would result, and the satellite would *spiral down toward the Earth.*

64. **2** Write out the problem.

Given: v_i = 500. m/s at 30° to horizontal Find: v_{i_y} = ?

Refer to the *Motion in a Plane* section in the *Reference Tables* to find an equation(s) that relates velocity and its vector components.

Solution:
$$v_{i_y} = v_i \sin\theta$$
$$= (500.\ \text{m/s})(\sin 30°)$$
$$= 250.\ \text{m/s}$$

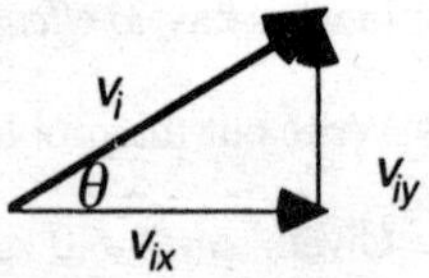

V_i θ V_{iy} V_{ix}

65. **2** *Kepler's third law of planetary motion* states that the ratio of the squares of the periods of any two planets or satellites orbiting a single body is equal to the ratio of the cubes of their average distance from that body. Thus, if T_A and T_B are the respective periods of satellites A and B, and r_A and r_B are their respective average distances from the Earth,

$$\left(\frac{T_A}{T_B}\right)^2 = \left(\frac{r_A}{r_B}\right)^3$$

Since the average distance from the Earth of satellite B is greater than the average distance of satellite A, the orbital period of B must be *longer* than the orbital period of A.

GROUP 2—**Internal Energy**

66. **1** The Kelvin temperature is an absolute scale that is directly proportional to the average kinetic energy of a substance. Absolute zero represents the *minimum kinetic energy of the particles in the molecule.*

Wrong Choices Explained:

(2) *Gravitational potential energy* is the change in energy of an object when moved in a gravitational field.

(3) *Specific heat* is the amount of heat required to raise the temperature of 1 gram of a substance by 1 Celsius degree.

(4) *Heat of fusion* is the amount of heat required to change the phase of 1 gram of a substance from solid to liquid at the melting point of the substance.

67. **3** To answer this question, refer to the table of *Heat Constants* in the *Reference Tables.* Look up the melting points and boiling points of the choices to determine what phases they are in at the temperatures given.

Mercury melts at –39°C and thus is a solid at –51°C; it boils at 357°C and thus is a liquid at –33°C.

Wrong Choices Explained:

(1) *Ammonia* melts at –78°C and thus is a liquid at –51°C; it boils at –33°C and thus is in a phase change from liquid to gas at –33°C.

(2) *Ethyl alcohol* melts at –117°C and thus is a liquid at –51°C; it boils at 79°C and thus is still a liquid at –33°C.

(4) *Lead* melts at 328°C and thus is a solid at –51°C; it boils at 1740°C and thus is still a solid at –33°C.

68. **1** *Specific heat* is the amount of heat required to raise the temperature of 1 gram of a substance by 1 Celsius degree. On the graph, substance *A* has the smallest slope, indicating that a relatively large amount of heat must be added to raise the temperature a relatively small amount. In other words, substance *A* has a relatively large specific heat.

69. **1** Write out the problem.

Given: $\Delta T = 10.\ C°$ Find: What could this substance be?

$Q = 2.4\ kJ$

$m = 1.0\ kg$

Refer to the *Internal Energy* section in the *Reference Tables* to find an equation(s) that relates temperature change, heat, and mass.

Solution:

$$Q = mc\Delta T_c$$

$$c = \frac{Q}{m\Delta T_c}$$

$$= \frac{2.4\ \text{kJ}}{(1.0\ \text{kg})(10.\ \text{C°})}$$

$$= 0.24 \frac{\text{kJ}}{\text{kg} \cdot \text{C°}}$$

Refer to the *Heat Constants* table in the *Reference Tables* to find a possible match. The value calculated for the specific heat, 0.24 kJ/kg · C°, matches the value for *silver*.

70 **4** Write out the problem.

Given: m = 0.50 kg   Find: Q needed to melt

The value for the heat of fusion of iron can be found in the *Heat Constants* table in the *Reference Tables.* The value obtained, H_f = 267 kJ/kg, is the amount of heat in kilojoules required to change the phase of 1 kilogram of iron from a solid to a liquid at its melting point.

Now refer to the *Internal Energy* section to locate an equation that relates amount of heat with mass and heat of fusion.

Solution: $Q = H_f m$

$$= \left(267 \frac{\text{kJ}}{\cancel{\text{kg}}}\right)(0.50 \cancel{\text{kg}})$$

$$= 133.5 \text{ kJ} = 130 \text{ kJ}$$

(Note that the value is reported to two significant figures because the worst measurement, that of mass, has two significant figures.)

71 **2** A liquid boils when its vapor pressure equals atmospheric pressure. The vapor pressure of water increases with increasing temperature; therefore, if the external pressure is raised, water will *boil at a higher temperature.* Ice is less dense than liquid water; therefore, an increase in pressure will favor the liquid phase and the water will *freeze at a lower temperature.*

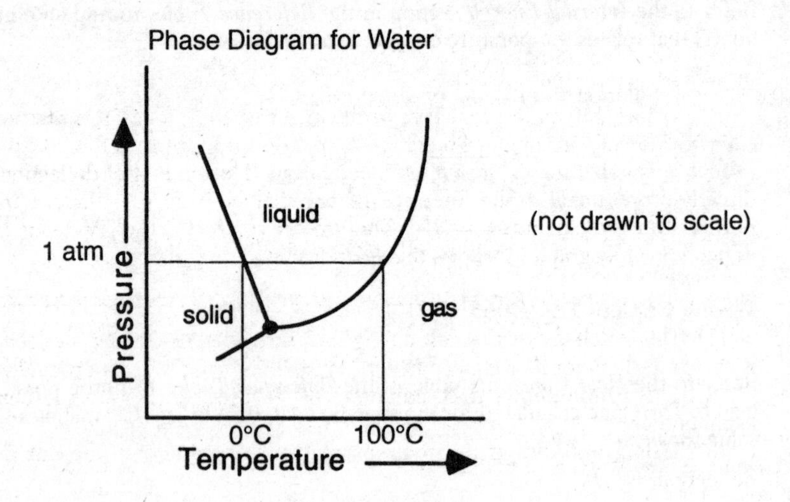

72. **4** According to the *kinetic molecular theory*, ideal gas particles (molecules or atoms) have negligible volumes and are *far away from one another.* Between the particles is empty space.

The collisions between the particles and the walls of the container, and the collisions between the particles, are assumed to be perfectly elastic collisions (there is no loss of kinetic energy). It is assumed that there are no forces of attraction or repulsion between the particles.

73. **3** According to *Boyle's law*, the pressure of a sample of gas at constant temperature is inversely proportional to the volume of the gas. In other words,

$$PV = k \text{ (a constant)}$$

Therefore, if the pressure is *doubled*, the volume is *halved*.

74. **2** The *second law of thermodynamics* states that natural processes proceed in a direction that *increases the total entropy of the universe.* Entropy is best described as a measure of disorder in a system.

Entropy, like thermal energy, is contained in an object. If heat is added to an object, entropy is increased. If heat is removed from an object, entropy is decreased.

75. **3** During a phase change such as boiling, heat energy is being added, yet the temperature remains constant; therefore, the average kinetic energy of the molecules is not increasing. The heat energy raises the potential energy of the particles in order to overcome the forces of attraction between them. The temperature of the liquid, however, *remains the same.*

GROUP 3—**Electromagnetic Applications**

76. **4** The galvanometer is an application of the fact that a current-carrying loop of wire in a magnetic field experiences a torque (a force applied in such a way that it produces rotational motion). The amount of deflection is directly proportional to the force on the wire, which in turn, is directly proportional to the current: $F = BI\ell$. The greater the force, the greater is the deflection. Only graph (4) shows this relationship.

Wrong Choices Explained:

(1) This graph shows that, for a range of currents up to a point, deflection is constant; then, for increasing currents, deflection increases proportionally.

(2) This graph shows that, as current increases, deflection increases exponentially.

(3) This graph shows an inverse relationship between current and deflection.

77. **3** *Increasing the potential difference applied to the armature* will increase the current that flows through it proportionally, according to Ohm's law ($V = IR$). The torque is directly proportional to the magnetic field strength, current in the loop, and length of the wire in the magnetic field: $F = BI\ell$; therefore, increasing the current will *increase the torque.*

Wrong Choices Explained:

(1) *Decreasing the current* would decrease the torque.

(2) *Decreasing the magnetic field* would decrease the torque.

(4) *Increasing the distance between the armature and the magnetic field poles* would decrease the strength of the magnetic field around the armature and thus decrease the torque.

78. **4** Iron is highly permeable to magnetic flux, and the *iron core* acts to strengthen and concentrate the magnetic field produced in the coil.

Wrong Choices Explained:

(1) The *split-ring commutator* provides the means by which the current in the wire loop is reversed every half-turn (180°).

(2) The *back emf* is the induced potential difference in the armature coil of the motor.

79. **3** Since electrons are negative, they will be attracted toward the positive plate, that is, *toward the top of the page.*

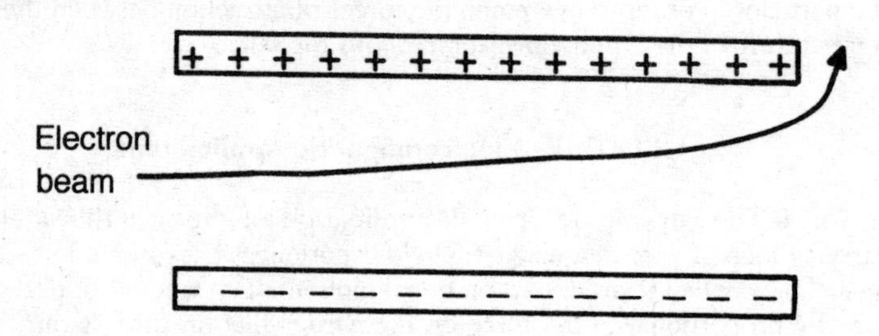

80. **2** Write out the problem.

Given: $v = 1.5 \times 10^6$ m/s Find: F = ?

$B = 3.0$ T

Refer to the *Electromagnetic Applications* section in the *Reference Tables* to find an equation(s) that relates force with velocity and magnetic field strength.

Solution: $F = qvB$, but a value for variable q, which represents the charge of a particle measured in coulombs, is needed. This problem deals with a proton, which carries a +1 elementary electric charge. Refer to the *List of Physical Constants* in the *Reference Tables* to find the value for q.

$$\begin{aligned} F &= qvB \\ &= (1.6 \times 10^{-19}\ \text{C})(1.5 \times 10^{6}\ \text{m/s})(3.0\ \text{T}) \\ &= 7.2 \times 10^{-13}\ \text{N} \end{aligned}$$

81. **1** To answer this question, use the following *left-hand rule*: The fingers of the left hand should point in the direction of the magnetic field, which in this case is directed out of the plane of the paper. The palm of the hand should be pointing in the direction of the force that the electron will experience. Since the electrons are to be deflected toward the top of the page, the palm should point toward the top of the page. Now the thumb should be pointing in the direction of the velocity of the electron. In this instance the thumb points toward the *right side* of the page, matching the path of electron A.

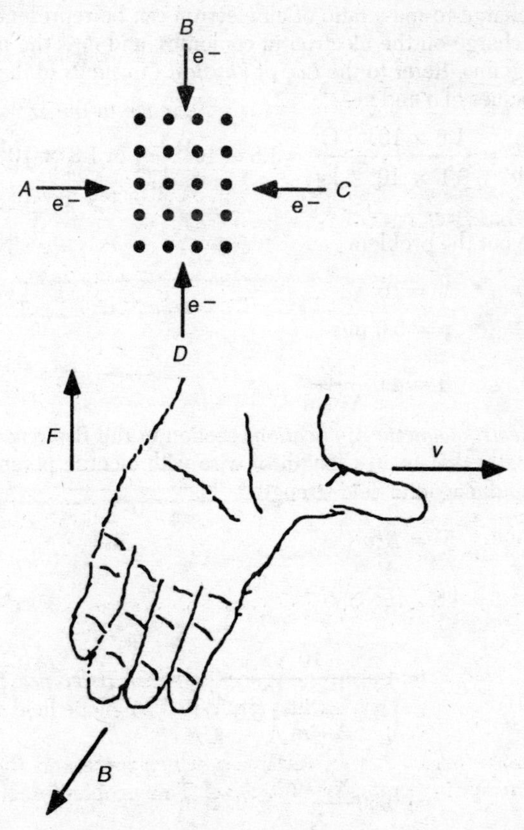

Wrong Choices Explained:

(2) *B* would be deflected toward the *right side* of the page.
(3) *C* would be deflected toward the *bottom* of the page.
(4) *D* would be deflected toward the *left side* of the page.

82. **1** The deflecting force, *F*, is directly proportional to the charge on the particle, the speed of the particle, and the magnetic field strength. Refer to the *Electromagnetic Applications* section of the *Reference Tables* for the equation:

$$F = qvB$$

If the speed of the particle is doubled, the force will be doubled. If the magnetic field strength is halved, the force will be halved also. If both these changes take place simultaneously, the effects will cancel each other and the magnitude of the deflecting force will remain *unchanged.*

83. **4** The charge-to-mass ratio of an electron can be represented as q/m, where q is the charge on the electron in coulombs, and m is the mass of the electron in kilograms. Refer to the *List of Physical Constants* in the *Reference Tables* for the values of q and m.

Solution: $$\frac{q}{m} = \frac{1.6 \times 10^{-19}\ \text{C}}{9.1 \times 10^{-31}\ \text{kg}} = 1.8 \times 10^{11}\ \frac{\text{C}}{\text{kg}},\ \text{or}\ 1.8 \times 10^{11}\ \text{C/kg}$$

84. **1** Write out the problem.

Given: $V = 10.\ \text{V}$ Find: $\ell = ?$

$v = 5.0\ \text{m/s}$

$B = 4.0 \frac{\text{N}}{\text{A} \cdot \text{m}}$

Refer to the *Electromagnetic Applications* section in the *Reference Tables* to find an equation(s) that relates length of wire with electric potential difference, velocity, and magnetic field strength.

Solution: $$V = B\ell v$$

$$\ell = \frac{V}{Bv}$$

$$= \frac{10.\ \text{V}}{\left(4.0 \frac{\text{N}}{\text{A} \cdot \cancel{\text{m}}}\right)\left(5.0 \frac{\cancel{\text{m}}}{\text{s}}\right)} = \frac{10. \frac{\text{J}}{\cancel{\text{C}}}}{20. \frac{\text{N}}{\frac{\cancel{\text{C}}}{\cancel{\text{s}}}} \cdot \cancel{\text{s}}}$$

$$= 0.50 \frac{\cancel{\text{N}} \cdot \text{m}}{\cancel{\text{N}}} = 0.50\ \text{m}$$

Wrong Choices Explained:

Choices (2), (3), (4) These results are obtained if the values are multiplied and/or divided incorrectly. These mistakes can be avoided if you keep track of the units during the calculation. Only if the values are correctly multiplied and/or divided will the units cancel properly, leaving the correct units for the variable for which you are solving.

85. **2** The ratio of the number of turns in the primary coil to the number of turns in the secondary coil is equal to the ratio of the voltage in the primary coil to the voltage in the secondary coil of a transformer. Refer to equation

$$\frac{N_p}{N_s} = \frac{V_p}{V_s}$$

in the *Electromagnetic Applications* section in the *Reference Tables*. The number of turns from primary to secondary coils is decreasing; therefore, the *voltage will decrease* from the primary to the secondary coil.

In an ideal transformer, power input into the primary coil equals power output in the secondary coil. Refer to equation

$$V_p I_p = V_s I_s$$

also in the *Electromagnetic Applications* section in the *Reference Tables*. Voltage and current are inversely proportional to each other; therefore, if voltage decreases, *current will increase*.

GROUP 4—**Geometric Optics**

86. **1** Plane mirrors always produce *virtual*, *erect*, and *reversed* images. To write the image, keep the letters right side up, but write them from right to left instead of left to right. In other words, reverse the order and obtain **WOWO**.

Wrong Choices Explained:

(2) This choice shows an image *identical* to the object.
(3) This choice shows an *upside-down* image.
(4) This choice shows an *inverted* (reversed and upside-down) image.

87. **4** Write out the problem.

Given: C (radius of curvature) = 0.60 m Find: d_i = ?

$C = 2f$ $\therefore f = 0.30$ m (Note that f has a positive value because this mirror is concave and has a real focus.)

$d_o = 0.40$ m

Refer to the *Geometric Optics* section in the *Reference Tables* to find an equation(s) that relates image distance with focal length and object distance.

Solution: $$\frac{1}{f} = \frac{1}{d_o} + \frac{1}{d_i}$$

$$\frac{1}{d_i} = \frac{1}{f} - \frac{1}{d_o}$$

$$= \frac{1}{0.30 \text{ m}} - \frac{1}{0.40 \text{ m}} = \frac{4}{1.2 \text{ m}} - \frac{3}{1.2 \text{ m}}$$

$$= \frac{1}{1.2 \text{ m}}$$

$$d_i = 1.2 \text{ m}$$

88. **2** *Spherical aberration* is the inability of spherical lenses and mirrors to bring all parallel rays to the same focus. The diagram illustrates this phenomenon by showing two different foci for a set of incoming parallel rays.

Wrong Choices Explained:

(1) *Chromatic aberration* occurs because different colors of light have different absolute indices of refraction in a dispersive medium such as glass. It is a property of lenses, not of mirrors.

(3) *Refraction* is the bending of a wave front as it enters a new medium obliquely, due to a change in the speed of the wave.

(4) *Dispersion* is the process of spreading light into its component colors.

89. **2** A lens that is thicker in its midsection *converges* parallel rays of light to a small point.

Wrong Choices Explained:

(1) This lens will *bend* parallel rays of light equally. The rays leaving this lens will still be parallel to each other but will be shifted from their original positions.

(3) A lens that is thinner in its midsection will *diverge* parallel rays of light.

(4) This piece of glass acts as a prism and will *disperse* light into its component colors.

90. **1** *Convex* mirrors are diverging mirrors. All convex mirrors have a virtual focus and always form virtual images smaller than the object.

Wrong Choices Explained:

(2) This image is the same size as the object and could be formed by a *plane* mirror, but not a convex mirror.

(3), (4) These images could be formed by a *concave* mirror if the object were located between one focal length of the mirror and the mirror.

91. **3** An object located any distance beyond one focal length from a converging lens will form a *real* image. If an object is located between the lens and one focal length, a *virtual* image will be formed.

92. **3** Write out the problem.

Given: $d_i = x$ Find: $S_i = ?$

$d_o = 2x$

$S_o = 0.050$ m

Refer to the *Geometric Optics* section in the *Reference Tables* to find an equation(s) that relates image size with image distance, object distance, and object size.

Solution:
$$\frac{S_o}{S_i} = \frac{d_o}{d_i}$$
$$S_i = \frac{S_o d_i}{d_o}$$
$$= \frac{(0.050 \text{ m})(\cancel{x})}{2\cancel{x}}$$
$$= 0.025 \text{ m}$$

93. **1** A diverging lens always produces a *virtual*, erect image that is *smaller* than the object.

94. **4** The absolute index of refraction of a lens depends on the color of the light; therefore, different colors will focus at different distances from the lens. This phenomenon of lenses is known as chromatic aberration. The use of filters to restrict light to a single wavelength will result in a single focus, eliminating *chromatic abberation.*

Wrong Choices Explained:

(1) *Diffusion* of light occurs as a result of the reflection of parallel light rays off a rough or uneven surface.

(2) *Diffraction* of light is the bending of light rays around an object or barrier in their path.

(3) *Polarization* of light restricts the plane of the electric and magnetic fields of the light waves to a single direction.

95. **2** The geometry of plane mirrors shows us that, for all object distances, the distance of the object from the mirror is always equal to the distance that the image appears behind the mirror. In this case, as the distance of the object from the mirror increases from 0.2 to 0.3 m, the image distance also *increases* from 0.2 to 0.3 m.

GROUP 5—**Solid State**

96. **4** In a semiconductor, resistance decreases with *increasing temperature* because more electrons are promoted to the conduction band from the valence band. A decreased resistance will result in an increased current.

Wrong Choices Explained:

(1) *Increasing the resistance* will decrease the current.

(2) *Reversing the battery polarity* will simply change the direction of current flow; it will have no effect on the magnitude of the current.

(3) *Reversing the connections* to the semiconductor will have no effect on either the direction or the magnitude of the current.

97. **2** An *insulator* has the largest energy gap between the valence band and the conduction band. The gap is too large for any significant number of electrons to bridge and thus be free to conduct in the conducting band. As a result, insulators are poor conductors of electricity.

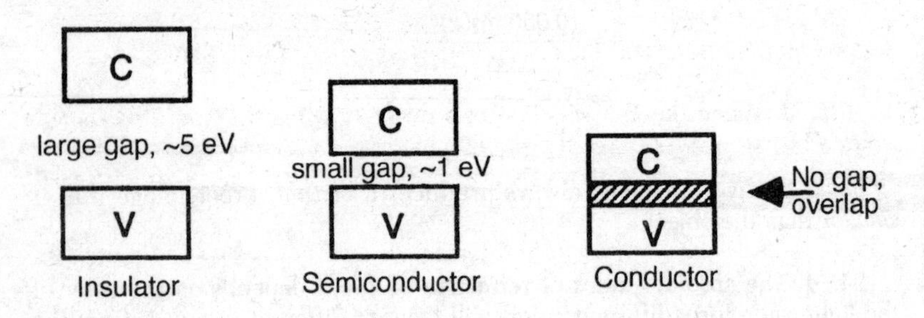

Wrong Choices Explained:

(1) A *conductor* (see the diagram above) has overlapping bands.

(3), (4) A *semiconductor* (see the diagram above) has a small gap that can be bridged by increasing the temperature. The difference between an *extrinsic* and an *intrinsic* semiconductor is that extrinsic semiconductors are doped in order to increase conduction whereas intrinsic semiconductors are not doped.

98. **4** In a circuit that contains a single type of semiconducting material, either *N*-type or *P*-type, reversing the battery polarity *will simply reverse* the direction of current in the circuit. It will have no effect on the magnitude of the current.

99. **1** Electrons move toward the positive terminal (because opposite charges attract), by moving into a hole ahead and leaving a hole behind them. Holes can be treated as positive charges, which flow toward the *negative terminal.*

100. **2** A *diode* contains an *N*-type semiconductor joined to a *P*-type semiconductor. If the diode is connected to an outside source of potential difference in such a way that the *N*-side is connected to the negative terminal of the potential source and the *P*-side is connected to the positive terminal, the diode is said to be forward biased and current will flow because the source of potential acts in a direction that reduces the potential barrier at the *P-N* junction.

If, however, the diode is connected so that the *N*-side is connected to the positive terminal and the *P*-side is connected to the negative terminal, the diode is said to be reversed biased and almost no current will flow because the source of potential acts to reinforce and enlarge the potential barrier at the *P-N* junction.

Wrong Choices Explained:

(1) A *transistor* contains three pieces of semiconducting material adjoining each other. Therefore, although there is forward and reverse biasing, a transistor is not the simplest circuit element that allows current to flow in only one direction.

(3), (4) A single piece of *N*-type or *P*-type *semiconducting material* has no potential barriers, and the direction of the current is irrelevant.

101. **3** When the *P-N junction* of a diode is *forward biased*, the outside source of potential opposes the potential barrier within the diode. When the applied potential is large enough to overcome the potential barrier, current flows and increased potential will produce increased current flow.

Wrong Choices Explained:

(1), (2) A single piece of *semiconducting material* cannot be biased because no potential barrier can be created to hinder current flow.

(4) A *reverse biased P-N junction* will hinder current flow; therefore, increased potential will have little effect on the magnitude of the current until the potential becomes very large (200–300 V). At this point, the diode breaks down, a phenomenon known as avalanching.

102. **2** The symbol 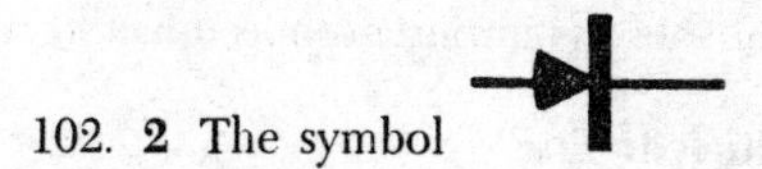 is used to represent a *P-N* junction; the arrow points from *P* to *N*. The symbol  represents a *P-N-P transistor.*

Wrong Choices Explained:

(1) An *N-P-N transistor* has a symbol identical to the *P-N-P* transistor except that the direction of the arrow in the symbol is reversed:

The arrow points in the direction of positive hole flow.

(3) A *zener diode* symbol looks very similar to the diode symbol for a *P-N* junction.

(4) An *integrated circuit* is the name for any number of tiny electronic devices that control electronic signals. An integrated circuit can include thousands of capacitors, diodes, resistors, and transistors, all on a single thin silicon layer.

103. **3** The base is the middle section of a transistor. Since this is an *N-P-N* transistor, the *P-type semiconductor material* is the center section and therefore the base.

104. **1** An *N-P-N transistor* acts as an amplifier. Small varying potential differences in the emitter base circuit produce corresponding variations of much larger amplitude in the collector circuit.

Wrong Choices Explained:

(2) A *diode* can act as a rectifier since it allows current flow in only one direction.

(3), (4) *N-type* and *P-type semiconductors* conduct electricity differently from each other and from conductors; however, neither can do anything else alone.

105. **2** An *integrated circuit* is the name for a large group of electronic components that are interconnected on a single block of silicon inside a computer. See also explanation for question 102, choice (4).

GROUP 6—**Nuclear Energy**

106. **3** All atomic nuclei may be represented by the same general symbol: ${}^{A}_{Z}X$, where X represents the particle; Z, called the atomic number, represents the number of protons; and A, called the mass number, is equal to the sum of neutrons and protons present in the nucleus. The number of neutrons (N) present in a nucleus is indicated by the expression: $N = A - Z$.

The atom $^{16}_{8}O$ has 16 – 8 = 8 neutrons, and $^{15}_{7}N$ also has 8 neutrons (15 – 7).

Wrong Choices Explained:

(1) $^{16}_{7}N$ has 16 – 7 = 9 neutrons.

(2) $^{17}_{8}O$ has 17 – 8 = 9 neutrons.

(4) $^{15}_{8}O$ has 15 – 8 = 7 neutrons.

107. **3** Protons, which are positively charged particles, are all concentrated in the nucleus of atoms, despite the electrostatic force of repulsion. Therefore, there must be a *strong* attractive nuclear force that holds the protons together. This force is also responsible for holding neutrons and protons together and neutrons and neutrons together. In short, it is responsible for the stability of nuclei. Within the nucleus it is about 100 times stronger than the electrostatic force. Beyond a few nucleon diameters, however, the nuclear force diminishes rapidly and becomes much less powerful than the electrostatic and gravitational forces. Although the nuclear force is the strongest force known to exist, it has a very *short range.*

108. **4** To solve this problem, use the mass-energy relationship, which can be found in the *List of Physical Constants* in the *Reference Tables*.

$$0.5\ \text{u}\left(\frac{9.3\times10^2\ \text{MeV}}{1\ \text{u}}\right) = 4.7\times10^2\ \text{MeV}$$

109. **4** The number of protons in the nucleus of an atom determines the nature of the element. Atoms with different numbers of protons represent different elements. Atoms of different isotopes of the same element contain the *same number of protons, but different numbers of neutrons.*

110. **3** Naturally occurring radioactive elements can decay via alpha, beta, or gamma decay. In beta decay, a neutron in the nucleus decays to a proton and an electron:

$$^{1}_{0}n \rightarrow {}^{1}_{1}p + {}^{0}_{-1}e$$

The mass number does not change since it is the sum of the neutrons and protons present in the nucleus; a neutron was lost, but a *proton was gained.* The electron does not remain in the nucleus; it is emitted and is known as the beta (β^-) particle.

111. **1** A nuclear equation must be balanced for both mass and charge.

$$^{14}_{6}C \rightarrow ^{14}_{7}N + X$$

Balancing for mass reveals that particle X has a mass number of 0 (14 = 14 + 0). Balancing for charge reveals that particle X has a charge of −1 (6 = 7 + (−1)). This mass number and charge match the description of an electron, which is also known as a *beta* (β^-) *particle.*

$$^{14}_{6}C \rightarrow ^{14}_{7}N + ^{0}_{-1}e$$

Wrong Choices Explained:

(2) A *gamma ray* has a charge of 0 and a mass number of 0: $^{0}_{0}\gamma$.

(3) A *neutron* has a charge of 0 and a mass number of 1: $^{1}_{0}n$.

(4) A *positron* has a charge of +1 and a mass number of 0: $^{0}_{+1}e$. Positrons are the antiparticles of electrons and are an example of antimatter. They are also known as positive beta particles (β^+).

112. **2** Write out the problem.

Given: half-life = 1,600 yr
total time = 4,800 yr
m_i = 10.0 kg

Find: m_f = ?

Refer to the *Nuclear Energy* section in the *Reference Tables* to find an equation(s) that relates final mass with half-life and initial mass.

Solution: $m_f = \dfrac{m_i}{2^n}$

To use this equation, you must first determine the total number of half-lives that have passed:

$$\frac{4{,}800\ \cancel{yr}}{1{,}600 \frac{\cancel{yr}}{\text{half-life}}} = 3 \text{ half-lives}$$

Now solve for total remaining mass:

$$m_f = \frac{10.0\ \text{kg}}{2^3} = \frac{10.0\ \text{kg}}{8}$$
$$= 1.25\ \text{kg}$$

113. **1** The function of a *moderator* in a nuclear reactor is to slow down the neutrons in order to control the rate of the reaction.

Wrong Choices Explained:

(2) The function of the *control rods* is to absorb the neutrons, either to control the rate of the reaction or to completely halt the reaction quickly.

(3) The *fuel rods* contain the material to be fissioned.

(4) *Accelerators* use electric and magnetic fields to accelerate charged particles, but they have no effect on neutrons, which carry no charge.

114. **2** The reacting nuclei need *large kinetic energies* in order to overcome the electrostatic repulsion between them.

115. **2** The difference between the mass of the nucleus and the mass of its nucleons is known as the mass defect of the nucleus. The energy equivalent of the mass defect, obtained by using the mass-energy relationship, $E = mc^2$ (refer to the *Nuclear Energy* section of the *Reference Tables*), is known as the binding energy of the nucleus. The *greater the mass defect,* the *greater the binding energy.*

PART III

116. Write out the problem.

Given: $v_i = 0$ m/s Find: $a = ?$
$v_f = 40.$ m/s
$\Delta t = 20.$ s

Refer to the *Mechanics* section in the *Reference Tables* to find an equation(s) that relates acceleration with velocity and time.

Solution:
$$\bar{a} = \frac{\Delta v}{\Delta t} = \frac{v_f - v_i}{\Delta t}$$
$$= \frac{40.\text{ m/s} - 0\text{ m/s}}{20.\text{ s}} = \frac{40.\text{ m/s}}{20.\text{ s}}$$
$$= 2.0\text{ m/s}^2$$

Alternatively, you can solve this problem using the slope of line *AB* on the graph. The slope of a line is defined as the change in y over the change in x. In this graph, y is speed and x is time; thus, the slope on a speed versus time graph is equal to the change in speed over the change in time, which is equal to the acceleration.

$$\text{slope} = \frac{\Delta y}{\Delta x} = \frac{\Delta v}{\Delta t} = \bar{a}$$
$$= \frac{40.\text{ m/s}}{20.\text{ s}}$$
$$= 2.0\text{ m/s}^2$$

NOTE: One point is awarded for writing the equation and for the subsitution of values with units. If the equation and/or units are not shown, you do not receive this point. One point is awarded for the correct answer (number and units). If there are no units, you do not receive this point. Note also that significant figures and scientific notation are not required to obtain credit.

117. Write out the problem.

Given: $\overline{v} = 40.\ \text{m/s}$ Find: $\Delta s = ?$

$\Delta t = 20.\ \text{s}$

Refer to the *Mechanics* section in the *Reference Tables* to find an equation(s) that relates displacement with velocity and time.

Solution: $$\overline{v} = \frac{\Delta s}{\Delta t}$$

$$\Delta s = \overline{v}\,\Delta t$$

$$= \left(40.\frac{\text{m}}{\not{\text{s}}}\right)(20.\ \not{\text{s}})$$

$$= 800\ \text{m} \quad \text{or} \quad 8.0 \times 10^2\ \text{m}$$

Alternatively, you can solve this problem directly from the graph. The area of a square or rectangle is calculated by multiplying the base times the height. In this case the base is the time, and the height is the speed; therefore, the area under the curve is equal to the displacement.

$$A = bh = \Delta t\overline{v}$$

$$= (20.\ \not{\text{s}})\left(40.\frac{\text{m}}{\not{\text{s}}}\right)$$

$$= 800\ \text{m} \quad \text{or} \quad 8.0 \times 10^2\ \text{m}$$

NOTE: One point is awarded for writing the equation and for the subsitution of values with units. If the equation and/or units are not shown, you do not receive this point. One point is awarded for the correct answer (number and units). If there are no units, you do not receive this point. Note also that significant figures and scientific notation are not required to obtain credit.

118. Write out the problem.

Given: $v_i = 40.\ \text{m/s}$ Find: $v = ?$

$v_f = 0\ \text{m/s}$

$\Delta t = 10.\ \text{s}$

Refer to the *Mechanics* section in the *Reference Tables* to find an equation(s) that relates average velocity to the variables we are given.

Solution: $$\overline{v} = \frac{v_f + v_i}{2}$$

$$= \frac{40.\ \text{m/s} + 0\ \text{m/s}}{2} = \frac{40.\ \text{m/s}}{2}$$

$$= 20.\ \text{m/s} \quad \text{or} \quad 20\ \text{m/s}$$

Note that time is not required for this calculation.

NOTE: One point is awarded for the correct answer. To receive this point, you must include units. Note also that significant figures and scientific notation are not required to obtain credit.

119. Electric field lines are drawn in such a way that the arrows of the lines point in the direction in which a small positive test charge would move if it were placed in the field. Since like charges repel and opposite charges attract, the field lines will point from the positive plate toward the negative plate. See the diagram below.

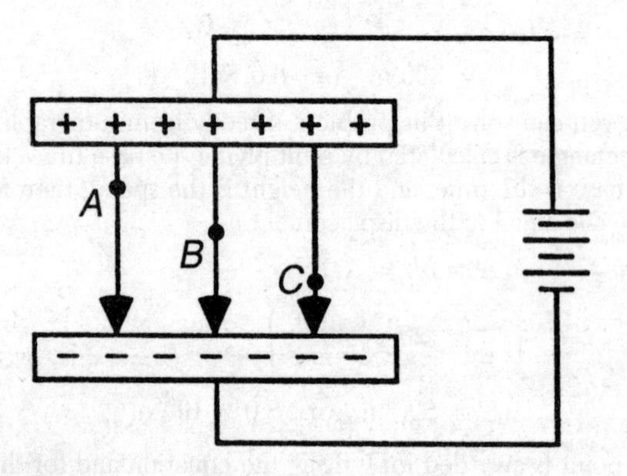

NOTE: One point is given for drawing three parallel lines passing through points *A*, *B*, and *C* as long as the lines do not pass through either of the plates.

One point is awarded for drawing three arrowheads, one on each line, pointing in the correct direction. Additional field lines may be added with no penalty as long as they are drawn correctly.

120. Write out the problem.

Given: $d = 2.0 \times 10^{-2}$ m $\quad$ Find: $E = ?$
$V = 1.0 \times 10^{-2}$ V

Refer to the *Electricity and Magnetism* section in the *Reference Tables* to find an equation(s) that relates electric field strength to the variables given.

Solution:
$$E = \frac{V}{d}$$
$$= \frac{1.0 \times 10^{2}\ \text{V}}{2.0 \times 10^{-2}\ \text{m}}$$
$$= 0.5 \times 10^{4}\ \frac{\text{V}}{\text{m}} \quad \text{or} \quad 5.0 \times 10^{3}\ \frac{\text{V}}{\text{m}}$$

Alternatively,
$$\frac{\text{V}}{\text{m}} = \frac{\frac{\text{J}}{\text{C}}}{\text{m}} = \frac{\frac{\text{N} \cdot \cancel{\text{m}}}{\text{C}}}{\cancel{\text{m}}},$$
$$E = 5000\frac{\text{N}}{\text{C}}$$

NOTE: One point is awarded for writing the equation and for the subsitution of values with units. If the equation and/or units are not shown, you do not receive this point. One point is awarded for the correct answer (number and units). If there are no units, you do not receive this point. Note also that significant figures and scientific notation are not required to obtain credit, and any correct units are accepted.

121. Write out the problem.

Given: $\theta_1 = 30°$ $\quad$ Find: $\theta_2 = ?$
medium 1 = Lucite
medium 2 = air

Refer to the *Wave Phenomena* section in the *Reference Tables* to find an equation(s) that relates angle of refraction to angle of incidence.

Solution: $n_1 \sin \theta_1 = n_2 \sin \theta_2$

To calculate the angle of refraction from the angle of incidence, you need to know the refractive indices of the two media. You can easily obtain this information from the *Absolute Indices of Refraction* list in the *Reference Tables*. The refractive indices for Lucite and air are 1.50 and 1.00, respectively.

$$\sin \theta_2 = \frac{n_1 \sin \theta_1}{n_2}$$

$$\theta_2 = \sin^{-1}\left(\frac{n_1 \sin \theta_1}{n_2}\right)$$

$$= \sin^{-1}\left(\frac{1.50 \sin 30.°}{1.00}\right)$$

$$= \sin^{-1}(1.50 \times 0.5) = \sin^{-1} (0.750)$$

$$= 49°$$

(If you are using a scientific calculator and are not getting the correct value for the angle, check to see whether you are in radian mode (RAD) or in gradient mode (GRAD). If you are in either of these, change the mode to degree mode (DEG). Make sure you are in this mode for all calculations involving angles measured in degrees.)

NOTE: One point is awarded for writing the equation and for the subsitution of values with units. If the equation and/or units are not shown, you do not receive this point. One point is awarded for the correct answer (number and units). If there are no units, you do not receive this point. Note also that significant figures and scientific notation are not required to obtain credit.

122. Angles of refraction, like angles of incidence, are measured with respect to the normal. Using a protractor, measure 49° to the right of the normal in the air and mark the spot. Draw a line from point *O* to the spot marked off, and add an arrowhead facing away from point *O*. See the diagram below.

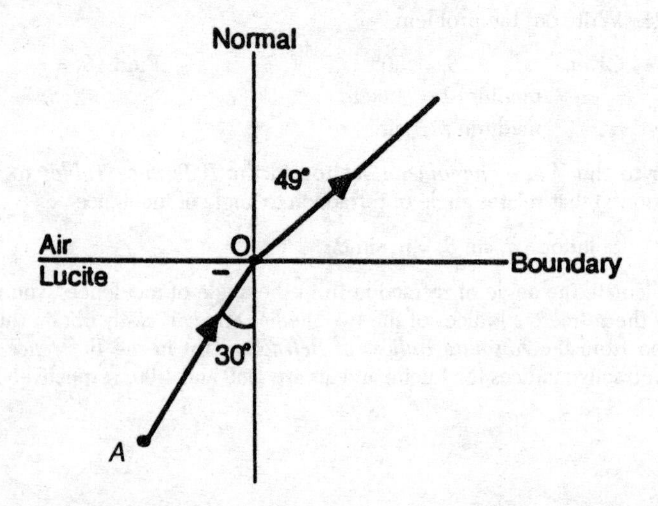

NOTE: One point will be awarded if the angle between the normal and the refracted ray is equal (±2°) to the value calculated in question 121. One point will be awarded if the refracted ray is drawn as shown (straight line, originating at point *O*, drawn to the right of the normal and having an arrow directed away from point *O*).

123. Refer to the *Wave Phenomena* section of the *Reference Tables* to find an equation that will allow you to calculate the critical angle of incidence for Lucite relative to air.

Solution:

$$\sin \theta_c = \frac{1}{n}$$

$$\theta_c = \sin^{-1}\left(\frac{1}{n}\right)$$

$$= \sin^{-1}\left(\frac{1}{1.50}\right) = \sin^{-1}(0.667)$$

$$= 42°$$

NOTE: One point is awarded for writing the equation and for the subsitution of values with units. If the equation and/or units are not shown, you do not receive this point. One point is awarded for the correct answer (number and units). If there are no units, you do not receive this point. Note also that significant figures and scientific notation are not required to obtain credit.

Topic	Question Numbers (total)	Wrong Answers (x)	Grade
Mechanics	1–14, 52–54, 116–118: (20)		$\frac{100(20-x)}{20} = \%$
Energy	15–20, 29: (7)		$\frac{100(7-x)}{7} = \%$
Electricity and Magnetism	21–28, 30–32, 35, 36, 119, 120: (15)		$\frac{100(15-x)}{15} = \%$
Wave Phenomena	33, 34, 37–45, 47, 55, 121–123: (16)		$\frac{100(16-x)}{16} = \%$
Modern Physics	46, 48–51: (5)		$\frac{100(5-x)}{5} = \%$
Motion in a Plane	56–65: (10)		$\frac{100(10-x)}{10} = \%$
Internal Energy	66–75: (10)		$\frac{100(10-x)}{10} = \%$
Electromagnetic Applications	76–85: (10)		$\frac{100(10-x)}{10} = \%$
Geometrical Optics	86–95: (10)		$\frac{100(10-x)}{10} = \%$
Solid State Physics	96–105: (10)		$\frac{100(10-x)}{10} = \%$
Nuclear Energy	106–115: (10)		$\frac{100(10-x)}{10} = \%$

To further pinpoint your weak areas, use the Topic Outline in the front of the book.

Examination June 1996

Physics

PART I

Answer all 55 questions in this part. [65]

Directions (1–55): For *each* statement or question, select the word or expression that, of those given, best completes the statement or answers the question. Record the answers to these questions in the spaces provided.

1 A car travels between the 100.-meter and 250.-meter highway markers in 10. seconds. The average speed of the car during this interval is

(1) 10. m/s
(2) 15 m/s
(3) 25 m/s
(4) 35 m/s

1 ______

2 A student walks 40. meters along a hallway that heads due north, then turns and walks 30. meters along another hallway that heads due east. What is the magnitude of the student's resultant displacement?

(1) 10. m
(2) 35 m
(3) 50. m
(4) 70. m

2 ______

3 The graph below represents the relationship between speed and time for a car moving in a straight line.

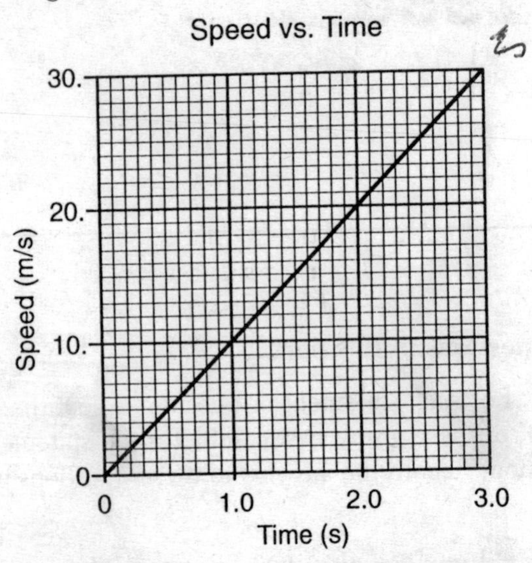

The magnitude of the car's acceleration is

(1) 1.0 m/s^2 (3) 10. m/s^2
(2) 0.10 m/s^2 (4) 0.0 m/s^2

3 ____

4 Oil drips at 0.4-second intervals from a car that has an oil leak. Which pattern best represents the spacing of oil drops as the car accelerates uniformly from rest?

(1) . . . .

(2) . .   .   .

(3) . . . .

(4) .   .. .

4 ____

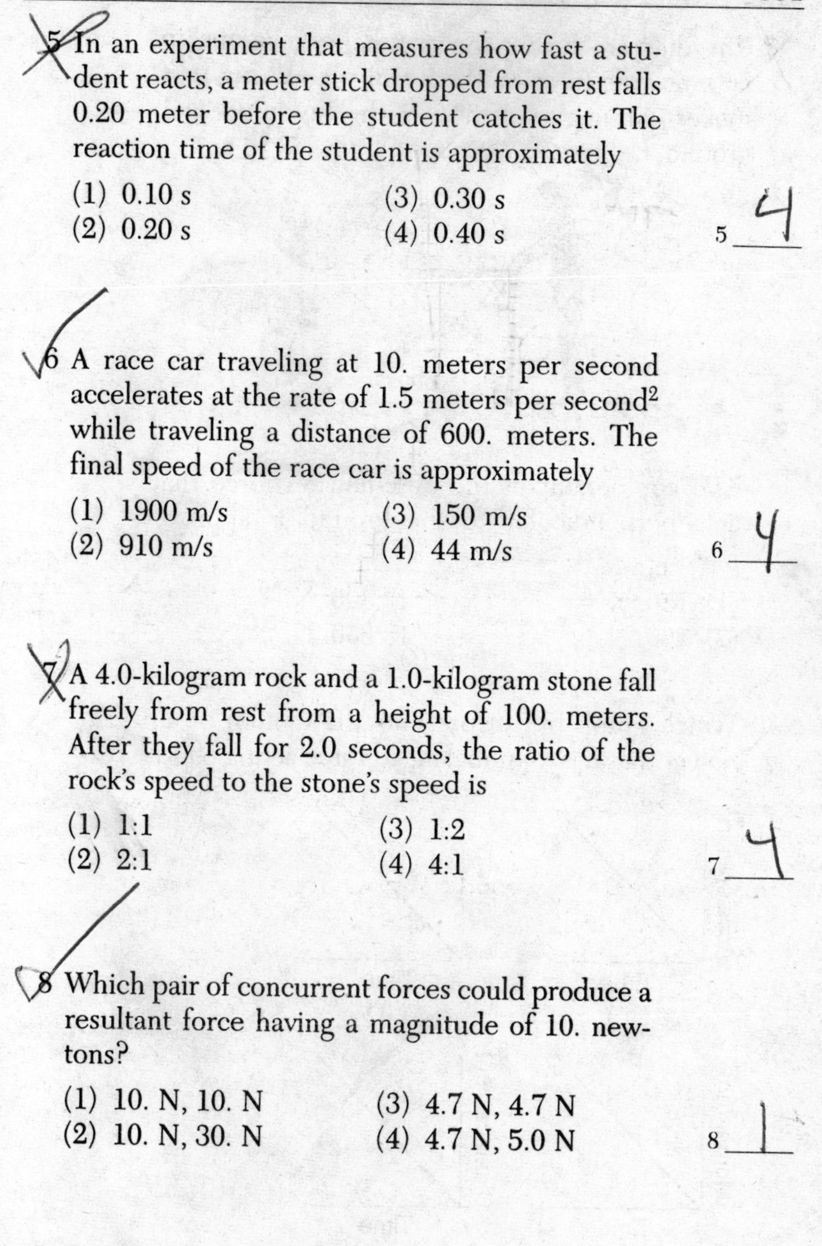

5 In an experiment that measures how fast a student reacts, a meter stick dropped from rest falls 0.20 meter before the student catches it. The reaction time of the student is approximately

(1) 0.10 s
(2) 0.20 s
(3) 0.30 s
(4) 0.40 s

5 ____

6 A race car traveling at 10. meters per second accelerates at the rate of 1.5 meters per second^2 while traveling a distance of 600. meters. The final speed of the race car is approximately

(1) 1900 m/s
(2) 910 m/s
(3) 150 m/s
(4) 44 m/s

6 ____

7 A 4.0-kilogram rock and a 1.0-kilogram stone fall freely from rest from a height of 100. meters. After they fall for 2.0 seconds, the ratio of the rock's speed to the stone's speed is

(1) 1:1
(2) 2:1
(3) 1:2
(4) 4:1

7 ____

8 Which pair of concurrent forces could produce a resultant force having a magnitude of 10. newtons?

(1) 10. N, 10. N
(2) 10. N, 30. N
(3) 4.7 N, 4.7 N
(4) 4.7 N, 5.0 N

8 ____

9 The diagram below shows a person exerting a 300.-newton force on the handle of a shovel that makes an angle of 60.° with the horizontal ground.

300. N

60.°

Shovel

Ground

The component of the 300.-newton force that acts perpendicular to the ground is approximately

(1) 150. N (3) 300. N
(2) 260. N (4) 350. N

9 _____

10 Which graph best represents the motion of an object that has *no* unbalanced force acting on it?

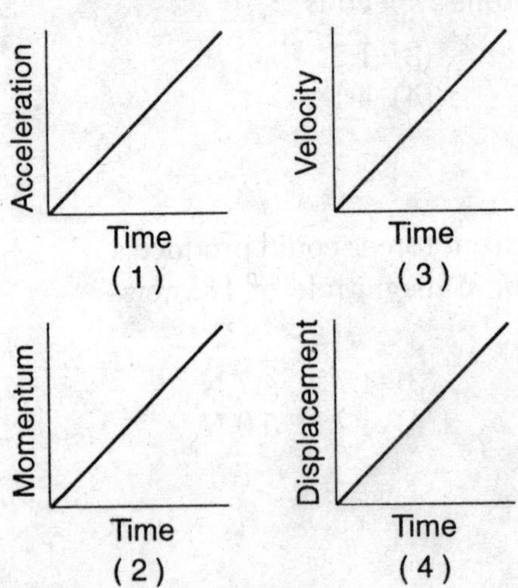

10 _____

11 The weight of an apple is closest to

(1) 10^{-2} N
(2) 10^{0} N
(3) 10^{2} N
(4) 10^{4} N

11 1

12 Two forces are applied to a 2.0-kilogram block on a frictionless, horizontal surface, as shown in the diagram below.

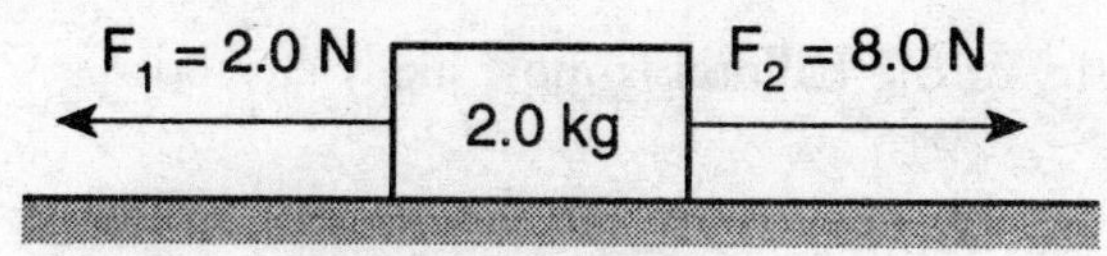

The acceleration of the block is

(1) 5.0 m/s^2 to the right
(2) 5.0 m/s^2 to the left
(3) 3.0 m/s^2 to the right
(4) 3.0 m/s^2 to the left

12 3

13 Compared to the inertia of a 0.10-kilogram steel ball, the inertia of a 0.20-kilogram Styrofoam ball is

1 one-half as great
2 twice as great
3 the same
4 four times as great

13 2

14 A 3.0-kilogram mass weighs 15 newtons at a given point in the Earth's gravitational field. What is the magnitude of the acceleration due to the gravity at this point?

(1) 45 m/s^2
(2) 9.8 m/s^2
(3) 5.0 m/s^2
(4) 0.20 m/s^2

14 3

15 As shown in the diagram below, an inflated balloon released from rest moves horizontally with velocity v.

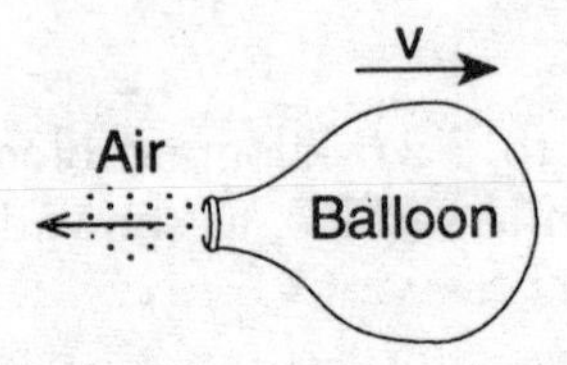

The velocity of the balloon is most likely caused by

1 action-reaction
2 centripetal force
3 gravitational attraction
4 rolling friction

15 1

16 A horizontal force is used to pull a 5.0-kilogram cart at a constant speed of 5.0 meters per second across the floor, as shown in the diagram below.

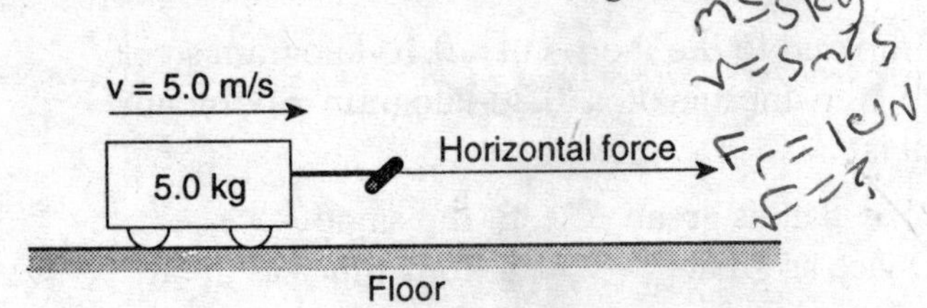

If the force of friction between the cart and the floor is 10. newtons, the magnitude of the horizontal force along the handle of the cart is

(1) 5.0 N
(2) 10. N
(3) 25 N
(4) 50. N

16 3

17 A bullet traveling at 5.0×10^2 meters per second is brought to rest by an impulse of 50. newton-seconds. What is the mass of the bullet?

(1) 2.5×10^4 kg
(2) 1.0×10^1 kg
(3) 1.0×10^{-1} kg
(4) 1.0×10^{-2} kg

17 3

18 A box weighing 1.0×10^2 newtons is dragged to the top of an incline, as shown in the diagram below.

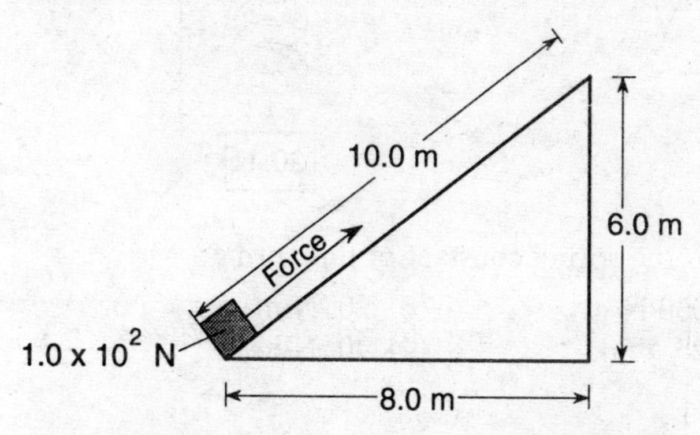

The gravitational potential energy of the box at the top of the incline is approximately

(1) 1.0×10^2 J
(2) 6.0×10^2 J
(3) 8.0×10^2 J
(4) 1.0×10^3 J

18 2

19 A 10.-newton force is required to move a 3.0-kilogram box at constant speed. How much power is required to move the box 8.0 meters in 2.0 seconds?

(1) 40. W
(2) 20. W
(3) 15 W
(4) 12 W

19 1

20 A 20.-newton weight is attached to a spring, causing it to stretch, as shown in the diagram below.

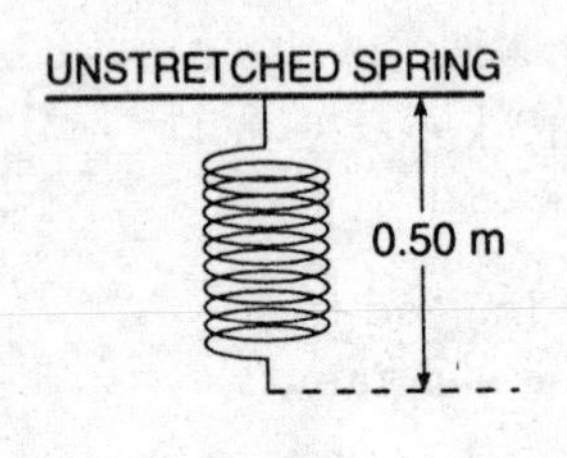

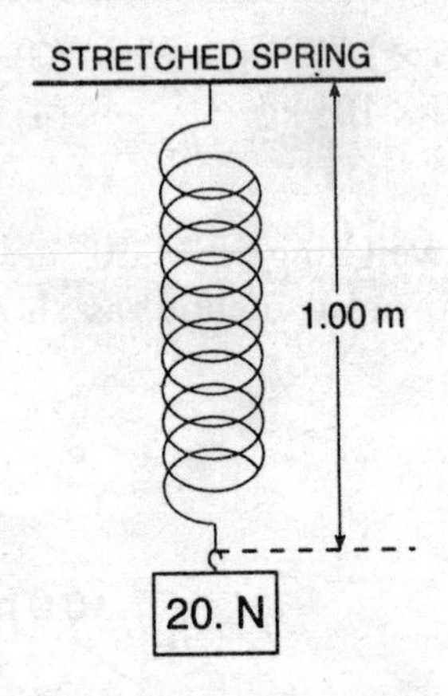

What is the spring constant of this spring?

(1) 0.050 N/m
(2) 0.25 N/m
(3) 20. N/m
(4) 40. N/m

20 4

21 Spring *A* has a spring constant of 140 newtons per meter, and spring *B* has a spring constant of 280 newtons per meter. Both springs are stretched the same distance. Compared to the potential energy stored in spring *A*, the potential energy stored in spring *B* is

1 the same
2 twice as great
3 half as great
4 four times as great

21 2

22 A cart of mass *m* traveling at speed *v* has kinetic energy KE. If the mass of the cart is doubled and its speed is halved, the kinetic energy of the cart will be

1 half as great
2 twice as great
3 one-fourth as great
4 four times as great

22 1

23 A repulsive electrostatic force of magnitude F exists between two metal spheres having identical charge q. The distance between their centers is r. Which combination of changes would produce *no* change in the electrostatic force between the spheres?

1 doubling q on one sphere while doubling r
2 doubling q on both spheres while doubling r
3 doubling q on one sphere while halving r
4 doubling q on both spheres while halving r

23 ____

24 An inflated balloon which has been rubbed against a person's hair is touched to a neutral wall and remains attracted to it. Which diagram best represents the charge distribution on the balloon and wall?

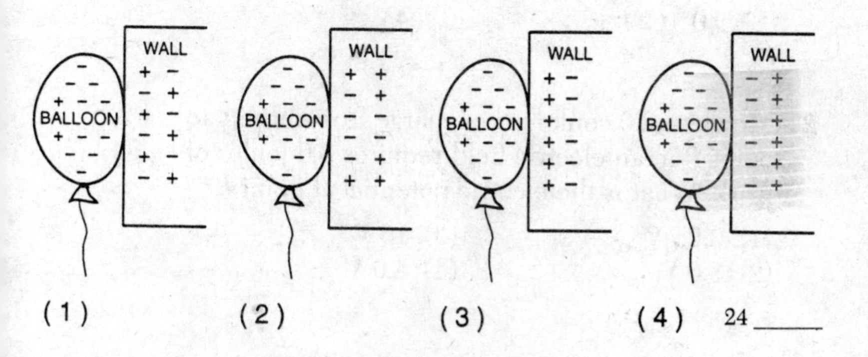

24 ____

25 Two metal spheres having charges of $+4.0 \times 10^{-6}$ coulomb and $+2.0 \times 10^{-5}$ coulomb, respectively, are brought into contact and then separated. After separation, the charge on each sphere is

(1) 8.0×10^{-11} C
(2) 8.0×10^{-6} C
(3) 2.1×10^{-6} C
(4) 1.2×10^{-5} C

25 ____

26 Which graph best represents the relationship between electric field intensity and distance from a point charge?

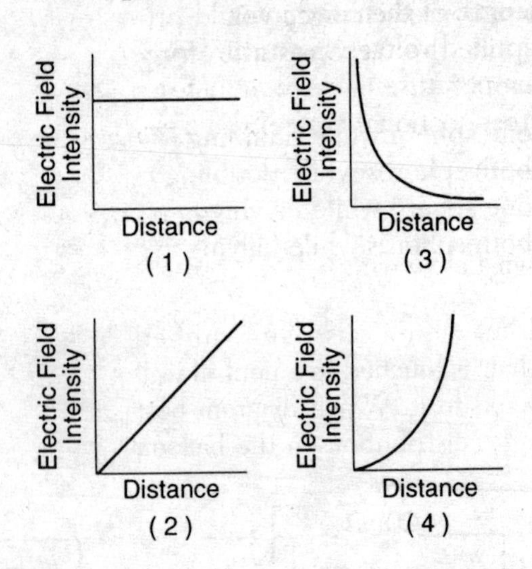

26 ____

27 Moving +2.0 coulombs of charge from infinity to point *P* in an electric field requires 8.0 joules of work. What is the electric potential at point *P*?

(1) 0.25 V (3) 16 V
(2) 8.0 V (4) 4.0 V

27 ____

28 What is the magnitude of the electrostatic force experienced by one elementary charge at a point in an electric field where the electric field intensity is 3.0×10^3 newtons per coulomb?

(1) 1.0×10^3 N (3) 3.0×10^3 N
(2) 1.6×10^{-19} N (4) 4.8×10^{-16} N

28 ____

29 A metal conductor is used in an electric circuit. The electrical resistance provided by the conductor could be increased by

1 decreasing the length of the conductor
2 decreasing the applied voltage in the circuit
3 increasing the temperature of the conductor
4 increasing the cross-sectional area of the conductor

29 _____

30 In the circuit shown below, voltmeter V_2 reads 80. volts.

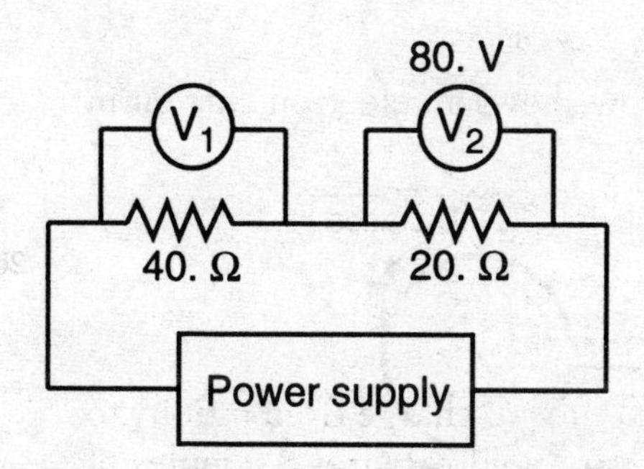

What is the reading of voltmeter V_1?

(1) 160 V
(2) 80. V
(3) 40. V
(4) 20. V

30 _____

31 In a lightning strike, a charge of 18 coulombs is transferred between a cloud and the ground in 2.0×10^{-2} second at a potential difference of 1.5×10^{6} volts. What is the average current produced by this strike?

(1) 3.6×10^{-1} A
(2) 9.0×10^{2} A
(3) 3.0×10^{4} A
(4) 7.5×10^{7} A

31 _____

32 The diagram below shows the current in a segment of a direct current circuit.

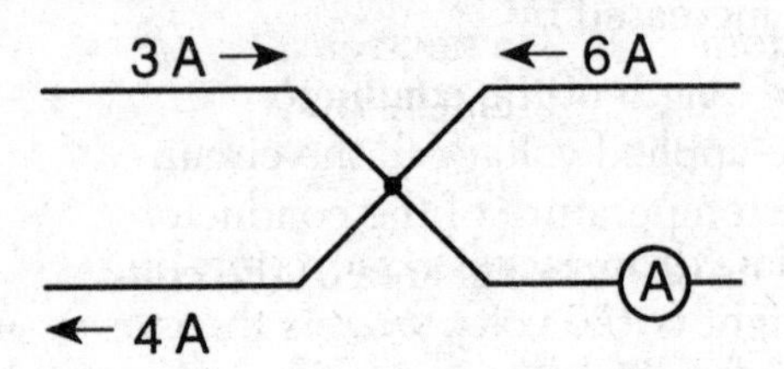

What is the reading of ammeter *A*?

(1) 1 A
(2) 5 A
(3) 7 A
(4) 8 A

32 _____

33 The diagram below shows an electron current in a wire loop.

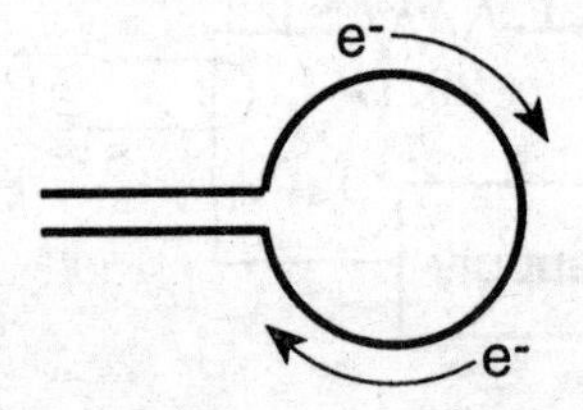

What is the direction of the magnetic field at the center of the loop?

1 out·of the page
2 into the page
3 clockwise
4 counterclockwise

33 _____

34 A charged particle is moving with a constant velocity. On entering a uniform magnetic field, the particle

1 must decrease in speed
2 must change the magnitude of its momentum
3 may change its direction of motion
4 may increase in kinetic energy

34 _____

35 Electromagnetic waves can be generated by accelerating

1 a hydrogen atom
2 a photon
3 a neutron
4 an electron

35 _____

36 If the potential drop across an operating 300.-watt floodlight is 120 volts, what is the current through the floodlight?

(1) 0.40 A
(2) 2.5 A
(3) 7.5 A
(4) 4.8 A

36 _____

37 Which diagram correctly shows a magnetic field configuration?

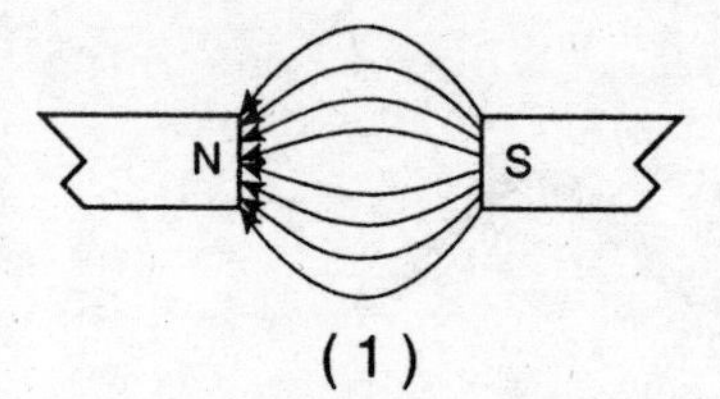

(1)

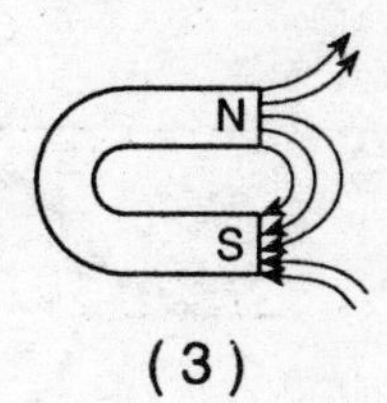

(3)

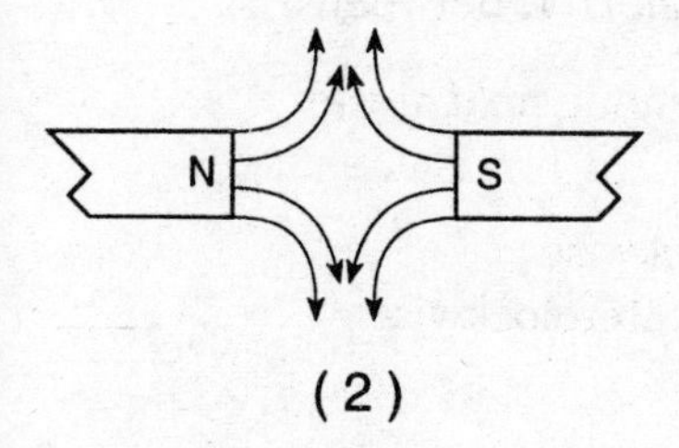

(2)

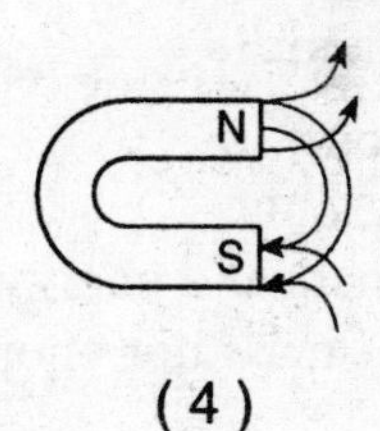

(4)

37 _____

38 Two points on a transverse wave that have the same magnitude of displacement from equilibrium are in phase if the points also have the

1 same direction of displacement and the same direction of motion
2 same direction of displacement and the opposite direction of motion
3 opposite direction of displacement and the same direction of motion
4 opposite direction of displacement and the opposite direction of motion

38 _____

39 A periodic wave travels through a rope, as shown in the diagram below.

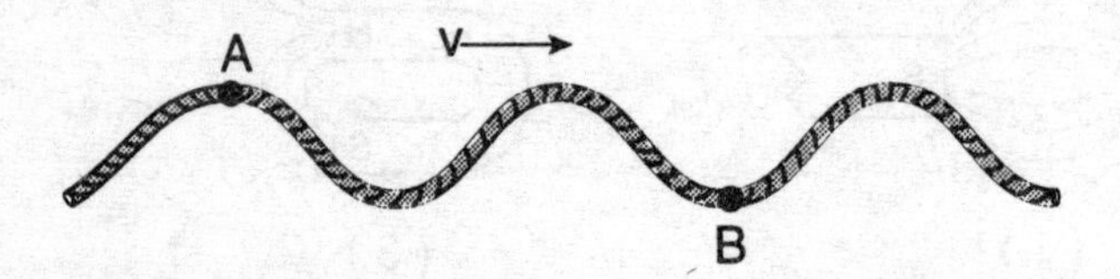

As the wave travels, what is transferred between points *A* and *B*?

1 mass, only
2 energy, only
3 both mass and energy
4 neither mass nor energy

39 _____

40 Which graph best represents the relationship between the frequency and period of a wave?

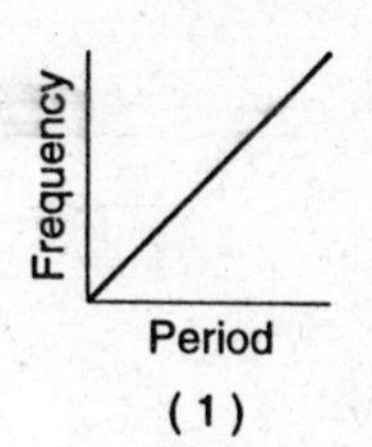

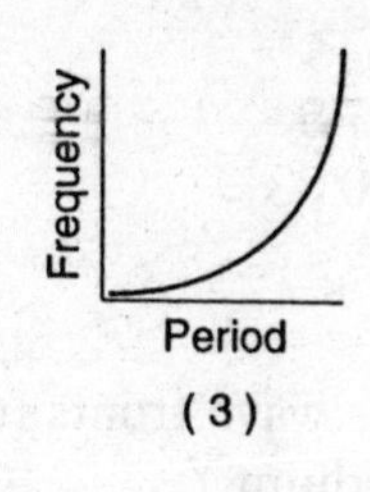

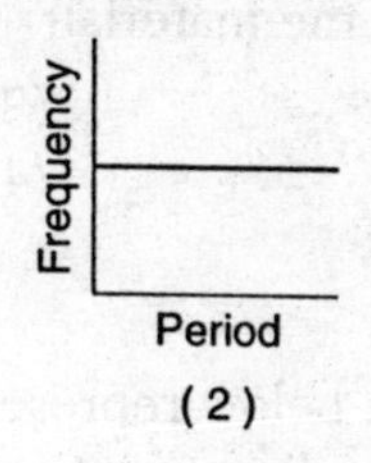

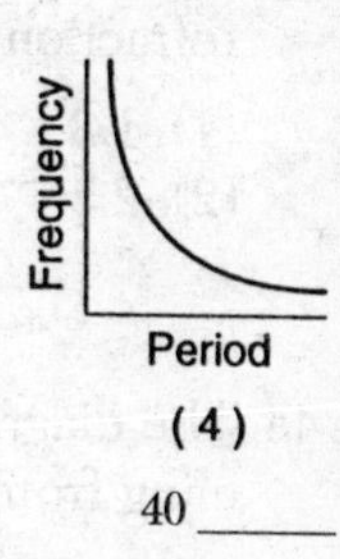

40 _____

41 A beam of monochromatic light ($\lambda = 5.9 \times 10^{-7}$ meter) crosses a boundary from air into Lucite at an angle of incidence of 45°. The angle of refraction is approximately

(1) 63° (2) 56° (3) 37° (4) 28°

41 _____

42 The diagram below represents shallow water waves of wavelength λ passing through two small openings, *A* and *B*, in a barrier.

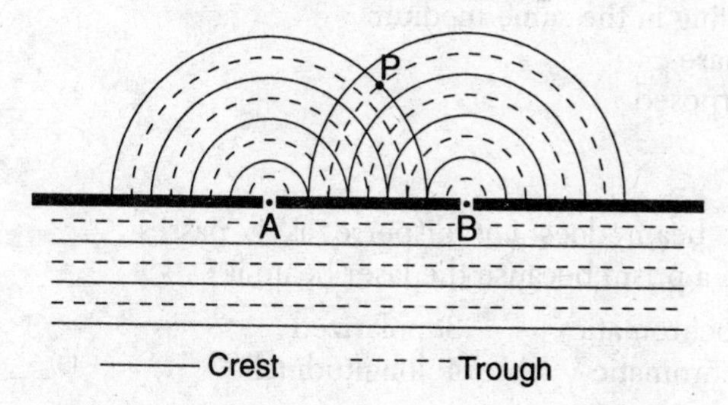

Compared to the length of path *BP*, the length of path *AP* is

(1) 1λ longer
(2) 2λ longer
(3) $\frac{1}{2}\lambda$ longer
(4) the same

42 _____

43 The speed of light in a material is 2.5×10^8 meters per second. What is the absolute index of refraction of the material?

(1) 1.2
(2) 2.5
(3) 7.5
(4) 0.83

43 _____

44 The diagram below represents wave fronts traveling from medium *X* into medium *Y*.

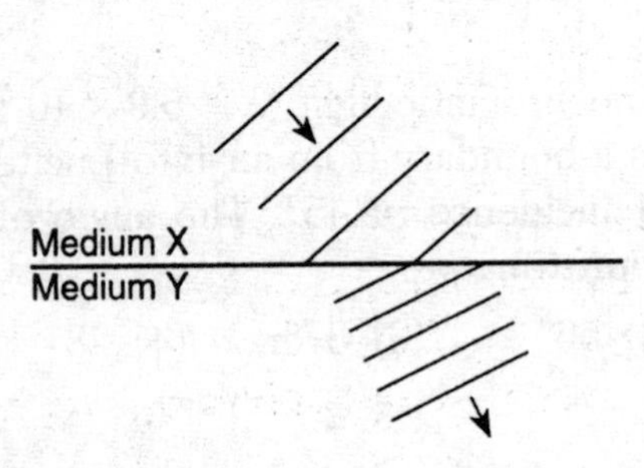

All points on any one wave front shown must be

1 traveling with the same speed
2 traveling in the same medium
3 in phase
4 superposed

44 _____

45 A laser beam does *not* disperse as it passes through a prism because the laser beam is

1 monochromatic
2 polychromatic
3 polarized
4 longitudinal

45 _____

46 The diagram below shows a wave phenomenon.

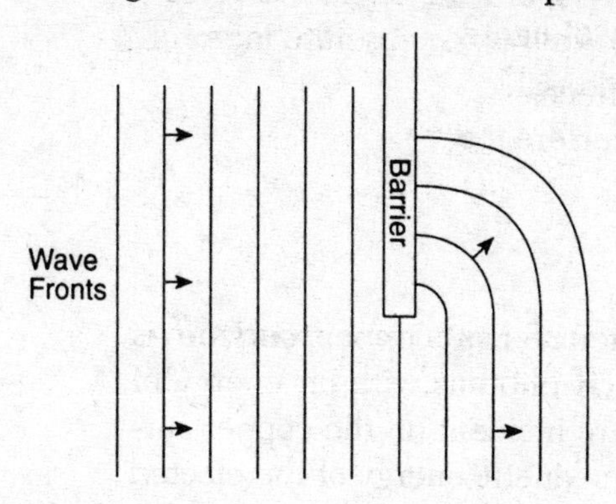

The pattern of waves shown behind the barrier is the result of

1 reflection
2 refraction
3 diffraction
4 interference

46 _____

47 In the diagram below, monochromatic light ($\lambda = 5.9 \times 10^{-7}$ meter) in air is about to travel through crown glass, water, and diamond.

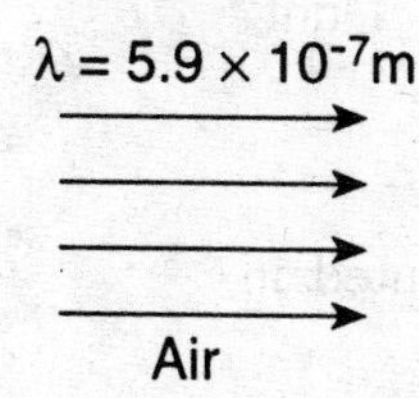

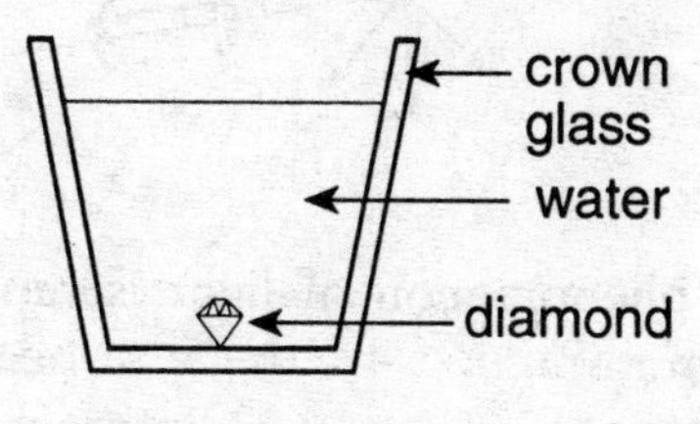

(not drawn to scale)

In which substance does the light travel the slowest?

1 air
2 diamond
3 water
4 crown glass

47 _____

48 Which phenomenon is most easily explained by the particle theory of light?

1 photoelectric effect
2 constructive interference
3 polarization
4 diffraction

48 ____

49 The work function for a copper surface is 7.3×10^{-19} joule. If photons with an energy of 9.9×10^{-19} joule are incident on the copper surface, the maximum kinetic energy of the ejected photoelectrons is

(1) 2.6×10^{-19} J
(2) 7.3×10^{-19} J
(3) 9.9×10^{-19} J
(4) 1.7×10^{30} J

49 ____

50 The diagram below shows sunglasses being used to eliminate glare.

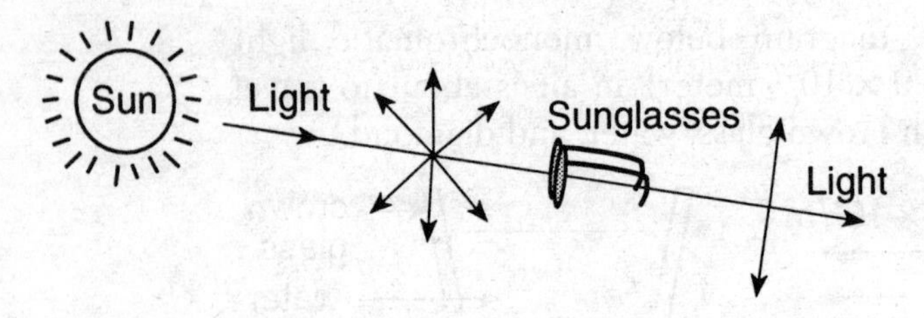

Which phenomenon of light is represented in the diagram?

1 dispersion
2 diffraction
3 internal reflection
4 polarization

50 ____

51 What is the minimum energy required to ionize a hydrogen atom in the $n = 3$ state?

(1) 13.60 eV
(2) 12.09 eV
(3) 5.52 eV
(4) 1.51 eV

51 ____

52 When a source of dim orange light shines on a photosensitive metal, no photoelectrons are ejected from its surface. What could be done to increase the likelihood of producing photoelectrons?

1 Replace the orange light source with a red light source.
2 Replace the orange light source with a higher frequency light source.
3 Increase the brightness of the orange light source.
4 Increase the angle at which the photons of orange light strike the metal.

52 _____

53 In Rutherford's model of the atom, the positive charge

1 is distributed throughout the atom's volume
2 revolves about the nucleus in specific orbits
3 is concentrated at the center of the atom
4 occupies most of the space in the atom

53 _____

Note that questions 54 and 55 have only three choices.

54 Light ($\lambda = 5.9 \times 10^{-7}$ meter) travels through a solution. If the absolute index of refraction of the solution is increased, the critical angle will

1 decrease
2 increase
3 remain the same

54 _____

55 An astronomer on Earth studying light coming from a star notes that the observed light frequencies are lower than the actual emitted frequencies. The astronomer concludes that the distance between the star and Earth is

1 decreasing
2 increasing
3 not changing

55 ____

PART II

This part consists of six groups, each containing ten questions. Each group tests an optional area of the course. Choose two of these six groups. Be sure that you answer all ten questions in each group chosen. Record the answers to these questions in the spaces provided. [20]

GROUP 1—**Motion in a Plane**

If you choose this group, be sure to answer questions **56–65**.

Base your answers to questions 56 through 58 on the information and diagram below.

A cannon elevated at an angle of 35° to the horizontal fires a cannonball, which travels the path shown in the diagram below. [Neglect air resistance and assume the ball lands at the same height above the ground from which it was launched.]

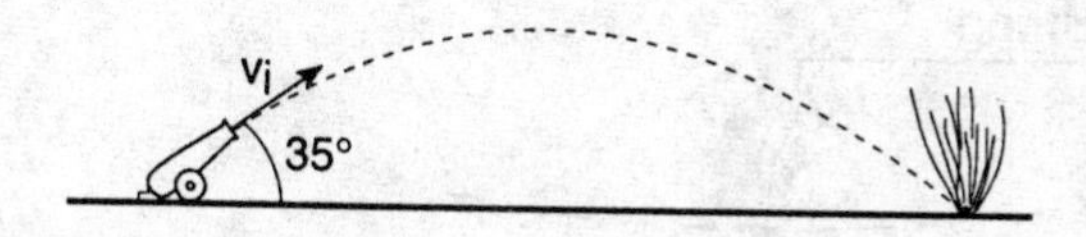

56 If the ball lands 7.0×10^2 meters from the cannon 10. seconds after it was fired, what is the horizontal component of its initial velocity?

(1) 70. m/s
(2) 49 m/s
(3) 35 m/s
(4) 7.0 m/s

56 _____

57 If the ball's time of flight is 10. seconds, what is the vertical component of its initial velocity?

(1) 9.8 m/s
(2) 49 m/s
(3) 70. m/s
(4) 98 m/s

57 _____

58 If the angle of elevation of the cannon is decreased from 35° to 30.°, the vertical component of the ball's initial velocity will

1 decrease and its horizontal component will decrease
2 decrease and its horizontal component will increase
3 increase and its horizontal component will decrease
4 increase and its horizontal component will increase

58 _____

Base your answers to questions 59 through 61 on the diagram below. The diagram shows a student spinning a 0.10-kilogram ball at the end of a 0.50-meter string in a horizontal circle at a constant speed of 10. meters per second. [Neglect air resistance.]

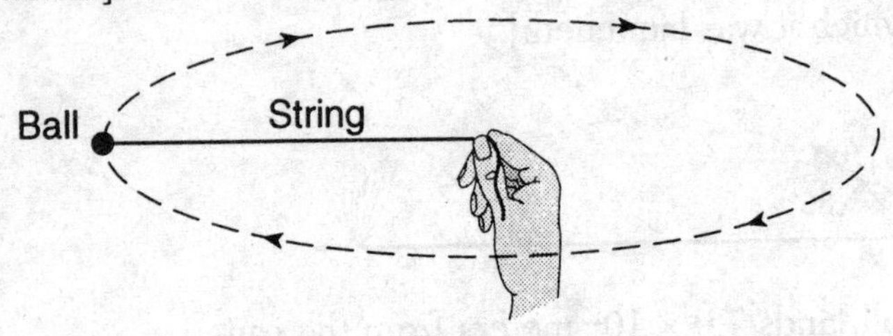

Note that question 59 has only three choices.

59 If the magnitude of the force applied to the string by the student's hand is increased, the magnitude of the acceleration of the ball in its circular path will

1 decrease 2 increase 3 remain the same

59 _____

60 The magnitude of the centripetal force required to keep the ball in this circular path is

(1) 5.0 N (2) 10. N (3) 20. N (4) 200 N 60 _____

61 Which is the best description of the force keeping the ball in the circular path?

1 perpendicular to the circle and directed toward the center of the circle
2 perpendicular to the circle and directed away from the center of the circle
3 tangent to the circle and directed in the same direction that the ball is moving
4 tangent to the circle and directed opposite to the direction that the ball is moving 61 _____

---

62 A convertible car with its top down is traveling at constant speed around a circular track, as shown in the diagram below.

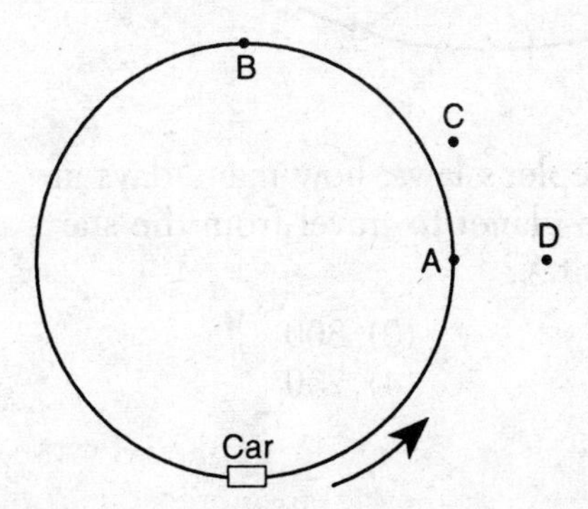

When the car is at point A, if a passenger in the car throws a ball straight up, the ball could land at point

(1) A (3) C
(2) B (4) D 62 _____

63 The comet Hyakutake, seen in the Earth's sky this year, will take more than 10,000 years to complete its orbit. Which object is at a focus of the comet's orbit?

1 Earth
2 Sun
3 Moon
4 Jupiter

63 _____

64 The diagram below represents the path of a planet moving in an elliptical orbit around a star. The orbital period of the planet is 1,000 days.

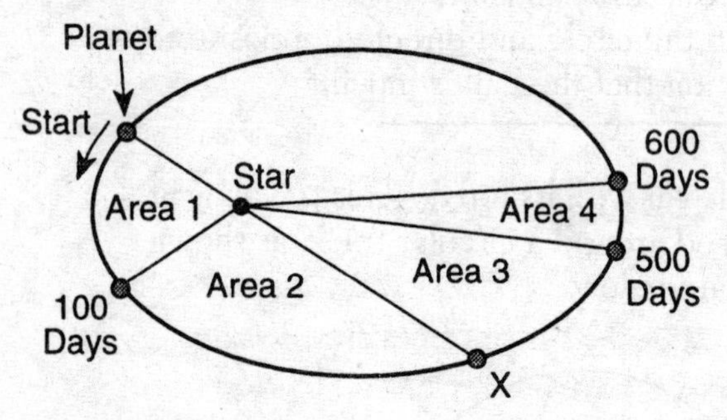

According to Kepler's laws, how many days are required for the planet to travel from the starting point to point *X*?

(1) 400
(2) 350
(3) 300
(4) 250

64 _____

65 A ball is projected horizontally to the right from a height of 50. meters, as shown in the diagram below.

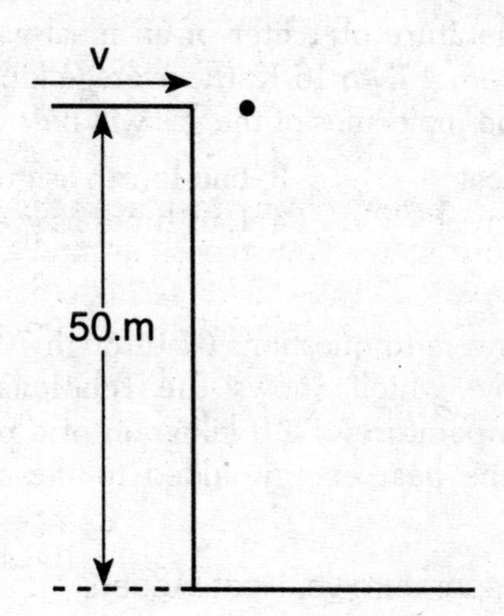

Which diagram best represents the position of the ball at 1.0-second intervals? [Neglect air resistance.]

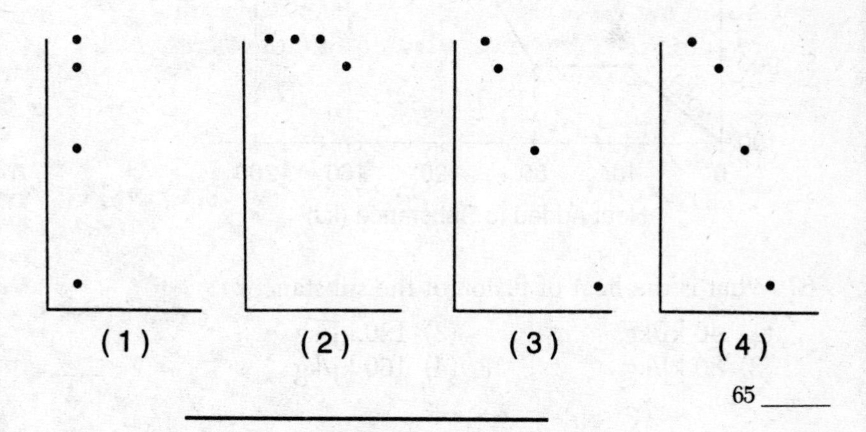

65 ______

GROUP 2—**Internal Energy**

If you choose this group, be sure to answer questions **66–75**.

66 If the temperature of 1 liter of an ideal gas is increased from 4 K to 16 K, the average kinetic energy of the molecules of the gas will be

1 half as great
2 twice as great
3 one-fourth as great
4 four times as great

66 _____

Base your answers to questions 67 through 70 on the graph below which shows the relationship between the temperature of 1.0 kilogram of a pure substance and the heat energy added to the substance.

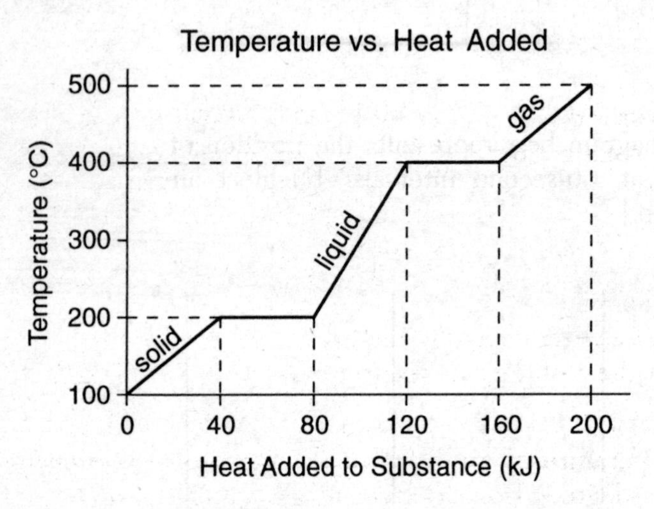

67 What is the heat of fusion of the substance?

(1) 40 kJ/kg
(2) 80 kJ/kg
(3) 120 kJ/kg
(4) 160 kJ/kg

67 _____

68 The freezing point of the substance is

(1) 400°C (3) 200°C
(2) 300°C (4) 100°C 68 _____

69 What is the specific heat of the substance as a gas?

(1) 0.04 kJ/kg•C° (3) 40 kJ/kg•C°
(2) 0.4 kJ/kg•C° (4) 4 kJ/kg•C° 69 _____

70 If the initial temperature of the sample is 100°C, the total amount of heat required to convert all of the substance to a gas is

(1) 40 kJ (3) 120 kJ
(2) 80 kJ (4) 160 kJ 70 _____

---

71 When a car is driven over snow, the snow under the tires may melt because the

1 pressure of the tires lowers the melting point of the snow
2 pressure of the tires raises the melting point of the snow
3 snow loses heat energy to the tires
4 specific heat of the snow is decreased 71 _____

72 The temperature of a water sample increased 5 Celsius degrees from its freezing point. The sample's rise in temperature on the Kelvin scale was

(1) 5 K (3) 278 K
(2) 9 K (4) 378 K 72 _____

73 Which graph best represents the relationship between pressure (P) and absolute temperature (T) for a fixed mass of an ideal gas in a rigid container?

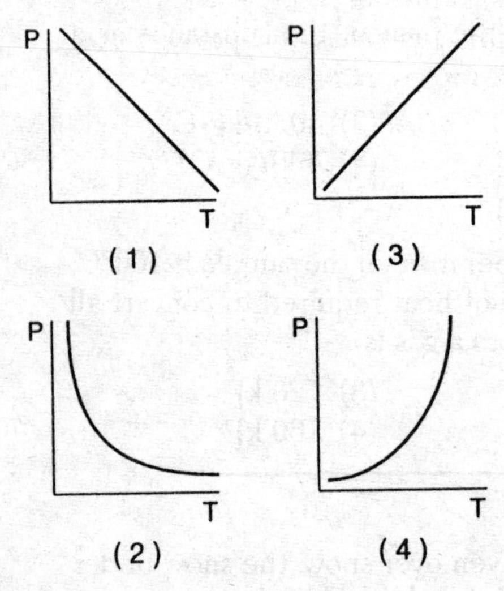

73 ____

74 Which characteristic of a gas sample results from the collision of gas molecules with the walls of its container?

1 specific heat
2 condensation point
3 temperature
4 pressure

74 ____

Note that question 75 has only three choices.

75 As the absolute temperature of a fixed mass of an ideal gas is increased at constant pressure, the volume occupied by the gas

1 decreases
2 increases
3 remains the same

75 ____

GROUP 3—**Electromagnetic Applications**

If you choose this group, be sure to answer questions **76–85**.

76 Which diagram best represents how galvanometer G can be modified to make it a voltmeter? [In the diagrams, R represents resistance.]

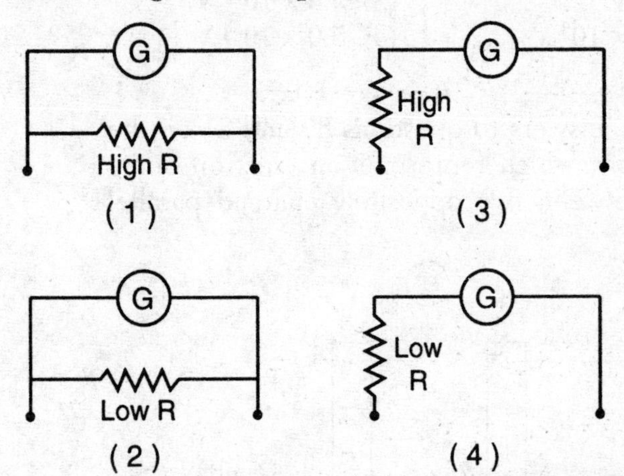

76 _____

77 A motor is to rotational mechanical energy as a generator is to

1 chemical potential energy
2 induced electrical energy
3 thermal internal energy
4 elastic potential energy

77 _____

78 A split-ring commutator is used to

1 reduce the voltage in a transformer
2 reduce the resistance of the shunt in an ammeter
3 make the light waves coherent in a laser
4 keep the torque acting in the same direction in a motor

78 _____

79 A straight conductor 1.0 meter long is moved at a constant speed of 10. meters per second perpendicular to a magnetic field. If the flux density of the field is 5.0×10^{-3} tesla, what is the magnitude of the electromotive force induced in the conductor?

(1) 0.0 V
(2) 2.0×10^{3} V
(3) 5.0×10^{-2} V
(4) 5.0×10^{-4} V

79 _____

Base your answers to questions 80 and 81 on the diagram below which represents an electron being projected between two oppositely charged parallel plates.

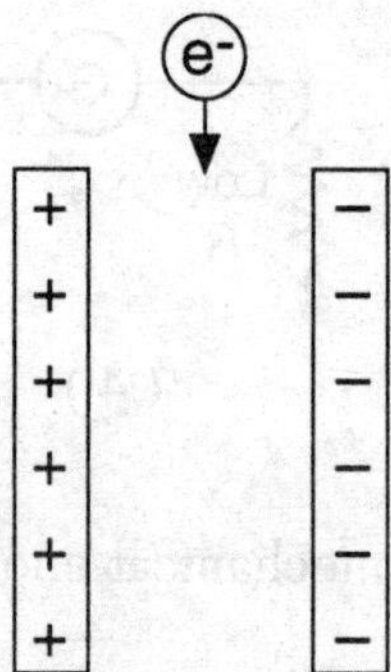

80 In which direction will the electric field deflect the electron?

1 into the page
2 out of the page
3 to the right
4 to the left

80 _____

Note that question 81 has only three choices.

81 As the electron moves through the electric field, the magnitude of the electric force on the electron

1 decreases 2 increases 3 remains the same

81 _____

82 The diagram below represents a negatively charged oil drop between two oppositely charged parallel plates. The forces acting on the oil drop are in equilibrium.

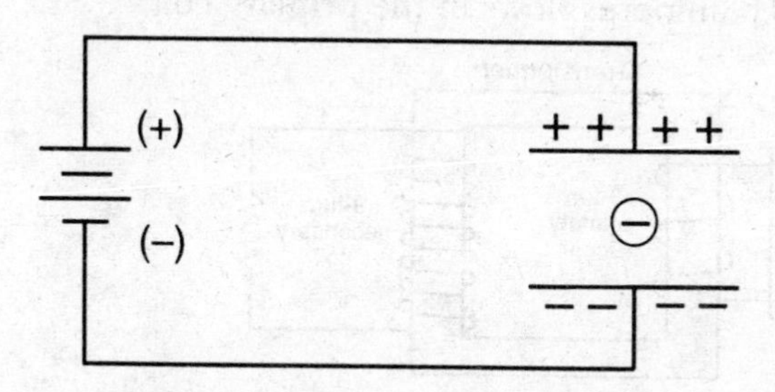

The oil drop could have a charge of

(1) 6.4×10^{-19} C
(2) 2.0×10^{-19} C
(3) 1.6×10^{-38} C
(4) 3.2×10^{-50} C

82 _____

83 As a charged particle moves through a magnetic field, the particle is deflected. The magnitude of the magnetic force acting on the particle is directly proportional to the

1 mass of the particle
2 electric charge on the particle
3 polarity of the magnetic field
4 work done on the charge by the magnetic field

83 _____

84 The 100% efficient transformer in the diagram below has three turns in its primary coil and nine turns in its secondary coil. When a 12-volt alternating current source is connected to the primary coil, 3.0 amperes flows in the primary coil.

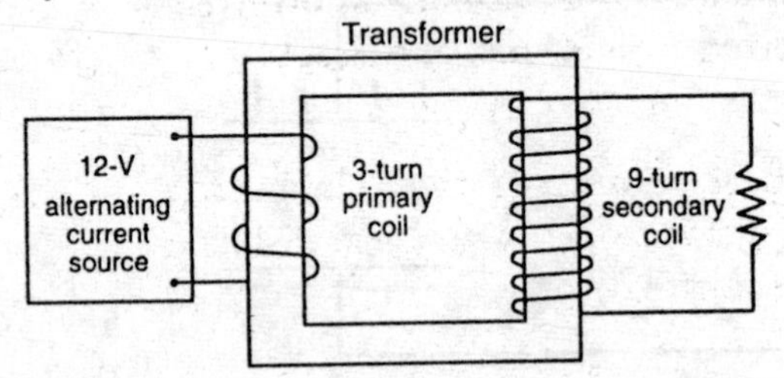

What potential difference and current are induced in the secondary coil?

(1) 36 V and 1.0 A
(2) 36 V and 9.0 A
(3) 4.0 V and 1.0 A
(4) 4.0 V and 9.0 A

84 _____

85 The diagram below shows particles produced by thermionic emission at the end of a heater element about to enter a magnetic field directed into the page.

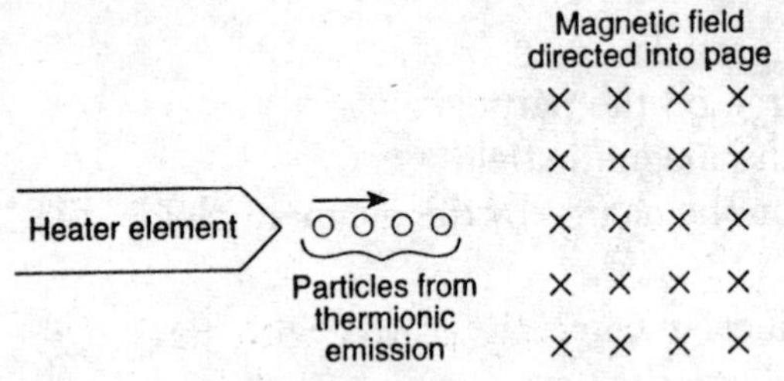

Upon entering the magnetic field, the particles will be deflected

1 toward the top of the page
2 toward the bottom of the page
3 into the page
4 out of the page

85 _____

GROUP 4—Geometric Optics

If you choose this group, be sure to answer questions **86–95**.

86 A plane mirror will form an image that is

1 virtual and erect
2 real and inverted
3 virtual and inverted
4 real and erect

86 _____

87 Which graph best represents the relationship between image distance (d_i) and object distance (d_o) for a plane mirror?

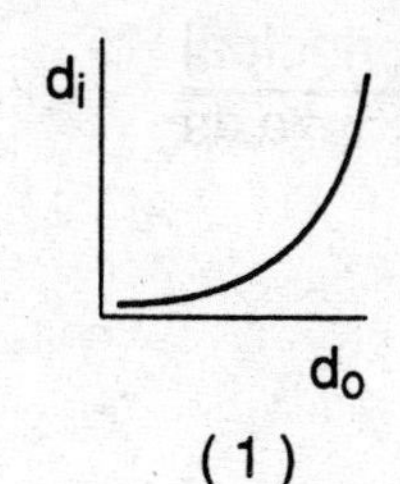

(1)

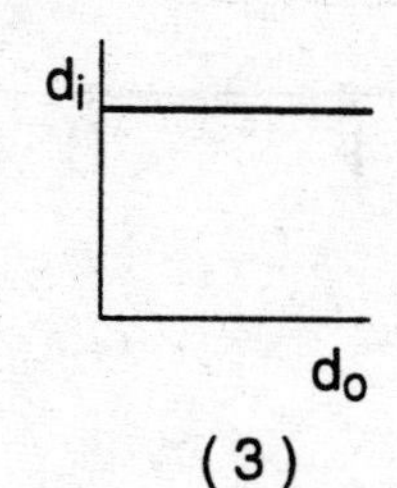

(3)

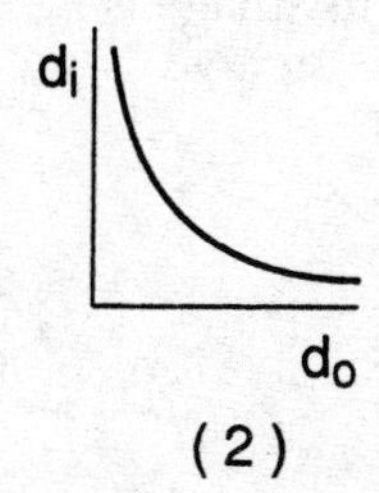

(2)

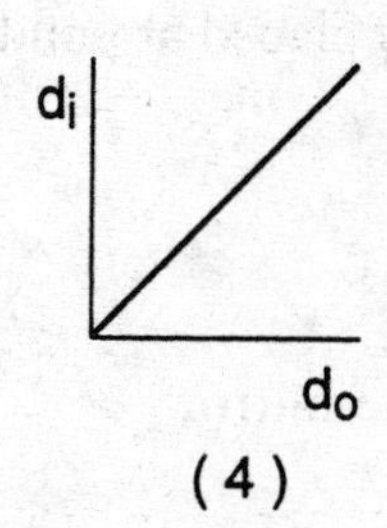

(4)

87 _____

88 Which lens defect is correctly paired with its cause?

1 chromatic aberration, caused by refraction
2 chromatic aberration, caused by diffraction
3 spherical aberration, caused by wave interference
4 spherical aberration, caused by wave polarization

88 _____

Base your answers to questions 89 and 90 on the information and diagram below. The diagram shows a concave (converging) spherical mirror having principal focus *F* and center of curvature *C*. Point *A* lies on the principal axis.

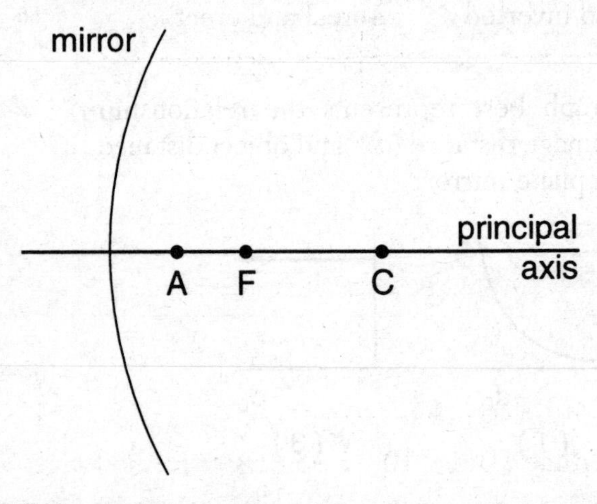

89 When an object is placed at point *A*, its image is observed

1 at *F*
2 between *F* and *C*
3 to the right of *C*
4 to the left of the mirror

89 _____

90 If an object is located at point *A*, its image is

1 virtual and inverted
2 virtual and erect
3 real and inverted
4 real and erect

90 _____

---

91 Which phenomenon allows a lens to focus light?

1 diffraction
2 refraction
3 interference
4 polarization

91 _____

92 The diagram below represents an object placed two focal lengths from a converging lens.

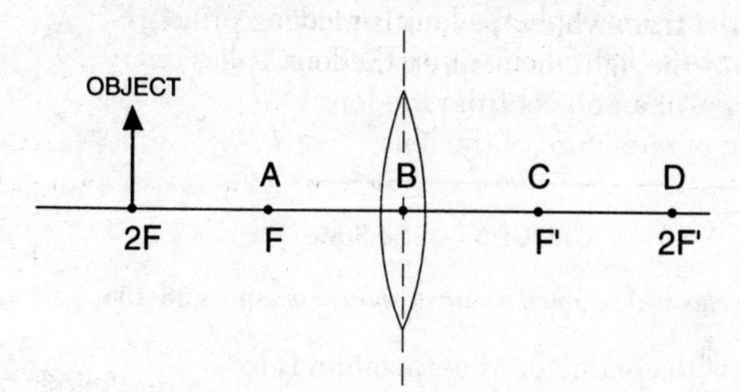

At which point will the image be located?

(1) *A* (3) *C*
(2) *B* (4) *D* 92 _____

93 An image that is 1.0×10^{-2} meter tall is formed on a screen behind a converging lens when an object 2.0 meters tall is placed 8.0 meters in front of the lens. What is the distance from the lens to the screen?

(1) 2.5×10^{-3} m (3) 4.0×10^{-2} m
(2) 2.5×10^{-1} m (4) 4.0×10^{-1} m 93 _____

94 A student uses a magnifying glass to examine the crystals in a mineral specimen. The magnifying glass contains a

1 convex (diverging) mirror
2 convex (converging) lens
3 concave (diverging) lens
4 plane mirror 94 _____

95 The focal length of a lens is *not* dependent on the

1 material from which the lens is made
2 color of the light incident on the lens
3 distance of an object from the lens
4 shape or curvature of the lens

95 _____

GROUP 5—**Solid State**

If you choose this group, be sure to answer questions **96–105**.

96 Copper is to conductor as germanium is to

1 insulator
2 fuse
3 resistor
4 semiconductor

96 _____

97 The diagram at the right shows the valence and conduction bands of an intrinsic semiconductor. Some of the electrons in the valence band have been promoted to the conduction band.

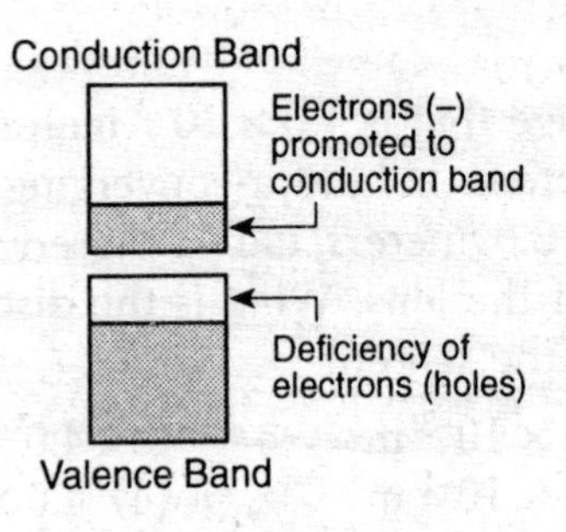

Which diagram below best represents the valence and conduction bands of the intrinsic semiconductor when its temperature is increased?

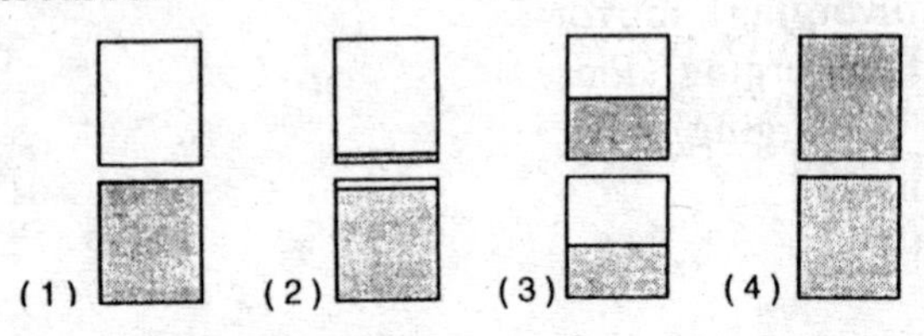

97 _____

98 What does the diagram below represent?

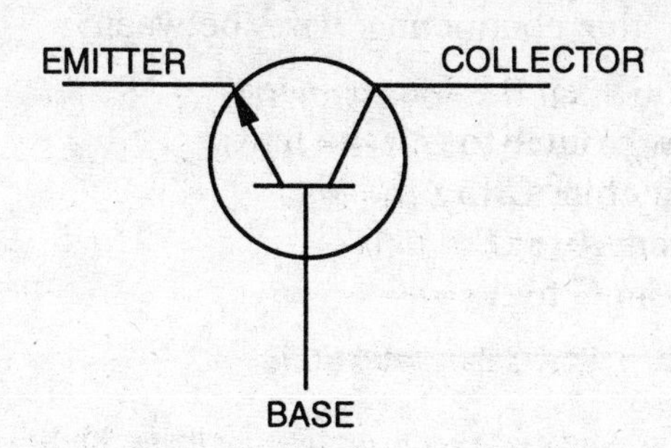

1 an *N-P-N* transistor
2 a *P-N-P* transistor
3 a zener diode
4 an avalanche region

98 _____

Base your answers to questions 99 and 100 on the diagram below which represents an *N*-type silicon semiconductor connected to a battery.

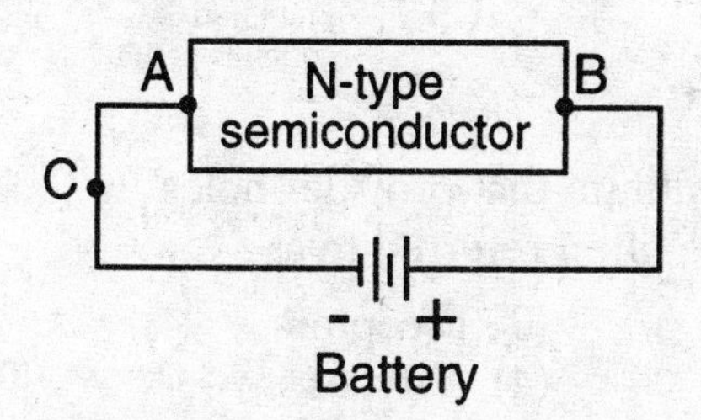

99 A very small amount of antimony, which has 5 valence electrons, had previously been added to the silicon crystal. This process produced

1 an excess of free electrons
2 an excess of free protons
3 more resistance
4 a higher emf

99 _____

100 Which statement best describes the flow of charge through the conducting wire between points *A* and *C*?

1 Electrons flow from *A* to *C*.
2 Electrons flow from *C* to *A*.
3 Holes flow from *A* to *C*.
4 Holes flow from *C* to *A*. 100 ____

---

101 Which diagram below represents a *P-N* junction with a forward bias?

+ P N −
(1)

+ P N +
(3)

− P N +
(2)

− P N −
(4)

101 ____

102 Materials such as indium that provide "holes" to semiconductors are often referred to as

(1) holistic
(2) donors
(3) acceptors
(4) *N*-types

102 ____

103 The diagram below shows alternating current input to a black box containing diodes.

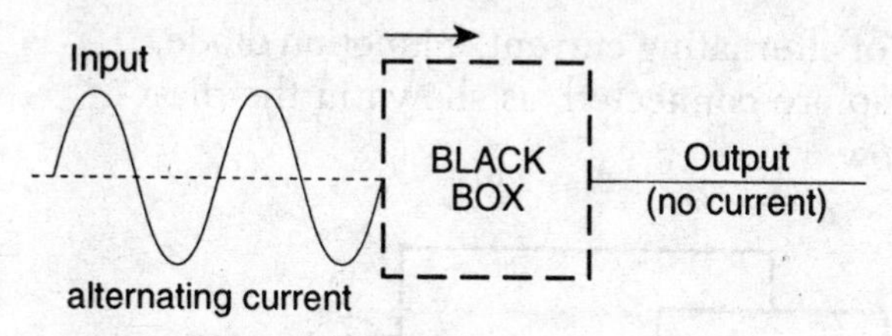

Which diode configuration in the black box accounts for the output (no current)?

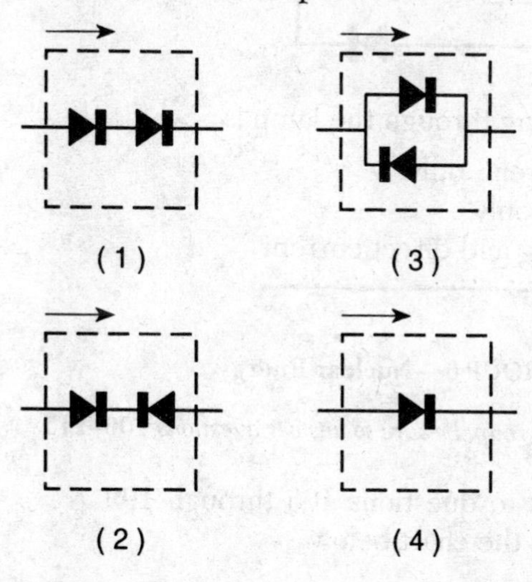

103 _____

104 As a result of transistor amplification, small increases in the emitter-base current will bring about large

1 decreases in the emitter-base voltage
2 decreases in the collector current
3 increases in the emitter-base voltage
4 increases in the collector current

104 _____

Note that question 105 has only three choices.

105 A source of alternating current, a junction diode, and a lamp are connected, as shown in the diagram below.

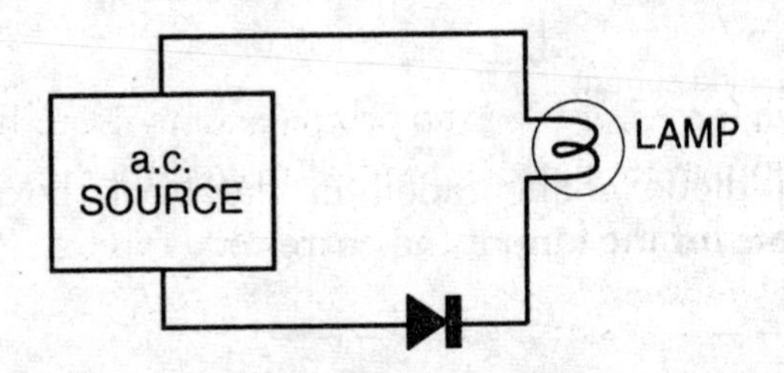

The current passing through the lamp is

1 alternating current, only
2 direct current, only
3 both alternating and direct current 105 _____

GROUP 6—**Nuclear Energy**

If you choose this group, be sure to answer questions **106–115**.

Base your answers to questions 106 through 109 on the information in the chart below.

Particle	Rest Mass
proton	1.0073 u
neutron	1.0087 u

106 The energy equivalent of the rest mass of a proton is approximately

(1) 9.4×10^2 MeV
(2) 1.9×10^3 MeV
(3) 9.1×10^{16} MeV
(4) 6.4×10^{18} MeV 106 _____

107 A tritium nucleus consists of one proton and two neutrons and has a total mass of 3.0170 atomic mass units. What is the mass defect of the tritium nucleus?

(1) 0.0014 u (3) 1.0010 u
(2) 0.0077 u (4) 2.0160 u 107 _____

108 Which force between the proton and neutrons in a tritium atom has the greatest magnitude?

1 electrostatic force 3 magnetic force
2 gravitational force 4 nuclear force 108 _____

109 Tritium would most likely be used as a

1 fuel in a fusion reaction
2 fuel in a fission reaction
3 coolant in a nuclear reactor
4 moderator in a nuclear reactor 109 _____

---

110 A nucleus having an odd number of protons and an odd number of neutrons is likely to be radioactive. Which nuclide matches this description?

(1) $^{29}_{14}Si$ (3) $^{32}_{16}S$

(2) $^{32}_{15}P$ (4) $^{35}_{17}Cl$ 110 _____

111 How do cloud chambers, spark chambers, and Geiger counters aid in the study of the nucleus?

1 They detect subatomic particles that exit the nucleus.
2 They detect the presence of a magnetic field around the nucleus.
3 They accelerate the nucleus before it collides with the particle beam.
4 They accelerate subatomic particles that exit the nucleus.

111 _____

112 Which nuclear particle is emitted as an atom of $^{238}_{92}U$ decays to $^{234}_{90}Th$?

1 neutron
2 positron
3 alpha particle
4 beta particle

112 _____

113 In the equation below, what is particle X?

$$^{9}_{4}Be + ^{4}_{2}He \rightarrow ^{12}_{6}C + X$$

1 an electron
2 a proton
3 a positron
4 a neutron

113 _____

114 In a nuclear reactor, the function of a control rod is to

1 slow down neutrons
2 speed up neutrons
3 absorb neutrons
4 produce neutrons

114 _____

115 The radioactive waste strontium-90 has a half-life of 28 years. How long must a sample of strontium-90 be stored to insure that only $\frac{1}{16}$ of the original sample remains as radioactive strontium-90?

(1) 28 years
(2) 56 years
(3) 84 years
(4) 112 years

115 _____

PART III

You must answer *all* questions in this part. [15]

Base your answers to questions 116 through 118 on the diagram and data table below. The diagram shows a worker moving a 50.0-kilogram safe up a ramp by applying a constant force of 300. newtons parallel to the ramp. The data table shows the position of the safe as a function of time.

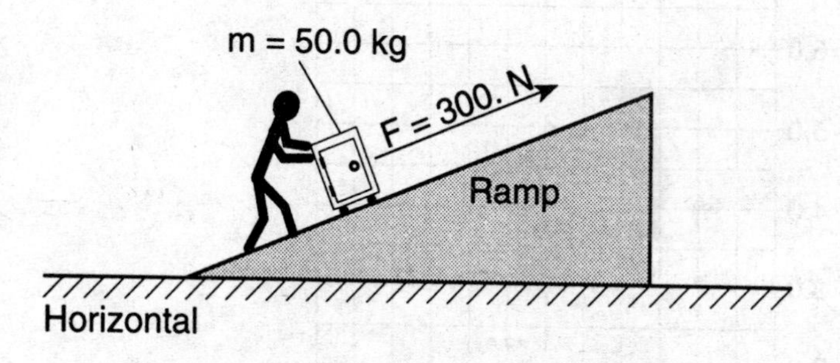

Time (s)	Distance Moved up the Ramp (m)
0.0	0.0
1.0	2.2
2.0	4.6
3.0	6.6
4.0	8.6
5.0	11.0

116 Using the information in the data table, construct a line graph on the grid provided. Plot the data points *and* draw the best-fit line. [2] 116 _____

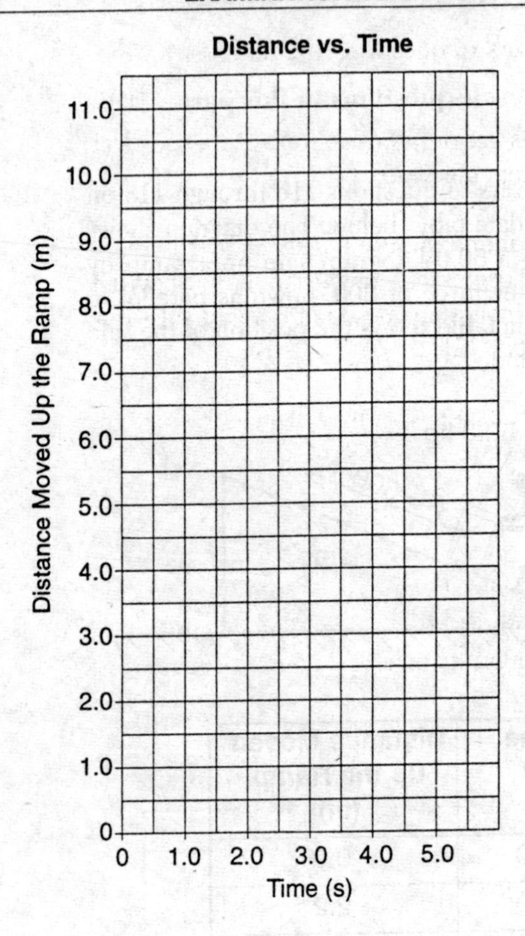

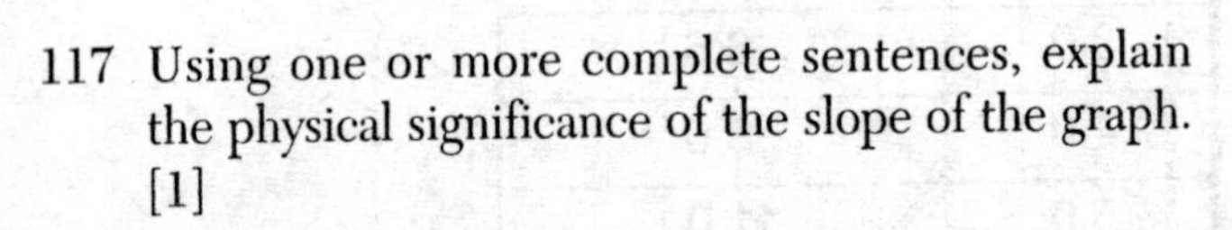

117 Using one or more complete sentences, explain the physical significance of the slope of the graph. [1]

117 ______

118 Calculate the work done by the worker in the first 3.0 seconds. [Show all calculations, including the equation and substitution with units.] [2]

118 ______

Base your answers to questions 119 and 120 on the information below.

An electron is accelerated from rest to a speed of 2.0×10^6 meters per second.

119 How much kinetic energy is gained by the electron as it is accelerated to this speed? [Show all calculations, including the equation and substitution with units.] [2]

120 What is the matter wavelength of the electron after it is accelerated to this speed? [Show all calculations, including the equations and substitution with units.] [3]

Base your answers to questions 121 through 124 on the information and diagram below. The diagram represents a wave generator having a constant frequency of 12 hertz producing parallel wave fronts in a ripple tank. The velocity of the waves is v.

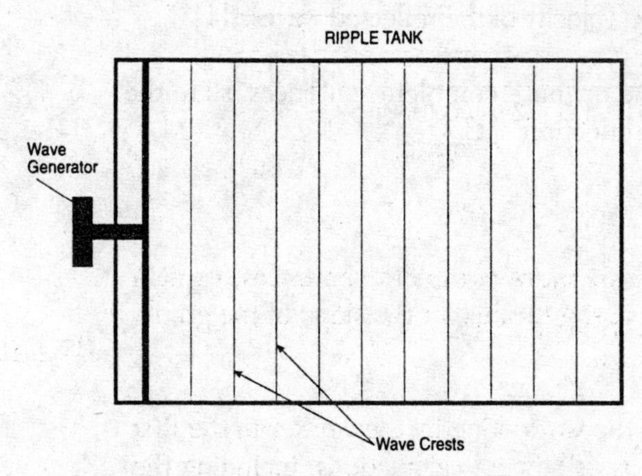

121 Using a ruler, measure the wavelength of the waves shown and record the value *to the nearest tenth of a centimeter.* [1]

121 ______

122 Determine the speed of the waves in the ripple tank. [Show all calculations, including the equation and substitution with units.] [2] 122 ______

123 A barrier is placed in the ripple tank as shown in the diagram below. 123 ______

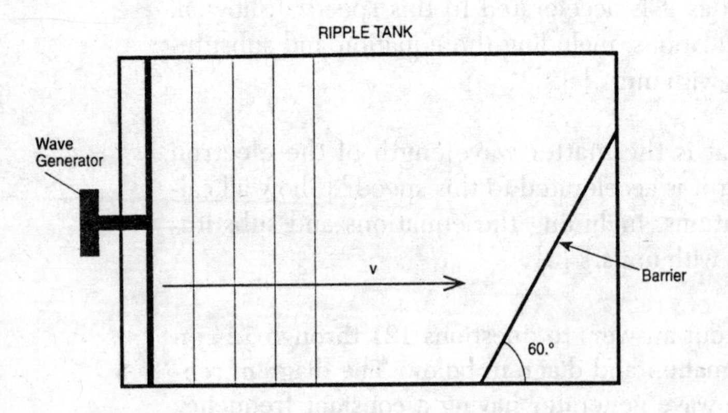

On the diagram, use a protractor and a straightedge to construct an arrow to represent the direction of the velocity of the reflected waves. [1]

124 Using one or more complete sentences, state the Law of Reflection. [1] 124 ______

Answers June 1996

Physics

Answer Key

PART I

1. 2	**11.** 2	**20.** 4	**29.** 3	**38.** 1	**47.** 2
2. 3	**12.** 3	**21.** 2	**30.** 1	**39.** 2	**48.** 1
3. 3	**13.** 2	**22.** 1	**31.** 2	**40.** 4	**49.** 1
4. 2	**14.** 3	**23.** 2	**32.** 2	**41.** 4	**50.** 4
5. 2	**15.** 1	**24.** 3	**33.** 1	**42.** 3	**51.** 4
6. 4	**16.** 2	**25.** 4	**34.** 3	**43.** 1	**52.** 2
7. 1	**17.** 3	**26.** 3	**35.** 4	**44.** 3	**53.** 3
8. 1	**18.** 2	**27.** 4	**36.** 2	**45.** 1	**54.** 1
9. 2	**19.** 1	**28.** 4	**37.** 3	**46.** 3	**55.** 2
10. 4					

PART II

Group 1	Group 2	Group 3	Group 4	Group 5	Group 6
56. 1	**66.** 4	**76.** 3	**86.** 1	**96.** 4	**106.** 1
57. 2	**67.** 1	**77.** 2	**87.** 4	**97.** 3	**107.** 2
58. 2	**68.** 3	**78.** 4	**88.** 1	**98.** 1	**108.** 4
59. 2	**69.** 2	**79.** 3	**89.** 4	**99.** 1	**109.** 1
60. 3	**70.** 4	**80.** 4	**90.** 2	**100.** 2	**110.** 2
61. 1	**71.** 1	**81.** 3	**91.** 2	**101.** 1	**111.** 1
62. 3	**72.** 1	**82.** 1	**92.** 4	**102.** 3	**112.** 3
63. 2	**73.** 3	**83.** 2	**93.** 3	**103.** 2	**113.** 4
64. 3	**74.** 4	**84.** 1	**94.** 2	**104.** 4	**114.** 3
65. 4	**75.** 2	**85.** 2	**95.** 3	**105.** 2	**115.** 4

PART III — See answers explained

Answers Explained

PART I

1. **2** The best way to approach any word problem is to correctly identify the quantities given and the quantity to be found.

Given: $s_i = 100.\text{ m}$ Find: $\overline{v} = ?$

$s_f = 250.\text{ m}$

$\Delta t = 10.\text{ s}$

Refer to the *Mechanics* section in the *Reference Tables* to find an equation that relates average speed with displacement and time.

Solution: $$\overline{v} = \frac{\Delta s}{\Delta t} = \frac{s_f - s_i}{\Delta t}$$

$$= \frac{250.\text{ m} - 100.\text{ m}}{10.\text{ s}}$$

$$= 15\text{ m/s}$$

2. **3** To find a *resultant vector*, place the given vectors head to tail, and draw the resultant vector from the tail of the first vector to the head of the last vector.

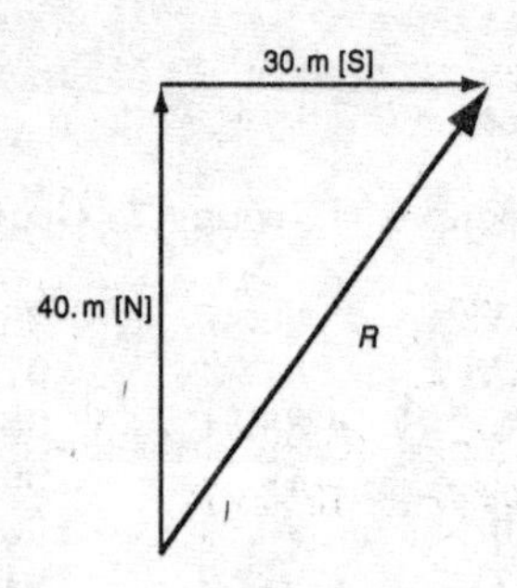

Since this is a right triangle, use the Pythagorean theorem:

$$a^2 + b^2 = c^2$$

to determine the magnitude of the resultant.

$$R^2 = (40.\text{ m})^2 + (30.\text{ m})^2 = 2500\text{ m}^2$$

$$R = 50.\text{ m}$$

3. **3** Write out the problem.

Given: From the graph: Find: $a = ?$

$v_f = 30.\ \text{m/s}$

$v_i = 0\ \text{m/s}$

$\Delta t = 3.0\ \text{s}$

Refer to the *Mechanics* section in the *Reference Tables* to find an equation that relates acceleration with velocity and time.

Solution: $$a = \frac{\Delta v}{\Delta t} = \frac{v_f - v_i}{\Delta t}$$

$$= \frac{30.\ \text{m/s} - 0\ \text{m/s}}{3.0\ \text{s}}$$

$$= 10.\ \text{m/s}^2$$

4. **2** Velocity is defined as the time rate change of displacement. The spacing between the dots, which represent oil drops, represents displacement, and the time intervals are equal; therefore, since the displacements are gradually increasing, the velocity of the car is gradually increasing from rest. Since acceleration is defined as the time rate change of velocity, the car has accelerated from rest.

Wrong Choices Explained:

(1) This pattern shows constant velocity because the displacements are equal.

(3) This pattern shows an initial constant velocity because the first few displacements are equal, followed by another set of equal displacements larger than the first set, indicating a new, larger constant velocity.

(4) This pattern shows varying velocities, increasing and decreasing because the displacements are increasing and decreasing.

5. **2** Write out the problem.

Given: $v_i = 0\ \text{m/s}$ Find: $\Delta t = ?$

$\Delta s = 0.20\ \text{m}$

$a = 9.8\ \text{m/s}^2$ (acceleration due to gravity; see the List of Physical Constants in the *Reference Tables*.)

Refer to the *Mechanics* section in the *Reference Tables* to find an equation that relates time with velocity, displacement, and acceleration.

Solution: $\Delta s = v_i \Delta t + \frac{1}{2} a(\Delta t)^2$

Then, since $v_i = 0$,

$$\Delta s = \frac{1}{2} a \Delta t^2$$

$$\Delta t = \sqrt{\frac{2\Delta s}{a}}$$

$$= \sqrt{\frac{2(0.20 \text{ m})}{9.8 \text{ m/s}^2}}$$

$$= 0.20 \text{ s}$$

6. **4** Write out the problem.

Given: $v_i = 10. \text{ m/s}$ Find: $v_f = ?$

$a = 1.5 \text{ m/s}^2$

$\Delta s = 600. \text{ m}$

Refer to the *Mechanics* section in the *Reference Tables* to find an equation that relates velocity with displacement and acceleration.

Solution: $v_f^{\,2} = v_i^{\,2} + 2a\Delta s$

$$v_f = \sqrt{v_i^{\,2} + 2a\Delta s}$$

$$= \sqrt{(10. \text{ m/s})^2 + 2(1.5 \text{ m/s}^2)(600. \text{ m})}$$

$$= 44. \text{ m/s}$$

7. **1** Free-fall motion is independent of mass; therefore the ratio of speeds is 1:1 because the rock and the stone fall from rest from the same height for the same amount of time with the same acceleration. As a result, the speed is the same for both.

8. **1** A maximum magnitude of a resultant vector occurs when the two component vectors are in the same direction and 0 degree apart, and a minimum occurs when they are 180 degrees apart, acting completely against each other. All other angles produce resultant vectors between this maximum and this minimum. The concurrent forces in choice (1) produce a resultant force with a maximum of 20 newtons and a minimum of 0 newton; 10 newtons falls within this range.

Wrong Choices Explained:

(2) The resultant force has a maximum of 40 N, a minimum of 20 N.

(3) The resultant force has a maximum of 9.4 N, a minimum of 0 N.
(4) The resultant force has a maximum of 9.7 N, a minimum of 0.3 N.

9. **2** Write out the problem.

Given: $F = 300.$ N Find: $F_{\perp} = ?$

$\theta = 60°$ to horizontal

To solve this problem, resolve vector *F* into its horizontal and vertical components.

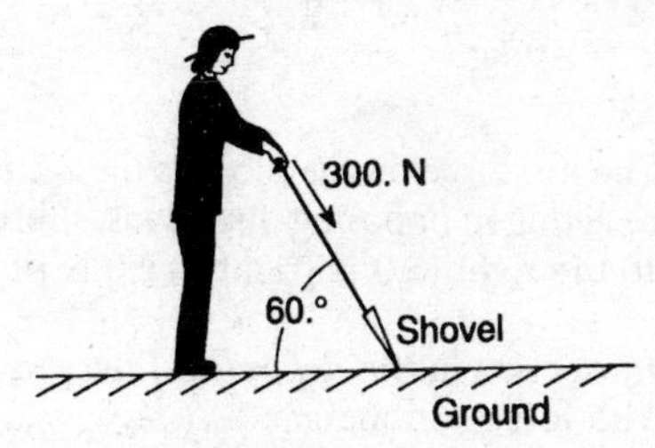

Solution:
$$\sin\theta = \frac{\text{opposite side}}{\text{hypotenuse}}$$
$$= \frac{F_{\perp}}{F}$$
$$F_{\perp} = F\sin\theta$$
$$= (300.\text{ N})(\sin 60°)$$
$$= 260.\text{ N}$$

10. **4** If an object has *no* unbalanced force acting on it, it experiences no acceleration. Graph (4) has a constant slope, which indicates a constant velocity, since the slope of a displacement versus time graph is equal to the velocity.

Wrong Choices Explained:

(1) This graph shows an increasing acceleration per unit time; therefore, there must be an unbalanced force.

(2) This graph shows an increasing momentum per unit time. The slope of this graph is equal to the change in momentum per unit time, which, by Newton's second law, is equal to the force. Since the slope is not zero, the force on the object is not zero.

(3) This graph shows an increasing velocity per unit time; therefore, the object experiences acceleration, indicating that it is experiencing an unbalanced force.

11. **2** The newton is the SI unit of force. A newton is approximately equal to 0.22 pound, a little less than a quarter pound. Of the four choices, 10^0 N, or 1 newton, is the only reasonable one for the weight of an apple.

Wrong Choices Explained:

(1) 10^{-2} N = 0.0022 lb or less than 0.04 oz.
(3) 10^{2} N = 22 lb.
(4) 10^{4} N = 2200 lb.

12. **3** Write out the problem.

Given: $F_1 = 2.0$ N to the left   Find: $a = ?$
$F_2 = 8.0$ N to the right
$m = 2.0$ kg

Solution: The net force on the block is the resultant of the two force vectors. Since they are acting in opposing directions, subtract their magnitudes to obtain 6 newtons to the right (8.0 N [right] – 2.0 N [left]).

Refer to the *Mechanics* section in the *Reference Tables* to find an equation that relates acceleration with force and mass.

$$F_{net} = ma$$

$$a = \frac{F_{net}}{m}$$

$$= \frac{6.0 \text{ N[right]}}{2.0 \text{ kg}}$$

$$= 3.0 \text{ m/s}^2 \text{[right]}$$

13. **2** Inertia is directly proportional to the mass of an object. Since the Styrofoam ball has twice the mass of the steel ball, its inertia is twice as great.

14. **3** Write out the problem.

Given: $m = 3.0$ kg   Find: $g = ?$
$w = 15$ N

Refer to the *Mechanics* section in the *Reference Tables* to find an equation that relates gravitational acceleration with mass and weight.

Solution: $w = mg$

$$g = \frac{w}{m}$$

$$= \frac{15 \text{ N}}{3.0 \text{ kg}}$$

$$= 5.0 \text{ m/s}^2$$

15. **1** Newton's third law states that, whenever one object exerts a force on a second object, the second object exerts on the first object an equal force that is opposite in direction. These equal and opposing forces are also known as action-reaction. Since the inflated balloon is moving, it must be overcoming the force of fluid friction due to the surrounding air and thus must be experiencing a force to the right since that is its direction of motion. Also, since the air from the balloon is moving to the left, it must be experiencing a force to the left. Choice (1) is the reasonable explanation for the the source of these opposite forces.

Wrong Choices Explained:

(2) Centripetal force is the force needed to keep an object moving in a circular path.

(3) Gravitational attraction is the force of attraction that exists between masses.

(4) Rolling friction is the force opposing the motion of one body that is rolling over another body.

16. **2** Since the cart is moving at constant velocity, there is no acceleration and, therefore, no net force on the cart. The force of friction always acts in the direction that opposes the motion of an object; in this case the force of friction, which is 10. newtons, is to the left. The horizontal force to the right along the handle of the cart must therefore also be equal to 10 newtons.

17. **3** Write out the problem.

Given: $v_i = 5.0 \times 10^2$ m/s   Find: $m = ?$

$v_f = 0$ m/s

$J = 50$ N·s

Refer to the *Mechanics* section in the *Reference Tables* to find equations that relate mass with velocity and impulse.

Solution:

$$J = F\Delta t$$

$$F\Delta t = m\Delta v$$

$$\therefore J = m\Delta v$$

$$m = \frac{J}{\Delta v}$$

$$= \frac{50 \text{ N}\cdot\text{s}}{5.0 \times 10^2 \text{ m/s}}$$

$$= 1.0 \times 10^{-1} \text{ kg}$$

18. **2** Write out the problem.

Given: $w = 1.0 \times 10^2$ N Find: PE = ?

$h = 6.0$ m

Refer to the *Energy* and *Mechanics* sections in the *Reference Tables* to find equations that relate potential energy with weight and height.

Solution: $\Delta PE = mg\Delta h$

$$w = mg$$

$$\therefore \Delta PE = wh$$

$$\Delta PE = (1.0 \times 10^2 \text{ N})(6.0 \text{ m})$$

$$= 6.0 \times 10^2 \text{ J}$$

19. **1** Write out the problem.

Given: $F = 10.$ N Find: $P = ?$

$m = 3.0$ kg

$\Delta s = 8.0$ m

$\Delta t = 2.0$ s

Refer to the *Energy* section in the *Reference Tables* to find an equation that relates power with force, displacement, and time.

Solution: $P = \dfrac{F\Delta s}{\Delta t}$

$$= \frac{(10. \text{ N})(8.0 \text{ m})}{2.0 \text{ s}}$$

$$= 40. \text{ W}$$

(Note that mass is not needed in this calculation)

20. **4** Write out the problem.

Given: $F = 20.$ N Find: $k = ?$

$x = 0.50$ m

Refer to the *Energy* section in the *Reference Tables* to find an equation that relates the spring constant with force and displacement from equilibrium.

Solution: $F = kx$

$$k = \frac{F}{x}$$

$$= \frac{20.\ \text{N}}{0.50\ \text{m}}$$

$$= 40.\ \text{N/m}$$

21. **2** In the *Energy* section in the *Reference Tables*, find that PE_S, the potential energy stored in a spring, is equal to $\frac{1}{2}kx^2$. Potential energy is directly proportional to the spring constant. Therefore, since the two springs, *A* and *B*, are stretched the same distance from equilibrium, and spring *B* has a spring constant twice as large as the spring constant of spring *A*, the potential energy stored in *B* will be twice as great as the potential energy stored in *A*.

22. **1** In the *Energy* section in the *Reference Tables*, find that kinetic energy, KE, is equal to $\frac{1}{2}mv^2$. Kinetic energy is directly proportional to mass and directly proportional to velocity squared. Therefore, doubling the mass would have the effect of doubling the kinetic energy, and halving the velocity would have the effect of quartering the kinetic energy. The product of these changes is $\frac{1}{2}$, and therefore the kinetic energy would be half as great.

23. **2** In the *Electricity and Magnetism* section in the *Reference Tables*, find that electrostatic force F, is equal to $\frac{kq_1q_2}{r_2}$. Electrostatic force is directly proportional to the magnitude of each of the two charges and inversely proportional to the distance squared. Therefore, doubling each of the two charges would have the effect of making the force 4 times as great, and doubling r would have the effect of quartering the force. The product of these changes is 1, and therefore there would be *no* net change in the force.

Wrong Choices Explained:

(1) This combination of changes would halve the force.

(3) This combination of changes would increase the force by a factor of 8.

(4) This combination of changes would increase the force by a factor of 16.

24. **3** Like charges repel and opposite charges attract. The inflated balloon has acquired a negative charge by being rubbed against the person's hair. It will repel negative charges in the wall away from the surface, leaving the wall's surface slightly positive, so that is attracts the negative balloon. Diagram (3) best shows the charge distribution on the balloon and wall.

Wrong Choices Explained:

(1) This diagram shows an equal distribution of charge in the wall; no separation of charge has occurred.

(2), (4) These diagrams show a separation of charge, but the surface of the wall is negative.

25. **4** Assume for this question that the two spheres are identical. That being the case, when they are brought into contact, charges will be transferred between them until the charge on both spheres *is the same*:

$$\frac{(4.0\times10^{-6}\text{C})+(2.0\times10^{-5}\text{C})}{2}=1.2\times10^{-5}\text{C}$$

26. **3** Refer to the *Electricity and Magnetism* section in the *Reference Tables* to find equations that relate electric field intensity with distance.

Solution: $$E=\frac{F}{q}$$

$$F=\frac{kQq}{r^2}$$

$$\therefore E=\frac{kQ}{r^2}$$

This equation shows that electric field intensity is inversely proportional to distance squared. Only graph (3) shows an inverse squared relationship.

Wrong Choices Explained:

(1) According to this graph, electric field intensity does not change with increasing distance.

(2) According to this graph, electric field intensity increases proportionally with increasing distance.

(4) According to this graph, electric field intensity increases exponentially with increasing distance.

27. **4** Write out the problem.

Given: $q=2.0\text{ C}$ Find: $V=?$

$W=8.0\text{ J}$

Refer to the *Electricity and Magnetism* section in the *Reference Tables* to find an equation that relates electric potential with charge and work.

Solution: $$V=\frac{W}{q}$$

$$=\frac{8.0\text{ J}}{2.0\text{ C}}$$

$$=4.0\text{ V}$$

28. **4** Write out the problem.

Given: $^{*}q = 1.6 \times 10^{-19}$ C Find: $F = ?$

$E = 3.0 \times 10^{3}$ N/C

*(one elementary charge; see the List of Physical Constants in the *Reference Tables*.)

Refer to the *Electricity and Magnetism* section in the *Reference Tables* to find an equation that relates force with charge and electric field intensity.

Solution: $E = \frac{F}{q}$

$$F = Eq$$

$$= (3.0 \times 10^{3} \text{ N/C})(1.6 \times 10^{-19} \text{ C})$$

$$= 4.8 \times 10^{-16} \text{ N}$$

29. **3** Resistance is directly proportional to the resistivity (ρ) of the conductor for a given temperature and to the length of the conductor, and inversely proportional to the cross-sectional area of the conductor:

$$R = \rho \frac{L}{A}$$

For a conductor, resistivity increases with increasing temperature; therefore resistance also increases with increasing temperature.

Wrong Choices Explained:

(1) A shorter conductor would have decreased resistance.

(2) For substances that obey Ohm's law, resistance is independent of the applied voltage. Therefore, changing the voltage in the circuit would have no effect.

(4) Increasing the cross-sectional area of the conductor would decrease resistance.

30. **1** Write out the problem.

Given: A series circuit Find: $V_1 = ?$

$R_1 = 40.\ \Omega$

$R_2 = 20.\ \Omega$

$V_2 = 80.$ V

Refer to the *Electricity and Magnetism* section in the *Reference Tables* to find equations that relate electric potential difference current and resistance for a series circuit.

Solution: $R = \frac{V}{I}$

$$I = \frac{V}{R}$$

$$I_1 = I_2 = I_3 = ...$$

$$\therefore \frac{V_1}{R_1} = \frac{V_2}{R_2}$$

$$V_1 = \frac{V_2R_1}{R_2}$$

$$= \frac{(80.\text{ V})(40.\ \Omega)}{20.\ \Omega}$$

$$= 160\text{ V}$$

31. **2** Write out the problem.

Given: $q = 18\text{ C}$ Find: $I = ?$

$\Delta t = 2.0 \times 10^{-2}\text{ s}$

$V = 1.5 \times 10^{6}\text{ V}$

Refer to the *Electricity and Magnetism* section in the *Reference Tables* to find an equation that relates current with charge and time.

Solution: $I = \frac{\Delta q}{\Delta t}$

$$= \frac{18\text{ C}}{2.0 \times 10^{-2}\text{ s}}$$

$$= 9.0 \times 10^{2}\text{ A}$$

(Note that electric potential difference is not needed in this calculation.)

32. **2** By the law of conservation of energy, the sum of all the currents entering a junction must equal the sum of all the currents leaving the junction. From the directions of the arrows, calculate that 9 amperes of current (3A + 6A) are flowing into the junction; therefore 9 amperes must flow out. Since 4 amperes are leaving in one branch, the second branch, which ammeter *A* is measuring, must contain 5 amperes (9 A – 4 A) of current.

33. **1** To solve this problem, use the following left-hand rule: Point the thumb of the left hand in the direction of electron flow. The fingers of the left hand will wrap around the wire (from wrist to fingertips) in the direction of the magnetic field around the wire.

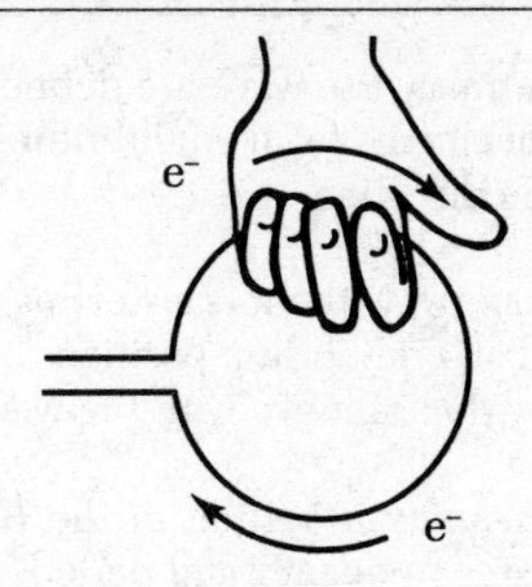

This method reveals that the magnetic field points out of the page at the center of the loop.

34. **3** A charged particle entering a magnetic field will experience a force unless it enters parallel to the magnetic field. If the force is present, it is always perpendicular to the particle's velocity. Consequently, the force, if present, can only change the *direction* of the particle's motion.

35. **4** Accelerating charges or changing magnetic fields produce electric and magnetic fields that move through space. These combined fields are known as electromagnetic waves. Choice (4), an electron, is the only choice that is a charged particle.

Wrong Choices Explained:

(1), (2), (3) All of these choices are particles that have no charge.

36. **2** Write out the problem.

Given: $P = 300.\text{ W}$ Find: $I = ?$

$V = 120\text{ V}$

Refer to the *Electricity and Magnetism* section in the *Reference Tables* to find an equation that relates current with power and electric potential difference.

Solution: $P = VI$

$$I = \frac{P}{V}$$

$$= \frac{300.\text{ W}}{120\text{ V}}$$

$$= 2.5\text{ A}$$

37. **3** The direction of a magnetic field line is defined as the direction that the north pole of a compass needle points to when it is placed in the magnetic field. On the outside of a magnet, the field lines leave from the north pole and enter through the south pole. Only in diagram (3) do the arrows of the field lines point in the correct direction.

38. **1** Two points on a transverse wave are defined as being in phase when they are at identical displacements from equilibrium and in the next instant of time will move in the same direction.

39. **2** Waves transfer energy without transferring mass. Individual particles of the rope vibrate around an equilibrium position as the periodic wave travels through the rope, but they do not move with the wave.

40. **4** In the *Wave Phenomena* section in the *Reference Tables*, find the equation $T = \frac{1}{f}$, which relates frequency and period. Period and frequency are inversely proportional to each other. Only graph (4) shows this relationship.

Wrong Choices Explained:

(1) This graph shows a direct relationship; as period increases, frequency also increases.

(2) According to this graph, a change in period has no effect on frequency.

(3) According to this graph, frequency increases exponentially with increasing period.

41. **4** Write out the problem.

Given: $\lambda = 5.9 \times 10^{-7}$ m   Find: $\theta_2 = ?$

medium 1 = air

medium 2 = Lucite

$\theta_2 = 45°$

Refer to the *Wave Phenomena* section in the *Reference Tables* to find an equation that relates angles of incidence and refraction.

Solution: $n_1 \sin\theta_1 = n_2 \sin\theta_2$

Refer to the *Absolute Indices of Refraction* table in the *Reference Tables* to find the values of n.

$$\sin\theta_2 = \frac{n_1 \sin\theta_1}{n_2}$$

$$\theta_2 = \sin^{-1}\frac{n_1 \sin\theta_1}{n_2}$$

$$= \sin^{-1}\frac{(1.00)(\sin 45°)}{1.50}$$

$$= 28°$$

42. **3** According to the diagram, point P is located where a crest meets a trough. Because of diffraction, that is, the bending of waves around an object, a wave passing through the two small openings became two new sources of circular waves. Since the two new waves came from the same source, left the barrier at the same instant, and travel in the same medium, they therefore have the same frequency and also the same wavelength. Consequently, if at some point a crest from one wave meets a trough of the other, the difference in their paths is some odd multiple of one-half wavelength. Specifically, in this diagram, from source A to point P there are four crests and four troughs, and from B there are three crests and four troughs, so path AP is one-half wavelength $\left(\frac{1}{2}\lambda\right)$ longer than path BP.

43. **1** Write out the problem.

Given: $v = 2.5 \times 10^8$ m/s   Find: $n = ?$

Refer to the *Wave Phenomena* section in the *Reference Tables* to find an equation(s) that relates absolute index of refraction and speed of light in a medium.

Solution: $n = \frac{c}{v}$

Refer to the *List of Physical Constants* in the *Reference Tables* to find the value for c, the speed of light in a vacuum.

$$n = \frac{3.0 \times 10^8 \text{ m/s}}{2.5 \times 10^8 \text{ m/s}}$$
$$= 1.2$$

44. **3** A wave front is defined as being all points on a three-dimensional wave that are in phase with each other.

Wrong Choices Explained:

(1) As can be seen in the diagram, a wave front can be in two different media; therefore, all points are not traveling with the same speed. The change in speed from one medium to another causes the bending of the wave front known as refraction.

(2) As can be seen from the diagram, when a wave front hits a boundary between two media obliquely, part of the wave front enters the new medium before the rest of the wave front.

(4) Superposition occurs when two different waves occupy the same region at the same time.

45. **1** Dispersion is the process of spreading light into its component colors. Monochromatic light is light of a single color. Light produced by a laser is coherent, that is, in phase, meaning that it has to be monochromatic.

46. **3** Diffraction is defined as the bending or spreading of waves into the region behind an obstacle. The pattern of waves behind the barrier in the diagram is due to diffraction.

Wrong Choices Explained:

(1) Reflection is the bouncing back of a wave that occurs when the wave reaches a barrier or the interface between two different media.

(2) Refraction is the bending of a wave from its original direction as it obliquely enters a second medium in which its speed changes.

(4) Interference is the reinforcement or cancellation of amplitude that occurs when two or more waves pass through the same region at the same time.

47. **2** The ratio of the speed of light in a vacuum to the speed of light in a medium is known as the absolute index of refraction of that medium. The slower the speed of light in the medium, the larger the value of its absolute index of refraction. Refer to the *Absolute Indices of Refraction* table in the *Reference Tables* to find the indices of refraction for air, crown glass, water, and diamond. Diamond has the largest absolute index of refraction (2.42).

Wrong Choices Explained:

(1) Air = 1.00
(3) Water = 1.33
(4) Crown glass = 1.52

48. **1** The photoelectric effect is the ejection of electrons from a photoemissive surface caused by the absorption of photons of sufficient energy. A photon is a bundle of light, also known as a quanta of light, and its energy is proportional to the frequency of the light used and is not proportional to the intensity of the light, which would correspond to the amplitude of a wave. Thus, the photoelectric effect can be explained by the particle theory of light but not by the wave theory of light.

Wrong Choices Explained:

(2) Constructive interference occurs as a result of the superposition of two waves occupying the same region at the same time and having amplitudes in the same direction of displacement from equilibrium.

(3) Polarization is the confinement of the vibrations of light waves to a single plane perpendicular to the direction of the wave's motion.

(4) Diffraction is the bending or spreading of waves into the region behind an obstacle.

49. **1** Write out the problem.

Given: $W_0 = 7.3 \times 10^{-19}$ J   Find: $KE_{max} = ?$

$E_{photon} = 9.9 \times 10^{-19}$ J

Refer to the *Modern Physics* section in the *Reference Tables* to find equations that relate maximum kinetic energy with the energy of an incident photon and the work function.

Solution:
$$E_{photon} = hf$$
$$KE_{max} = hf - W_0$$
$$\therefore KE_{max} = E_{photon} - W_0$$
$$= (9.9 \times 10^{-19}\ \text{J}) - (7.3 \times 10^{-19}\ \text{J})$$
$$= 2.6 \times 10^{-19}\ \text{J}$$

50. **4** Polarization is the confinement of the vibrations of a light wave to a single plane perpendicular to the direction of the wave's motion. This is what the sunglasses have done in the diagram.

Wrong Choices Explained:

(1) Dispersion is the process of spreading light into its component colors.

(2) Diffraction is the bending or spreading of waves into the region behind an obstacle.

(3) Internal reflection occurs when light passes from a medium with a higher index of refraction to a medium with a lower index of refraction and the angle of the wave with the boundary exceeds the critical angle.

51. **4** Refer to the *Energy Level Diagrams for Mercury and Hydrogen* in the *Reference Tables* to find the required information. According to the diagram on the right, the energy level for the n = 3 state is –1.51 eV. Therefore, 1.51 electron volts is the minimum amount of energy required to ionize the hydrogen atom in this state.

52. **2** For a photoelectron to be emitted from a photoemissive surface, the work function of the surface has to be met. The energy of a photon of light is directly proportional to the frequency of the light. Therefore, increasing the frequency of the light would increase the energy of the photon, enabling it to meet the work function of the surface and eject electrons.

Wrong Choices Explained:

(1) Red light has a lower frequency than orange light.

(3) Increasing brightness increases only the number of photons ejected, but not the energy they contain. Once the work function of a light source has been met, increasing the brightness will cause more electrons to be ejected at a specific kinetic energy but will have no effect if the work function hasn't been met.

(4) The angle of incidence is not related to the ejection of photoelectrons.

53. **3** Rutherford's gold foil experiment led him to conclude that the atom consists of mostly empty space, that all the positive charge is concentrated in a

small, dense nucleus, and that the electrons (negative charges) orbit in the empty space around the nucleus.

54. **1** Refer to the *Wave Phenomena* section in the *Reference Tables* to find an equation that relates critical angle with absolute index of refraction.

Solution: $\sin\theta_c = \frac{1}{n}$

$$\theta_c = \sin^{-1}\left(\frac{1}{n}\right)$$

If the absolute index of refraction (n) of the solution is increased, the critical angle (θ_c) will decrease.

55. **2** The Doppler effect is the variation in observed frequency when there is relative motion between the source of the light waves and the observer. When the source of light (here, a star) is moving away from the observer, the observed frequency is less than the actual frequency emitted. The astronomer concludes that the distance between the star and Earth is increasing.

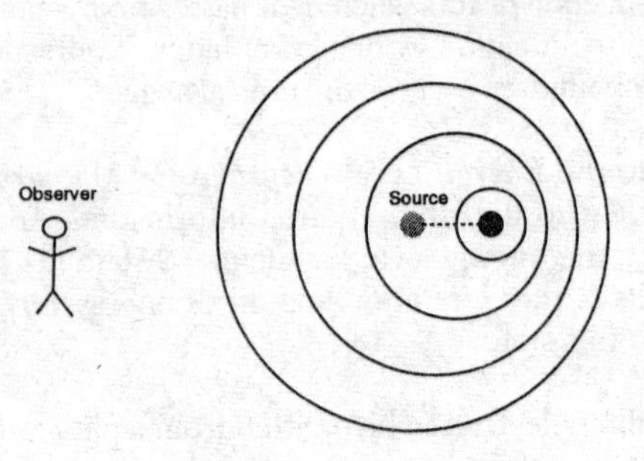

PART II

GROUP 1—Motion in a Plane

56. **1** Write out the problem.

Given: $\Delta s_x = 7.0 \times 10^2$ m   Find: $v_x = ?$

$\Delta t = 10.$ s

Refer to the *Mechanics* section in the *Reference Tables* to find an equation that relates velocity with displacement and time.

Solution: $v_x = \frac{\Delta s_x}{\Delta t}$

$= \frac{7.0 \times 10^2 \text{ m}}{10. \text{ s}}$

$= 70. \text{ m/s}$

57. **2** When the cannonball reaches its maximum height, the vertical component of its velocity will be equal to zero. Since acceleration in the vertical direction is constant and is equal to the acceleration due to gravity, and since it is assumed that the cannonball will land at the same height above the ground as it started, it will reach its maximum height halfway through its flight.

Given: $\Delta t = 10. \text{ s}$ Find: $v_{iy} = ?$

$v_{fy} = 0 \text{ m/s}$

$a = -9.8 \text{ m/s}^2$ (acceleration due to gravity; see the List of Physical Constants in the *Reference Tables*.)

Refer to the *Mechanics* section in the *Reference Tables* to find an equation that relates velocity with acceleration and time.

Solution: $a = \frac{\Delta v_y}{\Delta t}$

$v_{iy} = v_{fy} - a\Delta t$

$= 0 - (-9.8\text{m/s}^2)(5.0 \text{ s})$

$= 49 \text{ m/s}$

58. **2** Refer to the *Motion in a Plane* section in the *Reference Tables* to find equations that relate velocity with its vector components.

Solution: $v_{iy} = v_i \sin\theta$

$v_{ix} = v_i \cos\theta$

As the angle of elevation decreases, sin θ also decreases, therefore, the vertical component of the initial velocity will decrease. Also, cos θ increases; therefore the horizontal component of the velocity will increase.

59. **2** Refer to the *Motion in a Plane* section in the *Reference Tables* to find equations that relate force with centripetal acceleration.

Solution:
$$a_c = \frac{v^2}{r}$$
$$F_c = \frac{mv^2}{r}$$
$$\therefore F_c = ma_c$$

The last equation shows that force is directly proportional to centripetal acceleration, therefore increasing the applied force will increase the acceleration of the ball.

60. **3** Write out the problem.

Given: $m = 0.10$ kg   Find: $F_c = ?$

$r = 0.50$ m

$v = 10.$ m/s

Refer to the *Motion in a Plane* section in the *Reference Tables* to find an equation that relates centripetal force with mass, radius, and velocity.

Solution:
$$F_c = \frac{mv^2}{r}$$
$$= \frac{(0.10 \text{ kg})(10. \text{ m/s})^2}{0.50 \text{ m}}$$
$$= 20. \text{ N}$$

61. **1** Centripetal force is defined as the force required to keep an object moving in a circular path. Centripetal force, like any other force, is a vector quantity, and its direction is perpendicular to the circle and toward the center of the circular path.

62. **3** According to Newton's first law, an object in constant straight-line motion remains in constant straight-line motion unless acted upon by an unbalanced force. The direction of the velocity of an object moving in circular motion is tangent to the circle at any point, and the velocity points in the direction the object would take if the centripetal force were suddenly removed. At point *A*, the velocity of the car is directed toward point *C*. If a ball were thrown upward from the car at point *A*, it would also have a horizontal velocity relative to the ground and equal, in magnitude and direction, to the velocity of the car at the instant it was released. Free from an unbalanced horizontal force, the ball would move toward *C*.

63. **2** The comet is a satellite orbiting the Sun. According to Kepler's first law, all such satellites orbit the Sun in elliptical paths with the Sun located at one focus.

64. **3** Write out the problem.

Given: Area 1 + Area 2 = Area 3 + Area 4

Area 1 = 100 days

Area 4 = 100 days (600 days − 500 days)

Area 1 + Area 2 + Area 3 + Area 4 = 600 days

Find: Area 1 + Area 2 = ?

Solution:

100 days + Area 2 = Area 3 + 100 days

Area 2 = Area 3

100 days + Area 2 + Area 3 + 100 days = 600 days

Area 2 + Area 3 = 400 days

$\therefore 2 \times$ Area 2 = 400 days

Area 2 = 200 days

Area 1 + Area 2 = 100 days + 200 days = 300 days to reach point X

65. **4** A projectile moves simultaneously in the horizontal and vertical directions independently of each other. With air resistance neglected, it moves at a constant speed in the horizontal direction and experiences a downward acceleration due to gravity in the vertical direction. Only diagram (4) accurately depicts this type of motion.

Wrong Choices Explained:

(1) This diagram shows only free-fall motion in the vertical direction; there is no horizontal velocity component.

(2) This diagram shows a constant horizontal velocity for the first 3 s; there is no vertical component.

(3) This diagram shows the horizontal and vertical motions as not being independent of each other.

GROUP 2—**Internal Energy**

66. **4** Temperature is a measure of the average kinetic energy of the molecules of an ideal gas. Kelvin temperature is directly proportional to the average kinetic energy of the molecules; therefore, if the Kelvin temperature is increasing by a factor of 4, the average kinetic energy of the molecules must also be four times as great.

67. **1** Write out the problem.

Given: $m = 1.0$ kg Find: $H_f = ?$

Refer to the *Internal Energy* section in the *Reference Tables* to find an equation that relates heat of fusion with mass.

Solution: $Q = mH_f$

$$H_f = \frac{Q}{m}$$

Heat of fusion is the quantity of heat necessary to melt a unit amount of a substance at its melting point. According to the graph, 40 kilojoules are required to completely change 1 kilogram of the substance from a solid into a liquid.

$$H_f = \frac{40 \text{ kJ}}{1.0 \text{ kg}}$$

$$= 40 \text{ kJ/kg}$$

68. **3** The freezing point is the temperature at which freezing occurs, that is, the liquid phase changes to the solid phase. According to the graph, this occurs at 200°C.

69. **2** Write out the problem.

Given: $m = 1.0$ kg Find: $c = ?$

From the graph :

$Q = 40$ kJ

$\Delta T = 100$ C°

Refer to the *Internal Energy* section in the *Reference Tables* to find an equation that relates specific heat with mass, heat, and temperature change.

Solution: $Q = mc\Delta T$

$$c = \frac{Q}{m\Delta T}$$

$$= \frac{40 \text{ kJ}}{(1.0 \text{ kg})(100 \text{ C}^\circ)}$$

$$= 0.4 \text{ kJ/kg} \cdot \text{C}^\circ$$

70. **4** According to the graph, the boiling point is 400°C. This is the temperature at which the substance changes phase from liquid to gas. The temperature begins to rise again only when all the substance has changed phase. This point is reached at the 160 kJ mark.

71. **1** Snow is loosely packed ice crystals. Ice is less dense than liquid water; therefore, an increase in pressure will favor the more dense liquid phase and the snow will melt. As a result, the melting point of the snow is lowered.

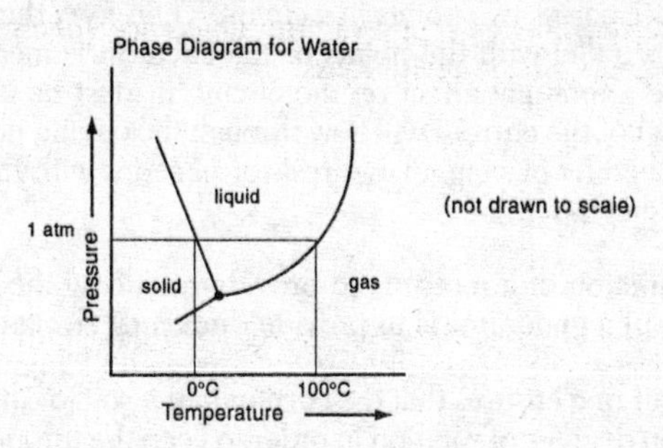

72. **1** The magnitude of a Celsius degree is equal to the magnitude of a kelvin; therefore, an increase of 5 Celsius degrees is equal to a rise in temperature of 5 kelvins.

73. **3** Pressure varies directly with absolute temperature for an ideal gas of fixed volume. Only graph (3) accurately portrays this relationship.

Wrong Choices Explained:

(1) This graph shows an inverse relationship; pressure decreases with increasing temperature.

(2) This graph shows an inverse proportion relationship.

(4) This graph indicates that pressure increases exponentially with increasing temperature.

74. **4** Pressure is defined as force per unit area. The collisions of the gas molecules with the walls of the container produce the forces, and the container wall is the area on which the forces act.

Wrong Choices Explained:

(1) Specific heat is the amount of heat required to raise the temperature of a unit mass of a substance by 1 Celsius degree.

(2) The condensation point is the temperature at which a gas changes phase into a liquid. This is equal to the boiling point of a substance.

(3) Temperature is the measure of the average kinetic energy of the molecules of the gas.

75. **2** The volume of an ideal gas at constant pressure varies directly with its Kelvin temperature; therefore, if the temperature of the sample is increasing, the volume is also increasing.

GROUP 3—**Electromagnetic Applications**

76. **3** A voltmeter is a galvanometer that has been modified to measure the potential difference across two points in a circuit. Therefore, the voltmeter must be connected in parallel with the points of the circuit it is measuring. For the voltmeter to have a minimal effect on the circuit, it must be a high-resistance device so that most of the current will flow through the circuit, not the voltmeter. This is accomplished by placing a large resistor in series with the galvanometer coil, as shown in diagram (3).

77. **2** The function of a motor is to provide rotational mechanical energy, and the function of a generator is to provide (induced) electrical energy.

78. **4** The split ring ensures that the current in the coil to which it is attached reverses every 180 degrees of rotation in order to keep the torque on the coil acting in the same direction in a motor.

79. **3** Write out the problem.

Given: $\ell = 1.0\text{ m}$ Find: $V = ?$

$v = 10.\text{ m/s}$

$B = 5.0 \times 10^{-3}\text{ T}$

Refer to the *Electromagnetic Applications* section in the *Reference Tables* to find an equation that relates electric potential difference with length, velocity, and flux density.

Solution: $V = B\ell v$

$= (5.0 \times 10^{-3}\text{ T})(1.0\text{ m})(10.\text{ m/s})$

$= 5.0 \times 10^{-2}\text{ V}$

80. **4** Like charges repel and opposite charges attract; therefore, the electron will be repelled from the negative plate and attracted toward the positive plate. In other words, the electric field will deflect the electron to the left.

81. **3** The force that a charged particle experiences in an electric field is equal to Eq, where E is the electric field intensity and q is the charge on the particle. Since the field is uniform, E doesn't change and therefore the magnitude of the force remains the same as the electron moves through the electric field.

82. **1** The elementary electric charge is equal to $1.6 \times 10^{-19}\text{ C}$ (see the List of Physical Constants in the *Reference Tables*), which is the charge on a single electron or proton. A charge on an oil drop resulting from an excess of electrons would be a multiple of the elementary charge. Only choice (1) is a multiple of the elementary charge.

83. **2** The force on a charged particle moving through a magnetic field can be calculated using the equation $F = qvB$ (see the *Electromagnetic Applications* section of the *Reference Tables*). This equation indicates that the force is directly proportional to the charge on the particle, to its velocity, and to the flux density of the magnetic field.

84. **1** Write out the problem.

Given: 100% efficiency

$V_p = 12\text{ V}$

$I_p = 3.0\text{ A}$

$N_p = 3$

$N_s = 9$

Find: $V_s = ?$

$I_s = ?$

Refer to the *Electromagnetic Applications* section in the *Reference Tables* to find equations that relate potential difference and current of a secondary coil with those of a primary coil and the respective number of turns of wire in each coil.

Solution:

$$\frac{N_p}{N_s} = \frac{V_p}{V_s}$$

$$V_s = \frac{V_p N_s}{N_p}$$

$$= \frac{(12\text{ V})(9)}{3}$$

$$= 36\text{ V}$$

$$V_p I_p = V_s I_s$$

$$I_s = \frac{V_p I_p}{V_s}$$

$$= \frac{(12\text{ V})(3.0\text{ A})}{36\text{ V}}$$

$$= 1.0\text{ A}$$

85. **2** Particles produced by thermionic emission are electrons. To solve this problem, use the following left-hand rule: The thumb of the left hand points in the direction of the velocity of the electrons, which in this case is to the right, and the fingers of the left hand point in the direction of the magnetic field, which in this case is into the plane of the paper. The direction of the force experienced by the electrons entering the magnetic field points away from the palm of the hand, which in this instance is toward the bottom of the page. The electrons experiencing this force will be deflected in the same direction, that is, toward the bottom of the page.

GROUP 4—**Geometric Optics**

86. **1** All plane mirrors form images that are virtual, erect, and the same size as the object.

87. **4** For a plane mirror the distance of the object (d_o) from the mirror is equal to the distance of the image (d_i) from the mirror. This direct relationship is correctly represented by graph (4).

Wrong Choices Explained:

(1) According to this graph, image distance increases exponentially with increasing object distance.

(2) According to this graph, an inverse relationship exists between object distance and image distance.

(3) According to this graph, image distance remains the same regardless of object distance.

88. **1** Chromatic aberration is inability of a single lens to refract all of the colors of light to a single focus. This inability is due to the fact that all the different colors of light have different refractive indices.

Wrong Choices Explained:

(3), (4) Spherical aberration is failure of mirrors and lenses with a spherical surface to bring parallel rays of light striking all parts of the surface to a single focus, either by reflection (mirrors) or refraction (lenses).

89. **4** An object placed at point *A* is located between the surface of the mirror and its focus. The image produced is virtual, enlarged, erect, and *behind* the mirror's surface, which in this situation is to the left of the mirror.

Wrong Choices Explained:

(1) An image will be located at the principal focus (*F*) when the object is at infinity.

(2) An image will be located between the focus (*F*) and the center of curvature (*C*) when the object is placed beyond the center of curvature.

(3) An image will be located to the right of the center of curvature—in other words, beyond *C*—if the object is located between the focus and the center of curvature.

90. **2** An object placed at point *A* is located between the surface of the mirror and its focus. The image produced is *virtual*, enlarged, *erect*, and behind the mirror's surface.

91. **2** Refraction is the bending of a wave front as it enters a new medium obliquely; this bending is due to a change in the speed of the wave. As light enters a (converging) lens, the rays above and below the axis are bent in a fashion that causes them to converge to a point.

Wrong Choices Explained:

(1) Diffraction is the bending of light rays around an object or barrier in their path.

(3) Interference is the superposition of one wave on another when two or more waves pass through the same region at the same time.

(4) Polarization is the confinement of the vibration of light to a single plane perpendicular to the wave's direction of motion.

92. **4** When an object is placed at a distance of two focal lengths from a converging lens, a real, inverted image will be formed two focal lengths from the lens and on the opposite side of the lens, that is, at point *D* in the diagram. The image will be the same size as the object.

93. **3** Write out the problem.

Given: $S_i = 1.0 \times 10^{-2}$ m   Find: $d_i = ?$

$S_o = 2.0$ m

$d_o = 8.0$ m

Refer to the *Geometric Optics* section in the *Reference Tables* to find an equation that relates image distance to image size, object size, and object distance.

Solution:
$$\frac{S_o}{S_i} = \frac{d_o}{d_i}$$
$$d_i = \frac{d_o S_i}{S_o}$$
$$= \frac{(8.0\text{ m})(1.0 \times 10^{-2}\text{ m})}{2.0\text{ m}}$$
$$= 4.0 \times 10^{-2}\text{ m}$$

94. **2** A magnifying glass creates an erect, virtual, enlarged image located on the same side of the lens as the object. A convex (converging) lens at a distance less than one focal length away from the crystals will produce this kind of image.

Wrong Choices Explained:

(1), (3) A convex (diverging) mirror and a concave (diverging) lens are capable of producing only smaller, virtual, erect images.

(4) A plane mirror is capable of producing only virtual, erect images that are the same size as the objects.

95. **3** The distance of an object from a lens can be varied, and the image distance will change accordingly based on the focal length of the lens. The focal length of a lens does *not* depend on the distance of the object from the lens, nor does the distance of the object from the lens depend on the focal length.

Wrong Choices Explained:

(1) Different materials have different refractive indices, so the material from which the lens is made plays a part in determining the focal length.

(2) Different colors of light have different refractive indices; therefore the color of the light incident on the lens also plays a part in determining the focal length.

(4) The shape or curvature of the lens affects the angle of incidence of light on its surface, which in turn will affect the angle of refraction; therefore, lens shape or curvature also plays a part in determining the focal length of the lens.

GROUP 5—**Solid State**

96. **4** Copper is a metal and therefore by definition is a conductor. Germanium is a semimetal and is a semiconductor.

97. **3** Increasing temperature allows more electrons to have sufficient kinetic energy to bridge the energy gap between the valence and conduction bands; in other words, more electrons are promoted with increasing temperature. Diagram (3) shows this pattern of electron promotion.

Wrong Choices Explained:

(1) In this diagram all the electrons that had been in the conduction band have been demoted back to the valence band.

(2) In this diagram most of the electrons that were in the conduction band have been demoted back to the valence band.

(4) In this diagram the conduction band is filled with electrons, but the valence band hasn't lost any electrons—an impossible situation.

98. **1** The diagram accurately represents an *N-P-N* transistor. The arrow points in the direction of positive hole flow; in other words it points from *P* to *N* or from the base to the emitter.

Wrong Choices Explained:

(2) A *P-N-P* transistor, shown below, would look identical except that the arrow in the diagram would point from the emitter to the base, indicating that the direction of positive hole flow is from *P* to *N*.

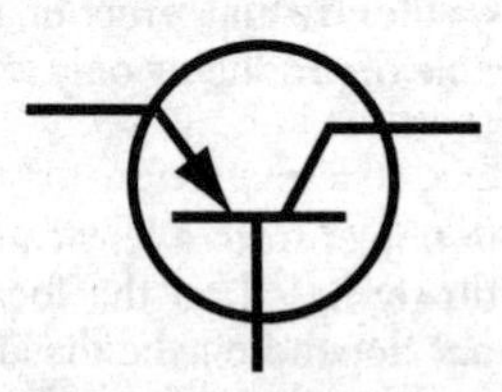

(3) A zener diode, shown below, is a reverse-biased diode used to control voltage output and is represented by the symbol used for a diode.

(4) The avalanche region, shown below, is the area where the breakdown voltage of a diode has been reached and a very large reverse current occurs.

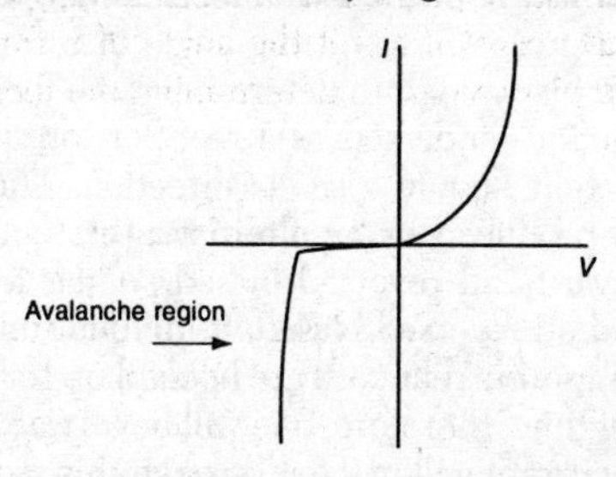

99. **1** Adding a small amount of impurity to an intrinsic semiconductor such as silicon is known as doping. Silicon contains 4 valence electrons. If the impurity (in this case, antimony) contains more valence electrons than the silicon, it provides an excess of free electrons, and since electrons are negative, the process is known as *N*-doping and the resulting extrinsic semiconducting material as an *N*-type semiconductor.

100. **2** The conducting wire is a metal, and its charge carrier is the electron. Electrons flow from the negative terminal of the battery toward the positive terminal; in this diagram the flow of charge is from point *C* to point *A*.

101. **1** A *P-N* junction diode is forward biased when the *P*-section is connected to the positive terminal and the *N*-section is connected to the negative terminal. This arrangement allows the potential barrier at the junction to be overcome and current to flow. Only diagram (1) accurately shows this situation.

Wrong Choices Explained:

(2) This diagram depicts a *P-N* junction, which is reverse biased. This arrangement reinforces the barrier at the junction and inhibits current flow.

(3), (4) These representations are not possible.

102. **3** When a material such as indium, the atoms of which contain only three valence electrons, is added to a semiconductor, the indium atoms take the place of the original atoms in the crystalline structure. Having only three valence electrons, these atoms can form only three valence bonds with the rest of the structure, leaving a hole in the valence shell. The hole will readily *accept* an electron from a neighboring atom to complete the fourth bond. Such materials as indium are therefore known as acceptors.

Wrong Choices Explained:

(2) Atoms that contain five valence electrons and are added to a semiconductor also take the place of original atoms in the crystalline structure. However, since only four electrons are needed to form the four covalent bonds to the structure, the fifth valence electron is free to be promoted to the conduction band to provide extra electrons. Therefore, these materials are known as donors.

(4) *N*-types are extrinsic semiconductors that have been doped with donor-type impurities.

103. **2** A *P-N* junction diode acts as a rectifier for alternating current because it limits the current flow to a single direction. This occurs because the alternating current is produced by an alternating potential, which makes the diode alternately forward and reversed biased. When it is forward biased, it allows current flow; when reversed biased, it inhibits current flow. In diagram (2) two diodes in the same current path are hooked up to be the reverse of each other. At any instant of time, therefore, one will be forward biased and the other reverse biased, so no current will make it through this segment of circuit.

Wrong Choices Explained:

(1) These two diodes are connected in the same direction. Therefore, this circuit has the same effect as one with a single diode, and the output will be a half-wave rectified signal, a pulsating DC output, as shown below.

(3) Although these two diodes are hooked up to be the opposite of each other, so that at any one time one will be forward biased while the other is reversed biased, they are connected in parallel. As a result, current will be able to flow through whichever one is forward biased, and thus the current output will be as if there were no diode at all, as shown below.

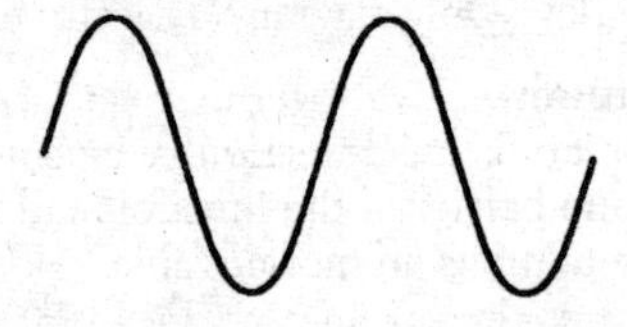

(4) The output from this single diode will be a half-wave rectified signal, a pulsating DC output, as shown in the diagram for choice (1).

104. **4** If a source of small, varying potential difference is applied to the emitter-base circuit, an amplified variation appears in the collector circuit. This change occurs because the ratio of the collector current to the base current is large and constant. Thus, small increases in the emitter-base current will produce large

increases in the collector current. A single transistor can provide an amplification of several hundred times the input.

105. **2** The circuit contains a junction diode. With an AC source, the diode is alternately forward and reversed biased, so the output current is a pulsating DC signal [see question 103, choice (4)]; therefore, the lamp receives pulsing direct current, only.

GROUP 6—**Nuclear Energy**

106. **1** Refer to the *List of Physical Constants* in the *Reference Tables* to find the mass-energy relationship. One atomic mass unit is equivalent to 9.3×10^2 MeV. Since a proton, at 1.0073 atomic mass units, is slightly larger than 1 atomic mass unit, choice (1) is the only reasonable answer for the energy equivalent of the rest mass of a proton.

107. **2** Write out the problem.

Given: 1 proton

2 neutrons

Total mass of nucleus = 3.0170 u

Find: Mass defect = ?

Solution: Total mass of nucleons = 1(1.0073 u) + 2(1.0087 u)

= 3.0247 u

Mass of nucleons – Mass of nucleus = Mass defect

3.0247 u – 3.0170 u = 0.00077 u

108. **4** The stability of nuclei is tied to the existence of nuclear force. The two nuclear forces, called the *strong* and *weak* interactions, are much more powerful at the very small distances present within the nucleus than are gravitational or electromagnetic forces. At larger distances, however, the strong and weak interactions lose their effectiveness; for this reason they are called *short-range* forces.

109. **1** The bringing together of light nuclei to form heavier, more stable nuclei is known as nuclear fusion and is the process by which stars, including our own Sun, produce their energy. Tritium, a hydrogen isotope that has two neutrons and would be considered a small, light nucleus, is used as a fuel in a fusion reaction.

Wrong Choices Explained:

(2) Fission is the splitting apart of a large nucleus by neutron absorption, with an accompanying release of energy and two or more neutrons.

(3) A coolant is used in a nuclear reactor to remove thermal energy from the core and is typically water or liquid sodium.

(4) A moderator is used in a nuclear reactor to slow down neutrons and is typically heavy water (D_2O, T_2O), graphite, or beryllium.

110. **2** All atomic nuclei (also called *nuclides*) and their component nucleons may be represented by the same general symbol:

$$\boxed{{}^{A}_{Z}X}$$

The letter X represents the letter(s) used to identify the particle; the letter Z, called the *atomic number*, indicates the number of elementary charges present (assumed to be positive unless a negative sign is written); the letter A, called the *mass number*, is equal to the sum of neutrons and protons present. The number of neutrons (N) present in an atomic nucleus is given by this equation:

$$N = A - Z$$

Only choice (2), ${}^{32}_{15}P$, represents a nuclide with an odd number (15) of protons and an odd number (32 – 15 = 17) of neutrons.

Wrong Choices Explained:

(1), (3) These choices have even atomic numbers.

(4) This choice has an even number of neutrons (35 – 17 = 18).

111. **1** To study nuclear reactions and their products, a number of *detection devices* have been invented. The principle underlying all detection devices is that the subatomic particles produced in nuclear reactions leave their "fingerprints" as they pass through the cloud chamber, spark chamber, or Geiger counter and that the device can detect these particles exiting the nucleus.

112. **3** Nuclear reactions, like any other chemical reactions, obey the laws of conservation of mass, energy, and electric charge.

Write out the problem.

$${}^{238}_{92}U \rightarrow {}^{234}_{90}Th + X$$

Particle X's mass and charge must be such that the sum of all the masses and charges on the right side of the equation is equal to the sum of all the masses and charges on the left side of the equation.

$$\text{Solution: } {}^{238}_{92}U \rightarrow {}^{234}_{90}Th + {}^{4}_{2}X$$

The mass and charge of particle X correspond to those of an alpha particle, ${}^{4}_{2}He$.

Wrong Choices Explained:

(1), (2), (4) These choices can be represented as ${}^{1}_{0}n$, ${}^{0}_{+1}e$, and ${}^{0}_{-1}e$, respectively.

113. **4** Nuclear reactions, like any other chemical reactions, obey the laws of conservation of mass, energy, and electric charge. Particle *X*'s mass and charge must be such that the sum of all the masses and charges on the right side of the equation is equal to the sum of all the masses and charges on the left side of the equation.

$$\text{Solution: } {}^{9}_{4}\text{Be} + {}^{4}_{2}\text{He} \rightarrow {}^{12}_{6}\text{C} + {}^{1}_{0}X$$

The mass and charge of particle *X* correspond to those of a neutron, ${}^{1}_{0}n$.

Wrong Choices Explained:

(1), (2), (3) These choices can be represented as ${}^{0}_{-1}e$, ${}^{1}_{1}H$, and ${}^{0}_{+1}e$, respectively.

114. **3** Control rods are inserted into the core of a nuclear reactor to regulate the rate of fission by absorbing neutrons. Control rods are usually made of cadmium or boron.

115. **4** The rate of decay of a radioactive isotope is measured in terms of a quantity called the *half-life*. The half-life is defined as the time in which a sample of the isotope decays to one-half of its original mass. The half-life of strontium-90 is 28 years. The decay of amount *x* of this isotope over a period of time is shown below:

$$x \xrightarrow{28y} \frac{1}{2}x \xrightarrow{28y} \frac{1}{4}x \xrightarrow{28y} \frac{1}{8}x \xrightarrow{28y} \frac{1}{16}x$$

Four half-lives or 112 years ($28 \times 4 = 112$) would have to pass before only one-sixteenth of the original sample would remain as radioactive strontium-90.

PART III

116.

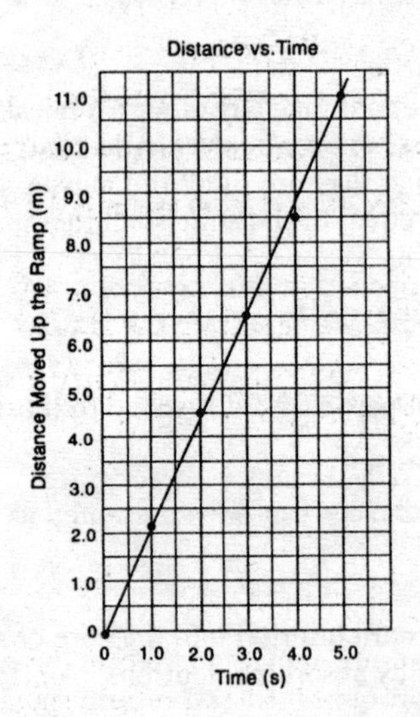

Note: One credit is awarded for plotting all points accurately (± 0.3 grid space). One credit is awarded for drawing a best-fit straight line. If one or more points are plotted incorrectly but a best-fit straight is drawn, this credit is still granted.

117. The slope of a graph is defined as the change in y over the change in x. In this graph, y is distance and x is time; thus, the slope of a distance-versus-time graph is equal to the change in distance over the change in time, which is equal to velocity.

Examples of acceptable responses therefore include:

The slope is the average speed of the safe.
The slope is the velocity.
The change in distance divided by the change in time is equal to the velocity.
The slope is the speed of the safe.

Note: One credit is granted for an acceptable response.

118. Write out the problem.

Given: $m = 50.0$ kg　　　Find: $W = ?$

$F = 300.$ N

$\Delta t = 3.0$ s

$\Delta s = 6.6$ m

Refer to the *Energy* section in the *Reference Tables* to find an equation that relates work with mass, force, time, and displacement.

Solution:

$$W = F\Delta s$$
$$= (300.\ \text{N})(6.6\ \text{m})$$
$$= 1980\ \text{N}\cdot\text{m} = 1980\ \text{J}$$
$$\text{or } 2.0\times10^3\ \text{J} = 2000\ \text{kg}\cdot\text{m}^2/\text{s}^2$$

(Note that mass is not needed in this calculation and that time is needed only to obtain the value of displacement from the graph.)

The following answers are acceptable: W = 1980 J, W = 1980 N • m, W = 2.0 $\times 10^3$ J, W = 2000 kg • m^2/s^2.

Note: One credit is awarded for writing the equation and for the substitution of values with units. If the equation and/or units are not shown, you do not receive this credit. One credit is awarded for the correct answer (number and units). If no units are shown, you do not receive this credit. Significant figures and scientific notation are not required to obtain this credit. The value used for distance in the calculation must be the value corresponding to t = 3.0 seconds on the graph.

119. Write out the problem.

Given: An electron — Find: KE = ?

$v = 2.0\times10^6$ m/s

Refer to the *Energy* section in the *Reference Tables* to find an equation that relates kinetic energy with velocity.

Solution: $\text{KE} = \frac{1}{2}mv^2$

Refer to the *List of Physical Constants* in the *Reference Tables* to find a value for the rest mass of the electron.

$$\text{KE} = \frac{1}{2}(9.1\times10^{-31}\ \text{kg})(2.0\times10^6\ \text{m/s})^2$$
$$= 1.8\times10^{-18}\ \text{J or } 1.82\times10^{-18}\ \text{kg}\cdot\text{m}^2/\text{s}^2$$

Note: One credit is awarded for writing the equation and for the substitution of values with units. If the equation and/or units are not shown, you do not receive this credit. One point is awarded for the correct answer (number and units).

If no units are shown, you do not receive this credit. Significant figures and scientific notation are not required to obtain this credit.

120. Write out the problem.

Given: $v = 2.0 \times 10^6$ m/s Find: $\lambda = ?$

$m = 9.1 \times 10^{-31}$ kg

Refer to the *Modern Physics* and *Mechanics* sections in the *Reference Tables* to find equations that relate wavelength with velocity and mass.

Solution:

$$p = \frac{h}{\lambda}$$

$$p = mv$$

$$\therefore mv = \frac{h}{\lambda}$$

$$\lambda = \frac{h}{mv}$$

Refer to the *List of Physical Constants* in the *Reference Tables* to find the value for h, Planck's constant.

$$\lambda = \frac{6.6 \times 10^{-34} \text{ J}\cdot\text{s}}{(9.1 \times 10^{-31} \text{ kg})(2.0 \times 10^{6} \text{ m/s})}$$

$$= 3.6 \times 10^{-10} \text{ m}$$

Alternatively, solve for momentum first and then substitute the answer into the second equation. Doing so will round the answer to 3.7×10^{-10} m, which is also acceptable.

Note: One credit is awarded for the correct combination of the two equations and one credit for the substitution of values with units, or one credit each for writing the equations with the substitution of values with units. If the equation and/or units are not shown, you do not receive these credits. A third credit is awarded for the correct answer (number and units). If no units are shown, you do not receive this credit. Significant figures and scientific notation are not required to obtain this credit.

121. Examples of acceptable responses include the following:

0.9 cm (± 0.1 cm)
9 mm (± 1 mm)
0.009 m (± 0.001 m)

Note: Unit must be included to receive this credit.

122. Write out the problem.

Given: $f = 12$ Hz Find: $v = ?$
$\lambda = 0.9$ cm (0.009 m)

Refer to the *Wave Phenomena* section in the *Reference Tables* to find an equation that relates velocity with frequency and wavelength.

Solution: $v = f\lambda$
$= (12 \text{ Hz})(0.009 \text{ m})$
$= 0.1 \text{ m/s}$

Other acceptable responses are 10.8 cm/s and 100 mm/s.

Note: One credit is awarded for writing the equation and for the substitution of values with units. If the equation and/or units are not shown, you do not receive this credit. One credit is awarded for the correct answer (number and units). If no units are shown, you do not receive this credit. Significant figures and scientific notation are not required to obtain this credit. Credit will be granted for any answer that correctly uses the response to question 121.

123.

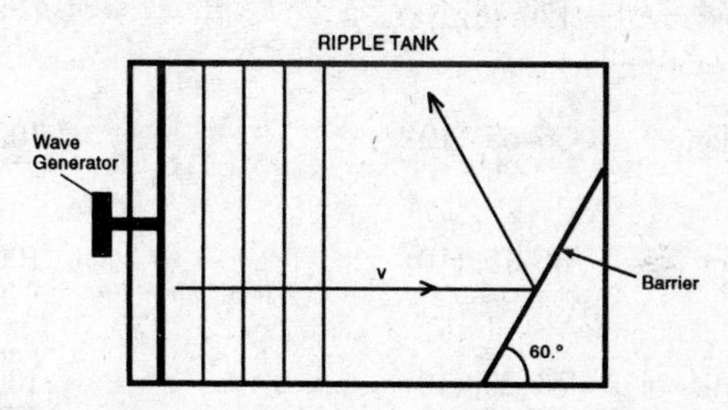

Note: One credit will be granted if the arrow forms an angle of 60.° ± 2° with the barrier and is directed away from the barrier as shown.

124. Examples of acceptable responses for one credit:

The angle of incidence is equal to the angle of reflection.
∠ of incidence = ∠ of reflection

Examples of *unacceptable* responses:

The angle that the incoming waves make with the barrier is equal to the angle that the reflected waves make with the barrier.

$\angle i = \angle r$

Topic	Question Numbers (total)	Wrong Answers (x)	Grade
Mechanics	1–17, 20, 116, 117: (20)		$\frac{100(20-x)}{20} = \%$
Energy	18, 19, 21, 22, 118, 119: (6)		$\frac{100(6-x)}{6} = \%$
Electricity and Magnetism	23–37: (15)		$\frac{100(15-x)}{15} = \%$
Wave Phenomena	38–47, 50, 54, 55, 121–124: (17)		$\frac{100(17-x)}{17} = \%$
Modern Physics	48, 49, 51–53 120: (6)		$\frac{100(6-x)}{6} = \%$
Motion in a Plane	56–65: (10)		$\frac{100(10-x)}{10} = \%$
Internal Energy	66–75: (10)		$\frac{100(10-x)}{10} = \%$
Electromagnetic Applications	76–85: (10)		$\frac{100(10-x)}{10} = \%$
Geometrical Optics	86–95: (10)		$\frac{100(10-x)}{10} = \%$
Solid State Physics	96–105: (10)		$\frac{100(10-x)}{10} = \%$
Nuclear Energy	106–115: (10)		$\frac{100(10-x)}{10} = \%$

To further pinpoint your weak areas, use the Topic Outline in the front of the book.

Examination June 1997

Physics

PART I

Answer all 55 questions in this part. [65]

Directions (1–55): For *each* statement or question, select the word or expression that, of those given, best completes the statement or answers the question. Record the answers to these questions in the spaces provided.

1 What is the total displacement of a student who walks 3 blocks east, 2 blocks north, 1 block west, and then 2 blocks south?

(1) 0
(2) 2 blocks east
(3) 2 blocks west
(4) 8 blocks

1 _____

2 A baseball pitcher throws a fastball at 42 meters per second. If the batter is 18 meters from the pitcher, approximately how much time does it take for the ball to reach the batter?

(1) 1.9 s
(2) 2.3 s
(3) 0.86 s
(4) 0.43 s

2 _____

3 The length of a high school physics classroom is probably closest to

(1) 10^{-2} m
(2) 10^{-1} m
(3) 10^{1} m
(4) 10^{4} m

3 _____

4 A stone is dropped from a bridge 45 meters above the surface of a river. Approximately how many seconds does the stone take to reach the water's surface?

(1) 1.0 s
(2) 10. s
(3) 3.0 s
(4) 22 s

4 _____

5 A 150.-newton force, F_1, and a 200.-newton force, F_2, are applied simultaneously to the same point on a large crate resting on a frictionless, horizontal surface. Which diagram shows the forces positioned to give the crate the greatest acceleration?

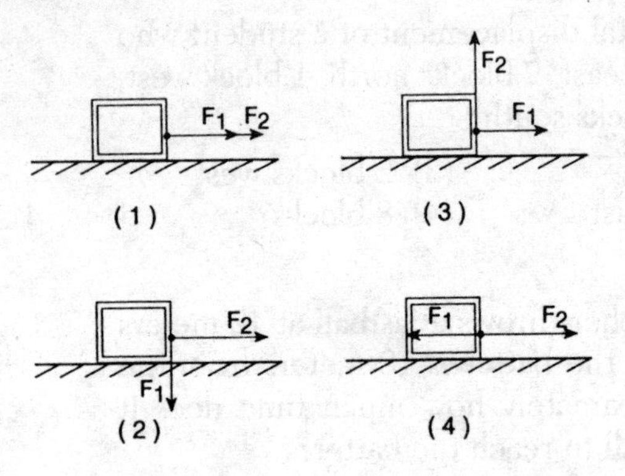

5 _____

6 The displacement-time graph below represents the motion of a cart along a straight line.

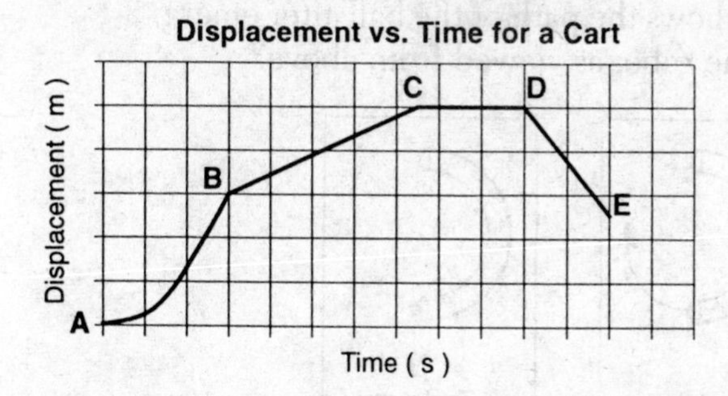

During which interval was the cart accelerating?

(1) *AB*
(2) *BC*
(3) *CD*
(4) *DE*

6 ______

7 A ball rolls through a hollow semicircular tube lying flat on a horizontal tabletop. Which diagram best shows the path of the ball after emerging from the tube, as viewed from above?

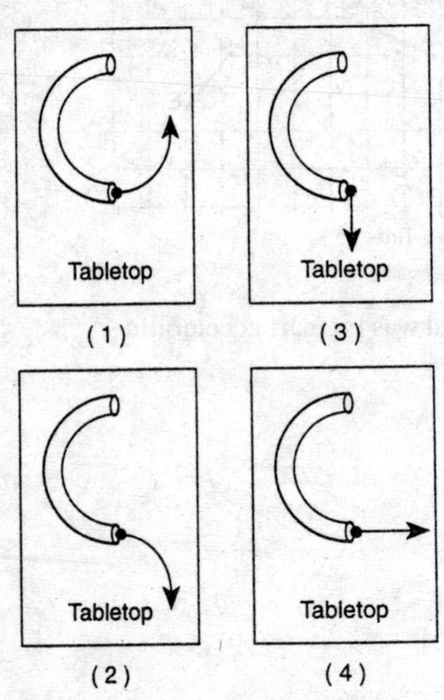

7 ____

8 A 100.-newton force acts on point P, as shown in the diagram below.

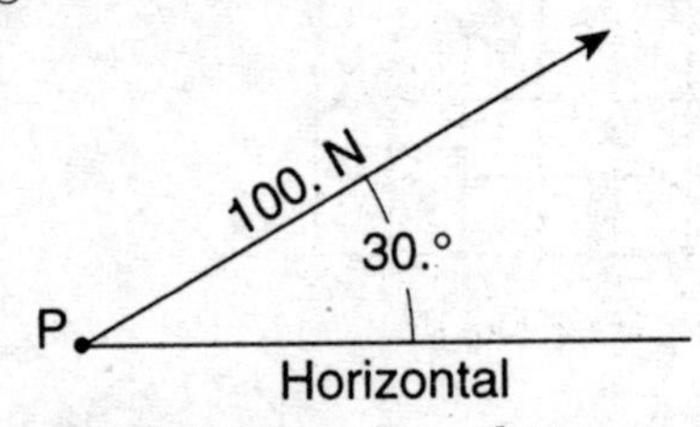

The magnitude of the vertical component of this force is approximately

(1) 30. N
(2) 50. N
(3) 71 N
(4) 87 N

8 ____

Base your answers to questions 9 and 10 on the information below.

A 1,000-kilogram car traveling with a velocity of +20. meters per second decelerates uniformly at –5.0 meters per second^2 until it comes to rest.

9 What is the total distance the car travels as it decelerates to rest?

(1) 10. m
(2) 20. m
(3) 40. m
(4) 80. m

9 _____

10 What is the magnitude of the impulse applied to the car to bring it to rest?

(1) 1.0×10^4 N•s
(2) 2.0×10^4 N•s
(3) 3.9×10^4 N•s
(4) 4.3×10^4 N•s

10 _____

---

11 The graph below shows the weight of three objects on planet *X* as a function of their mass.

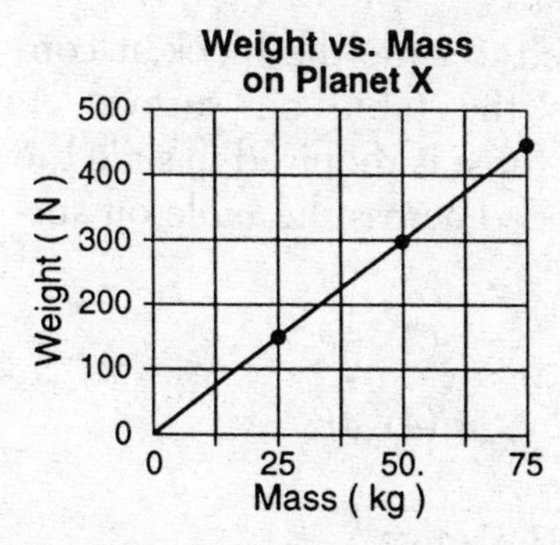

The acceleration due to gravity on planet *X* is approximately

(1) 0.17 m/s^2
(2) 6.0 m/s^2
(3) 9.8 m/s^2
(4) 50. m/s^2

11 _____

12 What is the magnitude of the net force acting on a 2.0×10^3-kilogram car as it accelerates from rest to a speed of 15 meters per second in 5.0 seconds?

(1) 6.0×10^3 N
(2) 2.0×10^4 N
(3) 3.0×10^4 N
(4) 6.0×10^4 N

12 _____

13 In the diagram below, surface *B* of the wooden block has the same texture as surface *A*, but twice the area of surface *A*.

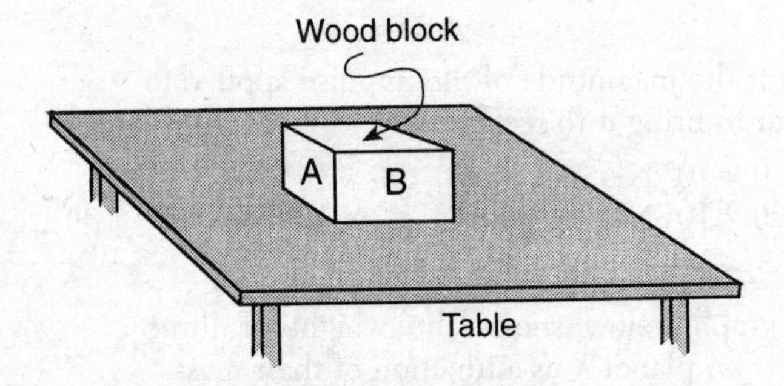

If force *F* is required to slide the block at constant speed across the table on surface *A*, approximately what force is required to slide the block at constant speed across the table on surface *B*?

(1) *F*
(2) 2*F*
(3) $\frac{1}{2}F$
(4) 4*F*

13 _____

14 The diagram below shows two carts on a horizontal, frictionless surface being pushed apart when a compressed spring attached to one of the carts is released. Cart *A* has a mass of 3.0 kilograms and cart *B* has a mass of 5.0 kilograms. The speed of cart *A* is 0.33 meter per second after the spring is released.

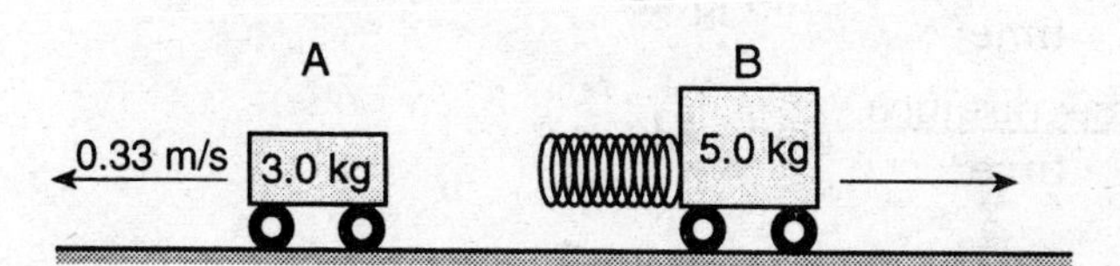

If the carts are initially at rest, what is the approximate speed of cart *B* after the spring is released?

(1) 0.12 m/s
(2) 0.20 m/s
(3) 0.33 m/s
(4) 0.55 m/s

14 _____

15 The magnitude of the gravitational force of attraction between Earth and the Moon is approximately

(1) 2.1×10^{20} N
(2) 6.0×10^{24} N
(3) 6.7×10^{-11} N
(4) 7.8×10^{28} N

15 _____

16 How much work is done on a downhill skier by an average braking force of 9.8×10^2 newtons to stop her in a distance of 10. meters?

(1) 1.0×10^1 J
(2) 9.8×10^1 J
(3) 1.0×10^3 J
(4) 9.8×10^3 J

16 _____

17 Which variable expression is paired with a corresponding unit?

(1) $\frac{\text{mass} \bullet \text{distance}}{\text{time}}$ and watt

(2) $\frac{\text{mass} \bullet \text{distance}^2}{\text{time}}$ and watt

(3) $\frac{\text{mass} \bullet \text{distance}^2}{\text{time}^2}$ and joule

(4) $\frac{\text{mass} \bullet \text{distance}}{\text{time}^3}$ and joule

17 ____

18 A spring has a spring constant of 120 newtons per meter. How much potential energy is stored in the spring as it is stretched 0.20 meter?

(1) 2.4 J
(2) 4.8 J
(3) 12 J
(4) 24 J

18 ____

19 The graph below shows the relationship between the elongation of a spring and the force applied to the spring causing it to stretch.

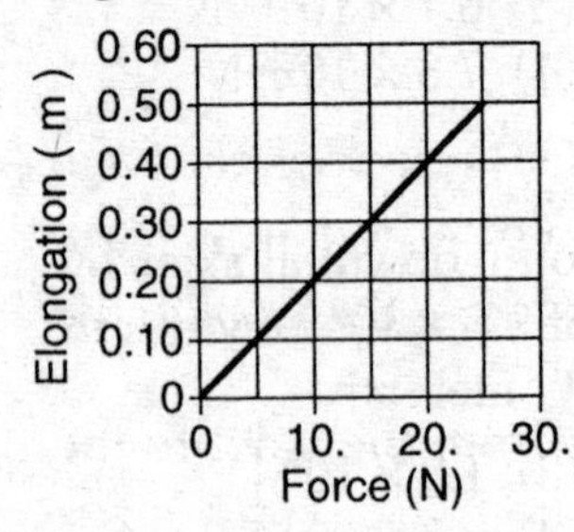

What is the spring constant for this spring?

(1) 0.020 N/m
(2) 2.0 N/m
(3) 25 N/m
(4) 50. N/m

19 ____

20 A motor having a maximum power rating of 8.1×10^4 watts is used to operate an elevator with a weight of 1.8×10^4 newtons. What is the maximum weight this motor can lift at an average speed of 3.0 meters per second?

(1) 6.0×10^3 N
(2) 1.8×10^4 N
(3) 2.4×10^4 N
(4) 2.7×10^4 N

20 _____

21 A cart of mass M on a frictionless track starts from rest at the top of a hill having height h_1, as shown in the diagram below.

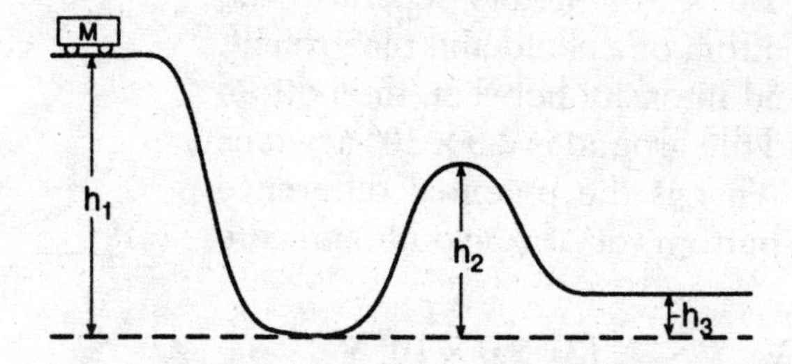

What is the kinetic energy of the cart when it reaches the top of the next hill, having height h_2?

(1) Mgh_1
(2) $Mg(h_1 - h_2)$
(3) $Mg(h_2 - h_3)$
(4) 0

21 _____

22 When a plastic rod is rubbed with wool, the wool acquires a positive charge because

1 electrons are transferred from the wool to the rod
2 protons are transferred from the wool to the rod
3 electrons are transferred from the rod to the wool
4 protons are transferred from the rod to the wool

22 _____

23 Three identical metal spheres are mounted on insulating stands. Initially, sphere A has a net charge of q and spheres B and C are uncharged. Sphere A is touched to sphere B and removed. Then sphere A is touched to sphere C and removed. What is the final charge on sphere A?

(1) q (3) $\frac{q}{3}$

(2) $\frac{q}{2}$ (4) $\frac{q}{4}$

23 ____

24 A distance of 1.0×10^3 meters separates the charge at the bottom of a cloud and the ground. The electric field intensity between the bottom of the cloud and the ground is 2.0×10^4 newtons per coulomb. What is the potential difference between the bottom of the cloud and the ground?

(1) 1.3×10^{23} V (3) 2.0×10^7 V

(2) 2.0×10^1 V (4) 5.0×10^{-2} V

24 ____

25 The diagram below shows proton P located at point A near a positively charged sphere.

B

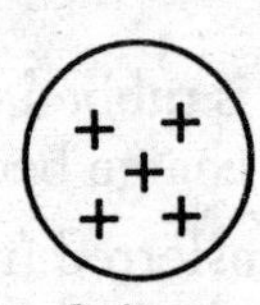

Sphere

If 6.4×10^{-19} joule of work is required to move the proton from point A to point B, the potential difference between A and B is

(1) 6.4×10^{-19} V (3) 6.4 V

(2) 4.0×10^{-19} V (4) 4.0 V

25 ____

26 What is the net static electric charge on a metal sphere having an excess of +3 elementary charges?

(1) 1.6×10^{-19} C
(2) 4.8×10^{-19} C
(3) 3.0×10^{0}C
(4) 4.8×10^{19} C

26____

27 The diagram below shows currents in a segment of an electric circuit.

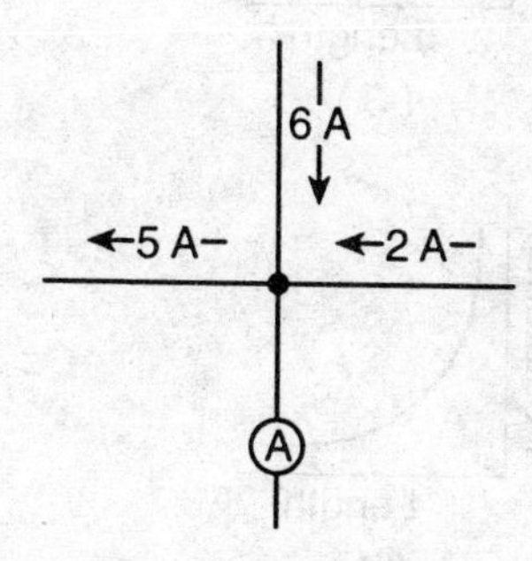

What is the reading of ammeter *A*?

(1) 8 A
(2) 2 A
(3) 3 A
(4) 13 A

27____

28 An operating lamp draws a current of 0.50 ampere. The amount of charge passing through the lamp in 10. seconds is

(1) 0.050 C
(2) 2.0 C
(3) 5.0 C
(4) 20. C

28____

29 Which graph best represents the relationship between the resistance of a copper wire of uniform cross-sectional area and the wire's length at constant temperature?

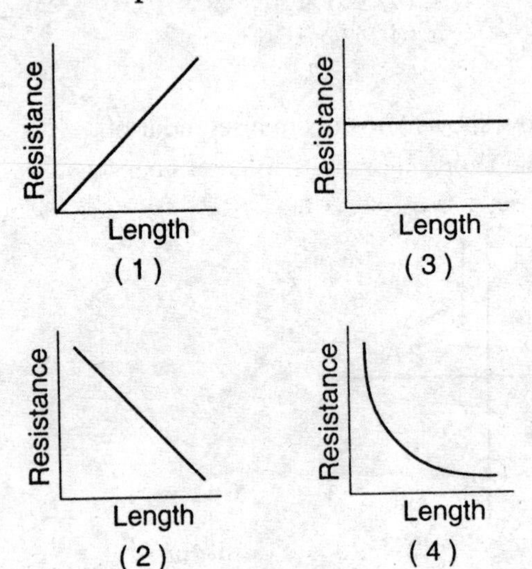

29_____

30 Which is a vector quantity?

1 electric charge
2 electrical resistance
3 electrical potential difference
4 electrical field intensity

30_____

31 To increase the brightness of a desk lamp, a student replaces a 60-watt light bulb with a 100-watt bulb. Compared to the 60-watt bulb, the 100-watt bulb has

1 less resistance and draws more current
2 less resistance and draws less current
3 more resistance and draws more current
4 more resistance and draws less current

31_____

32 An electric dryer consumes 6.0×10^6 joules of energy when operating at 220 volts for 30. minutes (1800 seconds). During operation, the dryer draws a current of approximately

(1) 10. A
(2) 15 A
(3) 20. A
(4) 25 A

32 ____

33 The diagram below shows a compass placed near the north pole, *N*, of a bar magnet.

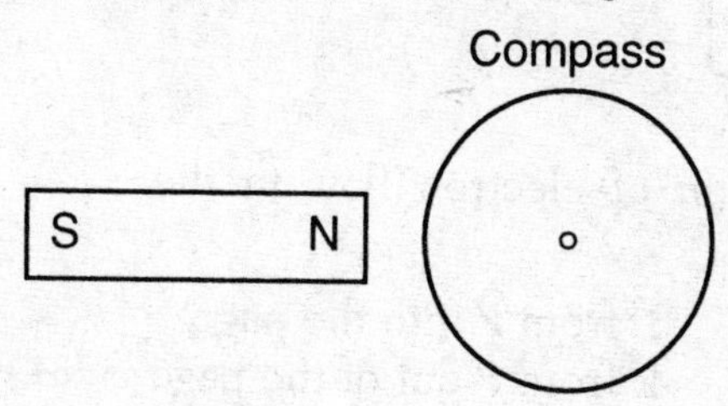

Which diagram best represents the position of the needle of the compass as it responds to the magnetic field of the bar magnet?

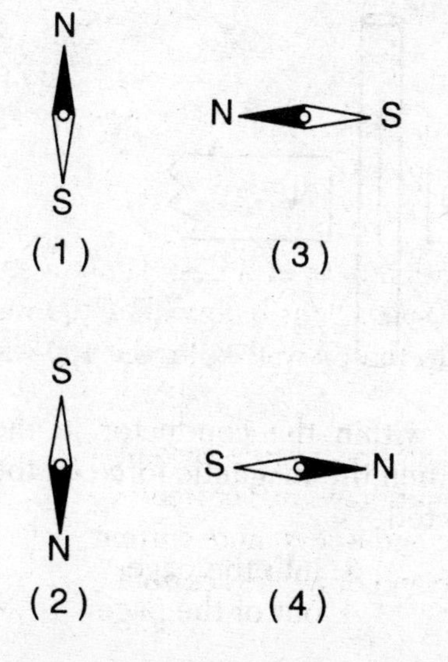

33 ____

34 The diagram below represents the magnetic field around point *P*, at the center of a current-carrying wire.

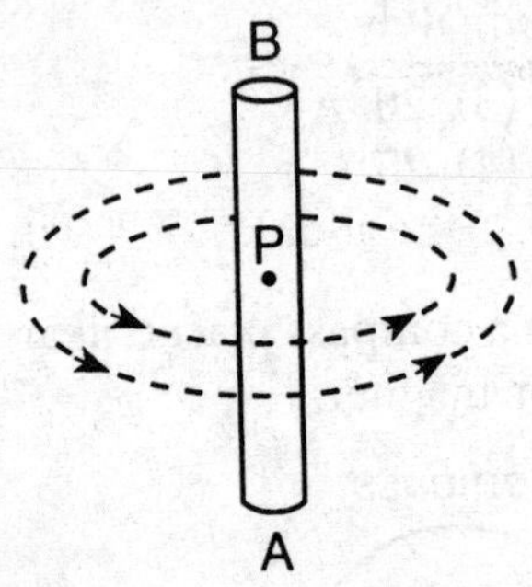

What is the direction of electron flow in the wire?

1 from *A* to *B*
2 from *B* to *A*
3 from *P* into the page
4 from *P* out of the page

34 ______

35 The diagram below represents a straight conductor between the poles of a permanent magnet.

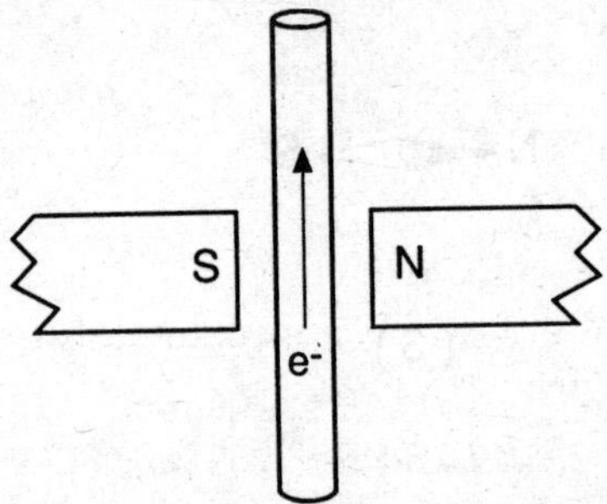

If electrons flow within the conductor in the direction shown, then the magnetic force on the conductor is directed

1 toward *N*
2 toward *S*
3 into the page
4 out of the page

35 ______

36 The diagram below shows a transverse wave moving to the right along a rope.

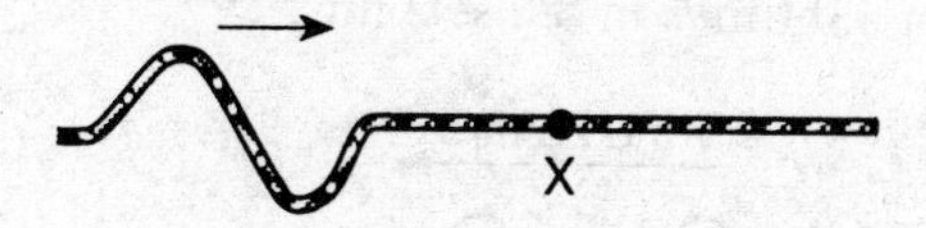

As the wave passes point *X*, the motion of *X* will be

1 up, then down
2 down, then up
3 left, then right
4 in a circle

36 _____

37 The frequency of a light wave is 5.0×10^{14} hertz. What is the period of the wave?

(1) 1.7×10^{6} s
(2) 2.0×10^{-15} s
(3) 6.0×10^{-7} s
(4) 5.0×10^{-14} s

37 _____

38 The amplitude of a sound wave is to its loudness as the amplitude of a light wave is to its

1 brightness
2 frequency
3 color
4 speed

38 _____

39 The speed of light in glycerol is approximately

(1) 1.0×10^{7} m/s
(2) 2.0×10^{8} m/s
(3) 3.0×10^{8} m/s
(4) 4.4×10^{8} m/s

39 _____

40 In the diagram below, a water wave having a speed of 0.25 meter per second causes a cork to move up and down 4.0 times in 8.0 seconds.

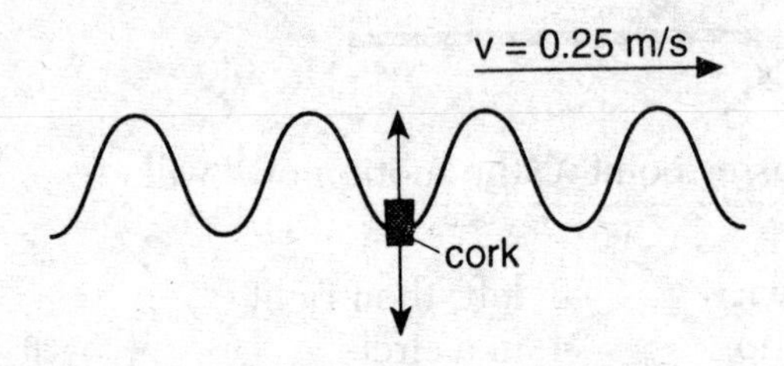

What is the wavelength of the water wave?

(1) 1.0 m
(2) 2.0 m
(3) 8.0 m
(4) 0.50 m

40 ____

41 The driver of a car sounds the horn while traveling toward a stationary person. Compared to the sound of the horn heard by the driver, the sound heard by the stationary person has

1 lower pitch and shorter wavelength
2 lower pitch and longer wavelength
3 higher pitch and shorter wavelength
4 higher pitch and longer wavelength

41 ____

42 The diagram below shows a ray of monochromatic light incident on an alcohol-flint glass interface.

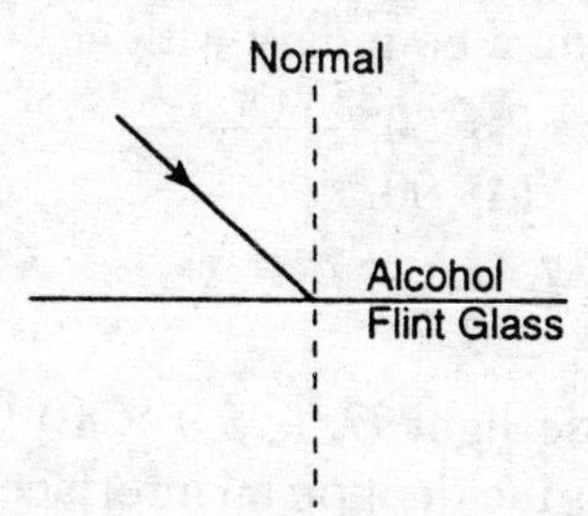

What occurs as the light travels from alcohol into flint glass?

1 The speed of the light decreases and the ray bends toward the normal.
2 The speed of the light decreases and the ray bends away from the normal.
3 The speed of the light increases and the ray bends toward the normal.
4 The speed of the light increases and the ray bends away from the normal.

42 ______

43 The diagram below shows straight wave fronts passing through an opening in a barrier.

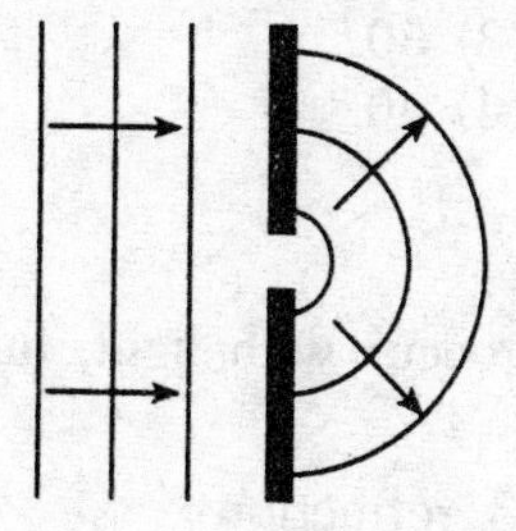

This wave phenomenon is called

1 reflection
2 refraction
3 polarization
4 diffraction

43 ______

44 The absolute index of refraction for a substance is 2.0 for light having a wavelength of 5.9×10^{-7} meter. In this substance, what is the critical angle for light incident on a boundary with air?

(1) 30.° (3) 60.°
(2) 45° (4) 90.° 44 ____

45 A ray of monochromatic light ($\lambda = 5.9 \times 10^{-7}$ meter) traveling in air is incident on an interface with a liquid at an angle of 45°, as shown in the diagram below.

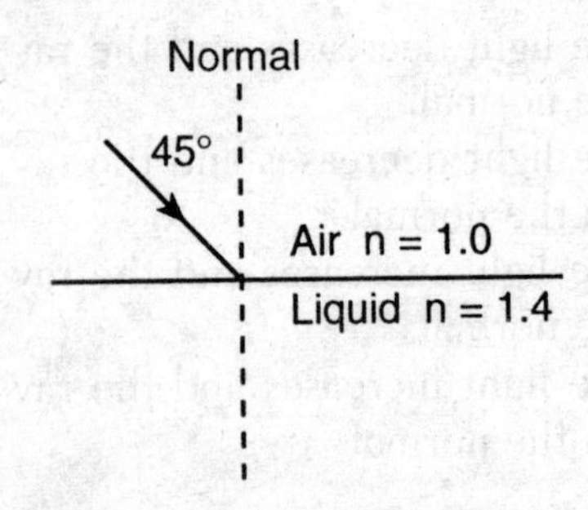

If the absolute index of refraction of the liquid is 1.4, the angle of refraction for the light ray is closest to

(1) 10.° (3) 30.°
(2) 20.° (4) 40.° 45 ____

46 Which phenomenon can occur with light, but *not* with sound?

1 interference 3 refraction
2 polarization 4 the Doppler effect 46 ____

47 As shown in the diagram below, speaker, S_1 and S_2, separated by a distance of 0.50 meter, are producing sound of the same constant frequency. A person walking along a path 4.0 meters in front of the speakers hears the sound reach a maximum intensity every 2.0 meters.

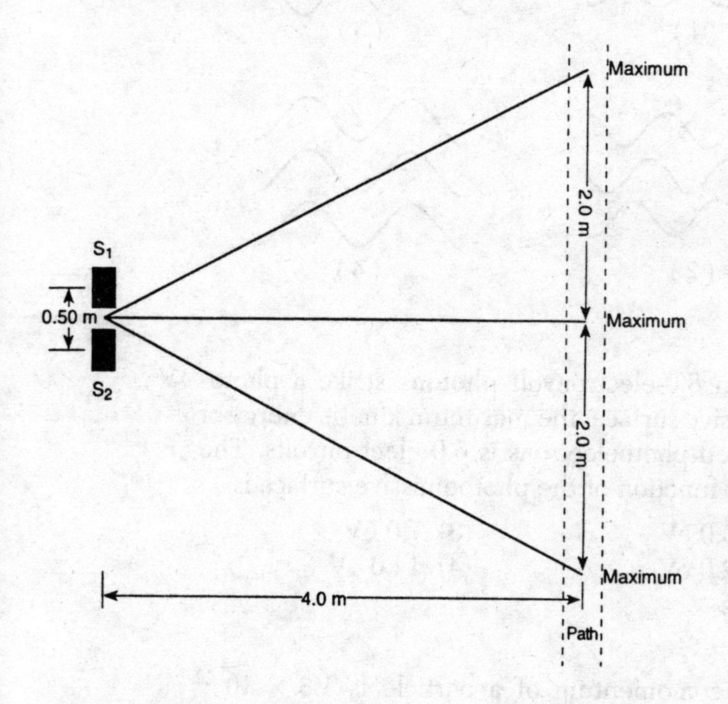

What is the wavelength of the sound produced by the speakers?

(1) 1.0 m
(2) 0.063 m
(3) 0.25 m
(4) 4.0 m

47 _____

48 Which diagram best represents light emitted from a coherent light source?

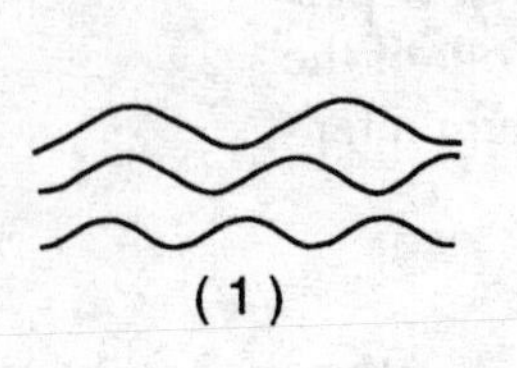
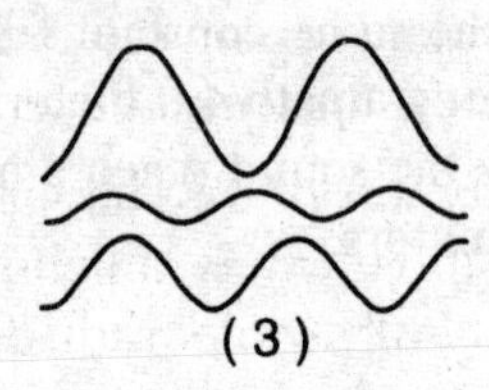
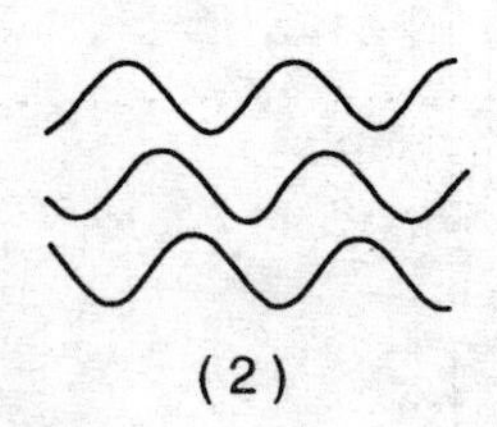
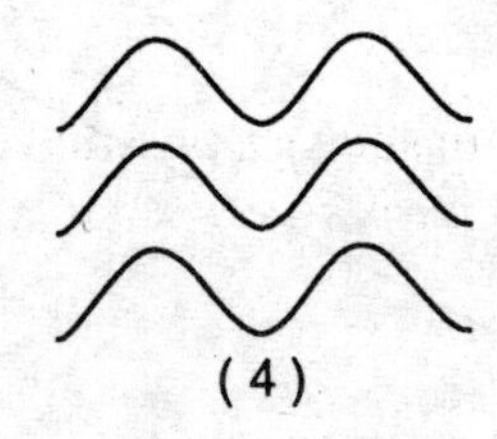

48 _____

49 When 8.0-electronvolt photons strike a photoemissive surface, the maximum kinetic energy of ejected photoelectrons is 6.0 electronvolts. The work function of the photoemissive surface is

(1) 0.0 eV
(2) 2.0 eV
(3) 7.0 eV
(4) 14.0 eV

49 _____

50 If the momentum of a particle is 1.8×10^{-22} kilogram-meter per second, its matter wavelength is approximately

(1) 1.2×10^{-55} m
(2) 2.7×10^{11} m
(3) 3.7×10^{-12} m
(4) 5.0×10^{-7} m

50 _____

51 The threshold frequency of a photoemissive surface is 7.1×10^{14} hertz. Which electromagnetic radiation, incident upon the surface, will produce the greatest amount of current?

1 low-intensity infrared radiation
2 high-intensity infrared radiation
3 low-intensity ultraviolet radiation
4 high-intensity ultraviolet radiation

51 _____

52 Which diagram shows a possible path of an alpha particle as it passes very near the nucleus of a gold atom?

(1) 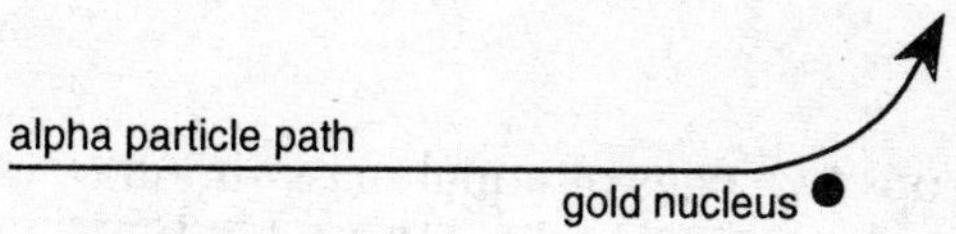 alpha particle path

gold nucleus

(2) 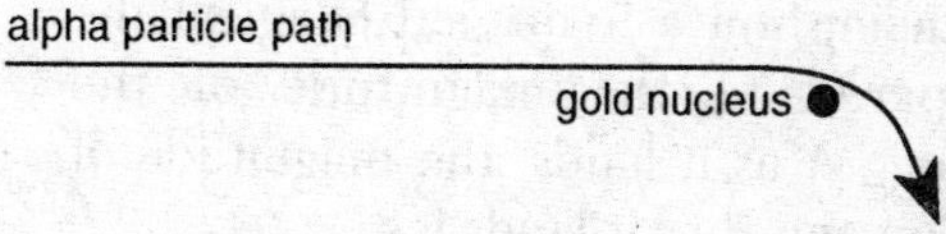 alpha particle path

gold nucleus

(3)  alpha particle path

gold nucleus

(4) alpha particle path

gold nucleus

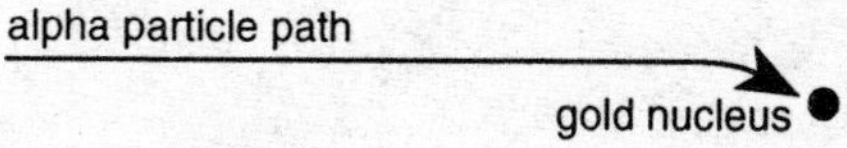

52 _____

53 A hydrogen atom could have an electron energy-level transition from $n = 2$ to $n = 3$ by absorbing a photon having an energy of

(1) 1.51 eV (3) 4.91 eV
(2) 1.89 eV (4) 10.20 eV

53 _____

Note that questions 54 and 55 have only three choices.

54 A bicyclist accelerates from rest to a speed of 5.0 meters per second in 10. seconds. During the same 10. seconds, a car accelerates from a speed of 22 meters per second to a speed of 27 meters per second. Compared to the acceleration of the bicycle, the acceleration of the car is

1 less
2 greater
3 the same

54 _____

55 A student drops two eggs of equal mass simultaneously from the same height. Egg *A* lands on the tile floor and breaks. Egg *B* lands intact, without bouncing, on a foam pad lying on the floor. Compared to the magnitude of the impulse on egg *A* as it lands, the magnitude of the impulse on egg *B* as it lands is

1 less
2 greater
3 the same

55 _____

PART II

This part consists of six groups, each containing ten questions. Each group tests an optional area of the course. Choose two of these six groups. Be sure that you answer all ten questions in each group chosen. Record the answers to these questions in the spaces provided. [20]

GROUP 1—Motion in a Plane

If you choose this group, be sure to answer questions **56–65**.

56 In which diagram do the arrows best represent the path of a satellite in a geosynchronous orbit?

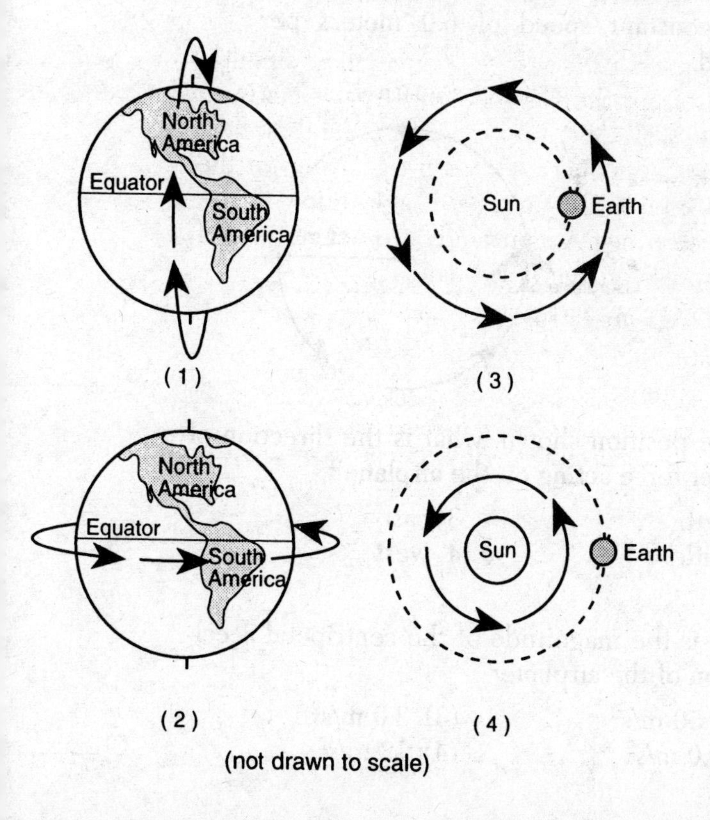

(not drawn to scale)

56 _____

57 A soccer ball travels the path shown in the diagram at the right. Which vector best represents the direction of the force of air friction on the ball at point *P*?

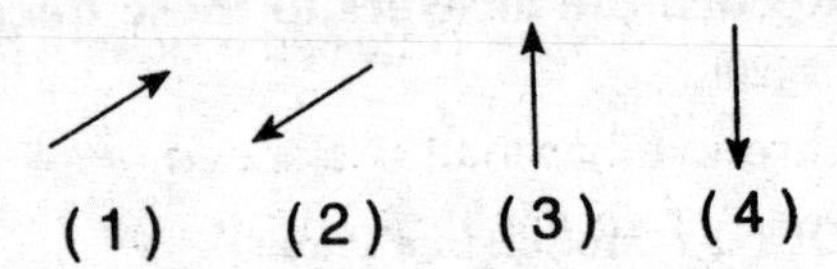

57 _____

Base your answers to questions 58 through 60 on the information and diagram below.

A 4.0-kilogram model airplane travels in a horizontal circular path of radius 12 meters at a constant speed of 6.0 meters per second.

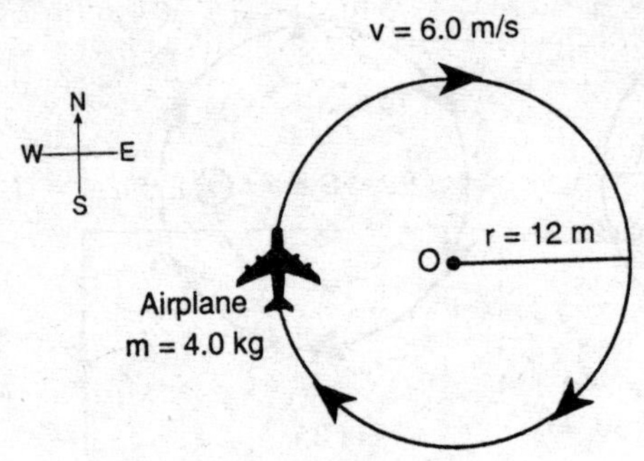

58 At the position shown, what is the direction of the net force acting on the airplane?

1 north
2 south
3 east
4 west

58 _____

59 What is the magnitude of the centripetal acceleration of the airplane?

(1) 0.50 m/s^2
(2) 2.0 m/s^2
(3) 3.0 m/s^2
(4) 12 m/s^2

59 _____

60 If the speed of the airplane is doubled and the radius of the path remains unchanged, the magnitude of the centripetal force acting on the airplane will be

1 half as much
2 twice as much
3 one-fourth as much
4 four times as much

60 ____

61 A baseball player throws a baseball at a speed of 40. meters per second at an angle of 30.° to the ground. The horizontal component of the baseball's speed is approximately

(1) 15 m/s
(2) 20. m/s
(3) 30. m/s
(4) 35 m/s

61 ____

62 Projectiles are fired from different angles with the same initial speed of 14 meters per second. The graph below shows the range of the projectiles as a function of the original angle of inclination to the ground, neglecting air resistance.

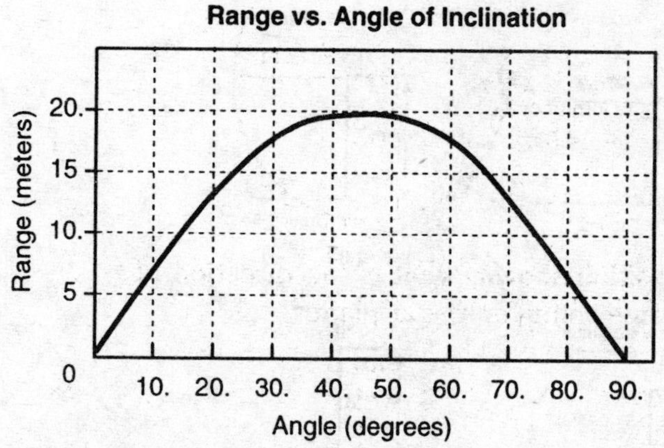

The graph shows that the range of the projectiles is

1 the same for all angles
2 the same for angles of 20.° and 80.°
3 greatest for an angle of 45°
4 greatest for an angle of 90.°

62 ____

63 The data table below gives the mean radius of orbit (R) and the period (T) of some planets orbiting the Sun.

Planet	Mean Radius of Orbit (R) ($\times 10^6$ kilometers)	Orbital Period (T) (days)
Mercury	58	88
Venus	108	225
Earth	150.	365
Mars	228	687

Which ratio is constant for these planets?

(1) $\frac{R}{T}$

(2) $\frac{R^2}{T}$

(3) $\frac{R^2}{T^2}$

(4) $\frac{R^3}{T^2}$

63 ______

64 Four different balls are thrown horizontally off the top of four cliffs. In which diagram does the ball have the shortest time of flight?

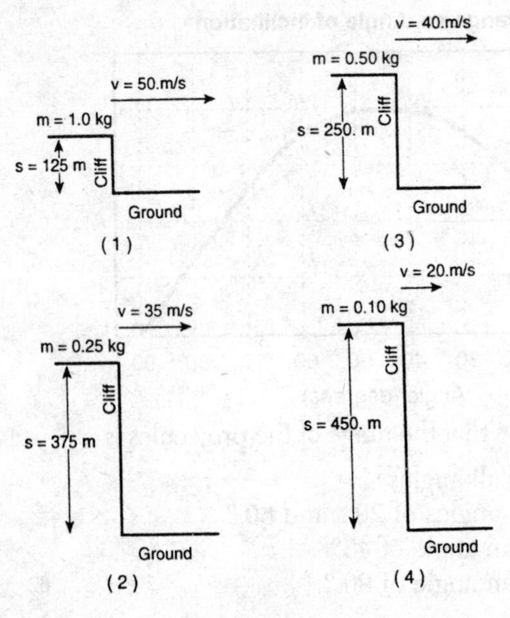

64 ______

65 The diagram below represents the path of a planet in an elliptical orbit around the Sun.

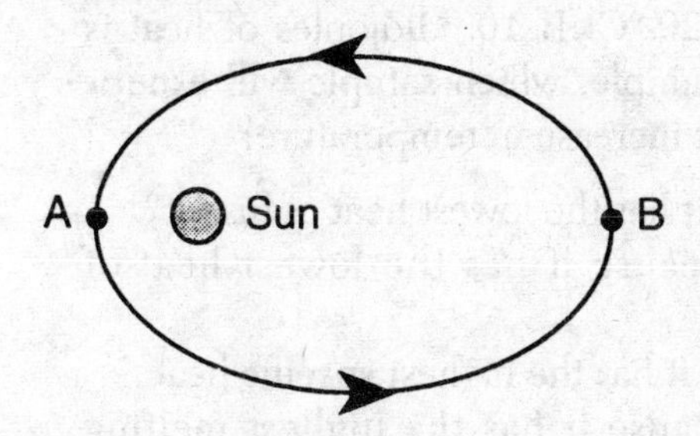

As the planet moves from point *A* to point *B*, what changes occur in its speed and kinetic energy?

1 Both speed and kinetic energy decrease.
2 Both speed and kinetic energy increase.
3 Speed decreases and kinetic energy increases.
4 Speed increases and kinetic energy decreases. 65_____

GROUP 2—**Internal Energy**

If you choose this group, be sure to answer questions **66–75.**

66 What is the boiling point of water at standard pressure on the Kelvin scale?

(1) 100. K (3) 273 K
(2) 212 K (4) 373 K 66_____

67 Two solid metal blocks are placed in an insulated container. If there is a net flow of heat between the blocks, they must have different

1 initial temperatures 3 specific heats
2 melting points 4 heats of fusion 67_____

68 Samples of lead, platinum, silver, and tungsten each have a mass of 1.0 kilogram and an initial temperature of 20.°C. If 10. kilojoules of heat is added to each sample, which sample will experience the smallest increase in temperature?

1 lead, because it has the lowest heat of fusion
2 platinum, because it has the lowest heat of vaporization
3 silver, because it has the highest specific heat
4 tungsten, because it has the highest melting point

68_____

69 When Juliann drinks cold water, her body warms the water until thermal equilibrium is reached. If she drinks 6 glasses (2.5 kilograms) of water at 0°C in a day, approximately how much energy must her body expend to raise the temperature of 37°C?

(1) 190 kJ (3) 840 kJ
(2) 390 kJ (4) 2300 kJ

69_____

70 Which statement is consistent with the kinetic theory of ideal gases?

1 Molecules are always stationary.
2 The force of attraction between molecules is large.
3 Molecules transfer energy through collisions.
4 The size of molecules is large compared to the distance that separates them.

70_____

71 The graph below shows temperature versus time for 1.0 kilogram of unknown material as heat is added at a constant rate.

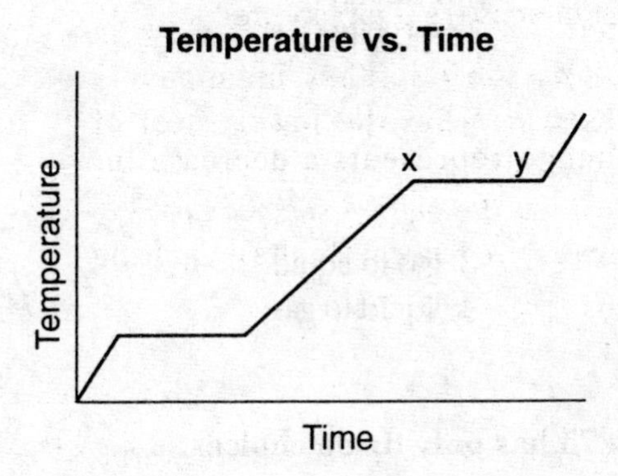

During interval *xy*, the material experiences

1 a decrease in internal energy and a phase change
2 an increase in internal energy and a phase change
3 no change in internal energy and a phase change
4 no change in internal energy and no phase change 71_____

72 Which graph best represents the relationship between volume (V) and absolute temperature (T_K) for a fixed mass of an ideal gas at constant pressure?

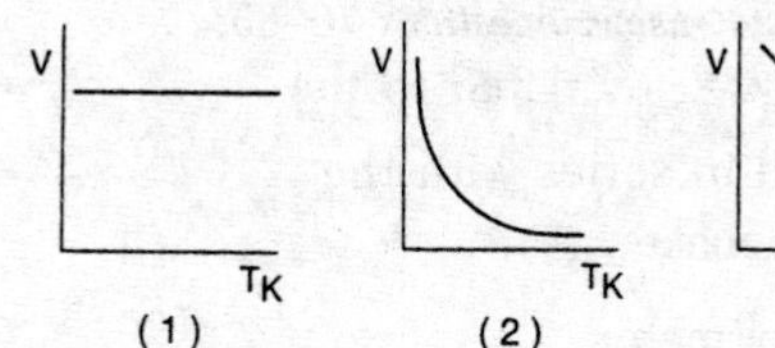

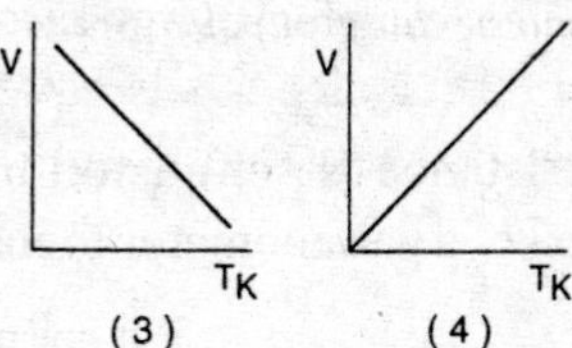

(1) (2) (3) (4) 72_____

73 How much heat must be removed from 0.50 kilogram of mercury at –39°C to change it from a liquid to a solid?

(1) 1.1 kJ
(2) 5.5 kJ
(3) 150 kJ
(4) 180 kJ

73____

74 Which phase change represents a decrease in entropy?

1 solid to gas
2 solid to liquid
3 gas to liquid
4 liquid to gas

74____

Note that question 75 has only three choices.

75 In an operating automobile, the pressure on the coolant in the radiator is greater than 1.0 atmosphere and its temperature is above 100.°C. If the radiator's pressure cap is removed, the pressure of the coolant is lowered to 1.0 atmosphere, causing the boiling point of the coolant to

1 decrease
2 increase
3 remain the same

75____

GROUP 3—Electromagnetic Applications

If you choose this group, be sure to answer questions **76–85**.

76 A high resistance is connected in series with the internal coil of a galvanometer to make

1 a motor
2 an ammeter
3 a voltmeter
4 a generator

76____

77 A student uses a voltmeter to measure the potential difference across a circuit resistor. To obtain a correct reading, the student must connect the voltmeter

1 in parallel with the circuit resistor
2 in series with the circuit resistor
3 before connecting the other circuit components
4 after connecting the other circuit components

77 _____

Base your answers to questions 78 and 79 on the diagram below which represents an electron moving with speed v to the right and about to enter a uniform magnetic field acting into the page.

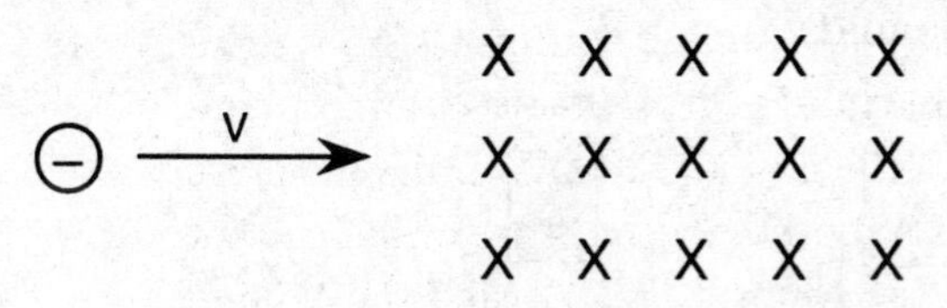

78 Upon entering the magnetic field, the electron will be deflected

1 into the page
2 out of the page
3 toward the top of the page
4 toward the bottom of the page

78 _____

79 If the speed of the electron is 3.0×10^3 meters per second and the magnitude of the magnetic field is 3.0×10^{-5} tesla, the magnitude of the magnetic force on the electron is approximately

(1) 4.8×10^{-24} N
(2) 1.4×10^{-20} N
(3) 4.8×10^{-16} N
(4) 9.0×10^{-2} N

79 _____

80 In a transformer, two coils of wire are wound around a common iron core. To operate properly, the transformer requires

1 an alternating-current source connected to the primary coil
2 a direct-current source connected to the secondary coil
3 more turns in the primary coil than in the secondary coil
4 more turns in the secondary coil than in the primary coil

80 _____

81 Two hollow-core solenoids, *A* and *B*, are connected by a wire, as shown in the diagram below. Two bar magnets, 1 and 2, are suspended just above the solenoids.

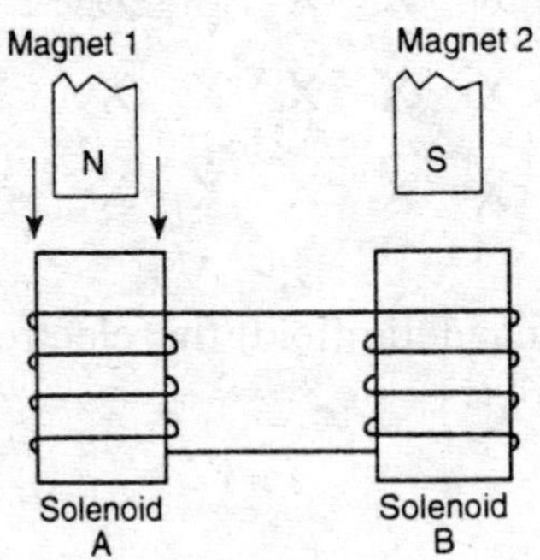

If the north pole of magnet 1 is dropped through solenoid *A*, the south pole of magnet 2 will simultaneously be

1 attracted by a magnetic force toward solenoid *B*
2 repelled by a magnetic force away from solenoid *B*
3 repelled by an electric force away from solenoid *B*
4 unaffected by solenoid *B*

81 _____

82 Which device can be used to increase the voltage from a source of direct current?

1 electroscope
2 mass spectrometer
3 induction coil
4 generator

82_____

83 The graph below shows induced potential difference versus time for the rotating armature of a generator.

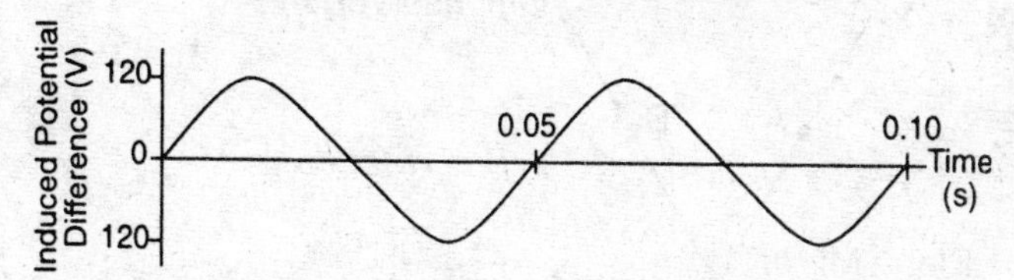

What is the frequency of armature rotation?

(1) 10 Hz
(2) 20 Hz
(3) 40 Hz
(4) 60 Hz

83_____

84 The transformer on a power pole steps down the voltage from 10,800 volts to 120. volts. If the secondary coil contains 360 turns, how many turns are on the primary coil?

(1) 30
(2) 90
(3) 3600
(4) 32,400

84_____

Note that question 85 has only three choices.

85 An electron is located between a pair of oppositely charged parallel plates. As the electron approaches the positive plate, the kinetic energy of the electron

1 decreases
2 increases
3 remains the same

85_____

GROUP 4—**Geometric Optics**

If you choose this group, be sure to answer questions **86–95**.

86 In the diagram below, a source produces a light ray that is reflected from a plane mirror.

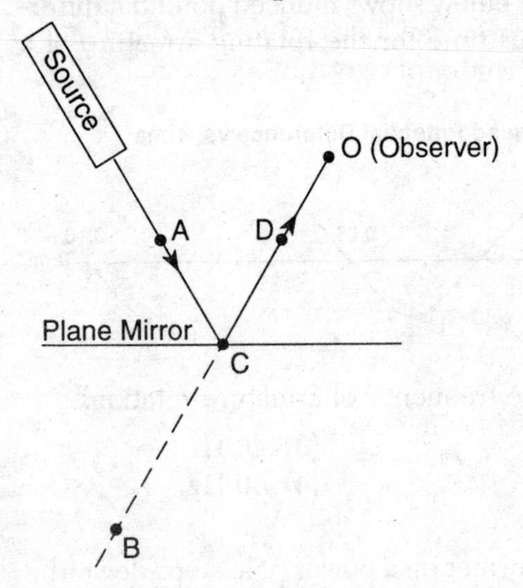

To an observer at point *O*, the light appears to originate from point

(1) *A*
(2) *B*
(3) *C*
(4) *D*

86 ____

87 A spherical mirror that forms only virtual images has a radius of curvature of 0.50 meter. The focal length of this mirror is

(1) –0.25 m
(2) +0.25 m
(3) –0.50 m
(4) +0.50 m

87 ____

88 A spherical concave mirror is used in the back of a car headlight. Where must the bulb of the headlight be located to produce a parallel beam of reflected light?

1 between the principal focus and the mirror
2 beyond the center of curvature of the mirror
3 at the principal focus of the mirror
4 at the center of curvature of the mirror

88 _____

Base your answers to questions 89 through 91 on the information and diagram below.

An object is located at the center of curvature *C* of a concave spherical mirror with principal focus *F*. The focal length of the mirror is 0.10 meter.

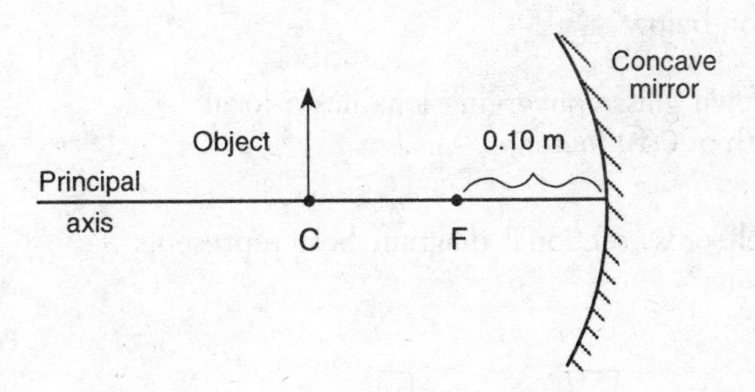

89 At what distance from the mirror is the image located?

(1) 0.10 m
(2) 0.20 m
(3) 0.30 m
(4) 0.40 m

89 _____

90 At what distance from the mirror could the object be placed to produce a virtual image of the object?

(1) 0.05 m (3) 0.30 m
(2) 0.10 m (4) 0.50 m 90 _____

Note that question 91 has only three choices.

91 As the object is moved from point *C* toward point *F*, the size of its image

1 decreases
2 increases
3 remains the same 91 _____

Base your answers to questions 92 and 93 on the information below.

A crown glass converging lens has a focal length of 0.10 meter.

92 Which cross-sectional diagram best represents this lens?

(1)

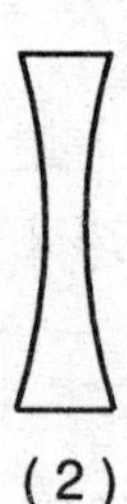

(2)

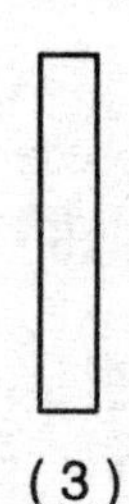

(3)

(4) 92 _____

93 An object is placed 0.30 meter from the lens. How far from the lens will an image of the object be formed?

(1) 0.30 m (3) 0.15 m
(2) 0.20 m (4) 0.10 m

93 _____

---

94 An object 0.080 meter high is placed 0.20 meter from a converging (convex) lens. If the distance of the image from the lens is 0.40 meter, the height of the image is

(1) 0.010 m (3) 0.080 m
(2) 0.040 m (4) 0.16 m

94 _____

95 A diverging (concave) lens can form images that are

1 virtual, only
2 inverted, only
3 either virtual or real
4 either inverted or erect

95 _____

GROUP 5—**Solid State**

If you choose this group, be sure to answer questions **96–105**.

96 Which graph best represents the relationship between conductivity and resistivity for a solid?

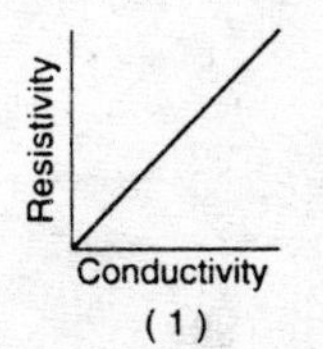

(1)

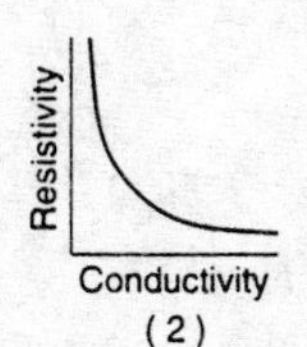

(2)

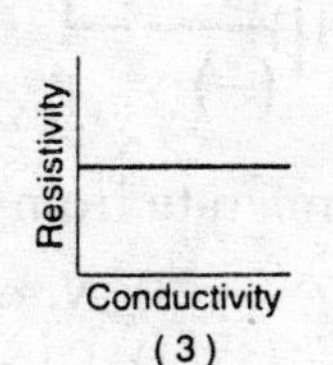

(3)

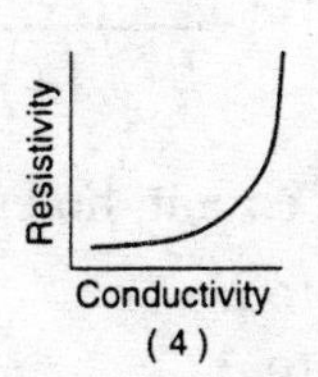

(4)

96 _____

97 The diagram below shows an input of alternating current to a "black box" and the resulting output current.

The "black box" most likely is

1 an LED
2 a transistor
3 a diode
4 an *N*-type semiconductor

97 _____

98 A *P*-type semiconductor is formed by adding impurities, which provide extra

1 electrons
2 protons
3 neutrons
4 holes

98 _____

Base your answers to questions 99 through 101 on the diagram below, which represents a semiconductor device connected to a battery.

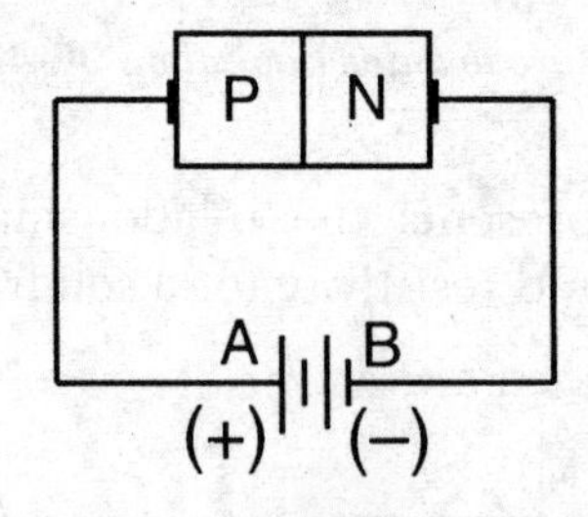

99 In the circuit, holes migrate from

(1) *A* to *B*
(2) *B* to *A*
(3) *N* to *P*
(4) *P* to *N*

99 _____

100 If an alternating potential difference is substituted for the battery, the result will be

1 an alternating current
2 a pulsating direct current
3 a steady direct current
4 no current flow

100 _____

101 Which symbol best represents the orientation of the device shown in the circuit diagram?

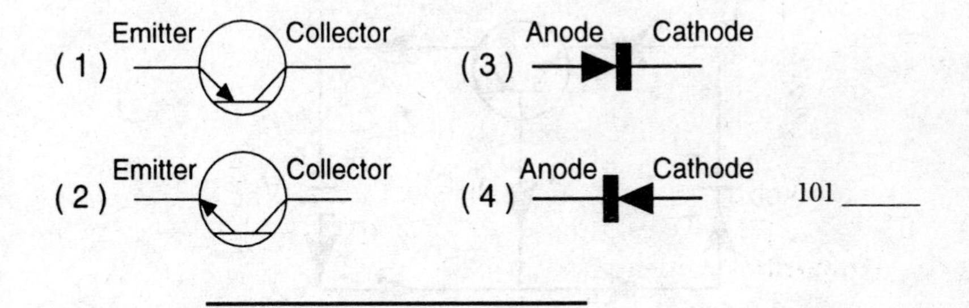

101 _____

Base your answers to questions 102 and 103 on the diagram and graph below.

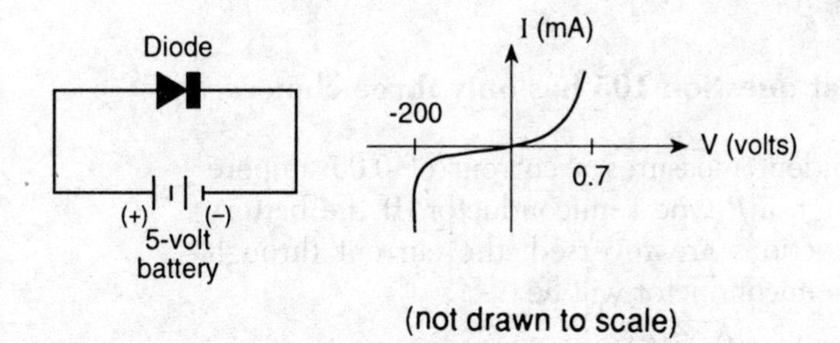

(not drawn to scale)

102 When connected as shown, the diode is

1 forward biased
2 reverse biased
3 open biased
4 closed biased

102 _____

103 Compared to the voltage at which this diode avalanches, a more heavily doped diode would

1 avalanche at a lower voltage, only
2 avalanche at a higher voltage, only
3 not avalanche at any voltage
4 avalanche at any voltage

103 _____

---

104 The diagram below represents an operating transistor circuit.

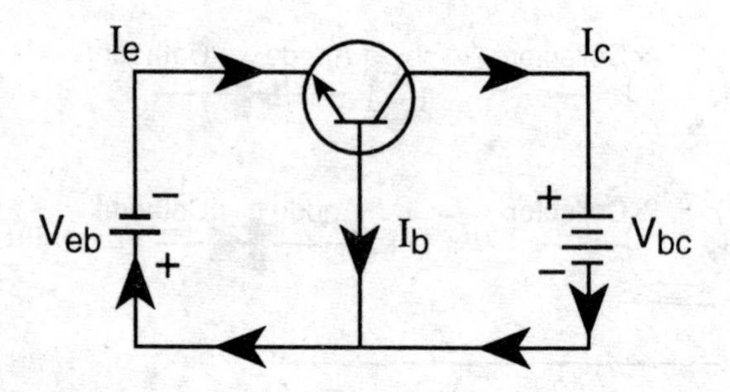

Which currents are approximately equal?

(1) I_b and I_e, only
(2) I_b and I_c, only
(3) I_c and I_e, only
(4) I_b, I_c, and I_e

104 _____

Note that question 105 has only three choices.

105 A student measures a current of 0.05 ampere through a *P*-type semiconductor. If the battery connections are reversed, the current through the semiconductor will be

1 less than 0.05 A
2 greater than 0.05 A
3 equal to 0.05 A

105 _____

GROUP 6—**Nuclear Energy**

If you choose this group, be sure to answer questions **106–115**.

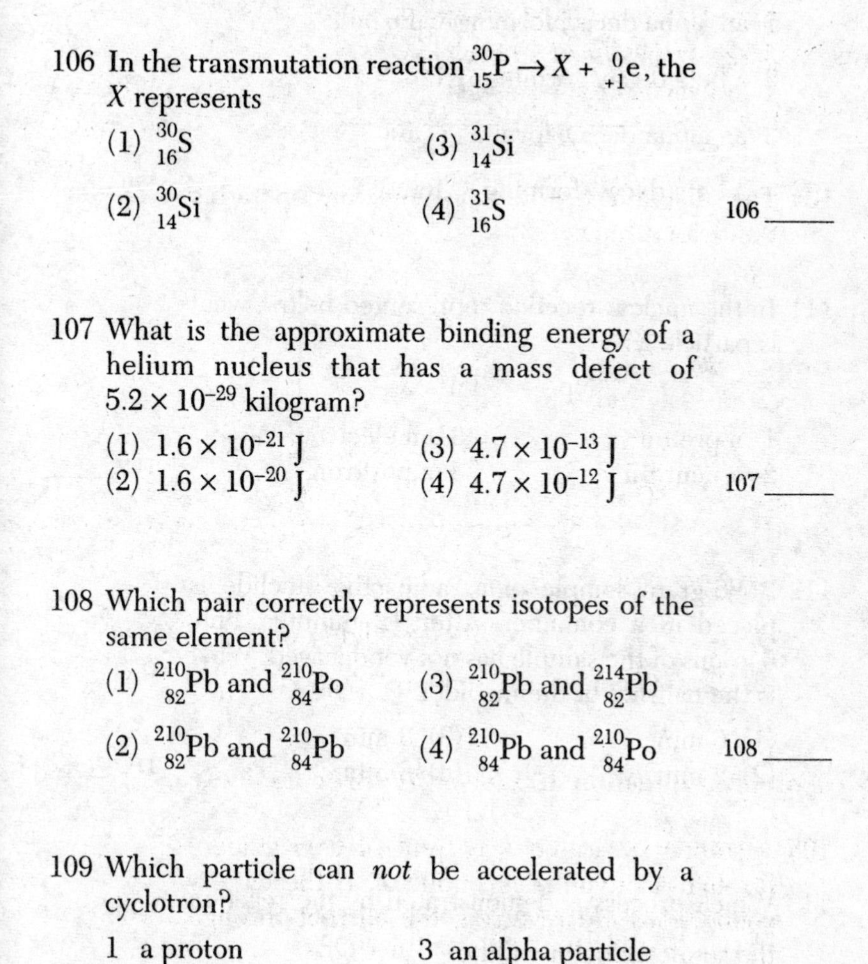

106 In the transmutation reaction $^{30}_{15}\text{P} \rightarrow X + \, ^{0}_{+1}\text{e}$, the X represents

(1) $^{30}_{16}\text{S}$
(2) $^{30}_{14}\text{Si}$
(3) $^{31}_{14}\text{Si}$
(4) $^{31}_{16}\text{S}$

106 _____

107 What is the approximate binding energy of a helium nucleus that has a mass defect of 5.2×10^{-29} kilogram?

(1) 1.6×10^{-21} J
(2) 1.6×10^{-20} J
(3) 4.7×10^{-13} J
(4) 4.7×10^{-12} J

107 _____

108 Which pair correctly represents isotopes of the same element?

(1) $^{210}_{82}\text{Pb}$ and $^{210}_{84}\text{Po}$
(2) $^{210}_{82}\text{Pb}$ and $^{210}_{84}\text{Pb}$
(3) $^{210}_{82}\text{Pb}$ and $^{214}_{82}\text{Pb}$
(4) $^{210}_{84}\text{Pb}$ and $^{210}_{84}\text{Po}$

108 _____

109 Which particle can *not* be accelerated by a cyclotron?

1 a proton
2 a neutron
3 an alpha particle
4 an electron

109 _____

110 According to the Uranium Disintegration Series, $^{222}_{86}Rn$ undergoes

1 an alpha decay, forming $^{218}_{84}Po$
2 a beta decay, forming $^{218}_{84}Po$
3 an alpha decay, forming $^{226}_{88}Ra$
4 a beta decay, forming $^{226}_{88}Ra$

110 _____

111 In the nuclear reaction represented below, what is particle *X*?

$$^{238}_{93}Np \rightarrow ^{238}_{94}Pu + X$$

1 a proton
2 a neutron
3 an electron
4 a positron

111 _____

112 A 96-gram sample of a radioactive nuclide is placed in a container. After 12 minutes, only 6 grams of the sample has not yet decayed. What is the half-life of the nuclide?

(1) 6 min
(2) 2 min
(3) 3 min
(4) 8 min

112 _____

113 Which process is demonstrated by the reaction $^{235}_{92}U + ^{1}_{0}n \rightarrow ^{141}_{56}Ba + ^{92}_{36}Kr + 3^{1}_{0}n + Q$?

1 nuclear fission
2 nuclear fusion
3 alpha decay
4 beta decay

113 _____

114 The principal reason for using neutrons to bombard a nucleus is that neutrons

1 have a relatively low atomic mass
2 have a very high kinetic energy
3 can be easily accelerated
4 are not repelled by the nucleus 114 _____

115 Which process is the source of the Sun's energy?

1 natural radioactive decay
2 electron capture
3 fission
4 fusion 115 _____

PART III

You must answer *all* questions in this part. [15]

Base your answers to questions 116 through 120 on the information and vector diagram below.

A 20.-newton force due north and a 40.-newton force due east act concurrently on a 10.-kilogram object, located at point *P*.

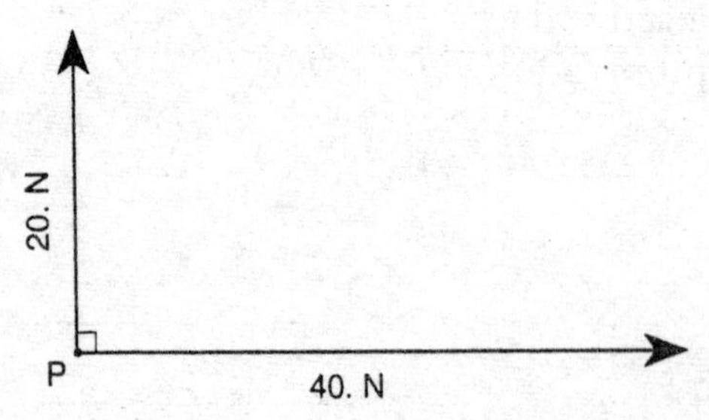

116 Using a ruler, determine the scale used in the vector diagram by finding the number of newtons represented by each centimeter. [1]

117 On the vector diagram provided, use a ruler and protractor to construct the vector that represents the resultant force. [1]

118 What is the magnitude of the resultant force? [1]

____

119 What is the measure of the angle (in degrees) between east and the resultant force? [1]

____

120 Calculate the magnitude of the acceleration of the object. [Show all calculations, including the equation and substitiution with units.] [2]

____

____

____

Base your answers to questions 121 through 123 on the information and diagram below.

Two waves, *A* and *B*, travel in the same direction in the same medium at the same time.

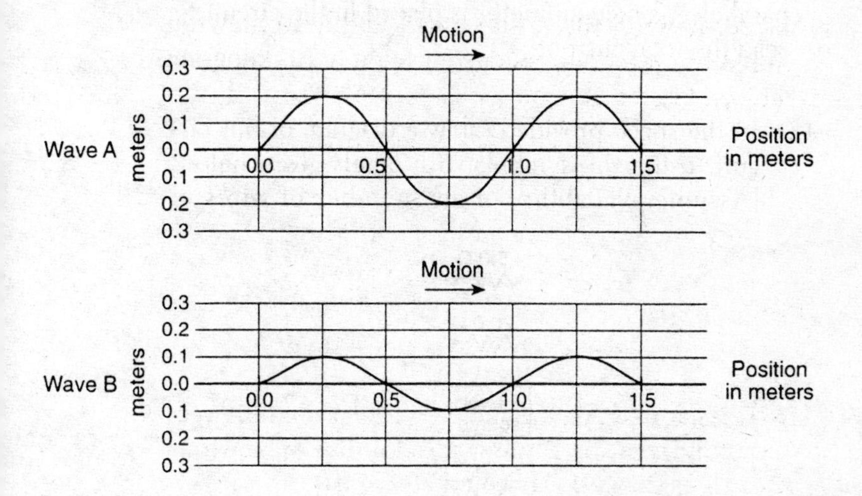

121 On the grid provided, draw the resultant wave produced by the superposition of waves *A* and *B*. [1]

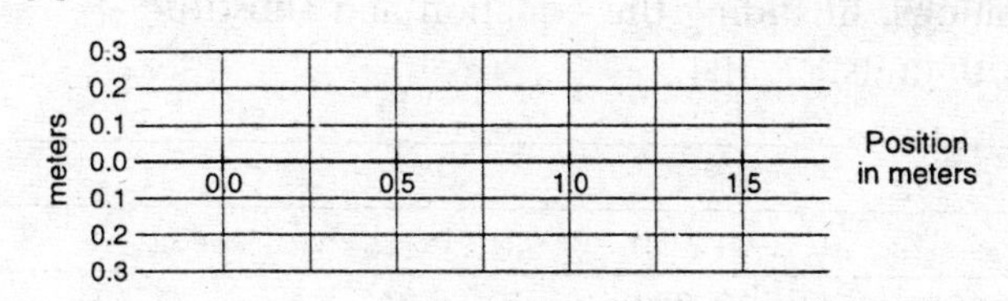

122 What is the amplitude of the resultant wave? [1]

____

123 What is the wavelength of the resultant wave? [1]

____

Base your answers to questions 124 through 126 on the information below.

A 5.0-ohm resistor, a 20.0-ohm resistor, and a 24-volt source of potential difference are connected in parallel. A single ammeter is placed in the circuit to read the total current.

124 In the space provided, draw a diagram of this circuit, using the symbols with labels given below. [Assume availability of any number of wires of

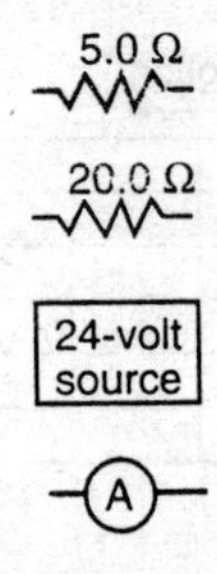

125 Determine the total circuit resistance. [Show all calculations, including the equation and substitution with units.] [2]

____

____

____

126 Determine the total circuit current. [Show all calculations, including the equation and substitution with units.] [2]

____

____

____

Answers June 1997

Physics

Answer Key

PART I

1. 2	**11.** 2	**20.** 4	**29.** 1	**38.** 1	**47.** 3
2. 4	**12.** 1	**21.** 2	**30.** 4	**39.** 2	**48.** 4
3. 3	**13.** 1	**22.** 1	**31.** 1	**40.** 4	**49.** 2
4. 3	**14.** 2	**23.** 4	**32.** 2	**41.** 3	**50.** 3
5. 1	**15.** 1	**24.** 3	**33.** 4	**42.** 1	**51.** 4
6. 1	**16.** 4	**25.** 4	**34.** 2	**43.** 4	**52.** 1
7. 4	**17.** 3	**26.** 2	**35.** 3	**44.** 1	**53.** 2
8. 2	**18.** 1	**27.** 3	**36.** 2	**45.** 3	**54.** 3
9. 3	**19.** 4	**28.** 3	**37.** 2	**46.** 2	**55.** 3
10. 2					

PART II

Group 1	Group 2	Group 3	Group 4	Group 5	Group 6
56. 2	**66.** 4	**76.** 3	**86.** 2	**96.** 2	**106.** 2
57. 2	**67.** 1	**77.** 1	**87.** 1	**97.** 2	**107.** 4
58. 3	**68.** 3	**78.** 4	**88.** 3	**98.** 4	**108.** 3
59. 3	**69.** 2	**79.** 2	**89.** 2	**99.** 4	**109.** 2
60. 4	**70.** 3	**80.** 1	**90.** 1	**100.** 2	**110.** 1
61. 4	**71.** 2	**81.** 1	**91.** 2	**101.** 3	**111.** 3
62. 3	**72.** 4	**82.** 3	**92.** 4	**102.** 1	**112.** 3
63. 4	**73.** 2	**83.** 2	**93.** 3	**103.** 1	**113.** 1
64. 1	**74.** 3	**84.** 4	**94.** 4	**104.** 3	**114.** 4
65. 1	**75.** 1	**85.** 2	**95.** 1	**105.** 3	**115.** 4

PART III — See answers explained

Answers Explained

PART I

1. **2** Displacement is a vector quantity. To find the resultant vector, simply sketch the vectors to scale, place them head to tail, and draw the resultant vector from the tail of the first vector to the head of the last vector.

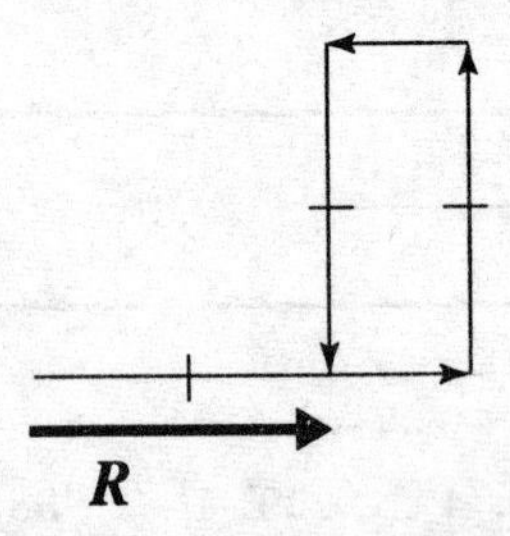

From the sketch it is clear that the total displacement of the student is equal to 2 blocks east.

2. **4** The best way to approach any word problem is to correctly identify the quantities given and the quantity to be found.

Given: $\bar{v} = 42$ m/s Find: $\Delta t = ?$

$\Delta s = 18$ m

Refer to the *Mechanics* section in the *Reference Tables* to find an equation(s) that relates time with average speed and displacement.

Solution:

$$\bar{v} = \frac{\Delta s}{\Delta t}$$

$$\Delta t = \frac{\Delta s}{\bar{v}}$$

$$= \frac{18m}{42 \text{ m/s}}$$

$$= 0.43 \text{ s}$$

3. **3** The *meter* is the SI unit of length. One meter is approximately equal to 3.3 feet. Of the four choices, 10^1 m, or 10 m, is the only reasonable one for the length of a high school physics classroom.

Wrong Choices Explained:

(1) 10^{-2} m = 1 cm, or a little less than 0.5 in.
(2) 10^{-1} m = 1 dm = 10 cm, or approximately 4 in.
(4) 10^4 m = 10 km, or approximately 5 mi.

4. **3** Write out the problem.

Given: $v_{i_y} = 0$ Find: $\Delta t = ?$

$\Delta s_y = 45$ m

$a_y = 9.8$ m/s² (Refer to List of Physical Constants for this value)

Refer to the *Mechanics* section in the *Reference Tables* to find an equation(s) that relates time with displacement, acceleration, and velocity.

Solution:
$$\Delta s_y = v_{i_y}\Delta t + \frac{1}{2}a_y(\Delta t)^2$$
$$\Delta t = \sqrt{\frac{2\Delta s}{a}}$$
$$= \sqrt{\frac{2(45\text{ m})}{9.8\text{ m/s}^2}}$$
$$= 3.0\text{ s}$$

5. **1** According to Newton's second law, $F = ma$, the greater the net force, the greater the acceleration. The net force on an object is the vector sum of all the forces acting on that object. Two forces acting in unison will produce the greatest net force when they are 0 degree apart. The figure in choice (1) correctly represents this situation.

Wrong Choices Explained:

(4) This diagram shows two forces applied in opposite directions (180 degrees apart). This will produce the smallest net force.

(2), (3) These diagrams show forces applied at 90 degrees to each other. This will produce a net force with a magnitude between that produced by choice (1) and that produced by choice (4).

6. **1** On a distance versus time graph the slope is equal to the average velocity. Section *AB* of the graph shows a changing slope; therefore, during this interval the velocity is changing and, hence, the cart is accelerating.

Wrong Choices Explained:

(2) Section *BC* has a constant positive slope, which indicates a constant positive velocity.

(3) Sections *CD* has a slope of 0, which indicates that the cart is at rest.

(4) Section *DE* has a constant negative slope, which indicates a constant velocity in the opposite direction.

7. **4** According to Newton's first law, an object in motion continues uniformly in a straight line unless acted upon by an unbalanced force. Once the ball exits the tube, it no longer experiences a centripetal force from the sides of the tube and continues in a straight line from its exit point. The diagram in choice (4) correctly illustrates this situation.

Wrong Choice Explained:

(2) This diagram would have been correct had the tube been placed vertically on the tabletop and not laid flat on the surface.

8. **2** Write out the problem.

Given: $\boldsymbol{F}$ = 100. N at 30° to horizontal   Find: $\boldsymbol{F}_{\text{vertical}} = ?$

To solve this problem, resolve the vector $\boldsymbol{F}$ into its horizontal and vertical components.

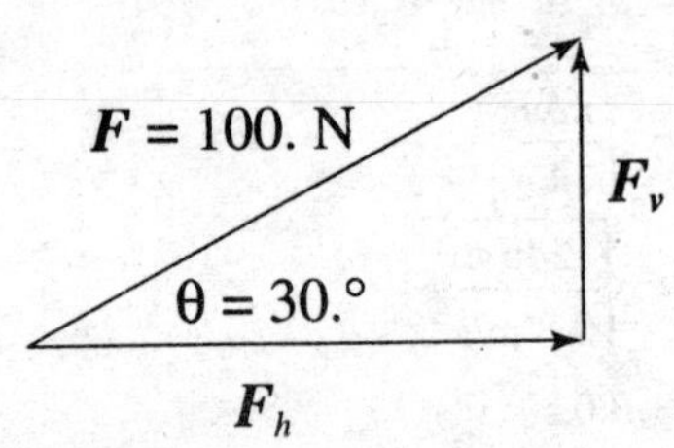

Solution:

$$\sin\theta = \frac{\text{opposite side}}{\text{hypotenuse}}$$

$$= \frac{\boldsymbol{F}_{\text{vertical}}}{\boldsymbol{F}}$$

$$\boldsymbol{F}_{\text{vertical}} = \boldsymbol{F}\cdot\sin\theta$$

$$= (100.\ \text{N})(\sin 30°)$$

$$= 50.\ \text{N}$$

9. **3** Write out the problem.

Given: m = 1,000. kg   Find: $\Delta \boldsymbol{s} = ?$

$\boldsymbol{v}_i$ = +20. m/s

$\boldsymbol{a}$ = –5.0 m/s²

$\boldsymbol{v}_f$ = 0

Refer to the *Mechanics* section in the *Reference Tables* to find an equation that relates displacement with mass, velocities, and acceleration.

Solution:

$$\boldsymbol{v}_f^{\,2} = \boldsymbol{v}_i^{\,2} + 2\boldsymbol{a}\,\Delta\boldsymbol{s}$$

$$\Delta\boldsymbol{s} = \frac{\boldsymbol{v}_f^{\,2} - \boldsymbol{v}_i^{\,2}}{2\boldsymbol{a}}$$

$$= \frac{0 - (20.\ \text{m/s})^2}{2(-5.0\ \text{m/s}^2)}$$

$$= 40.\ \text{m}$$

(Note that mass is not needed in this calculation.)

10. **2** Write out the problem.

Given: $m = 1{,}000.\text{ kg}$ Find: $\boldsymbol{J} = ?$

$\boldsymbol{v}_i = +20.\text{ m/s}$

$\boldsymbol{a} = 5.0\text{ m/s}^2$

$\boldsymbol{v}_f = 0$

Refer to the *Mechanics* section in the *Reference Tables* to find an equation(s) that relates impulse with mass, velocities, and acceleration.

Solution:

$$\begin{aligned}\boldsymbol{J} &= \boldsymbol{F}\,\Delta t\\ \boldsymbol{F}\,\Delta t &= m\,\Delta\boldsymbol{v}\\ \therefore \boldsymbol{J} &= m\,\Delta\boldsymbol{v}\\ &= m(\boldsymbol{v}_f - \boldsymbol{v}_i)\\ &= 1000.\text{ kg}(0 - 20.\text{ m})\\ &= -2.0\times10^4\text{ kg}\cdot\text{m/s}\\ &= -2.0\times10^4\text{ N}\cdot\text{s}\end{aligned}$$

(Note that acceleration is not needed in this calculation and that the negative sign simply indicates direction but is not needed in the answer.)

11. **2** The slope of a weight versus mass graph is equal to the acceleration due to gravity. Refer to the *Mechanics* section in the *Reference Tables*: $w = mg$; therefore $g = w/m$. The slope of the line in the graph is equal to 6.0 m/s^2.

12. **1** Write out the problem.

Given: $m = 2.0\times10^3\text{ kg}$ Find: $\boldsymbol{F} = ?$

$\boldsymbol{v}_i = 0$

$\boldsymbol{v}_f = 15\text{ m/s}$

$\Delta t = 5.0\text{ s}$

Refer to the *Mechanics* section in the *Reference Tables* to find an equation(s) that relates force with mass, velocities, and time.

Solution:

$$\begin{aligned}\boldsymbol{F}\Delta t &= m\Delta\boldsymbol{v}\\ \boldsymbol{F} &= \frac{m(\boldsymbol{v}_f - \boldsymbol{v}_i)}{\Delta t}\\ &= \frac{(2.0\times10^3\text{kg})(15\text{ m/s} - 0)}{5.0\text{ s}}\\ &= 6.0\times10^3\text{ N}\end{aligned}$$

13. **1** The force due to friction is equal to the product of μ (coefficient of friction between two surfaces, a constant) times the normal force. It does not depend on the surface area; therefore, since surfaces *A* and *B* are essentially the same, μ is the same. Since the block's mass and the acceleration due to gravity have not been changed by placing the block on its side, the normal force is also the same. Therefore, the force due to friction is the same, and the force required to slide the block at constant speed is the same, that is, ***F***.

14. **2** Write out the problem.

Given: $m_A = 3.0$ kg
$m_B = 5.0$ kg
$\boldsymbol{v}_{A_f} = 0.33$ m/s
$\boldsymbol{v}_{A_i} = 0$
$\boldsymbol{v}_{B_i} = 0$

Find: $\boldsymbol{v}_{B_f} = ?$

Refer to the *Mechanics* section in the *Reference Tables* to find an equation(s) that relates velocities with masses.

Solution: $\boldsymbol{p} = m\boldsymbol{v}$

This problem deals with conservation of momentum; in a closed system the total initial momentum of a system equals the total final momentum of a system:

$$\boldsymbol{p}_{\text{total}\,i} = \boldsymbol{p}_{\text{total}\,f}$$

$$m_A\boldsymbol{v}_{A_i} + m_B\boldsymbol{v}_{B_i} = m_A\boldsymbol{v}_{Af} + m_B\boldsymbol{v}_{Bf}$$

$$0 = m_A\boldsymbol{v}_{Af} + m_B\boldsymbol{v}_{Bf}$$

$$\boldsymbol{v}_{Bf} = \frac{-m_A\boldsymbol{v}_{Af}}{m_B}$$

$$= \frac{-(3.0 \text{ kg})(0.33 \text{ m/s})}{5.0 \text{ kg}}$$

$$= -0.20 \text{ m/s}$$

(Note that the negative sign simply indicates direction and is not needed in the answer.)

15. **1** Write out the problem.

Find: $\boldsymbol{F}_{G_{Earth\text{–}Moon}} = ?$

Refer to the *Mechanics* section in the *Reference Tables* to find an equation(s) that pertains to gravitational force.

$$F = \frac{Gm_1 m_2}{r^2}$$

Refer to the List of Physical Constants in the *Reference Tables* to find the value of G, the gravitational constant.

$$G = 6.7 \times 10^{-11}\ \text{N} \cdot \text{m}^2/\text{kg}^2$$
$$m_{\text{Earth}} = 6.0 \times 10^{24}\ \text{kg}$$
$$m_{\text{Moon}} = 7.4 \times 10^{22}\ \text{kg}$$
$$r_{\text{Earth–Moon}} = 3.8 \times 10^{8}\ \text{m}$$

Solution:

$$F_{G_{\text{Earth–Moon}}} = \frac{Gm_{\text{Earth}} m_{\text{Moon}}}{r_{\text{Earth–Moon}}{}^2}$$

$$= \frac{(6.7 \times 10^{-11}\ \text{N} \cdot \text{m}^2/\text{kg}^2)(6.0 \times 10^{24}\ \text{kg})(7.4 \times 10^{22}\ \text{kg})}{(3.8 \times 10^{8}\ \text{m})^2}$$

$$= 2.1 \times 10^{20}\ \text{N}$$

16. **4** Write out the problem.

Given: $\boldsymbol{F} = 9.8 \times 10^2$ N Find: W = ?
$\Delta \boldsymbol{s}$ = 10. m

Refer to the *Energy* section in the *Reference Tables* to find an equation(s) that relates work with force and displacement.

Solution: $W = \boldsymbol{F} \Delta \boldsymbol{s}$

$$= (9.8 \times 10^2\ \text{N})(10.\ \text{m})$$
$$= 9.8 \times 10^3\ \text{J}$$

17. **3** Translate the variables into appropriate units.

Solution: $$\frac{\text{mass} \cdot \text{distance}^2}{\text{time}^2} = \frac{\text{kg} \cdot \text{m}^2}{\text{s}^2} = \text{N} \cdot \text{m} = \text{J}$$

Wrong Choices Explained:

(1) $\frac{\text{mass} \cdot \text{distance}}{\text{time}} = \frac{\text{kg} \cdot \text{m}}{\text{s}}$, unit of momentum

(2) $\frac{\text{mass} \cdot \text{distance}^2}{\text{time}} = \frac{\text{kg} \cdot \text{m}^2}{\text{s}}$, not an SI unit of measure

(4) $\frac{\text{mass} \cdot \text{distance}}{\text{time}^3} = \frac{\text{kg} \cdot \text{m}}{\text{s}^3}$, not an SI unit of measure

18. **1** Write out the problem.

Given: $k = 120$ N/m Find: PE_s
$x = 0.20$ m

Refer to the *Energy* section in the *Reference Tables* to find an equation(s) that relates potential energy to a spring constant and a change in spring length as it is stretched.

Solution:
$$PE_s = \frac{1}{2}kx^2$$
$$= \frac{1}{2}(120 \text{ N/m})(0.20 \text{ m})$$
$$= 2.4 \text{ J}$$

19. **4** Refer to the *Energy* section in the *Reference Tables* to find an equation(s) that relates the spring constant of a spring with a change in length of the spring as it is stretched and with force.

Solution:
$$\boldsymbol{F} = kx$$
$$k = \frac{\boldsymbol{F}}{x}$$

The slope of this line is constant, indicating a direct relationship between force and elongation. Therefore, by choosing any point on the line and substituting the values into the equation above, you can obtain a value for k.

The value calculated from any point is equal to 50. N/m.

20. **4** Write out the problem.

Given: $P = 8.1 \times 10^4$ W Find: $\boldsymbol{F} = ?$
$w_{\text{elevator}} = 1.8 \times 10^4$ N
$\bar{v} = 3.0$ m/s

Refer to the *Energy* section in the *Reference Tables* to find an equation(s) that relates force with power and average speed.

Solution:
$$P = \boldsymbol{F}\bar{v}$$
$$\boldsymbol{F} = \frac{P}{\bar{v}}$$
$$= \frac{8.1 \times 10^4 \text{ W}}{3.0 \text{ m/s}}$$
$$= 2.7 \times 10^4 \text{ N}$$

(Note that the weight of the elevator is not necessary in this calculation.)

21. **2** Write out the problem.

Given: $m = M$ Find: $KE_f = ?$

$v_i = 0$

$h_i = h_1$

$h_f = h_2$

Refer to the *Energy* section in the *Reference Tables* to find an equation(s) that relates kinetic energy with mass, velocity, and heights.

Solution: $\Delta PE = mg\Delta h$

$$KE = \frac{1}{2}mv^2$$

This problem deals with conservation of energy. In a closed system, the total initial energy of the system is equal to the total final energy.

$$PE_i + KEi = PE_f + KE_f$$

$$mgh_i + \frac{1}{2}mv_i^{\ 2} = mgh_f + KE_f$$

$$KE_f = mgh_i + \frac{1}{2}mv_i^{\ 2} - mgh_f$$

$$= Mgh_1 + 0 - Mgh_2$$

$$= Mg(h_1 - h_2)$$

22. **1** The proton is nearly 2,000 times more massive than the electron and is tightly bound in the nucleus (along with the neutrons). As a result, ordinary objects become electrically charged by gaining or losing electrons. If an object gains electrons, the excess of electrons gives the object a negative charge. If an object loses electrons, the deficiency of electrons gives the object a positive charge.

Since electric charge must be conserved, electrons are transferred from one object to another. Therefore, in this example, the wool acquired a positive charge because electrons were transferred from the wool to the plastic rod.

23. **4** Since spheres *A*, *B*, and *C* are identical, charge will be transferred among them until the charge is the same for all three. Therefore, when sphere *A* makes contact with sphere *B*, each will have $q/2$ charge. When sphere *A* then subsequently makes contact with sphere *C*, each will have $q/4$ charge.

24. **3** Write out the problem.

Given: $d = 1.0 \times 10^3$ m Find: $V = ?$

$\mathbf{E} = 2.0 \times 10^4$ N/C

Refer to the *Electricity and Magnetism* section in the Reference Tables to find an equation(s) that relates potential difference with electric field intensity and distance.

Solution: $\mathbf{E} = \frac{V}{d}$

$$V = \mathbf{E}d$$
$$= (2.0 \times 10^4 \text{ N/C})(1.0 \times 10^3 \text{ M})$$
$$= 2.0 \times 10^7 \text{ V}$$

25. **4** Write out the problem.

Given: $W = 6.4 \times 10^{-19}$ J Find: $V = ?$
1 proton

Refer to the *Electricity and Magnetism* section in the *Reference Tables* to find an equation(s) that relates potential difference with work (energy).

Solution: $V = \frac{W}{q}$

Refer to the List of Physical Constants in the *Reference Tables* to obtain a value for q, the elementary electric charge.

$$V = \frac{6.4 \times 10^{-19} \text{ J}}{1.6 \times 10^{-19} \text{ C}}$$
$$= 4.0 \text{ V}$$

26. **2** Refer to the List of Physical Constants in the *Reference Tables* to obtain the value for q, the elementary electric charge.

Solution: $q_{\text{total}} = 3 \times q$
$$= 3(1.6 \times 10^{-19} \text{ C})$$
$$= 4.8 \times 10^{-19} \text{ C}$$

27. **3** As a consequence of the law of conservation of electric charge, the sum of the currents entering a junction must be equal to the sum of the currents leaving the junction.

Solution: $I_{\text{tot}_{\text{in}}} = I_{\text{tot}_{\text{out}}}$
$$6 \text{ A} + 2\text{A} = 5 \text{ A} + x\text{A}$$
$$x = 3 \text{ A}$$

28. **3** Write out the problem.

Given: $I = 0.50$ A Find: $q = ?$
$\Delta t = 10.$ s

Refer to the *Electricity and Magnetism* section in the *Reference Tables* to find an equation that relates charge with current and time.

Solution: $I = \frac{q}{\Delta t}$

$q = I\Delta t$

$= (0.50 \text{ A})(10. \text{ s})$

$= 5.0 \text{ C}$

29. **1** The resistance (R) of a wire is directly proportional to the resistivity (ρ) of the substance at a particular temperature, and to the length of the wire (L) and is inversely proportional to the cross-sectional area of the wire (A):

$$R = \rho\frac{L}{A}$$

The graph in choice (1) correctly represents a direct relationship between resistance and length when temperature and cross-sectional area are constant.

Wrong Choices Explained:

(2) This graph shows that resistance is constant for all lengths of wire.

(3) This graph shows an inverse relationship between resistance and length.

(4) This graph shows an inversely proportional relationship between resistance and length.

30. **4** A vector quantity has both magnitude and direction. A scalar quantity has only magnitude. Electric field intensity has both magnitude and direction and is therefore a vector quantity.

31. **1** Refer to the *Electricity and Magnetism* section in the *Reference Tables* to find an equation(s) that relates power with resistance and current.

Solution: $P = IV$
$V = IR$

Assuming a constant voltage source, if P is larger, then I is larger (power is directly proportional to current); if I is larger, then R is smaller (current is inversely proportional to resistance). Choice (1) correctly states these conclusions; the 100-W bulb has less resistance and draws more current than the 60-W bulb.

32. **2** Write out the problem.

Given: $W = 6.0 \times 10^6 \text{ J}$ Find: $I = ?$
$V = 220 \text{ V}$
$\Delta t = 1800 \text{ s}$

Refer to the *Electricity and Magnetism* section in the *Reference Tables* to find an equation(s) that relates current with work, potential difference, and time.

Solution: $W = VIt$

$$I = \frac{W}{Vt}$$

$$= \frac{6.0 \times 10^6 \text{ J}}{(220 \text{ V})(1800 \text{ s})}$$

$$= 15 \text{ A}$$

33. **4** The direction of a magnetic field line is defined as the direction to which the north pole of a compass needle points when the compass is placed in the magnetic field. On the outside of a magnet, the field lines leave from the north pole and enter through the south pole. Therefore, the diagram in choice (4) best represents the position of the compass needle as it responds to the magnetic field of the bar magnet.

34. **2** To solve this problem, use the following left-hand rule: Wrap the fingers of the left hand in the direction of the magnetic field, and the thumb will point in the direction of electron flow, that is, from *B* to *A* in the diagram.

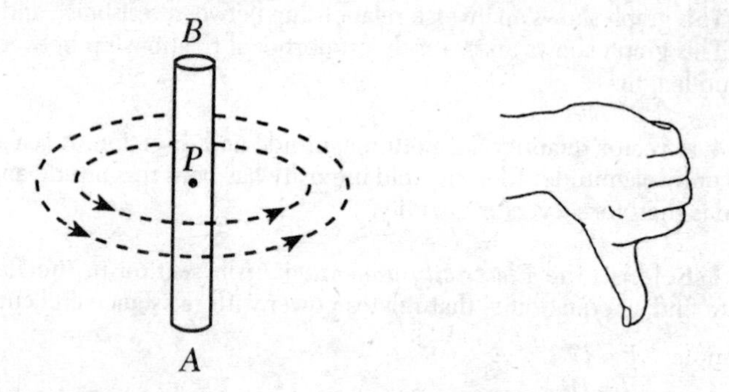

35. **3** To solve this problem, use the following left-hand rule: Point the thumb in the direction of the electron flow, and point the fingers in the direction of the magnetic field. The palm of the hand will point in the direction in which the magnetic force on the conductor is directed, that is, into the page.

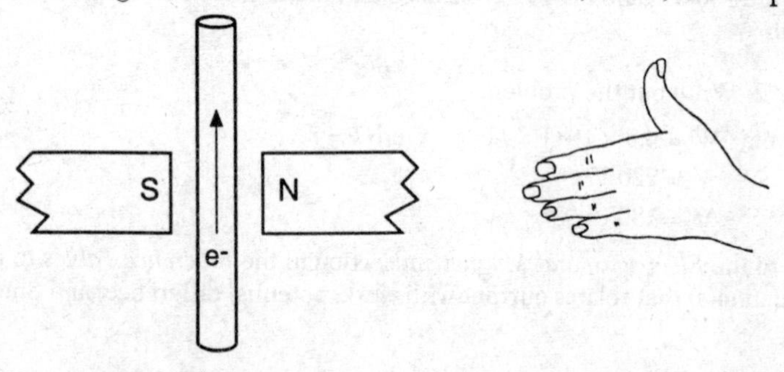

36. **2** A wave transfers energy, not mass. Individual particles in a medium move in simple harmonic motion about an equilibrium point. In a transverse wave, the particles vibrate in a direction perpendicular to the direction of motion of the wave. Since the wave front in the diagram is moving from left to right and a trough hits point X first, the motion of X will move down, then up as the crest passes by.

37. **2** Write out the problem.

Given: $f = 5.0 \times 10^{14}$ Hz   Find: $T = ?$

Refer to the *Wave Phenomena* section in the *Reference Tables* to find an equation(s) that relates period with frequency.

Solution:
$$T = \frac{1}{f}$$
$$= \frac{1}{5.0 \times 10^{14}\ \text{Hz}}$$
$$= 2.0 \times 10^{-15}\ \text{s}$$

38. **1** The amplitude of a wave is a measure of the energy contained in the wave. In sound, amplitude corresponds to loudness; in light, amplitude corresponds to brightness.

Wrong Choices Explained:

(2), (3) The frequency of a light wave corresponds to color.

39. **2** The speed of light in a vacuum, c, is constant under all circumstances. Refer to the List of Physical Constants in the *Reference Tables* to find its value.

The speed of light is less in a material medium than in a vacuum and depends on the nature of the medium and the frequency of the light.

Given: $c = 3.0 \times 10^8$ m/s   Find: $v_{glycerol} = ?$

Refer to the *Wave Phenomena* section in the *Reference Tables* to find an equation(s) that relates the speed of light in a medium with the speed of light in a vacuum.

Solution:
$$n = \frac{c}{v}$$

Refer to the Absolute Indices of Refraction table in the *Reference Tables* to obtain the value of n for glycerol.

$$v = \frac{c}{n}$$
$$= \frac{3.0 \times 10^8\ \text{m/s}}{1.47}$$
$$= 2.0 \times 10^8\ \text{m/s}$$

40. **4** Write out the problem.

Given: $v = 0.25$ m/s Find: $\lambda = ?$

$f = 4.0$ times in 8.0 s = 0.50 Hz

Refer to the *Wave Phenomena* section in the *Reference Tables* to find an equation(s) that relates wavelength with velocity and frequency.

Solution:
$$v = f\lambda$$
$$\lambda = \frac{v}{f}$$
$$= \frac{0.25 \text{ m/s}}{0.50 \text{ Hz}}$$
$$= 0.50 \text{ m}$$

41. **3** This question refers to the Doppler effect, which is the result of relative motion between a source of waves and an observer. As the distance between the source and the observer decreases, the frequency, which in sound corresponds to pitch, of the source, as perceived by the observer, is increased. This phenomenon is due to a series of sound waves being crowded together on the side nearest the observer.

When velocity is constant, frequency is inversely proportional to wavelength ($v = f\lambda$); therefore, wavelength is shorter. Choice (3) correctly states these conclusions; the sound heard by the stationary person has higher pitch and shorter wavelength.

42. **1** The absolute index of refraction of a material is defined as the ratio of the speed of light in a vacuum to the speed of light in the material:

$$n = \frac{c}{v}$$

The slower the speed of light in a material, the larger the absolute index of refraction. According to Snell's law ($n_1 \sin \theta_1 = n_2 \sin \theta_2$), the greater the absolute index of refraction, the smaller the angle of refraction.

Refer to the Absolute Indices of Refraction table in the *Reference Tables* to obtain the values for alcohol (1.36) and flint glass (1.61). Since flint glass has a larger index of refraction, the speed of light will decrease, and the angle of refraction will be smaller than the angle of incidence; therefore, the ray will bend toward the normal.

43. **4** Diffraction is defined as the bending or spreading of a wave into the region behind an obstacle. The diagram clearly shows this phenomenon.

Wrong Choices Explained:

(1) Reflection is the bouncing back of a wave when it reaches a barrier or an interface between two media.

(2) Refraction is the bending of a wave from its original direction as it obliquely enters a second medium in which its speed changes.

(3) Polarization is a process that produces transverse waves that vibrate in only one plane.

44. **1** Write out the problem.

Given: $n = 2.0$ Find: $\theta_c = ?$

$\lambda = 5.9 \times 10^{-7}$ m

Refer to the *Wave Phenomena* section in the *Reference Tables* to find an equation(s) that relates the critical angle with the absolute index of refraction and wavelength.

Solution:
$$\sin\theta_c = \frac{1}{n}$$
$$\theta_c = \sin^{-1}\left(\frac{1}{n}\right)$$
$$= 30°.$$

(Note that the wavelength is not needed in this calculation.)

45. **3** Write out the problem.

Given: $\lambda = 5.9 \times 10^{-7}$ m Find: $\theta_{liq} = ?$

$\theta_{air} = 45°$

$n_{air} = 1.0$

$n_{liq} = 1.4$

Refer to the *Wave Phenomena* section in the *Reference Tables* to find an equation(s) that relates angles with absolute indices of refraction and wavelength.

Solution:
$$n_1 \sin\theta_1 = n_2 \sin\theta_2$$
$$n_{air} \sin\theta_{air} = n_{liq} \sin\theta_{liq}$$
$$\theta_{liq} = \sin^{-1}\left(\frac{n_{air} \sin\theta_{air}}{n_{liq}}\right)$$
$$= \sin^{-1}\left[\frac{(1.0)(\sin 45°)}{1.4}\right]$$
$$= 30°.$$

(Note that the wavelength is not needed in this calculation, only as the reference for the refractive indices.)

46. **2** Polarization is a process that produces transverse waves that vibrate in only one plane. Polarization is limited to transverse waves. Since sound is a longitudinal wave, the phenomenon of polarization *cannot* occur.

Wrong Choices Explained:

(1) Interference is the reinforcement or the cancellation of amplitude that occurs when two or more waves pass through the same region at the same time.

(3) Refraction is the bending of a wave from its original direction as it obliquely enters a second medium in which its speed changes.

(4) The Doppler effect is the variation in observed frequency from actual frequency when there is a relative motion between the source of the waves and the observer.

47. **3** Write out the problem.

Given: $d = 0.50$ m Find: $\lambda = ?$

$L = 4.0$ m

$x = 2.0$ m

Refer to the *Wave Phenomena* section in the *Reference Tables* to find an equation(s) that relates wavelength with distance between slits, distance from slit to screen, and distance between maxima.

Solution:
$$\frac{\lambda}{d} = \frac{x}{L}$$
$$\lambda = \frac{xd}{L}$$
$$= \frac{(2.0 \text{ m})(0.50 \text{ m})}{4.0 \text{ m}}$$
$$= 0.25 \text{ m}$$

48. **4** Coherent light is defined as a series of light waves that have a fixed phase relationship—the type of light produced by a laser. The diagram in choice (4) accurately represents light waves that are completely in phase with each other, that is, light emitted from a coherent light source.

49. **2** Write out the problem.

Given: $E = 8.0$ eV Find: $W_0 = ?$

$KE = 6.0$ eV

Refer to the *Modern Physics* section in the *Reference Tables* to find an equation(s) that relates work function with energy of a photon and kinetic energy.

Solution:
$$E_{photon} = hf$$
$$KE_{max} = hf - W_0$$
$$\therefore KE_{max} = E_{photon} - W_0$$
$$= 8.0 \text{ eV} - 6.0 \text{ eV}$$
$$= 2.0 \text{ eV}$$

50. **3** Write out the problem.

Given: $p = 1.8 \times 10^{-22}$ kg · m/s   Find: $\lambda = ?$

Refer to the *Modern Physics* section in the *Reference Tables* to find an equation(s) that relates wavelength with momentum.

Solution: $p = \frac{h}{\lambda}$

Refer to the List of Physical Constants in the *Reference Tables* to obtain the value for *h*, Planck's constant.

$$\lambda = \frac{h}{p}$$
$$= \frac{6.6 \times 10^{-34} \text{ J} \cdot \text{s}}{1.8 \times 10^{-22} \text{ kg} \cdot \text{m/s}}$$
$$= 3.7 \times 10^{-12} \text{ m}$$

51. **4** According to photoelectric experiments, the frequency of light determines how energetic the electrons are; the intensity determines the rate at which the electrons are ejected (i.e., the current). Light with a frequency of 7.1×10^{14} Hz corresponds to blue light. Since the frequency of ultraviolet light exceeds the threshold frequency of the photoemissive surface in the question, electrons will be emitted and high-intensity ultraviolet radiation will produce a greater number of ejected electrons, hence a greater amount of current.

Wrong Choices Explained:

(1), (2) The frequency of infrared light does not exceed the threshold frequency of the photoemissive material; therefore, no electrons will be ejected, regardless of the intensity.

52. **1** An alpha particle is a helium nucleus and therefore carries a +2 elementary electric charge. The nucleus of a gold atom is made up of protons and neutrons and hence carries a large, positive electric charge. Since like charges repel, only the diagram in choice (1) correctly shows the path of an alpha particle being deflected by a repulsive force—here, the gold nucleus.

53. **2** Write out the problem.

Given: 1 hydrogen atom Find: $E_{\text{photon}} = ?$

$E_i \Rightarrow n = 2$

$E_f \Rightarrow n = 3$

Refer to the Energy Level Diagram for Hydrogen in the *Reference Tables* to obtain the values for the $n = 2$ and $n = 3$ energy level states.

Refer to the *Modern Physics* section in the *Reference Tables* to find an equation(s) that relates the energy of a photon with energy levels.

Solution:
$$E_{\text{photon}} = E_i - E_f$$
$$= -3.40 \text{ eV} - (-1.51 \text{ eV})$$
$$= -1.89 \text{ eV}$$

(Note that the negative sign simply indicates the absorption, rather than the release, of a photon.)

54. **3** Write out the problem.

Given: $\boldsymbol{v}_{b_i} = 0$ Find: $\boldsymbol{a}_c$ vs. $\boldsymbol{a}_b$

$\boldsymbol{v}_{b_f} = 5.0 \text{ m/s}$

$\Delta t = 10. \text{ s}$

$\boldsymbol{v}_{c_i} = 22 \text{ m/s}$

$\boldsymbol{v}_{c_f} = 27 \text{ m/s}$

Refer to the *Mechanics* section in the *Reference Tables* to find an equation(s) that relates accelerations with velocities and time.

Solution:
$$\boldsymbol{a} = \frac{\Delta \boldsymbol{v}}{\Delta t}$$

$$\boldsymbol{a}_c = \frac{\boldsymbol{v}_{c_f} - \boldsymbol{v}_{c_i}}{\Delta t} = \frac{27 \text{ m/s} - 22 \text{ m/s}}{10. \text{ s}} = 0.50 \text{ m/s}^2$$

$$\boldsymbol{a}_b = \frac{\boldsymbol{v}_{b_f} - \boldsymbol{v}_{b_i}}{\Delta t} = \frac{5.0 \text{ m/s} - 0}{10. \text{ s}} = 0.50 \text{ m/s}^2$$

The acceleration of the car is the same as the acceleration of the bicycle.

55. **3** Both eggs are dropped from rest; and since all objects experience the same acceleration due to gravity in free fall, regardless of mass, the eggs will have the same final velocity when they hit. Also, since both have the same mass, they will have the same momentum when they hit.

Since impulse, $\boldsymbol{J} = \boldsymbol{F}\,\Delta t$, and change in momentum, $\Delta \boldsymbol{p} = m\,\Delta \boldsymbol{v}$, are equal to each other: $\boldsymbol{F}\,\Delta t = m\,\Delta \boldsymbol{v}$, the eggs experience the same impulse. Egg *A*, which hits the floor, experiences a larger force for a shorter time, while egg *B*, which hits the foam, experiences a smaller force for a longer time.

PART II

GROUP 1—**Motion in a Plane**

56. **2** In a geosynchronous orbit the period of the satellite is equal to the period of the Earth's rotation (approximately 24 hours). A satellite in geosynchronous orbit remains continually in the same position above the Earth's surface. The arrows in the diagram in choice (2) most accurately represent the path of a satellite in this type of orbit.

57. **2** The force due to friction always acts in the direction directly opposite to the direction of motion. The direction of the arrow in choice (2) best represents the opposite direction of the arrow representing the direction of motion at point *P*; that is, this arrow best represents the direction of the force of air friction on the ball.

58. **3** The airplane is moving in a circular path, and by definition the force it experiences to keep it in a circular pathway is a centripetal force and is always directed toward the center of the circular path. At the point indicated, the center of the path—and hence the direction of the net force acting on the airplane—is toward the east.

59. **3** Write out the problem.

Given: $m = 4.0\text{ kg}$ Find: $\boldsymbol{a}_c = ?$

$r = 12\text{ m}$

$\boldsymbol{v} = 6.0\text{ m/s}$

Refer to the *Motion in a Plane* section in the *Reference Tables* to find an equation(s) that relates centripetal acceleration with mass, radius, and velocity.

Solution:
$$\boldsymbol{a}_c = \frac{\boldsymbol{v}^2}{r} = \frac{(6.0\text{ m/s})^2}{12\text{ m}} = 3.0\text{ m/s}^2$$

(Note that mass is not needed in this calculation.)

60. **4** Refer to the *Motion in a Plane* section in the *Reference Tables* to find an equation(s) that relates centripetal force with speed and radius.

Solution: $\boldsymbol{F}_c = \dfrac{m\boldsymbol{v}^2}{r}$

From this equation you can see that centripetal force is directly proportional to velocity squared. Therefore, if the speed is doubled and the mass and radius remain unchanged, the magnitude of the centripetal force acting on the airplane will be four times as much.

61. **4** Write out the problem.

Given: $\boldsymbol{v}_i = 40.$ m/s Find: $\boldsymbol{v}_{i_x} = ?$

$\theta = 30°$ to the ground

Refer to the *Motion in a Plane* section in the *Reference Tables* to find an equation(s) that relates the horizontal component of velocity with a velocity and an angle.

Solution: $\boldsymbol{v}_{i_x} = \boldsymbol{v}_i \cos\theta$

$= (40.\text{ m/s})(\cos 30°)$

$= 35.\text{ m/s}$

62. **3** The highest point on the curve on the graph represents the greatest range, 20 m, which corresponds to an angle of inclination of 45°.

Wrong Choices Explained:

(1) If this were true, the graph would have been a horizontal straight line
(2) 20° corresponds to a range of 13 m; 80°, to a range of 7 m.
(4) 90° corresponds to a range of 0, which is clearly not the greatest.

63. **4** Kepler's third law states that the ratio of the cube of the mean radius of a planet's orbit to the square of its period of revolution about the Sun—that is, the ratio $\dfrac{R^3}{T^2}$ —is the same for all planetary bodies in the solar system.

64. **1** Time of flight is the time an object takes to fall to the ground. All four balls, regardless of their masses, will fall with the same acceleration due to gravity. Since all the balls are thrown horizontally, their initial velocities in the vertical or y direction will all be 0. The only difference between the balls is the distance of their fall. Since

$$\Delta \boldsymbol{s}_y = \boldsymbol{v}_{i_y}\Delta t + \frac{1}{2}\boldsymbol{a}_y(\Delta t)^2$$

$$\Delta t = \sqrt{\frac{2\Delta \boldsymbol{s}_y}{\boldsymbol{a}_y}}$$

∴ The smaller the Δs, the shorter the time of flight.

The ball in choice (1) has the shortest distance, 125 m, to fall and, therefore, has the shortest flight time.

65. **1** According to Kepler's second law, a line from the Sun to a planet sweeps out equal areas of space in equal amounts of time. This means that the planet moves faster as it approaches a point closest to the Sun and moves slower as it moves further away from the Sun.

In the diagram, as the planet moves from *A* to *B*, it is moving away from the Sun, so its speed will decrease. Since kinetic energy is proportional to velocity squared, the planet's kinetic energy will also decrease.

GROUP 2—**Internal Energy**

66. **4** Standard pressure is equal to 1 atm or 101.3 kPa. At that pressure the boiling point of water is 100°C. Kelvin temperatures are obtained by adding 273 to Celsius temperatures; therefore, at standard pressure, water boils at 373 K.

67. **1** According to the second law of thermodynamics, heat flows spontaneously from a warmer body to a cooler body. In order to have a net flow of heat between the metal blocks, they must have different initial temperatures.

Wrong Choices Explained:

(2) The melting point of a substance is the temperature at which the solid and liquid phases can coexist.

(3) The specific heat of a substance is defined as the amount of heat required to raise the temperature of 1 gram of the substance by 1 Celsius degree.

(4) The heat of fusion of a substance is defined as the amount of heat required to change 1 gram of the substance from the solid phase to the liquid phase at the melting point of the substance.

68. **3** Refer to the *Internal Energy* section in the *Reference Tables* to find an equation(s) that relates mass with temperature and heat.

Solution: $Q = mc\,\Delta T$

Given that all four samples have the same mass and the same amount of heat is being applied to all samples, the smallest temperature change will result from the largest specific heat (c). Refer to the Heat Constants table in the *Reference Tables.* Silver has the highest specific heat (0.24 kJ/kg · C°) of all the samples; therefore it will experience the smallest increase in temperature.

69. **2** Write out the problem.

Given: $m = 2.5$ kg Find: $Q = ?$

$T_i = 0°C$

$T_f = 37°C$

Refer to the *Internal Energy* section in the *Reference Tables* to find an equation(s) that relates heat with temperature and mass.

Solution: $Q = mc\,\Delta T$

Refer to the Heat Constants table in the *Reference Tables* to obtain the value of c, the specific heat of water.

$$Q = mc(T_f - T_i)$$
$$= (2.5 \text{ kg})(4.19 \text{ kJ/kg} \cdot \text{C°})(37°\text{C} - 0°\text{C})$$
$$= 390 \text{ kJ}$$

70. **3** The kinetic theory of ideal gas behavior assumes that all collisions are perfectly elastic. When a molecule collides with the container wall, no energy is transferred. However, when two molecules collide, both momentum and energy are transferred between the molecules.

Wrong Choices Explained:

(1) Molecules are constantly in motion; they are never stationary.

(2) The kinetic theory assumes that no forces of attraction exist between molecules.

(4) Molecular size is assumed to be negligible.

71. **2** Since heat is being added and the temperature is not changing during interval *xy*, the heat is increasing the internal energy of the substance, overcoming the forces of attraction between the molecules, and a phase change is taking place.

72. **4** According to Charles's law, volume and absolute temperature are directly proportional to one another. The graph in choice (4) correctly represents the direct relationship between volume and Kelvin temperature for a fixed mass of an ideal gas at constant pressure.

Wrong Choices Explained:

(1) This graph shows that no relationship exists between volume and absolute temperature.

(2) This graph shows an inversely proportional relationship between volume and absolute temperature.

(3) This graph shows an inverse relationship between volume and temperature.

73. **2** Write out the problem.

Given: $m = 0.50\text{ kg}$ Find: $Q = ?$

$T = -39°\text{C}$

Refer to the *Internal Energy* section in the *Reference Tables* to find an equation(s) that relates heat with mass and constant temperature.

Solution: $Q = mH_f$

Refer to the Heat Constants table in the *Reference Tables* to obtain the value for the heat of fusion, H_f, of mercury.

$$Q = (0.50\text{ kg})(11\text{ kJ/kg})$$
$$= 5.5\text{ kJ}$$

74. **3** Entropy is a measure of the disorder or randomness of a system. A gas has more entropy than a liquid, which, in turn, has more entropy than a solid. Choice (3) is a phase change, gas to liquid, that represents a decrease in entropy.

75. **1** By definition, a liquid boils when its vapor pressure equals atmospheric pressure. Vapor pressure is directly proportional to temperature. Therefore, the boiling point of a liquid is higher at higher pressures. In this question, the pressure on the coolant is lowered, and as a result the boiling point of the coolant is decreased.

GROUP 3—**Electromagnetic Applications**

76. **3** A voltmeter is a galvanometer that has been modified to measure the potential difference across two points in a circuit. Therefore, the voltmeter must be connected in parallel with the two points in the circuit that it is measuring. For the voltmeter to have a minimal effect on the circuit, it must be a high-resistence device so that most of the current will flow through the circuit, not through the voltmeter. This high resistance is gained by placing a large resistance in series with the galvanometer coil.

Wrong Choices Explained:

(1), (4) Neither a motor nor a generator contains a galvanometer.

(2) An ammeter is a galvanometer that is placed in parallel with a low-resistance device known as a shunt; the ammeter must be in series with the circuit to produce a minimal effect on the circuit.

77. **1** A voltmeter is a modified galvanometer in which the galvanometer coil is connected in series with a high resistance device. In order for the voltmeter to have a nominal effect on the circuit, it must be connected in parallel with the circuit resistor it is measuring, so that only a very small amount of

current will flow through the galvanometer while the majority of the current remains in the circuit.

78. **4** To solve this problem, use the following left-hand rule: Place the thumb so that it points in the direction of the velocity of the electron, and the other fingers so that they point in the direction of the magnetic field. The palm of the hand will point in the direction of the deflecting force, that is, toward the bottom of the page.

79. **2** Write out the problem.

Given: 1 electron Find: $\boldsymbol{F}$ = ?

$\boldsymbol{v} = 3.0 \times 10^3$ m/s

$\boldsymbol{B} = 3.0 \times 10^{-5}$ T

Refer to the *Electromagnetic Applications* section in the *Reference Tables* to find an equation(s) that relates force with velocity and magnetic field intensity.

Solution: $\boldsymbol{F} = q\boldsymbol{vB}$

Refer to the List of Physical Constants in the *Reference Tables* to obtain the value for the charge, q, on an electron.

$$\boldsymbol{F} = (1.6 \times 10^{-19}\ \text{C})(3.0 \times 10^3\ \text{m/s})(3.0 \times 10^{-5}\ \text{T})$$

$$= 1.4 \times 10^{-20}\ \text{N}$$

80. **1** For current to be induced in a coil of wire, the wire must pass through magnetic field lines, or vice versa. Only when there is continuous relative movement between the two will a current exist. Only an alternating-current source connected to the primary coil generates an alternating magnetic field which is carried through the iron core and generates an alternating current in the secondary coil.

81. **1** According to Lenz's law, an induced current generates a magnetic field of its own that opposes the motion that is inducing the current. Therefore, since the north end of magnet 1 is dropped toward solenoid *A*, as shown in the diagram, the north end of the solenoid will be toward the top of the page. Use the following left-hand rule: Placing the thumb in the direction of the north pole of solenoid *A*; the fingers will wrap around solenoid *A* in the direction of the electron current.

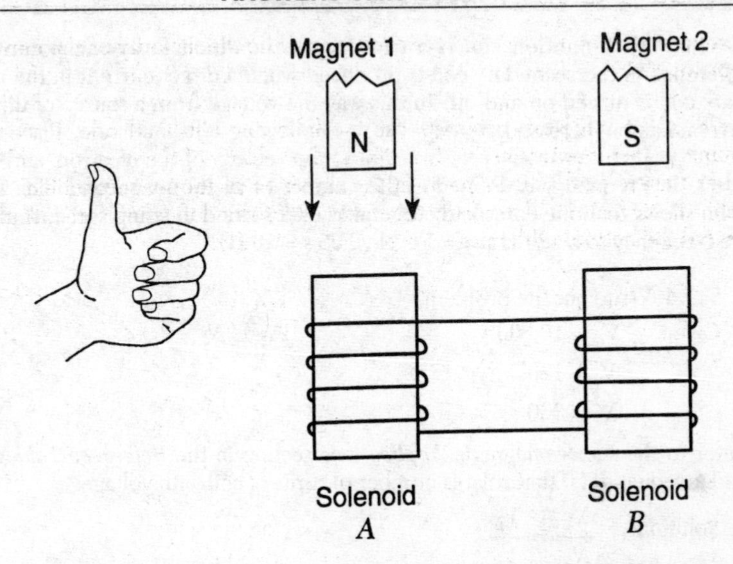

Use the same rule, but in reverse, on solenoid *B*: Wrap the fingers of the left hand around solenoid *B* in the direction of the electron flow in the coil; the thumb will point in the direction of the north pole of the solenoid. As the diagram shows, the north end of solenoid *B* faces the south pole of magnet 2. Therefore, the south pole of magnet 2 will be attracted by a magnetic force toward solenoind *B*.

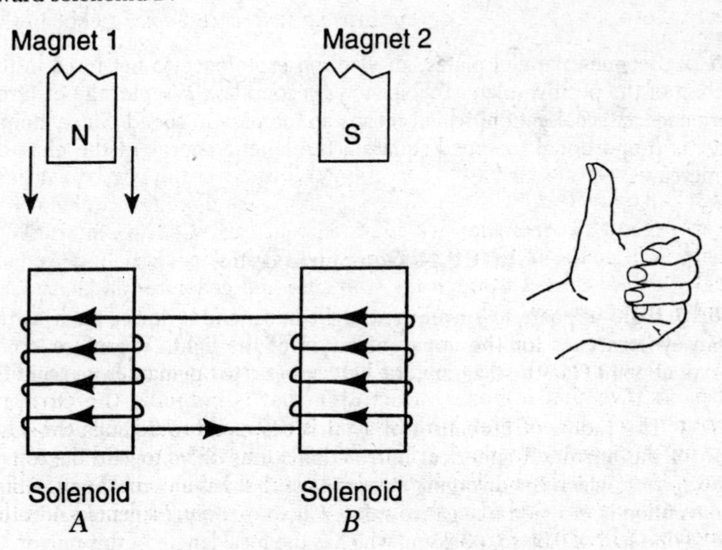

82. **3** An induction coil is a transformer in which a varying potential difference in the secondary coil is produced when a direct current in the primary coil is turned on and off. In this way the voltage from a source of direct current can be increased.

83. **2** Frequency is defined as the number of cycles per second. The diagram shows that one complete cycle takes 0.05 second to complete. Calculate the frequency to be 20 Hz. f = 1 cycle/0.05 s = 20 Hz.

84. **4** Write out the problem.

Given: $V_p = 10{,}800$ V Find: $N_p = ?$

$V_s = 120.$ V

$N_s = 360$

Refer to the *Electromagnetic Applications* section in the *Reference Tables* to find an equation(s) that relates number of turns of coil with voltage.

Solution:
$$\frac{N_p}{N_s} = \frac{V_p}{V_s}$$

$$N_p = \frac{N_s V_p}{V_s}$$

$$= \frac{(360)(10{,}800\text{ V})}{120.\text{ V}}$$

$$= 32{,}400$$

85. **2** Between parallel plates, an electron experiences a net force in the direction of the positive plate. By Newton's second law, $\boldsymbol{F} = m\boldsymbol{a}$, the electron experiences an acceleration, which means an increase in speed. Since kinetic energy is proportional to speed squared, the kinetic energy of the electron also increases.

GROUP 4—**Geometric Optics**

86. **2** Light appears to travel in straight lines, and a device such as the human eye searches for the apparent origin of the light. Therefore, to an observer at point *O* in the diagram, the light appears to originate from point *B*.

87. **1** The radius of curvature of a mirror is equal to 2 times the focal length of the mirror. A spherical mirror that forms only virtual images is a convex mirror, which is a diverging mirror. As such it has a virtual focus, which by convention is assigned a negative value. Choice (1) represents the negative value of one-half of 0.50 m, –0.25 m, which is the focal length of this mirror.

88. **3** Light rays that are angled so that they pass through the focus of a mirror will be reflected parallel to the principal axis. If the bulb of the headlight is placed at the principal focus of the mirror, all the light rays will come from the focus and will be reflected as parallel rays, which will produce a beam of light.

89. **2** Write out the problem.

Given: $f = 0.10$ m Find: $d_i = ?$

$d_o = C = 2f = 0.20$ m

Refer to the *Geometric Optics* section in the *Reference Tables* to find an equation(s) that relates image distance with focal length and object distance.

Solution:
$$\frac{1}{f} = \frac{1}{d_o} + \frac{1}{d_i}$$
$$\frac{1}{d_i} = \frac{1}{f} - \frac{1}{d_o}$$
$$= \frac{1}{0.10 \text{ m}} - \frac{1}{0.20 \text{ m}}$$
$$= \frac{2}{0.20 \text{ m}} - \frac{1}{0.20 \text{ m}}$$
$$= \frac{1}{0.20 \text{ m}}$$
$$d_i = 0.20 \text{ m}$$

90. **1** Concave spherical mirrors can produce both real and virtual images. Virtual images occur when the object is placed between the focal point of the mirror and the surface of the mirror. Only the distance in choice (1) falls within this range for this mirror.

91. **2** Refer to the *Geometric Optics* section in the *Reference Tables* to find an equation(s) that relates object and image distance with object and image size.

Solution:
$$\frac{S_o}{S_i} = \frac{d_o}{d_i}$$
$$S_i = \frac{S_o d_i}{d_o}$$

From the equation, as object distance decreases, image size increases.

92. **4** Converging lenses are thicker in the middle than at the ends. Only the lens in choice (4) fits this profile.

93. **3** Write out the problem.

Given: $f = 0.10$ m $\quad$ Find: $d_i = ?$

$d_o = 0.30$ m

Refer to the *Geometric Optics* section in the *Reference Tables* to find an equation(s) that relates image distance with focal length and object distance.

Solution:
$$\frac{1}{f} = \frac{1}{d_o} + \frac{1}{d_i}$$
$$\frac{1}{d_i} = \frac{1}{f} - \frac{1}{d_o}$$
$$= \frac{1}{0.10 \text{ m}} - \frac{1}{0.30 \text{ m}}$$
$$= \frac{3}{0.30 \text{ m}} - \frac{1}{0.30 \text{ m}}$$
$$= \frac{2}{0.30 \text{ m}}$$
$$= \frac{1}{0.15 \text{ m}}$$
$$d_i = 0.15 \text{ m}$$

94. **4** Write out the problem.

Given: $S_o = 0.080$ m $\quad$ Find: $S_i = ?$

$d_o = 0.20$ m

$d_o = 0.40$ m

Refer to the *Geometric Optics* section in the *Reference Tables* to find an equation(s) that relates object and image sizes with object and image distances.

Solution:
$$\frac{S_o}{S_i} = \frac{d_o}{d_i}$$
$$S_i = \frac{S_o d_i}{d_o}$$
$$= \frac{(0.080 \text{ m})(0.40 \text{ m})}{0.20 \text{ m}}$$
$$= 0.16 \text{ m}$$

95. **1** Diverging (concave) lenses can form only virtual erect images.

GROUP 5—**Solid State**

96. **2** Conductivity and resistivity are reciprocals of each other ($\sigma = 1/\rho$); in other words, they are inversely proportional to each other. Only the graph in choice (2) correctly represents this type of relationship for a solid.

Wrong Choices Explained:

(1) This graph shows a direct relationship.
(3) This graph indicates no relationship.
(4) This graph indicates an exponential relationship.

97. **2** A transistor serves as an amplification device. The diagram shows that the output current is an amplification of the input current; therefore, the "black box" most likely contains a transistor.

Wrong Choices Explained:

(1), (3) An LED (light-emitting diode) or a diode acts as a current rectifier and would produce a pulsating direct current.

(4) A current passing through an *N*-type semiconductor would emerge unchanged.

98. **4** By definition a *P*-type (positive type) semiconductor is formed by adding a doping agent which has 3 electrons in the valance shell and so provides extra (positive) holes for conduction.

99. **4** Holes can be considered as moving positive charges. As such, they are attracted toward the negative electrode and, in the circuit shown, would migrate from the *P*-type semiconducting material to the *N*-type semiconducting material.

100. **2** A *P-N* junction diode acts as a rectifier for alternating current because it limits the current flow to a single direction. This occurs because the alternating current is produced by an alternating potential, which alternately makes the diode forward and reversed biased. When the diode is forward biased, it allows current flow; when it is reversed biased, it inhibits current flow. The result is a pulsating direct current.

101. **3** The device shown in the diagram is a *P-N* junction diode. Choices (3) and (4) are the circuit symbols for a diode. Only choice (3), however, has the arrow pointing in the right direction (i.e., in the direction of hole flow, from *P* to *N*), for this forward-biased diode.

Wrong Choices Explained:

(1), (2) These are the symbols for a *P-N-P* and an *N-P-N* transistor, respectively.

(4) This would be the symbol for the diode if it were reverse biased (*N-P*) in this circuit.

102. **1** The diode symbol in the diagram represents a *P-N* junction diode that is forward biased. The *P*-type material is connected to the positive terminal; the *N*-type material, to the negative terminal. As a result the potential barrier of the junction is overcome and current flows.

103. **1** Doping the diode more heavily would make it more conductive and, therefore, more sensitive to reverse bias. Therefore, the avalanche would occur at a lower voltage, only.

104. **3** Since the base of the transistor is very thin, most (98%) of the charge carriers flow through the base (*b*) from emitter (*e*) to collector (*c*).

105. **3** A *P*-type semiconducting material simply conducts current, utilizing an excess of holes in the valence band. How the batteries are connected is irrelevant; the current through the semiconductor will be equal to 0.05 A.

GROUP 6—**Nuclear Energy**

106. **2** Nuclear reactions, like any other chemical reactions, obey the laws of conservation of mass and conservation of energy.

Write out the problem.

Given: $^{30}_{15}P \rightarrow X + {}^{0}_{+1}e$

The mass and charge of particle *X* must be such that the sum of all the masses and charges on the right side of the reaction are equal to the sum of all the masses and charges on the left side.

Solution: $^{30}_{15}P \rightarrow {}^{30}_{14}X + {}^{0}_{+1}e$

The mass and charge of particle *X* correspond to those of the particle in choice (2), $^{30}_{14}$Si.

107. **4** Write out the problem.

Given: $m = 5.2 \times 10^{-29}$ kg   Find: $E = ?$

Refer to the *Nuclear Energy* section in the *Reference Tables* to find an equation(s) that relates mass and energy.

Solution: $E = mc^2$

Refer to the List of Physical Constants in the *Reference Tables* to obtain the value for the speed of light in a vacuum.

$$E = (5.2 \times 10^{-29}\ \text{kg})(3.0 \times 10^{8}\ \text{m/s})^2$$
$$= 4.7 \times 10^{-12}\ \text{J}$$

108. **3** The number of protons in the nucleus of an atom determines the nature of the element. Atoms with different numbers of protons represent different elements. Atoms of different isotopes of the same element contain the same number of protons, but different numbers of neutrons. Only choice (3) correctly represents this situation.

109. **2** Neutrons carry no charge and therefore are not affected by the electric and magnetic fields that are the components in a cyclotron. In other words, neutrons can *not* be accelerated by a cyclotron.

110. **1** Nuclear reactions, like any other chemical reactions, obey the laws of conservation of mass and conservation of energy. Refer to the *Uranium Disintegration Series* diagram in the *Reference Tables.*

Write out the problem.

$$^{222}_{86}Rn \rightarrow ^{218}_{84}Po + X$$

The mass and charge of particle *X* must be such that the sum of all the masses and charges on the right side of the reaction are equal to the sum of all the masses and charges on the left side.

Solution: $^{222}_{86}Rn \rightarrow ^{218}_{84}Po + ^{4}_{2}X$

The mass and charge of particle *X* correspond to those of a helium nucleus, $^{4}_{2}He$, that is, an alpha particle. Choice (1) gives the correct mode of decay and the correct resulting atom.

111. **3** Nuclear reactions, like any other chemical reactions, obey the laws of conservation of mass and conservation of energy.

Write out the problem.

$$^{238}_{93}Np \rightarrow ^{238}_{94}Pu + X$$

The mass and charge of particle *X* must be such that the sum of all the masses and charges on the right side of the reaction are equal to the sum of all the masses and charges on the left side.

Solution: $^{238}_{93}Np \rightarrow ^{238}_{94}Pu + ^{0}_{-1}X$

The mass and charge of particle X correspond to those of an electron, $^{0}_{-1}e$.

Wrong Choices Explained:

(1) $^{1}_{1}H$ is the symbol for a proton.

(2) $^{1}_{0}n$ is the symbol for a neutron.

(4) $^{0}_{+1}e$ is the symbol for a positron.

112. **3** Write out the problem.

Given: $m_i = 96$ g   Find: half-life = ?

total time = 12 min

$m_f = 6$ g

Refer to the *Nuclear Energy* section in the *Reference Tables* to find an equation(s) that relates half-life with mass.

Solution:

$$m_f = \frac{m_i}{2^n}$$

$$2^n = \frac{m_i}{m_f} = \frac{96\text{ g}}{6\text{ g}} = 16$$

$$n = 4$$

$$\text{half-life} = \frac{\text{total time}}{n} = \frac{12\text{ min}}{4} = 3\text{ min}$$

113. **1** Fission is the process of splitting a heavy nucleus, such as U-235, into lighter fragments. Fission is accompanied by the release of large quantities of energy. This is the process represented by the given reaction.

Wrong Choices Explained:

(2) Nuclear fusion is the process of uniting lighter nuclei, such as deuterium, into a heavier nucleus.

(3) Alpha decay is a natural radioactive process that results in the emission of an alpha particle from a nuclide.

(4) Beta decay is a natural radioactive process that results in the emission of a beta particle from a nuclide.

114. **4** Neutrons carry no charge and therefore do not experience electrostatic forces of repulsion from the nucleus. The fact that neutrons are not repelled by the nucleus is the principal reason why they are used to bombard a nucleus.

115. **4** The Sun is composed primarily of hydrogen and uses the process of fusion to produce its energy. Spectroscopic evidence indicates the presence of helium (the product of fusion) and hydrogen in the Sun.

PART III

116. If you measure the 40.-N vector, you obtain a length of approximately 8.0 cm, which gives you the following scale: 1.0 cm is approximately equal to 5.0 N (±0.2 N).

Note: One credit is granted for an acceptable response with units.

117. The vector that represents the resultant force is as follows:

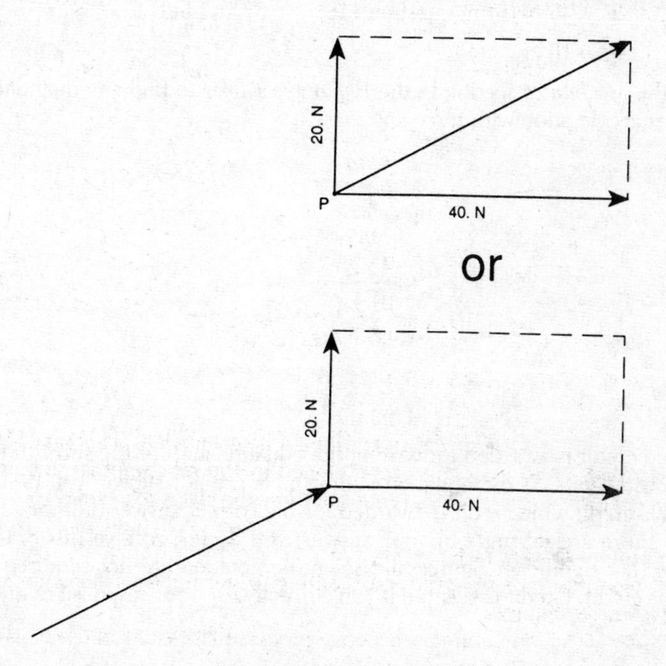

(**Note:** To receive one credit, the 8.9 cm ± 0.2-cm vector must include an arrowhead at the end. The resultant vector need not be labeled to receive this credit.)

118. If you measure the resultant vector, you obtain a magnitide of 45 N (±2 N) for the resultant force. Alternatively, the magnitude may be obtained by calculation using the Pythagorean theorem or trigonometry.

(**Note:** To receive one credit the correct unit must be included. The credit is granted also if the answer is based correctly on the answers to questions 116 and 117.)

119. Using a protractor, you obtain a degree measurement of 27° ± 2° for the angle between east and the resultant force.

(**Note:** The credit is granted also if the answer is based correctly on the answer to question 117 or is calculated using the tangent function tan θ = 20 N/40 N.)

120. Write out the problem.

Given: $\mathbf{F}$ = 45 N (from question 118) Find: $\boldsymbol{a}$ = ?

m = 10 kg

Refer to the *Mechanics* section in the *Reference Tables* to find an equation(s) that relates acceleration with force and mass.

$$\mathbf{F} = m\boldsymbol{a}$$

$$\boldsymbol{a} = \frac{\mathbf{F}}{m}$$

$$= \frac{45 \text{ N}}{10. \text{ kg}}$$

$$= 4.5 \text{ N/kg}$$

or

$$= 4.5 \text{ m/s}^2$$

Note: One credit is awarded for writing the equation and for the substitution of values with units. If the equation and/or units are not shown, you do not receive this credit. One credit is awarded for the correct answer (number and units). If there are no units in your answer, you do not receive this credit. Note also that significant figures and scientific notation are not required to obtain this credit. Credit is granted if you correctly use your response to question 118.

121. The resultant wave produced by the superposition of waves *A* and *B* is as follows:

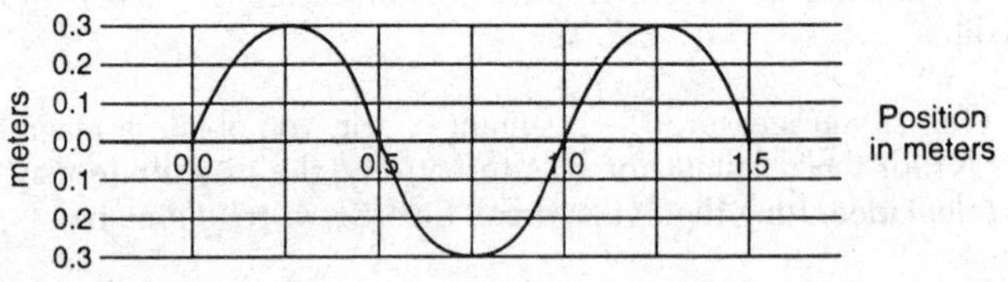

Note: One credit is awarded for the correct answer.

122. From the grid you obtain 0.3 m (±0.02 m) or 30 cm (±2 cm) as the amplitude of the resultant wave.

Note: To receive one credit, the correct unit must be included in your answer. Note also that one credit is granted if your answer is based correctly on the answer to question 121.

123. From the grid you obtain 1.0 m or 100 cm as the wavelength of the resultant wave.

Note: To receive one credit, the correct unit must be included in your answer. Note also that credit is granted if your answer is correct based on the answer to question 121.

124. The circuit diagram is as follows:

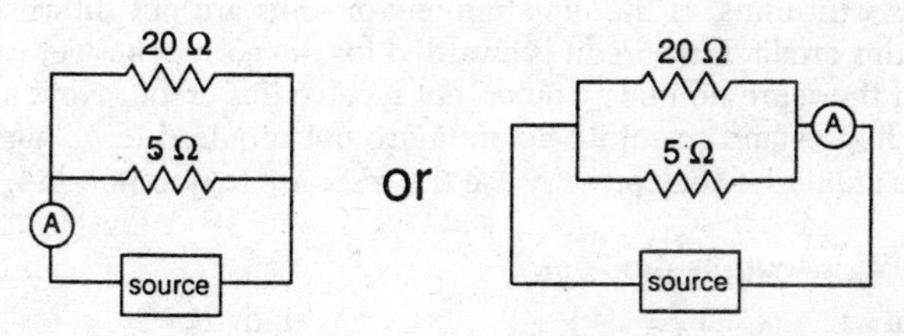

Note: One credit is granted for a circuit that contains two resistors, labeled "5 ohms" and "20 ohms," connected in parallel with a source that may be labeled "source," "24 V," or "24-V source." One credit is granted for a single ammeter properly placed to measure the total current. This credit is disallowed if more than one ammeter is present in the circuit. If the circuit drawn is a series circuit, and the first credit is not granted, this credit may still be granted if the ammeter is properly placed to measure the total current.

125. Write out the problem.

Given: R_1 = 50-Ω Find: $R_t = ?$

R_2 = 20-Ω

V = 24-V source

Refer to the *Electricity and Magnetism* section in the *Reference Tables* to find an equation(s) that relates total circuit resistance with individual resistance.

$$\frac{1}{R_t} = \frac{1}{R_1} + \frac{1}{R_2}$$

$$= \frac{1}{5.0\ \Omega} + \frac{1}{20.0\ \Omega}$$

$$= \frac{4}{20\ \Omega} + \frac{1}{200\ \Omega}$$

$$= \frac{5}{20\ \Omega}$$

$$R_t = \frac{20\ \Omega}{5}$$

$$= 4.0\ \Omega$$

Note: One credit is awarded for writing the equation and for the substitution of values with units. If the equation and/or units are not shown, you do not receive this credit. One credit is awarded for the correct answer (number and units). If there are no units, you do not receive this credit. Note also that significant figures and scientific notation are not required to obtain this credit. Credit is granted if you correctly use the response to question 124.

126. Write out the problem.

Given: $V_T = 24$ V Find: $I_T = ?$

$R_T = 4.0\text{-}\Omega$

Refer to the *Electricity and Magnetism* section in the *Reference Tables* to find an equation(s) that relates current with potential difference and resistance.

$$R_T = \frac{V_T}{I_T}$$

$$I_T = \frac{V_T}{R_T}$$

$$= \frac{24\ \text{V}}{4.0\ \Omega}$$

$$= 6.0\text{V}/\Omega \text{ or } 6.0\ \text{A}$$

Note: One credit is awarded for writing the equation and for the substitution of values with units. If the equation and/or units are not shown, you do not receive this credit. One credit is awarded for the correct answer (number and units). If there are no units, you do not receive this credit. Note also that significant figures and scientific notation are not required to obtain this credit. Credit is granted if you correctly use the response to question 125.

Topic	Question Numbers (total)	Wrong Answers (x)	Grade
Mechanics	1–15, 19, 54, 55, 116–120: (23)		$\frac{100(23-x)}{23} = \%$
Energy	16, 17, 18, 20, 21: (5)		$\frac{100(5-x)}{5} = \%$
Electricity and Magnetism	23–35, 124, 126, 156: (16)		$\frac{100(16-x)}{16} = \%$
Wave Phenomena	36–48, 121, 122 123: (16)		$\frac{100(16-x)}{16} = \%$
Modern Physics	49–53: (5)		$\frac{100(5-x)}{5} = \%$
Motion in a Plane	56–65: (10)		$\frac{100(10-x)}{10} = \%$
Internal Energy	66–75: (10)		$\frac{100(10-x)}{10} = \%$
Electromagnetic Applications	76–85: (10)		$\frac{100(10-x)}{10} = \%$
Geometrical Optics	86–95: (10)		$\frac{100(10-x)}{10} = \%$
Solid State Physics	96–105: (10)		$\frac{100(10-x)}{10} = \%$
Nuclear Energy	106–115: (10)		$\frac{100(10-x)}{10} = \%$

To further pinpoint your weak areas, use the Topic Outline in the front of the book.

Examination June 1998

Physics

PART I

Answer all 55 questions in this part. [65]

Directions (1–55): For *each* statement or question, select the word or expression that, of those given, best completes the statement or answers the question. Record the answers to these questions in the spaces provided.

1 A car travels 12 kilometers due north and then 8 kilometers due west going from town *A* to town *B*. What is the magnitude of the displacement of a helicopter that flies in a straight line from town *A* to town *B*?

(1) 20. km
(2) 14 km
(3) 10. km
(4) 4 km

1 ____

2 What is the approximate diameter of a dinner plate?

(1) 0.0025 m (3) 0.25 m
(2) 0.025 m (4) 2.5 m 2____

3 Which two graphs best represent the motion of an object falling freely from rest near Earth's surface?

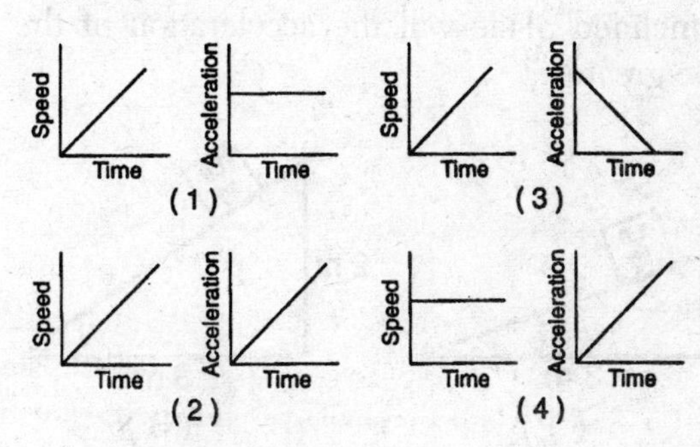

3____

4 Forces F_1 and F_2 act concurrently on point P, as shown in the diagram below.

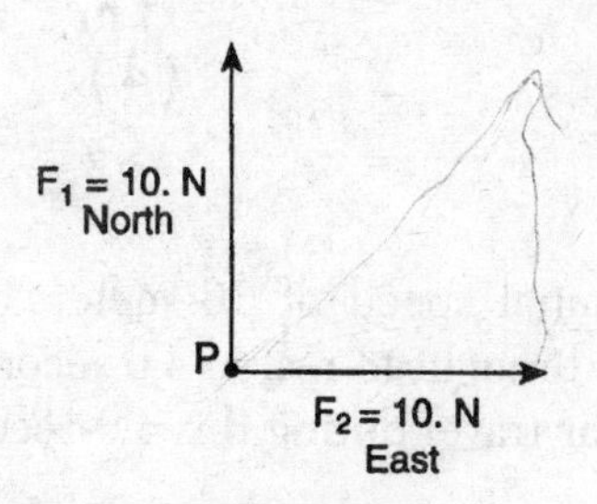

The equilibrant of F_1 and F_2 is

(1) 14 N southwest (3) 20. N southwest
(2) 14 N southeast (4) 20. N southeast 4____

5 What is the average velocity of a car that travels 30. kilometers due west in 0.50 hour?

(1) 15 km/hr
(2) 60. km/hr
(3) 15 km/hr west
(4) 60. km/hr west

5 4

6 A 1.0-kilogram block is placed on each of four frictionless planes inclined at different angles. On which inclined plane will the acceleration of the block be greatest?

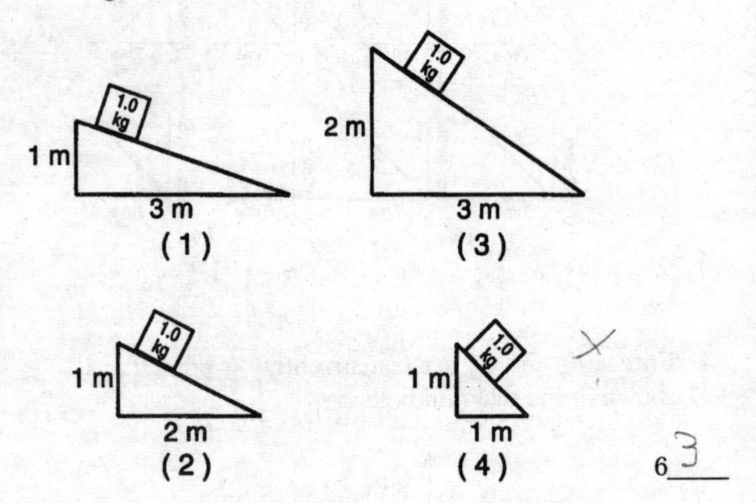

6 3

7 A car having an initial speed of 16 meters per second is uniformly brought to rest in 4.0 seconds. How far does the car travel during this 4.0-second interval?

(1) 32 m
(2) 82 m
(3) 96 m
(4) 4.0 m

7 1

8 A man weighs 900 newtons standing on a scale in a stationary elevator. If some time later the reading on the scale is 1200 newtons, the elevator must be moving with

1 constant acceleration downward
2 constant speed downward
3 constant acceleration upward
4 constant speed upward

8 2

9 Net force F causes mass m_1 to accelerate at rate a. A net force of $3F$ causes mass m_2 to accelerate at rate $2a$. What is the ratio of mass m_1 to mass m_2?

(1) 1:3
(2) 2:3
(3) 1:2
(4) 1:6

9 4

10 What is the magnitude of the gravitational force between an electron and a proton separated by a distance of 1.0×10^{-10} meter?

(1) 1.0×10^{-47} N
(2) 1.5×10^{-46} N
(3) 1.0×10^{-37} N
(4) 1.5×10^{-36} N

10 4

11 On the surface of planet X, the acceleration due to gravity is 16 meters per second^2. What is the weight of a 6.0-kilogram mass located on the surface of planet X?

(1) 2.7 N
(2) 59 N
(3) 96 N
(4) 940 N

11____

12 A book weighing 20. newtons slides at constant velocity down a ramp inclined 30.° to the horizontal as shown in the diagram below.

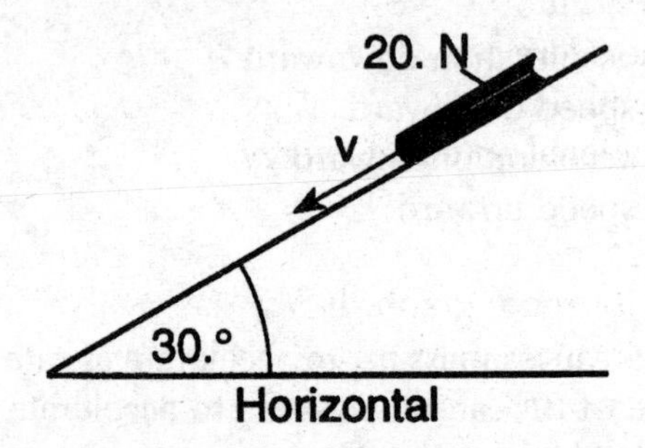

What is the force of friction between the book and the ramp?

(1) 10. N up the ramp
(2) 17 N up the ramp
(3) 10. N down the ramp
(4) 17 N down the ramp

12_____

13 A 0.60-kilogram softball initially at rest is hit with a bat. The ball is in contact with the bat for 0.20 second and leaves the bat with a speed of 25 meters per second. What is the magnitude of the average force exerted by the ball on the bat?

(1) 8.3 N
(2) 15 N
(3) 3.0 N
(4) 75 N

13_____

14 If the speed of a moving object is doubled, which quantity associated with the object must also double?

1 its momentum
2 its kinetic energy
3 its acceleration
4 its gravitational potential energy 14____

15 The velocity-time graph below represents the motion of a 3-kilogram cart along a straight line. The cart starts at $t = 0$ and initially moves north.

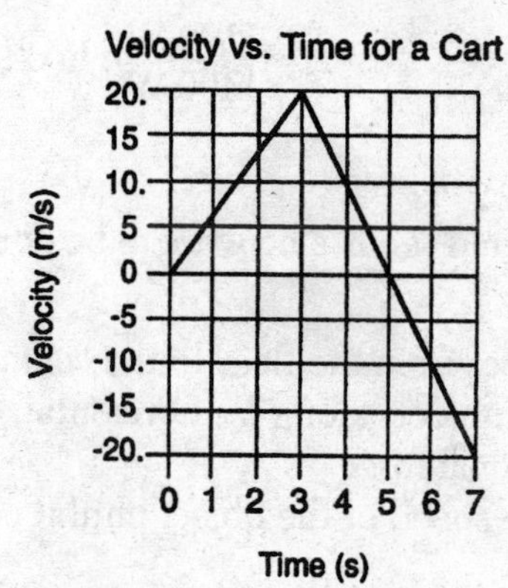

What is the magnitude of the change in momentum of the cart between $t = 0$ and $t = 3$ seconds?

(1) 20 kg•m/s
(2) 30 kg•m/s
(3) 60 kg•m/s
(4) 80 kg•m/s 15____

16 A person kicks a 4.0-kilogram door with a 48-newton force causing the door to accelerate at 12 meters per second^2. What is the magnitude of the force exerted by the door on the person?

(1) 48 N
(2) 24 N
(3) 12 N
(4) 4.0 N

16____

17 A 45-kilogram bicyclist climbs a hill at a constant speed of 2.5 meters per second by applying an average force of 85 newtons. Approximately how much power does the bicyclist develop?

(1) 110 W
(2) 210 W
(3) 1100 W
(4) 1400 W

17____

18 Which action would require no work to be done on an object?

1 lifting the object from the floor to the ceiling
2 pushing the object along a horizontal floor against a frictional force
3 decreasing the speed of the object until it comes to rest
4 holding the object stationary above the ground

18____

19 The diagram below shows a 1.5-kilogram kitten jumping from the top of a 1.80-meter-high refrigerator to a 0.90-meter-high counter.

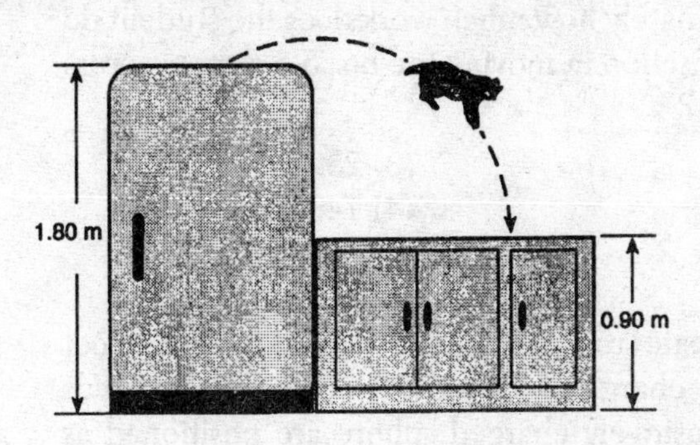

Compared to the kitten's gravitational potential energy on top of the refrigerator, the kitten's gravitational potential energy on top of the counter is

1 half as great
2 twice as great
3 one-fourth as great
4 four times as great

19____

20 A 60.-kilogram student running at 3.0 meters per second has a kinetic energy of

(1) 180 J
(2) 270 J
(3) 540 J
(4) 8100 J

20____

21 A student pulls a box across a horizontal floor at a constant speed of 4.0 meters per second by exerting a constant horizontal force of 45 newtons. Approximately how much work does the student do against friction in moving the box 5.5 meters across the floor?

(1) 45 J (3) 250 J
(2) 180 J (4) 740 J 21____

22 Two plastic rods, *A* and *B*, each possess a net negative charge of 1.0×10^{-3} coulomb. The rods and a positively charged sphere are positioned as shown below.

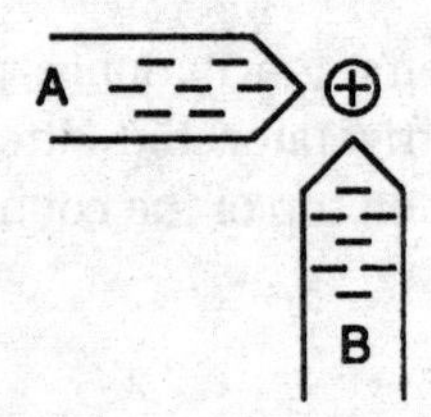

Which vector best represents the resultant electrostatic force on the sphere?

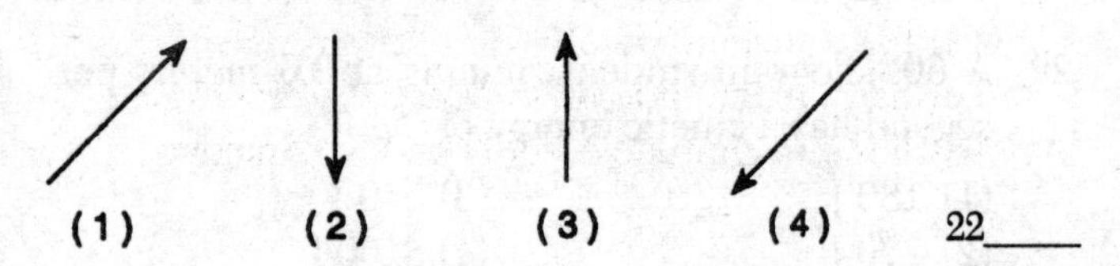

22____

23 A sphere has a net excess charge of -4.8×10^{-19} coulomb. The sphere must have an excess of

(1) 1 electron
(2) 1 proton
(3) 3 electrons
(4) 3 protons

23_____

24 The diagram below shows two metal spheres suspended by strings and separated by a distance of 3.0 meters. The charge on sphere *A* is $+5.0 \times 10^{-4}$ coulomb and the charge on sphere *B* is $+3.0 \times 10^{-5}$ coulomb.

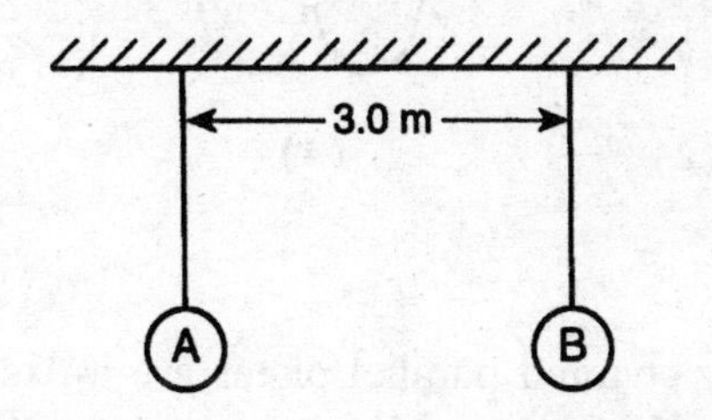

Which statement best describes the electrical force between the spheres?

1 It has a magnitude of 15 N and is repulsive.
2 It has a magnitude of 45 N and is repulsive.
3 It has a magnitude of 15 N and is attractive.
4 It has a magnitude of 45 N and is attractive.

24_____

25 Which diagram best represents the electric field near a positively charged conducting sphere?

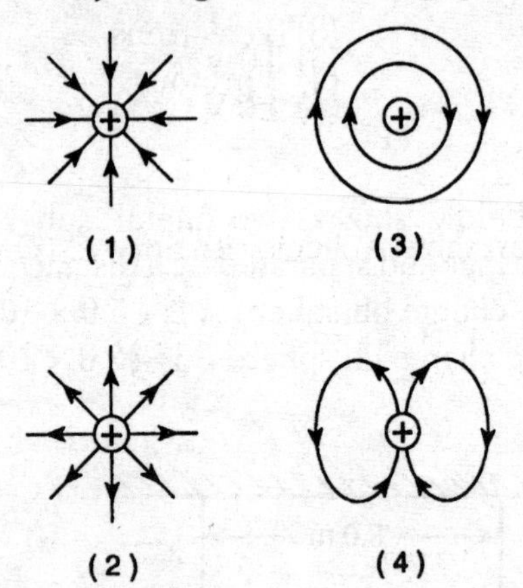

25____

26 Two oppositely charged parallel plates are a fixed distance apart. Which graph best represents the relationship between the electric field intensity (E) between the plates and the potential difference (V) across the plates?

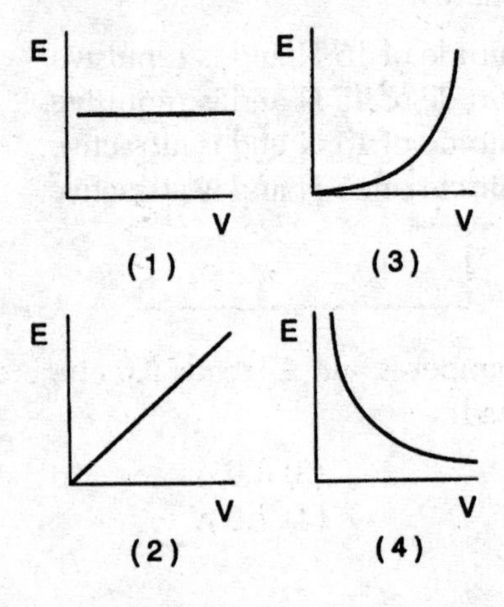

26____

27 What is the potential difference across a 2.0-ohm resistor that draws 2.0 coulombs of charge per second?

(1) 1.0 V (3) 3.0 V
(2) 2.0 V (4) 4.0 V

27____

28 The diagram below shows a circuit with three resistors.

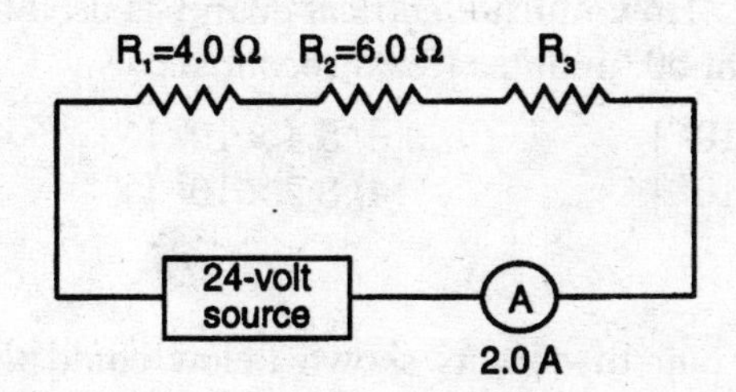

What is the resistance of resistor R_3?

(1) 6.0 Ω (3) 12 Ω
(2) 2.0 Ω (4) 4.0 Ω

28____

29 Three ammeters are placed in a circuit as shown below.

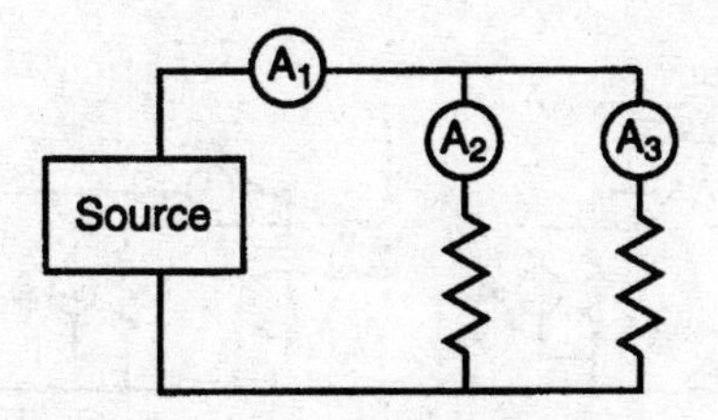

If A_1 reads 5.0 amperes and A_2 reads 2.0 amperes, what does A_3 read?

(1) 1.0 A (3) 3.0 A
(2) 2.0 A (4) 7.0 A

29____

30 An electric motor draws 150 amperes of current while operating at 240 volts. What is the power rating of this motor?

(1) 1.6 W
(2) 3.8×10^2 W
(3) 3.6×10^4 W
(4) 5.4×10^6 W

30____

31 An operating 75-watt lamp is connected to a 120-volt outlet. How much electrical energy is used by the lamp in 60. minutes (3600 seconds)?

(1) 4.5×10^3 J
(2) 2.7×10^5 J
(3) 5.4×10^5 J
(4) 3.2×10^7 J

31____

32 In which pair of circuits shown below could the readings of voltmeters V_1 and V_2 and ammeter A be correct?

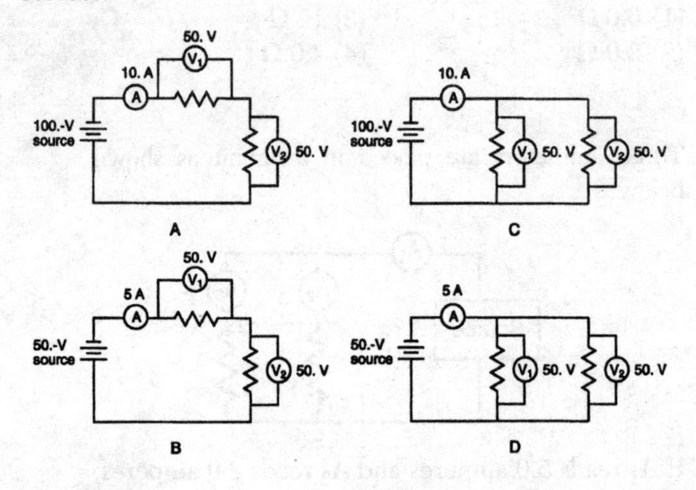

(1) A and B
(2) B and C
(3) C and D
(4) A and D

32____

33 The diagram below shows a coil of wire (solenoid) connected to a battery.

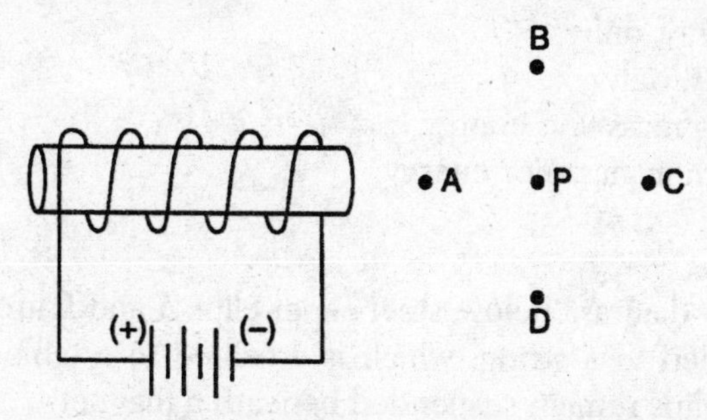

The north pole of a compass placed at point *P* would be directed toward point

(1) *A* (3) *C*
(2) *B* (4) *D* 33____

34 Two bar magnets of equal strength are positioned as shown.

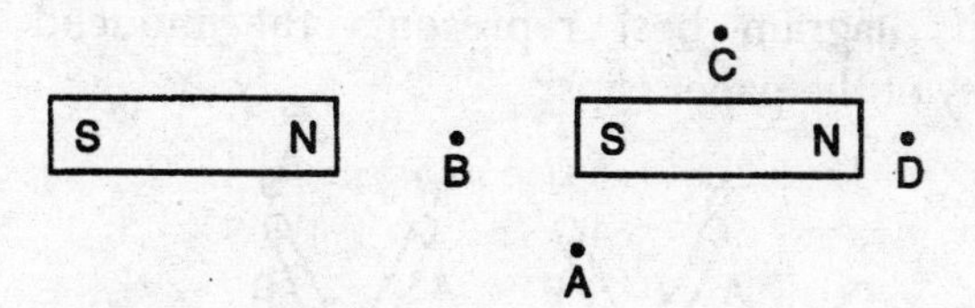

At which point is the magnetic flux density due to the two magnets greatest?

(1) *A* (3) *C*
(2) B (4) *D* 34____

35 As a sound wave travels through air, there is a net transfer of

1 energy, only
2 mass, only
3 both mass and energy
4 neither mass nor energy

35____

36 In the diagram below, steel paper clips *A* and *B* are attached to a string, which is attached to a table. The clips remain suspended beneath a magnet.

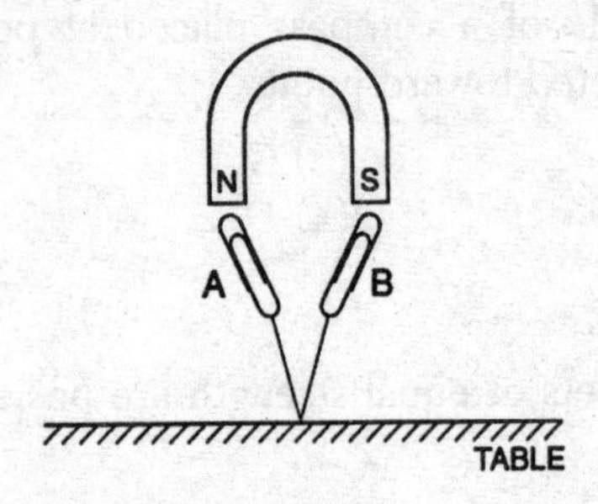

Which diagram best represents the induced polarity of the paper clips?

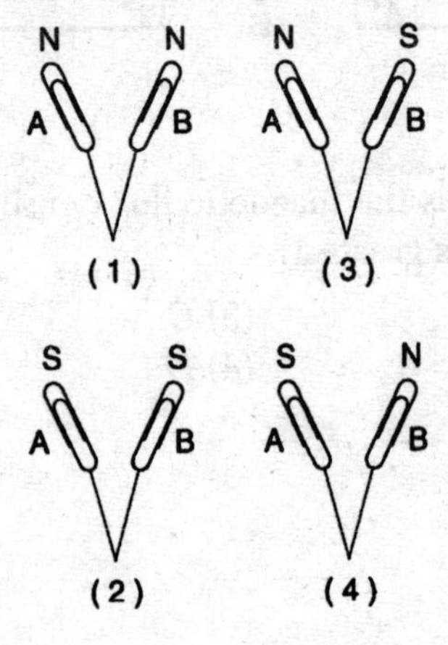

36____

37 The diagram below shows two pulses, *A* and *B*, moving to the right along a uniform rope.

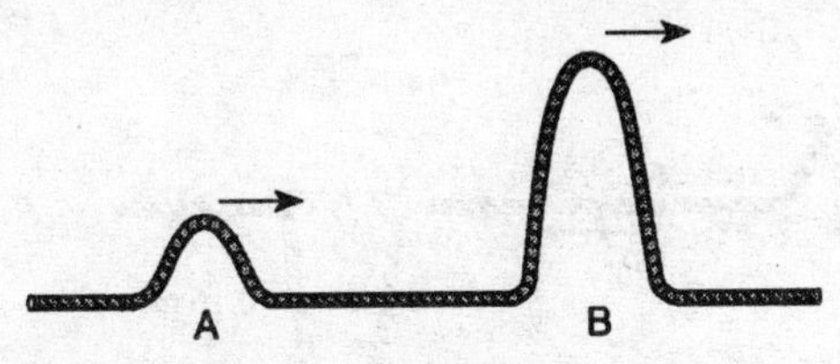

Compared to pulse *A*, pulse *B* has

1 a slower speed and more energy
2 a faster speed and less energy
3 a faster speed and the same energy
4 the same speed and more energy

37_____

38 A wave generator located 4.0 meters from a reflecting wall produces a standing wave in a string, as shown in the diagram below.

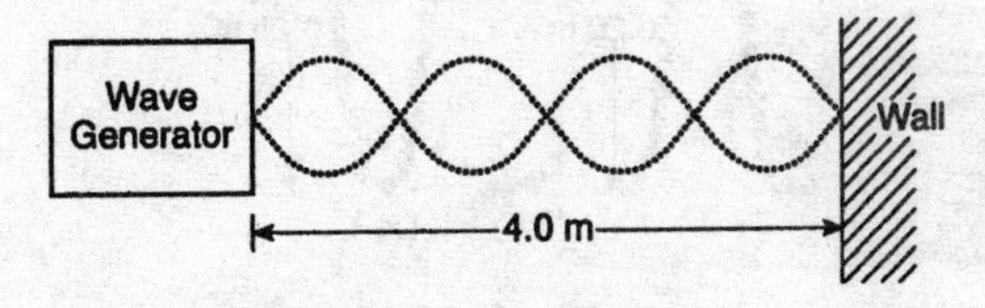

If the speed of the wave is 10. meters per second, what is its frequency?

(1) 0.40 Hz
(2) 5.0 Hz
(3) 10. Hz
(4) 40. Hz

38_____

39 The diagram below shows two pulses, each of length ℓ, traveling toward each other at equal speed in a rope.

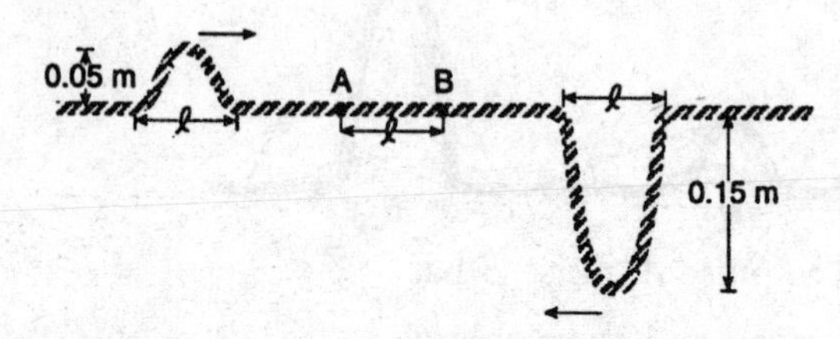

Which diagram best represents the shape of the rope when both pulses are in region *AB*?

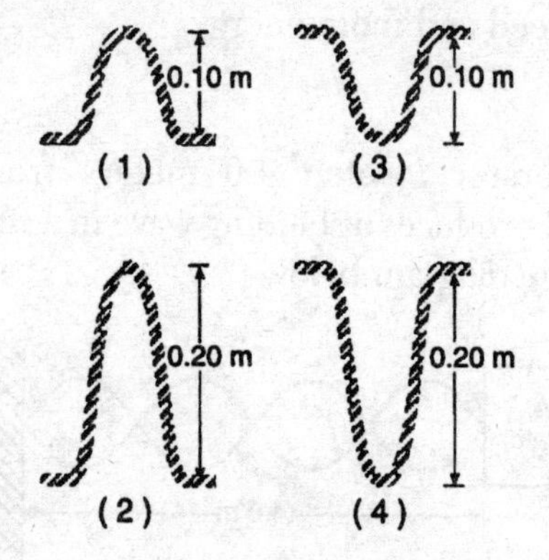

39_____

40 A nearby object may vibrate strongly when a specific frequency of sound is emitted from a loudspeaker. This phenomenon is called

1 resonance
2 the Doppler effect
3 reflection
4 interference

40_____

Base your answers to questions 41 and 42 on the diagram below, which represents waves *A*, *B*, *C*, and *D* traveling in the same medium.

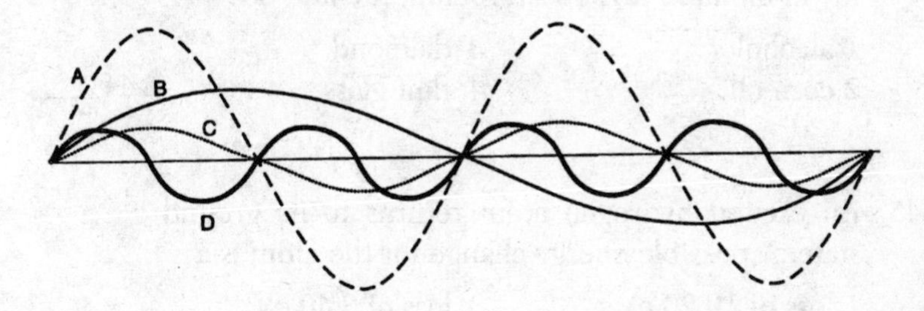

41 Which two waves have the same wavelength?

(1) *A* and *B*
(2) *A* and *C*
(3) *B* and *D*
(4) *C* and *D*

41_____

42 Which wave has the longest period?

(1) *A*
(2) *B*
(3) *C*
(4) *D*

42_____

---

43 A ray of light strikes a plane mirror at an angle of incidence equal to 35°. The angle between the incident ray and the reflected ray is

(1) 0°
(2) 35°
(3) 55°
(4) 70.°

43_____

44 A ray of light (λ= 5.9 × 10^{-7} meter) traveling in air is incident on an interface with medium *X* at an angle of 30.°. The angle of refraction for the light ray in medium *X* is 12.°. Medium *X* could be

1 alcohol
2 corn oil
3 diamond
4 flint glass

44____

45 An excited hydrogen atom returns to its ground state. A possible energy change for the atom is a

1 loss of 10.20 eV
2 gain of 10.20 eV
3 loss of 3.40 eV
4 gain of 3.40 eV

45____

46 In which diagram below could the light source and optical device be used to demonstrate the phenomenon of dispersion?

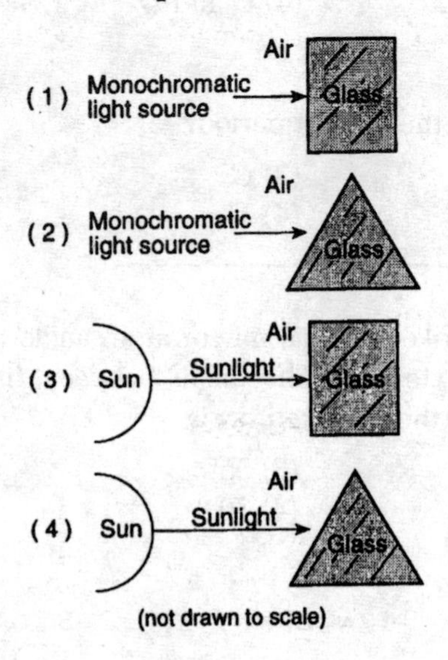

(not drawn to scale)

46____

47 The four-line Balmer series spectrum shown below is emitted by a hydrogen gas sample in a laboratory. A star moving away from Earth also emits a hydrogen spectrum.

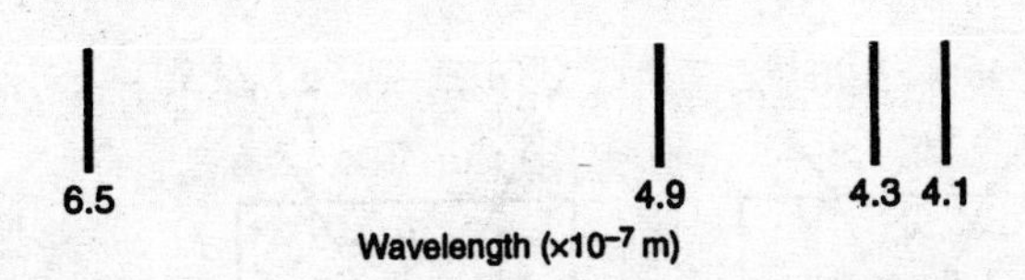

Which spectrum might be observed on Earth for this star?

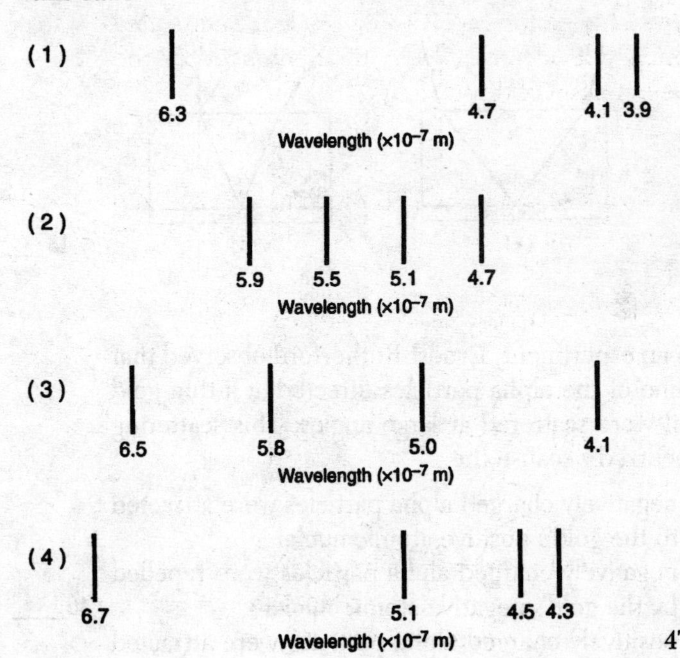

47_____

48 A ray of monochromatic light traveling in air enters a rectangular glass block obliquely and strikes a plane mirror at the bottom. Then the ray travels back through the glass and strikes the air-glass interface. Which diagram below best represents the path of this light ray? [*N* represents the normal to the surface.]

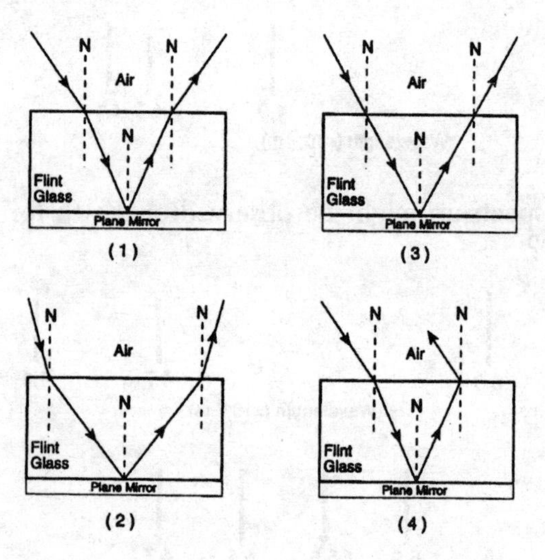

48_____

49 In an experiment, Ernest Rutherford observed that some of the alpha particles directed at a thin gold foil were scattered at large angles. This scattering occurred because the

1 negatively charged alpha particles were attracted to the gold's positive atomic nuclei
2 negatively charged alpha particles were repelled by the gold's negative atomic nuclei
3 positively charged alpha particles were attracted to the gold's negative atomic nuclei
4 positively charged alpha particles were repelled by the gold's positive atomic nuclei

49_____

50 The maximum kinetic energy of an electron ejected from a metal by a photon depends on

1 the photon's frequency, only
2 the metal's work function, only
3 both the photon's frequency and the metal's work function
4 neither the photon's frequency nor the metal's work function

50____

51 During a collision between a photon and an electron, there is conservation of

1 energy, only
2 momentum, only
3 both energy and momentum
4 neither energy nor momentum

51____

52 A ray of monochromatic light is traveling in flint glass. The ray strikes the flint glass–air interface at an angle of incidence greater than the critical angle for flint glass. Which diagram best represents the path of this light ray?

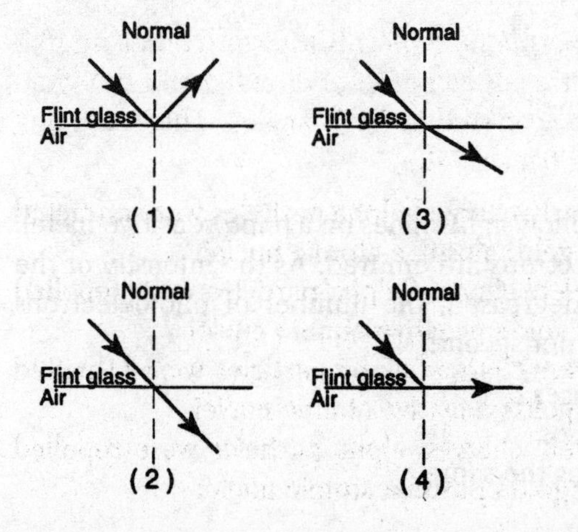

52____

Note that questions 53 through 55 have only three choices.

53 When an incandescent light bulb is turned on, its thin wire filament heats up quickly. As the temperature of this wire filament increases, its electrical resistance

1 decreases
2 increases
3 remains the same

53____

54 As shown in the diagrams below, a lump of clay travels horizontally to the right toward a block at rest on a frictionless surface. Upon collision, the clay and the block stick together and move to the right.

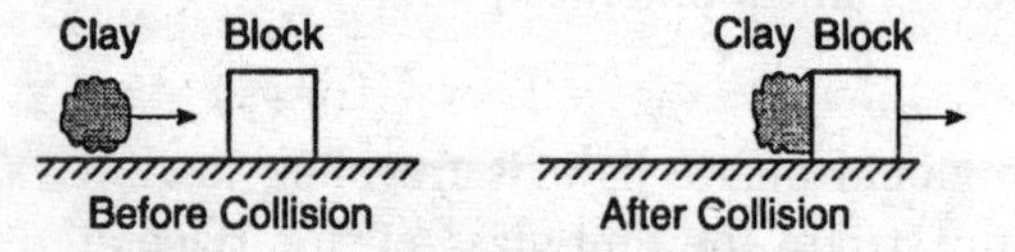

Compared to the total momentum of the clay and the block before the collision, the momentum of the clay-block system after the collision is

1 less
2 greater
3 the same

54____

55 When yellow light shines on a photosensitive metal, photoelectrons are emitted. As the intensity of the light is decreased, the number of photoelectrons emitted per second

1 decreases
2 increases
3 remains the same

55____

PART II

This part consists of six groups, each containing ten questions. Each group tests an optional area of the course. Choose two of these six groups. Be sure that you answer all ten questions in each group chosen. Record the answers to these questions in the spaces provided. [20]

GROUP 1—**Motion in a Plane**

If you choose this group, be sure to answer questions **56–65**.

Base your answers to questions 56 and 57 on the information and diagram below.

A student standing on a knoll throws a snowball horizontally 4.5 meters above the level ground toward a smokestack 15 meters away. The snowball hits the smokestack 0.65 second after being released. [Neglect air resistance.]

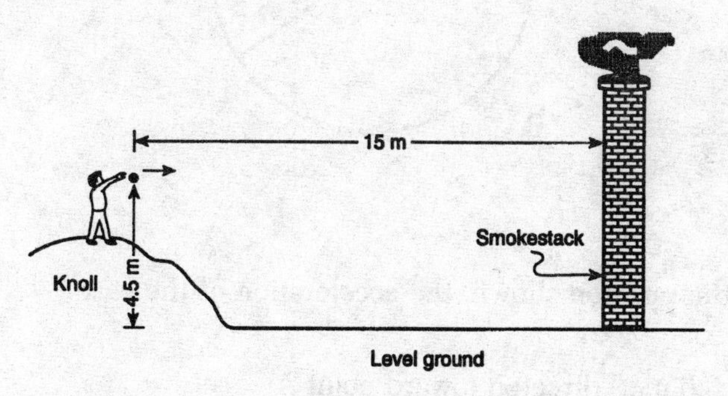

56 Approximately how far above the level ground does the snowball hit the smokestack?

(1) 0.0 m
(2) 0.4 m
(3) 2.4 m
(4) 4.5 m

56_____

57 At the instant the snowball is released, the horizontal component of its velocity is approximately

(1) 6.9 m/s
(2) 9.8 m/s
(3) 17 m/s
(4) 23 m/s

57____

Base your answers to questions 58 through 60 on the diagram below which shows a 2.0-kilogram cart traveling at a constant speed in a horizontal circle of radius 3.0 meters. The magnitude of the centripetal force of the cart is 24 newtons.

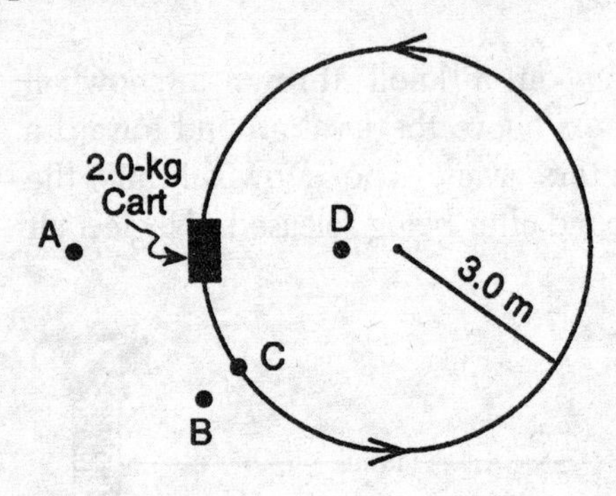

58 In the position shown, the acceleration of the cart is

(1) 8.0 m/s^2 directed toward point *A*
(2) 8.0 m/s^2 directed toward point *D*
(3) 12 m/s^2 directed toward point *A*
(4) 12 m/s^2 directed toward point *D*

58____

59 Which statement correctly describes the direction of the cart's velocity and centripetal force in the position shown?

1 Velocity is directed toward point *B*, and the centripetal force is directed toward point *A*.
2 Velocity is directed toward point *B*, and the centripetal force is directed toward point *D*.
3 Velocity is directed toward point *C*, and the centripetal force is directed toward point *A*.
4 Velocity is directed toward point *C*, and the centripetal force is directed toward point *D*. 59____

60 What is the speed of the cart?

(1) 6.0 m/s (3) 36 m/s
(2) 16 m/s (4) 4.0 m/s 60____

---

61 An artillery shell is fired at an angle to the horizontal. Its initial velocity has a vertical component of 150 meters per second and a horizontal component of 260 meters per second. What is the magnitude of the initial velocity of the shell?

(1) 9.0×10^4 m/s (3) 3.0×10^2 m/s
(2) 4.1×10^2 m/s (4) 1.1×10^2 m/s 61____

62 The diagram below shows a projectile moving with speed v at the top of its trajectory.

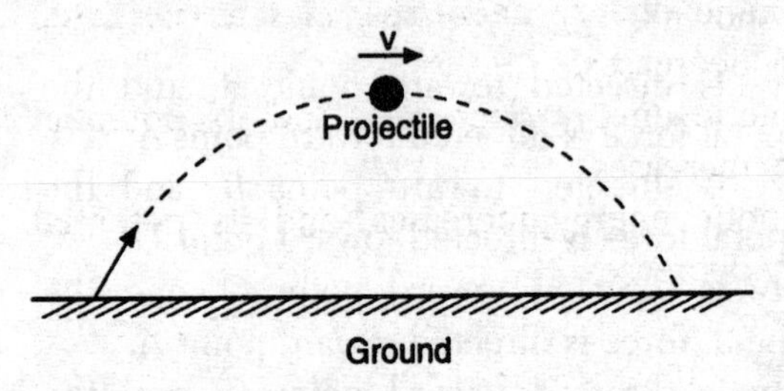

Which vector best represents the acceleration of the projectile in the position shown?

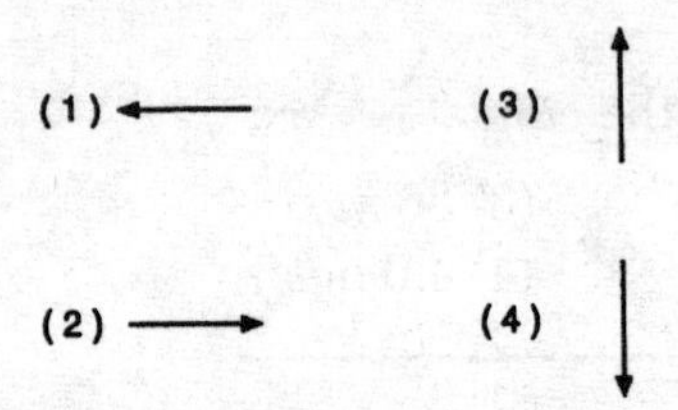

62_____

Base your answers to question 63 and 64 on the diagram below. A planet, *P*, moves around the Sun, *S*, in an elliptical orbit. The amount of time required for the planet to travel from point *A* to point *B* is equal to the amount of time required to travel from point *C* to point *D*.

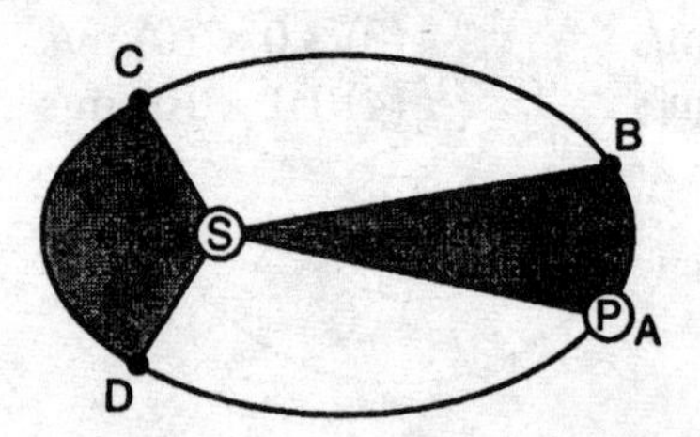

63 As the planet moves from point B to point C, how do its kinetic energy and potential energy change?

1 Its kinetic energy decreases, and its potential energy decreases.
2 Its kinetic energy decreases, and its potential energy increases.
3 Its kinetic energy increases, and its potential energy decreases.
4 Its kinetic energy increases, and its potential energy increases. 63_____

Note that question 64 has only three choices.

64 Compared to the area of region ABS, the area of region CDS is

1 smaller
2 larger
3 the same 64_____

---

65 The diagram below shows four planets, A, B, C, and D, orbiting a star.

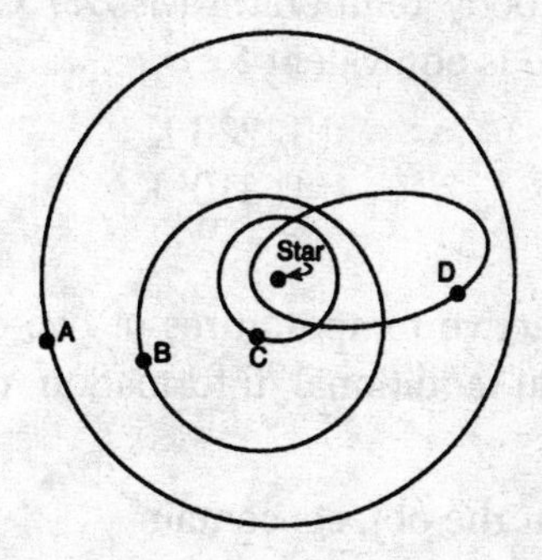

Which planet has the greatest orbital period?

(1) A (3) C
(2) B (4) D 65_____

GROUP 2—Internal Energy

If you choose this group, be sure to answer questions **66–75**.

66 Which graph best represents the relationship between the average kinetic energy ($\overline{KE}$) of the random motion of the molecules of an ideal gas and its absolute temperature (T)?

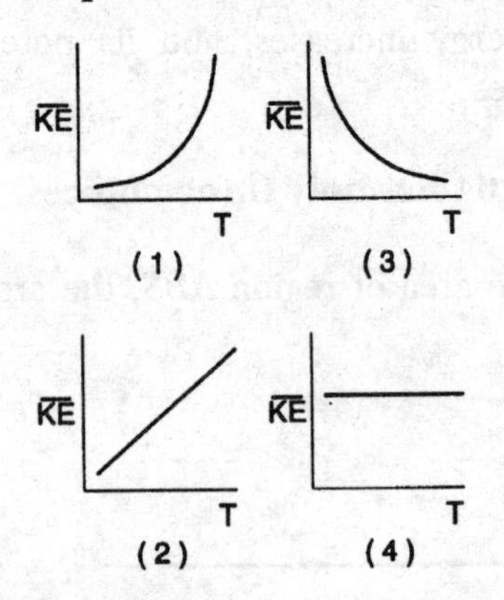

66_____

67 Normal human body temperature is 37° Celsius. This temperature is equivalent to

(1) 63 K
(2) 137 K
(3) 236 K
(4) 310. K

67_____

68 If only the respective temperatures of two objects are known, what additional information can be determined?

1 how much heat the objects contain
2 how much heat the warmer object can supply to the colder object
3 whether a heat exchange would take place if the objects were in contact
4 the total amount of energy the objects contain

68_____

69 Equal masses of platinum, iron, aluminum, and copper are at their respective boiling points. Which of these metals requires the greatest amount of heat to change from the liquid to the gaseous phase?

1 aluminum
2 iron
3 platinum
4 copper

69____

70 A mixture of ice and water is heated at a constant rate. Which graph best represents the relationship between the temperature of the mixture (T) and the heat added (Q)?

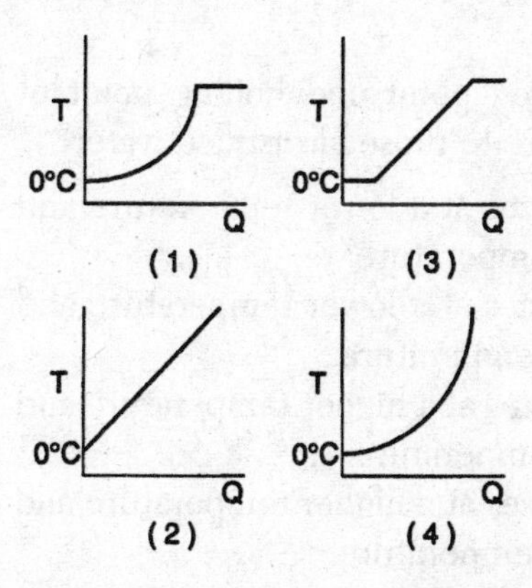

70____

71 A 0.060-kilogram ice cube at 0.0°C is placed in a glass containing 0.250 kilogram of water at 25°C. Which statement describes this system when equilibrium is reached? [Assume no external exchange of heat.]

1 The ice is completely melted and the water temperature is above 0°C.
2 The ice is completely melted and the water temperature is 0°C.
3 Part of the ice remains frozen and the water temperature is above 0°C.
4 Part of the ice remains frozen and the water temperature is 0°C.

71____

72 As 6.00 kilograms of a liquid substance at its freezing point completely freezes, it gives off enough heat to melt 3.00 kilograms of ice at 0°C. The heat of fusion of the substance is

(1) 2.05 kJ/kg
(2) 4.19 kJ/kg
(3) 167 kJ/kg
(4) 668 kJ/kg

72____

73 As lead melts, there is a change in its

1 temperature
2 heat of fusion
3 average molecular kinetic energy
4 average molecular potential energy

73____

74 How do the freezing point and boiling point of ocean water compare to those of distilled water?

1 Ocean water freezes at a lower temperature and boils at a lower temperature.
2 Ocean water freezes at a lower temperature and boils at a higher temperature.
3 Ocean water freezes at a higher temperature and boils at a lower temperature.
4 Ocean water freezes at a higher temperature and boils at a higher temperature.

74____

Note that question 75 has only three choices.

75 A cylinder fitted with a piston contains a fixed mass of an ideal gas. Heat is added to the gas, causing it to expand and raise the piston. If all the added heat is converted to work done in raising the piston, the internal energy of the gas will

1 decrease
2 increase
3 remain the same

75____

GROUP 3—**Electromagnetic Applications**

If you choose this group, be sure to answer questions **76–85**.

76 In the diagram below, a free electron is traveling upward at speed v parallel to a conductor. An electron current begins to flow upward in the conductor.

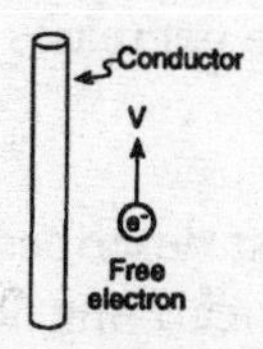

Which diagram best represents the resulting magnetic field, B, and the direction of the magnetic force, F, on the free electron?

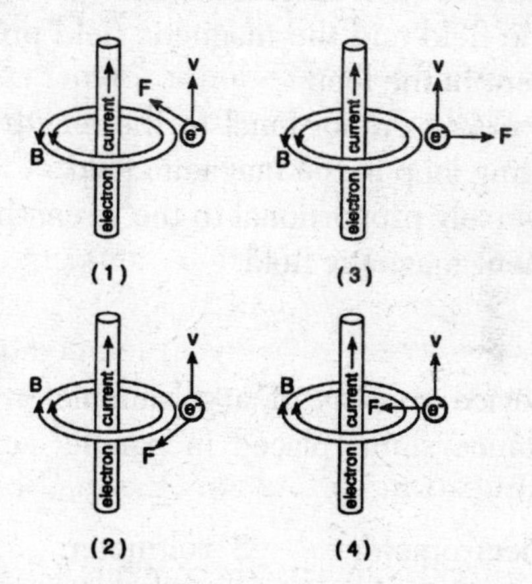

76_____

77 A magnetic force acts on a charged particle moving at a constant speed perpendicular to a uniform magnetic field. The magnitude of the magnetic force on the particle will increase if the

1 flux density of the magnetic field increases
2 time of travel of the charge increases
3 magnitude of the charge decreases
4 speed of the charge decreases

77____

78 Which statement best describes the torque experienced by a current-carrying loop of wire in an external magnetic field?

1 It is due to the current in the loop of wire, only.
2 It is due to the interaction of the external magnetic field and the magnetic field produced by current in the loop.
3 It is inversely proportional to the length of the conducting loop in the magnetic field.
4 It is inversely proportional to the strength of the permanent magnetic field.

78____

79 Which device consists of a galvanometer with a low-resistance shunt placed in parallel across its terminals?

1 mass spectrometer
2 transformer
3 voltmeter
4 ammeter

79____

80 An operating electric motor produces a back emf, which opposes the applied potential difference. As a result, the armature current

1 decreases
2 increases
3 changes from d.c. to a.c.
4 changes from a.c. to d.c. 80____

81 As the temperature of a surface increases, how does the rate of thermionic emission change?

1 Electrons are emitted at a lower rate.
2 Electrons are emitted at a higher rate.
3 Protons are emitted at a lower rate.
4 Protons are emitted at a higher rate. 81____

82 In an operating mass spectrometer, the motion of positive ion beams is influenced by

1 electric fields, only
2 magnetic fields, only
3 both electric and magnetic fields
4 neither electric nor magnetic fields 82____

83 The diagram below, which illustrates the Millikan oil drop experiment, shows a 3.2×10^{-14}-kilogram oil drop with a charge of -1.6×10^{-18} coulomb. The oil drop was in equilibrium when the upward electric force on the drop was equal in magnitude to the gravitational force on the drop.

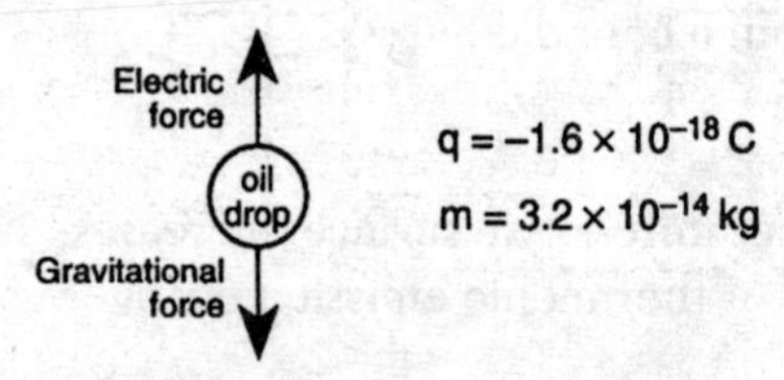

What was the magnitude of the electric field intensity when this oil drop was in equilibrium?

(1) 2.0×10^{-5} N/C
(2) 2.0×10^{5} N/C
(3) 5.0×10^{-5} N/C
(4) 5.0×10^{5} N/C

83_____

84 The diagram below shows an end view of a metal rod moving upward perpendicular to a uniform magnetic field having a flux density of 2.0×10^{-5} tesla. The 2.0-meter-long wire is moving at a constant speed of 3.0 meters per second.

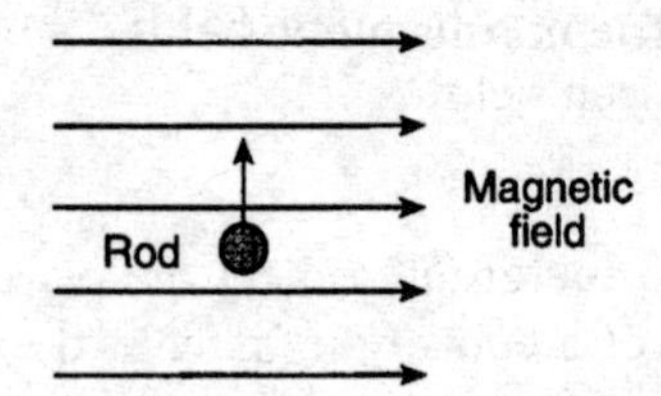

What is the emf induced across the rod?

(1) 0.060 V
(2) 0.12 V
(3) 1.2 V
(4) 6.0 V

84_____

85 The diagram below shows a step-up transformer having a primary coil with two windings and a secondary coil with four windings.

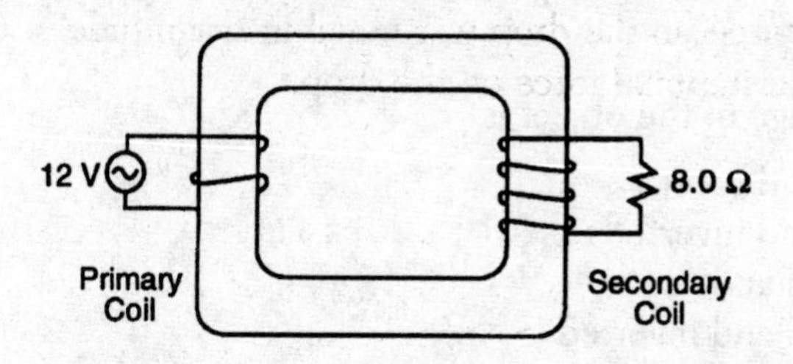

When a potential difference of 12 volts is applied to the primary coil, what is the current in a 8.0-ohm resistor connected to the secondary coil as shown?

(1) 0.33 A
(2) 0.75 A
(3) 3.0 A
(4) 4.5 A

85____

GROUP 4—**Geometric Optics**

If you choose this group, be sure to answer questions **86–95**.

Base your answers to questions 86 through 88 on the information below.

When a 0.020-meter-tall object is placed 0.15 meter in front of a converging mirror, the object's image appears 0.30 meter in front of the mirror.

86 The focal length of the mirror is

(1) 0.45 m
(2) 0.10 m
(3) –0.10 m
(4) –0.15 m

86____

87 The image of the object is

1 real and erect
2 real and inverted
3 virtual and erect
4 virtual and inverted

87____

88 How tall is the image?

(1) 0.010 m
(2) 0.020 m
(3) 0.030 m
(4) 0.040 m

88____

---

89 The diagram below shows a light ray parallel to the principal axis of a spherical convex (diverging) mirror. Point *F* is the virtual focal point of the mirror and *C* is the center of curvature.

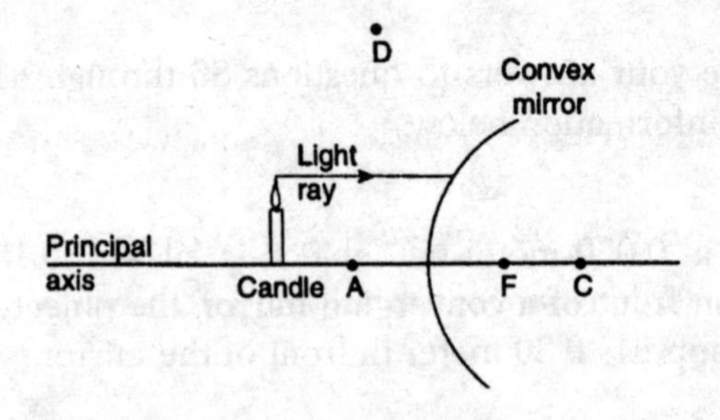

After the light ray is reflected, it will pass through point

(1) *A*
(2) *C*
(3) *D*
(4) *F*

89____

Base your answers to questions 90 through 92 on the information and diagram below. A convex lens having optical center *O* and principal focus *F* is used to produce an image of a candle. Ray *RF* is shown.

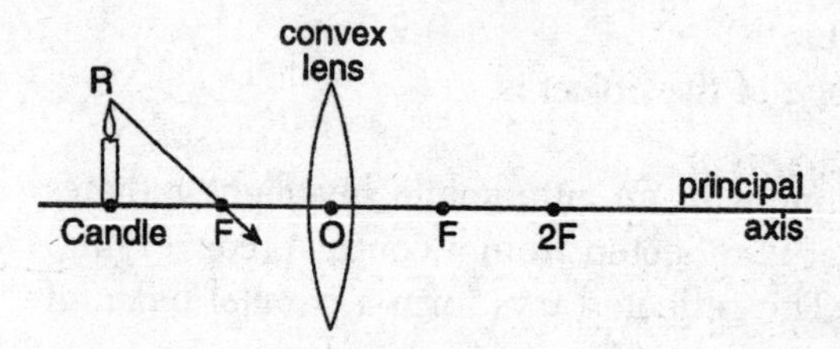

90 The lens is being used to produce an image of the candle that is

1 virtual and erect
2 virtual and inverted
3 real and erect
4 real and inverted

90____

91 When ray *RF* reaches the lens, the ray will

1 reflect back through point *R*
2 polarize and travel perpendicular to the principal axis
3 refract and pass through point 2*F*
4 refract and emerge parallel to the principal axis

91____

Note that question 92 has only three choices.

92 As the candle is moved toward the left, the size of its image will

1 decrease
2 increase
3 remain the same

92____

93 A 2.0-meter-tall student is able to view his entire body at once using a plane mirror. The minimum length of the mirror is

(1) 1.0 m (3) 1.5 m
(2) 0.50 m (4) 2.5 m

93____

94 The filament in an automobile headlight radiates light that is reflected from a concave (converging) mirror. The reflected rays form a parallel beam of light because the filament is placed

1 between the mirror and the principal focus
2 at the mirror's principal focus
3 at the mirror's center of curvature
4 beyond the mirror's center of curvature

94____

95 Which phenomenon may cause a concave mirror to form fuzzy, out-of-focus images?

1 spherical aberration
2 chromatic aberration
3 dispersion
4 refraction

95____

GROUP 5—**Solid State**

If you choose this group, be sure to answer questions **96–105**.

96 The diagram below represents the wave form and phase of an alternating current signal.

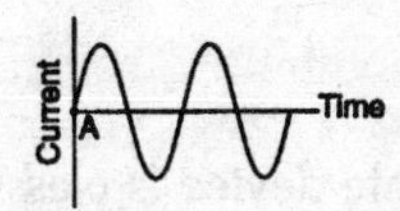

Which graph best represents the wave form and phase of this signal after it has passed through a diode rectifier?

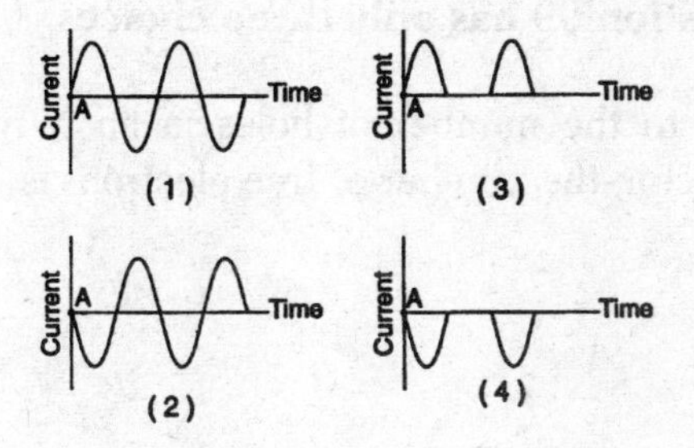

96_____

97 Which graph best represents the relationship between current and potential difference for a forward-biased diode?

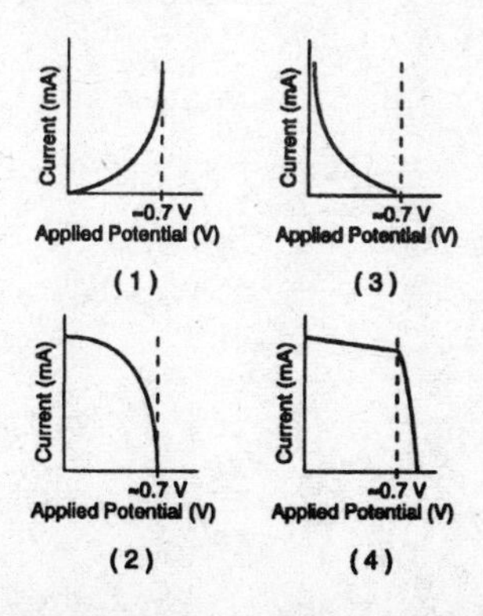

97_____

98 A circuit is shown in the diagram below.

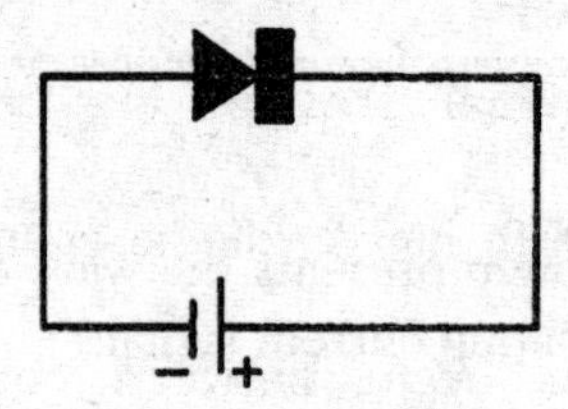

The current in the electronic device is classified as

1 reverse biased
2 forward biased
3 unbiased
4 transistorized

98____

Note that question 99 has only three choices.

99 Compared to the number of holes in an *N*-type semiconductor, the number of free electrons is

1 less
2 greater
3 the same

99____

100 The primary source of holes in *P-N-P* transistors is the

1 transmitter
2 collector
3 base
4 emitter

100____

101 What occurs as the temperature of a solid conductor increases?

1 The cross-sectional area of the conductor decreases.
2 The electrons with higher kinetic energy move slower.
3 More collisions occur between conduction electrons and atom kernels.
4 More electrons move from the conduction band into the valence band. 101____

102 Charge carriers in a semiconductor can be

1 electrons, only
2 holes, only
3 both electrons and holes
4 both electrons and protons 102____

103 What type of semiconductor device is formed by the two semiconductors shown below?

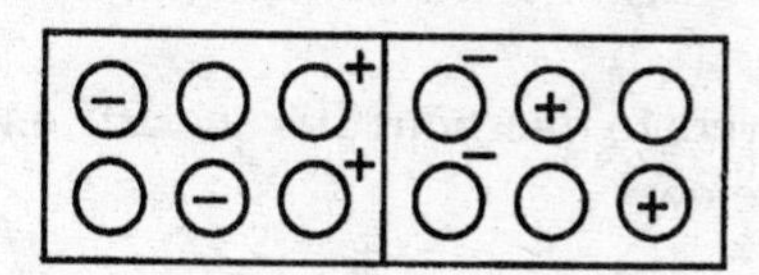

1 triode
2 transistor
3 donor
4 diode 103____

104 Indium is an element that has only three valence electrons. If a very small amount of indium was added to a germanium crystal, the resulting semiconductor would be

(1) *N*-type
(2) *P*-type
(3) negatively charged
(4) positively charged

104____

105 How should the junctions of an *N-P-N* transistor be biased?

1 The emitter-base is forward biased and the base-collector is reverse biased.
2 The emitter-base and collector-base are forward biased.
3 The emitter-base is reverse biased and the base-collector is forward biased.
4 The emitter-base and collector-base are reverse biased.

105____

GROUP 6—**Nuclear Energy**

If you choose this group, be sure to answer questions **106–115**.

Base your answers to questions 106 and 107 on the equation below.

$$^{32}_{16}S + ^{1}_{0}n \rightarrow ^{32}_{15}P + X$$

106 How many neutrons are in an atom of $^{32}_{15}P$?

(1) 15
(2) 16
(3) 17
(4) 32

106____

107 What is particle *X*?

1 a proton
2 a neutron
3 an alpha particle
4 a beta particle

107____

---

108 Compared to the gravitational force between two nucleons in an atom of helium, the nuclear force between the nucleons is

1 weaker and has a shorter range
2 weaker and has a longer range
3 stronger and has a shorter range
4 stronger and has a longer range

108____

109 The subatomic particles that make up protons are called

1 hyperons
2 baryons
3 positrons
4 quarks

109____

110 Which nuclear phenomenon produces a change in the mass number of a nucleus?

1 alpha decay
2 electron capture
3 gamma ray emission
4 positron emission

110____

111 According to the Uranium Disintegration Series, how many different isotopes of polonium (Po) are formed as $^{238}_{92}U$ decays to $^{206}_{82}Pb$?

(1) 1
(2) 2
(3) 3
(4) 0

111____

112 Gamma radiation consists of a stream of high-energy

1 photons
2 protons
3 neutrons
4 electrons

112____

113 A radioactive nuclide sample has a half-life of 3.0 days. If 2.0 kilograms of the sample remains unchanged after 9.0 days, what was the initial mass of the sample?

(1) 18kg
(2) 16 kg
(3) 8.0 kg
(4) 6.0 kg

113____

114 Uranium-235 and plutonium-239 are used as fuels in nuclear reactors because of their

1 ability to undergo fission
2 ability to undergo fusion
3 inability to absorb neutrons
4 inability to release neutrons

114____

115 What is represented by the nuclear reaction below?

$$^{32}_{16}S + ^{1}_{0}n \rightarrow ^{32}_{15}P + X$$

1 fusion
2 fission
3 alpha decay
4 beta decay

115____

PART III

You must answer *all* questions in this part. [15]

Base your answers to questions 116 and 117 on the information below.

A 5.0-kilogram block weighing 49 newtons sits on a frictionless, horizontal surface. A horizontal force of 20. newtons toward the right is applied to the block. [Neglect air resistance.]

116 On the diagram below, draw a vector to represent each of the three forces acting on the block. Use a ruler and a scale of 1.0 centimeter = 10. newtons. Begin each vector at point *C* and label its magnitude in newtons. [3]

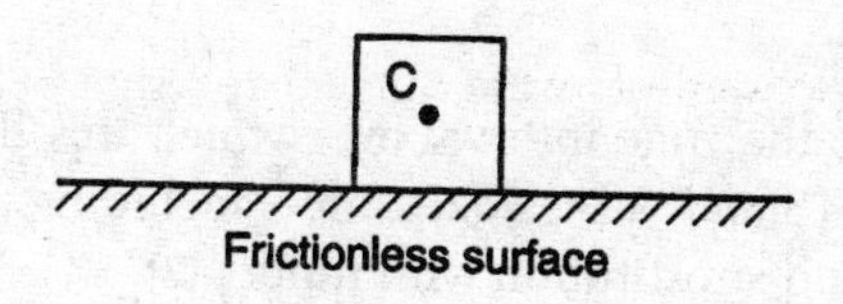

117 Calculate the magnitude of the acceleration of the block. [Show all calculations, including the equation and substitution with units.] [2]

Base your answers to questions 118 through 120 on the information below.

A scientist set up an experiment to collect data about lightning. In one lightning flash, a charge of 25 coulombs was transferred from the base of a cloud to the ground. The scientist measured a potential difference of 1.8×10^6 volts between the cloud and the ground and an average current of 2.0×10^4 amperes.

118 Determine the time interval over which this flash occurred. [Show all calculations, including the equation and substitution with units.] [2]

119 Determine the amount of energy, in joules, involved in the transfer of the electrons from the cloud to the ground. [Show all calculations, including the equation and substitution with units.] [2]

120 The scientist was several kilometers from the lightning flash. Using one or more complete sentences, explain why the scientist saw the lightning flash several seconds before he heard the sound of the thunderclap from that flash. [1]

Base your answers to questions 121 through 124 on the information in the data table below. The data were obtained by varying the force applied to a spring and measuring the corresponding elongation of the spring.

Applied Force (N)	Elongation of Spring (m)
0.0	0.00
4.0	0.16
8.0	0.27
12.0	0.42
16.0	0.54
20.0	0.71

Directions (121–123): Using the information in the table, construct a graph on the grid provided following the directions below.

121 Mark an appropriate scale on the axis labeled "Elongation (m)." [1]

122 Plot the data points for force versus elongation. [1]

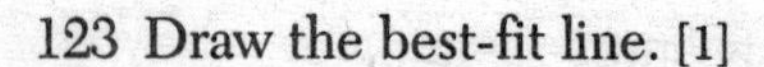

123 Draw the best-fit line. [1]

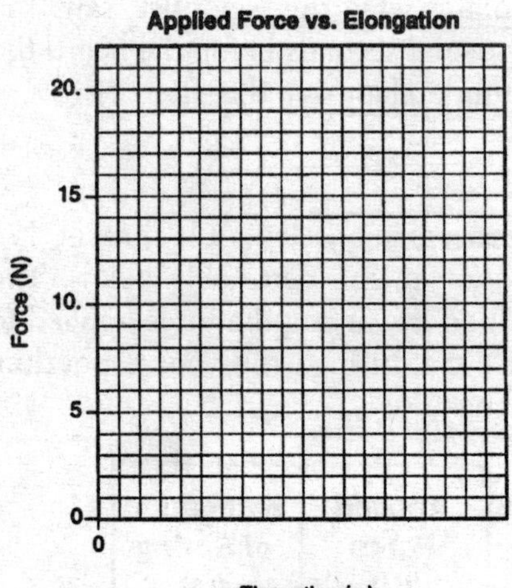

124 *Using the best-fit line*, determine the spring constant of the spring. [Show all calculations, including the equation and substitution with units.] [2]

____

____

____

Answers June 1998

Physics

Answer Key

PART I

1. 2	11. 3	20. 2	29. 3	38. 2	47. 4
2. 3	12. 1	21. 3	30. 3	39. 3	48. 1
3. 1	13. 4	22. 4	31. 2	40. 1	49. 4
4. 1	14. 1	23. 3	32. 4	41. 2	50. 3
5. 4	15. 3	24. 1	33. 1	42. 2	51. 3
6. 4	16. 1	25. 2	34. 2	43. 4	52. 1
7. 1	17. 2	26. 2	35. 1	44. 3	53. 2
8. 3	18. 4	27. 4	36. 4	45. 1	54. 3
9. 2	19. 1	28. 2	37. 4	46. 4	55. 1
10. 1					

PART II

Group 1	Group 2	Group 3	Group 4	Group 5	Group 6
56. 3	66. 2	76. 4	86. 2	96. 3	106. 3
57. 4	67. 4	77. 1	87. 2	97. 1	107. 1
58. 4	68. 3	78. 2	88. 4	98. 1	108. 3
59. 2	69. 1	79. 4	89. 3	99. 2	109. 4
60. 1	70. 3	80. 1	90. 4	100. 4	110. 1
61. 3	71. 1	81. 2	91. 4	101. 3	111. 3
62. 4	72. 3	82. 3	92. 1	102. 3	112. 1
63. 3	73. 4	83. 2	93. 1	103. 4	113. 2
64. 3	74. 2	84. 2	94. 2	104. 2	114. 1
65. 1	75. 3	85. 3	95. 1	105. 1	115. 1

PART III — See answers explained

Answers Explained

PART I

1. **2** To find a *resultant vector*, place the given vectors head to tail, and draw the resultant vector from the tail of the first vector to the head of the second vector.

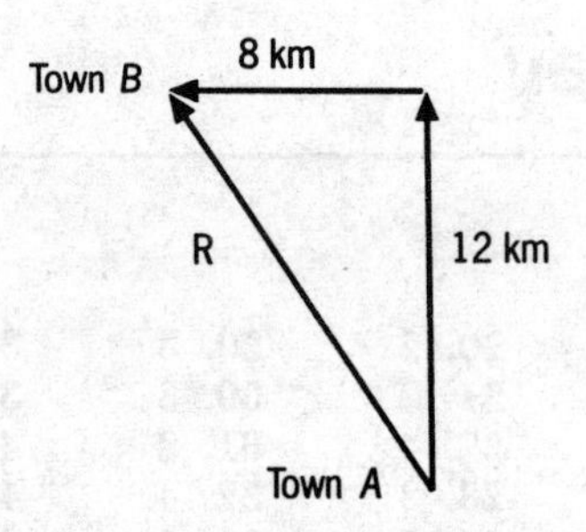

Since the result is a right triangle, use the Pythagorean theorem

$$a^2 + b^2 = c^2$$

to determine the magnitude of the resultant.

$$\begin{aligned} R^2 &= (12\text{ km})^2 + (8\text{ km})^2 \\ &= 144\text{ km}^2 + 64\text{ km}^2 \\ &= 208\text{ km}^2 \\ R &= 14\text{ km to the nearest whole number} \end{aligned}$$

2. **3** There are 100 cm in 1 m, and there are 2.54 cm in 1 in. If you convert 0.25 m into inches (see below), you obtain a value of 9.8 in. Thus 0.25 m is a reasonable size for the diameter of a dinner plate.

$$0.25\text{ m}\left(\frac{100.\text{ cm}}{1.0\text{ m}}\right)\left(\frac{1.0\text{ in.}}{2.54\text{ cm}}\right) = 9.8\text{ in.}$$

Wrong Choices Explained:

(1), (2), (4) 0.0025 m is approximately equal to 0.1 in., 0.025 m is approximately equal to 1 in., and 2.5 m is approximately equal to 98 in. All these sizes are ridiculous for the diameter of a dinner plate.

3. **1** An object falling freely near the surface of Earth would be subject to a constant acceleration due to Earth's gravity. Only the acceleration versus time graph in choice (1) represents a constant acceleration over time. With a constant acceleration, the speed of the object would be increasing, and the speed versus time graph in this choice correctly represents increasing speed over time.

Wrong Choices Explained:

(2) In this choice the acceleration versus time graph is incorrect. It represents an increasing acceleration over time.

(3) In this choice the acceleration versus time graph is incorrect. It represents a decreasing acceleration over time.

(4) In this choice both the speed and acceleration versus time graphs are incorrect. The speed graph represents a constant speed over time, and the acceleration graph represents an increasing acceleration over time.

4. **1** To solve this problem, you need to solve for the resultant force. The equilibrant or balancing force has a magnitude equal to the resultant force but has the opposite direction.

To find a *resultant vector*, place the given vectors head to tail, and draw the resultant vector from the tail of the first vector to the head of the second vector.

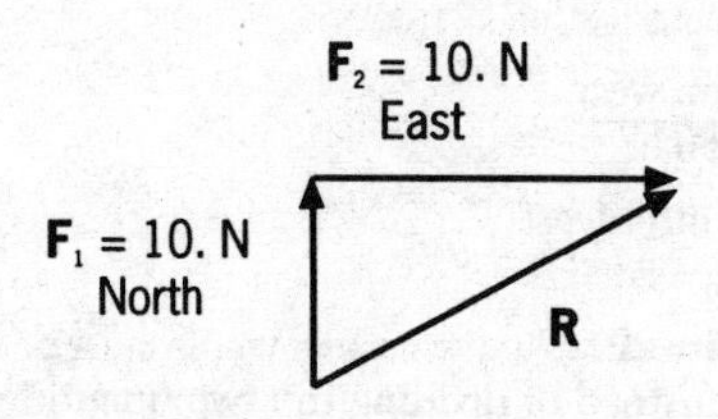

Since the result is a right triangle, use the Pythagorean theorem

$$a^2 + b^2 = c^2$$

to determine the magnitude of the resultant.

$$R^2 = (10.\ N)^2 + (10.\ N)^2$$
$$= 100\ N^2 + 100\ N^2$$
$$= 200\ N^2$$
$$R = 14\ N \text{ to the nearest whole number}$$

Since the resultant has a magnitude of 14 N and is in the northeasterly direction, the equilibrant will also have a magnitude of 14 N but will be in the southwesterly direction.

Wrong Choices Explained:

(2) Correct magnitude, but you forgot that the equilibrant vector is in the opposite direction from the resultant vector.

(3) Correct direction, but you added the two forces together as if they were in the same direction.

(4) Wrong direction, and you added the two forces together as if they were in the same direction.

5. **4** The best way to approach any word problem is to correctly identify the quantities given and the quantity to be found.

Given: $\Delta s = 30.\ \text{km west}$ Find: $\bar{v} = ?$

$\Delta t = 0.50\ \text{hr}$

Refer to the *Mechanics* section in the *Reference Tables* to find an equation(s) that relates average velocity with displacement and time.

Solution:

$$\bar{v} = \frac{\Delta s}{\Delta t}$$
$$= \frac{30.\ \text{km west}}{0.50\ \text{hr}}$$
$$= 60\ \text{km/hr west}$$

Wrong Choices Explained:

(1) You multiplied instead of dividing the two quantities, and you did not indicate the direction, an essential component of a vector quantity.

(2) Correct magnitude, but you did not indicate the direction, an essential component of a vector quantity.

(3) You multiplied instead of dividing the two quantities.

6. **4** The acceleration of an object down an inclined plane is due to the component of the object's weight, which is parallel to the surface of the inclined plane. The steeper the hill, the greater the parallel component of the object's weight, the greater the acceleration. (For more information regarding the parallel

component see the answer to question 12.) Since each of the four choices involves an object with a mass of 1.0 kg, the object weights will all be equal, so the angle of the slope is the only variable determining the parallel component.

Use the mathematical relationship

$$\tan\theta = \frac{\text{opposite}}{\text{adjacent}}$$

to find that choice (4) has the steepest slope: 45°.

Wrong Choices Explained:

(1), (2), (3) These choices have slopes of 18°, 27°, and 34°, respectively.

7. **1** Write out the problem.

Given: $v_i = 16$ m/s Find: $\Delta s = ?$

$v_f = 0$ m/s

$\Delta t = 4.0$s

Refer to the *Mechanics* section in the *Reference Tables* to find an equation(s) that relates displacement with velocity and time.

Solution:

$$\bar{v} = \frac{v_i + v_f}{2} = \frac{16\text{ m/s} + 0\text{ m/s}}{2} = 8.0\text{ m/s}$$

$$\bar{v} = \frac{\Delta s}{\Delta t}$$

$$\Delta s = \bar{v} \cdot \Delta t = (8.0\text{ m/s})(4.0\text{ s}) = 32\text{ m}$$

8. **3** According to Newton's third law, if object *A* exerts a force on object *B*, then object *B* exerts an equal and opposite force back on object *A*. When the man in the stationary elevator steps on the scale, the scale exerts an upward force that supports him (the normal force). He, in turn, places a downward force on the scale. It is this force that registers his weight: 900 N.

Since the later reading of 1200 N on the scale shows an increase in the downward force that the man is applying, it follows that the scale is, in turn, exerting a greater upward force on the man. The same force that is accelerating the elevator upward is responsible for the increased upward force on the man.

Wrong Choices Explained:

(1) A constant downward acceleration would decrease the upward force on the man, and the scale reading would be less than 900 N.

(2), (4) If the elevator were moving at constant speed, there would be no net force on the man regardless of direction.

9. **2** Write out the problem.
According to Newton's second law of motion, $F_{net} = ma$.

Given: $F = m_1 a$ Find: $\frac{m_2}{m_1}$

$3F = m_2 2a$

Solution: $m_1 = \frac{F}{a}$ $m_2 = \frac{3F}{2a}$

$$\frac{m_1}{m_2} = \frac{\frac{F}{a}}{\frac{3F}{2a}} = \frac{F}{a} \times \frac{2a}{3F}$$

$$= \frac{2}{3} \text{ or } 2{:}3$$

10. **1** Newton's law of universal gravitation

$$F = \frac{Gm_1 m_2}{r^2}$$

is used to calculate the magnitude of gravitational forces between two masses.

Refer to the *Mechanics* section in the *Reference Tables* to find this equation.

Refer to the *List of Physical Constants* in the *Reference Tables* to find the values of the gravitational constant and the masses of the electron and proton. Substitute these values into the equation and solve.

$$F = \frac{\left(6.7 \times 10^{-11} \frac{\text{N}\cdot\text{m}^2}{\text{kg}^2}\right)(9.1 \times 10^{-31} \text{ kg})(1.7 \times 10^{-27} \text{ kg})}{(1.0 \times 10^{-10} \text{ m})^2}$$

$$= \frac{1.0 \times 10^{-67} \text{N}\cdot\text{m}^2}{1.0 \times 10^{-20} \text{m}^2}$$

$$= 1.0 \times 10^{-47} \text{N}$$

11. **3** Write out the problem.

Given: $g = 16 \text{ m/s}^2$ Find: $w = ?$
$m = 6.0 \text{ kg}$

Refer to the *Mechanics* section in the *Reference Tables* to find an equation(s) that relates weight with gravitational acceleration and mass.

Solution: $w = mg$
$= (6.0 \text{ kg})(16 \text{ m/s}^2)$
$= 96 \text{ N}$

12. **1** Three forces are acting on the book: the force due to gravity (weight), the normal force, and the force due to friction. See diagram below.

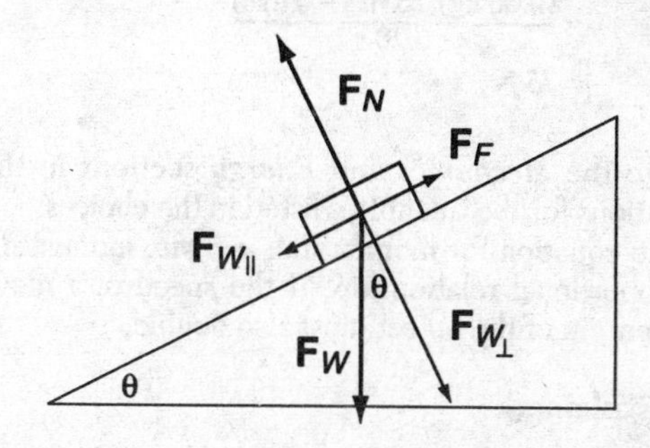

(Note that the force due to weight (gravity) has been resolved into its perpendicular and parallel components.)

According to Newton's first law, since the book is moving at constant velocity, there is no net force on the book. Therefore, the force of friction must be equal in magnitude and opposite in direction to the force pulling the book down the ramp($F_{w||}$) . Simple trigonometry yields the relationship $F_{||} = F \sin \theta$.

Solution: $F_{||} = (20. \text{ N})(\sin 30°)$
$= (20. \text{ N})(0.5000)$
$= 10. \text{ N}$

Wrong Choices Explained:

(2) You used cosine instead of sine.
(3) Correct magnitude, wrong direction.
(4) Wrong direction, and you used cosine instead of sine.

13. **4** Write out the problem.

Given: $m = 0.60$ kg Find: $F = ?$

$v_i = 0$ m/s

$\Delta t = 0.20$ s

$v_f = 25$ m/s

Refer to the *Mechanics* section in the *Reference Tables* to find an equation(s) that relates force with mass, speed, and time.

Solution:

$$F\,\Delta t = m\,\Delta v$$

$$F = \frac{m\,\Delta v}{\Delta t} = \frac{m(v_f - v_i)}{\Delta t}$$

$$= \frac{(0.60\text{ kg})(25\text{ m/s} - 0\text{ m/s})}{0.20\text{ s}}$$

$$= 75\text{ N}$$

14. **1** Refer to the *Mechanics* and *Energy* sections in the *Reference Tables* to find equations for the quantities listed in the choices.

According to the equation for momentum, $p = mv$, momentum and speed have a directly proportional relationship. If the speed of a moving object is doubled, the momentum of the object must also double.

Wrong Choices Explained:

(2) $KE = \frac{1}{2}mv^2$. If the speed of an object is doubled, the kinetic energy will quadruple.

(3) $\bar{a} = \frac{\Delta v}{\Delta t} = \frac{v_f - v_i}{\Delta t}$. The acceleration of an object need not double for the speed to double.

(4) $\Delta PE = mg\,\Delta h$. The speed of an object is not directly related to its potential energy.

15. **3** Write out the problem.

Given: $m = 3$ kg Find: $\Delta p = ?$

From the graph:

at $t = 0$ s, $v_i = 0$ m/s

at $t = 3$ s, $v_f = 20.$ m/s

Refer to the *Mechanics* section in the *Reference Tables* to find an equation(s) that relates momentum with mass and velocity.

Solution:
$$p = mv$$
$$\therefore \Delta p = m\,\Delta v$$
$$= m(v_j - v_i)$$
$$= (3\text{ kg})(20.\text{ m/s} - 0\text{ m/s})$$
$$= 60\frac{\text{kg}\cdot\text{m}}{\text{s}}$$

16. **1** According to Newton's third law, if object *A* exerts a force on object *B*, then object *B* exerts an equal and opposite force back on object *A*. Therefore, since the person in the question exerted a 48-N force on the door, the door exerts a 48-N force back on the person.

17. **2** Write out the problem.

Given: $m = 45$ kg Find: $P = ?$

$\bar{v} = 2.5$ m/s

$F = 85$ N

Refer to the *Energy* section in the *Reference Tables* to find an equation(s) that relates power with mass, velocity, and force.

Solution:
$$P = F\bar{v}$$
$$= (85\text{ N})(2.5\text{ m/s})$$
$$= 210\text{ W}$$

(Note that mass is not needed in this calculation.)

18. **4** Work is defined as the product of force and displacement, assuming that both are in the same direction. Holding an object stationary above the ground requires no work to be done on the object because there is zero displacement.

Wrong Choices Explained:

(1), (2), (3) In all three of these actions a force is being applied in the same direction as the displacement, so work is being done.

19. **1** Write out the problem.

Given: $m = 1.5$ kg
$h_{\text{refrigerator}} = 1.80$ m
$h_{\text{counter}} = 0.90$ m

Find: PE_{counter} compared to $PE_{\text{refrigerator}} = ?$

Refer to the *Energy* section in the *Reference Tables* to find an equation(s) that relates potential energy with mass and height.

Solution: $PE_1 = mgh_1$ $PE_2 = mgh_2$

Since the mass of the kitten and the acceleration due to gravity (g) do not change from the refrigerator to the counter, the only factor that changes the potential energy is the height. According to the equation, height is directly proportional to the potential energy; therefore, since the counter is half as high as the refrigerator, the kitten's gravitational potential energy on top of the counter is half as great.

Wrong Choice Explained:

(2) You compared the potential energy of the kitten on the refrigerator to the potential energy of the kitten on the counter.

20. **2** Write out the problem.

Given: $m = 60.$ kg
$v = 3.0$ m/s

Find: KE = ?

Refer to the *Energy* section in the *Reference Tables* to find an equation(s) that relates kinetic energy with mass and speed.

Solution:

$$\text{KE} = \frac{1}{2}mv^2$$
$$= \frac{1}{2}(60.\text{ kg})(3.0\text{ m/s})^2$$
$$= 270\text{ J}$$

21. **3** Write out the problem.

Given: $\bar{v} = 4.0$ m/s Find: $W = ?$
$F = 45$ N
$\Delta s = 5.5$ m

Refer to the *Energy* section in the *Reference Tables* to find an equation(s) that relates work with speed, force, and displacement.

Solution:
$$W = F\Delta s$$
$$= (45 \text{ N})(5.5 \text{ m})$$
$$= 250 \text{ J}$$

(Note that speed is not needed in this calculation.)

22. **4** Opposite charges attract. Draw vectors to represent the forces of attraction on the sphere from rod *A* and from rod *B*.

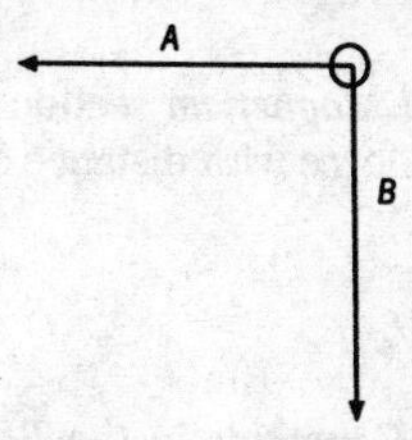

To find a *resultant vector*, place the vectors head to tail, and draw the resultant vector from the tail of the first vector to the head of the second vector.

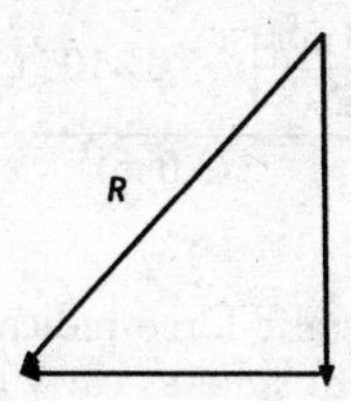

Vector (4) correctly represents the resultant electrostatic force on the sphere.

Wrong Choices Explained:

(1) You assumed a repulsive force instead of an attractive force.
(2) You forgot to include the force from rod *A*.
(3) You forgot to include the force from rod *A*, and you assumed a repulsive force instead of an attractive force.

23. **3** Refer to the *List of Physical Constants* in the *Reference Tables* to find the value of one elementary electric charge (1.6×10^{-19} C). Electrons carry negative elementary charges, and protons positive elementary charges. Since the total charge in this case is negative, it indicates an excess of electrons.

Solution:
$$-4.8\times10^{-19}\text{C}\left(\frac{1e^-}{-1.6\times10^{-19}\text{C}}\right)=3e^-$$

This sphere must have an excess of 3 electrons.

24. **1** Write out the problem.

Given: $r = 3.0$ m　　　Find: $F = ?$

$q_A = +5.0\times10^{-4}\text{C}$

$q_B = +3.0\times10^{-5}\text{C}$

Refer to the *Electricity and Magnetism* section in the *Reference Tables* to find an equation(s) that relates force with distance and charge.

Solution: $$F=\frac{kq_1q_2}{r^2}$$

Refer to the *List of Physical Constants* in the *Reference Tables* to find the value of k, the electrostatic constant.

$$F=\frac{\left(9.0\times10^9\,\frac{\text{N}\cdot\text{m}^2}{\text{C}^2}\right)(+5.0\times10^{-4}\text{C})(+3.0\times10^{-5}\text{C})}{(3.0\text{ m})^2}$$

$$=15\text{ N}$$

A positive value for an electrostatic force indicates a repulsive force. This is consistent with the fact that both spheres carry positive charges and that like charges repel.

Wrong Choices Explained:

(2) You forgot to square the distance.

(3) Correct magnitude; but like charges repel, opposites attract.

(4) You forgot to square the distance; and like charges repel, opposites attract.

25. **2** Field lines, lines drawn to visualize an electric field, represent the path that a very small positive test charge would take while in the field. Because in this problem the sphere creating the electric field is positive, and because like charges repel, the field lines should point away from the sphere. Diagram (2) correctly represents this situation.

26. **2** Refer to the *Electricity and Magnetism* section in the *Reference Tables* to find an equation that relates electric field intensity with potential difference.

$$E = \frac{V}{d}$$

This equation indicates a directly proportional relationship between E and V. Graph (2) correctly represents a directly proportional relationship between the electric field intensity between the plates and the potential difference across the plates.

Wrong Choices Explained:

(1) This graph indicates that increasing V has no effect on E.

(3) This graph indicates that E increases exponentially with increasing V.

(4) This graph shows an inversely proportional relationship between V and E.

27. **4** Write out the problem.

Given: $R = 2.0\ \Omega$ Find: $V = ?$

$q = 2.0\ \text{C}$

$\Delta t = 1.0\ \text{s}$

Refer to the *Electricity and Magnetism* section in the *Reference Tables* to find an equation(s) that relates potential difference with resistance, charge, and time.

Solution:

$$I = \frac{q}{\Delta t} = \frac{2.0\ \text{C}}{1.0\ \text{s}} = 2.0\ \text{A}$$

$$V = IR = (2.0\ \text{A})(2.0\ \Omega) = 4.0\ \text{V}$$

28. **2** Write out the problem.

Given: $R_1 = 4.0\ \Omega$ Find: $R_3 = ?$
$R_2 = 6.0\ \Omega$

$R_2 = 6.0\ \Omega$
series circuit
$V = 24\ V$
$I = 2.0\ A$

Refer to the *Electricity and Magnetism* section in the *Reference Tables* to find an equation(s) that relate total resistance in a series circuit with potential difference and current.

Solution:

$$V_t = I_t R_t$$
$$R_t = \frac{V_t}{I_t}$$
$$= \frac{24\ V}{2.0\ A}$$
$$= 12\ \Omega$$

$$R_t = R_1 + R_2 + R_3$$
$$R_3 = R_t - (R_1 + R_2)$$
$$= 12\ \Omega - (4.0\ \Omega + 6.0\ \Omega)$$
$$= 12\ \Omega - 10.\ \Omega$$
$$= 2\ \Omega$$

29. **3** The drawing in this question represents a parallel circuit. Refer to the *Electricity and Magnetism* section in the *Reference Tables* to find an equation for current in a parallel circuit and substitute the given values.

Solution:

$$I_t = I_1 + I_2$$
$$\therefore A_1 = A_2 + A_3$$
$$A_3 = A_1 - A_2$$
$$= 5.0\ A - 2.0\ A$$
$$= 3.0\ A$$

30. **3** Write out the problem.

Given: $I = 150\ A$ Find: $P = ?$
$V = 240\ V$

Refer to the *Electricity and Magnetism* section in the *Reference Tables* to find an equation(s) that relates power with current and potential difference.

Solution:

$$P = VI$$
$$= (240\ V)\ (150A)$$
$$= 3.6 \times 10^4\ W$$

31. **2** Write out the problem.

Given: $P = 75$ W   Find: $W = ?$
$V = 120$ V
$t = 60.$ min $= 3600$ s

Refer to the *Electricity and Magnetism* section in the *Reference Tables* to find an equation(s) that relates energy with power, potential difference and time.

Solution: $W = Pt$
$= (75 \text{ W})(3600 \text{ s})$
$= (75 \text{ J/s})(3600 \text{ s})$
$= 2.7 \times 10^5 \text{ J}$

(Note that potential difference (voltage) is not needed in this calculation.)

Wrong Choices Explained:

(1) You substituted 60 min for the time instead of 3600 s (the standard SI unit of time is the second).

(3) You substituted 60 min for the time instead of 3600 s (the standard SI unit of time is the second), and you multiplied by the voltage.

(4) You multiplied by the voltage.

32. **4** According to the law of conservation of energy, the potential difference across the entire circuit (V_t), supplied by the power source, is equal to the sum of the potential differences ($V_1, V_2, \ldots$) across all the resistances.

In a series circuit, therefore, all the potential differences across the resistors are added, and the sum must equal the potential difference of the source. Refer to the *Electricity and Magnetism* section in the *Reference Tables* to find the equation for potential difference in a series circuit. Circuit *A* is a correctly labeled possible series circuit.

In a parallel circuit, since each resistance comprises an independent path for the flowing charges so that, if one resistance ceases to operate, the others can continue to function, the potential difference in each path must equal the potential difference of the source. Refer to the *Electricity and Magnetism* section in the *Reference Tables* to find the equation for potential difference in a parallel circuit. Circuit *D* is a correctly labeled possible parallel circuit.

Therefore, in circuits *A* and *D* the readings of the voltmeters and the ammeter could be correct.

Wrong Choices Explained:

(1) Circuit *A* is correctly labeled; but in circuit *B*, a series circuit, the total potential across the resistances equals 100 V, while the source is providing only 50 V.

(2) In circuit B, a series circuit, the total potential across the resistances equals 100 V, while the source is providing only 50 V. In circuit C, a parallel circuit, each path indicates a potential difference of 50 V, yet the power source is providing 100 V.

(3) In circuit C, a parallel circuit, each path indicates a potential difference of 50 V, yet the power source is providing 100 V. Circuit D is represented correctly.

33. **1** To determine the location of the north pole of a solenoid, wrap the fingers of the left hand in the direction of the electron flow (use the right hand for conventional positive current). The thumb will point toward one end of the coil, which is the north pole. In this example, the north pole is located at the left end of the solenoid.

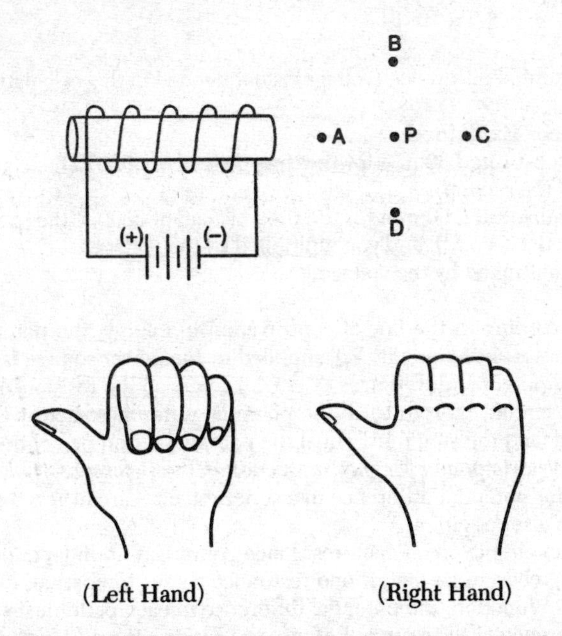

(Left Hand) (Right Hand)

Since opposite poles attract, the north pole of a compass placed at point P in the diagram will point toward the south pole of the solenoid, that is, toward point A.

34. **2** By convention, magnetic field lines used to represent magnetic flux are drawn so that they point away from the north and toward the south, and they form closed loops inside individual magnets. The diagram that follows shows the magnetic field around two bar magnets with opposite poles facing each other. The flux lines have the greatest density between the two facing poles.

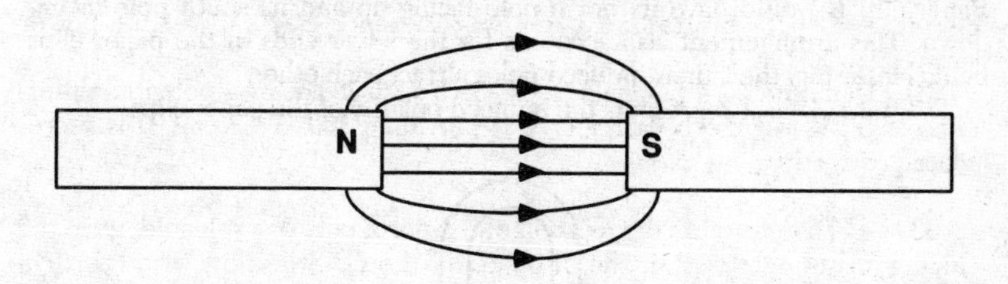

Therefore, the magnetic flux density due to the two magnets of equal strength is greatest at point *B* in the diagram in the question.

35. **1** A wave transfers energy, only. Although individual particles vibrate around their rest positions as the energy moves by, they do not move along with it, so there is no transfer of mass.

36. **4** The external magnetic field of the horseshoe magnet causes all the domains in the paper clips (areas where groups of atoms align their unpaired electrons so that they spin in the same direction) to align in the same direction.

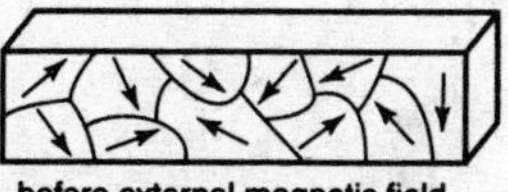
before external magnetic field

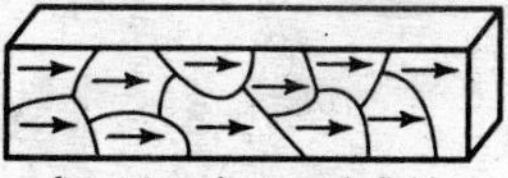
after external magnetic field

Since like poles repel and opposite poles attract, the induced field in paper clip *A* would have its south pole facing up and its north pole facing down. Paper clip *B* would have its north pole facing up and its south pole facing down. This arrangement also accounts for the lower ends of the paper clips being closer together; their induced poles attract each other.

Diagram (4) best represents the induced polarity of the paper clips.

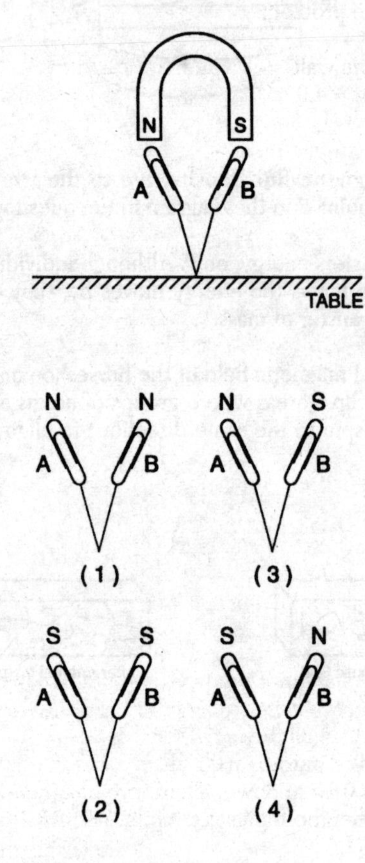

Wrong Choices Explained:

(1), (2), (3) The arrangements in these diagrams are not consistent with opposite poles attracting and like poles repelling.

37. **4** The speed of a mechanical wave depends solely on the medium it is traveling in; therefore, pulse *A* and pulse *B* are moving at the same speed. The amplitude of a wave is related to the energy carried by the wave; therefore, since pulse *B* has a larger amplitude than pulse *A*, it has more energy.
Compared to pulse *A*, pulse *B* has the same speed and more energy.

38. **2** Write out the problem.

Given: distance from wall to generator = 4.0 m Find: $f = ?$

Refer to the *Wave Phenomena* section in the *Reference Tables* to find an equation(s) that relates frequency with distance and speed.

Solution: $v = f\lambda$

The distance between two successive points on a periodic wave that are in phase is the wavelength, λ, of the wave. There are two sets of successive points in phase in the given diagram of a standing wave; therefore, since the distance from the wall to the generator is 4.0 m, the wavelength of a single wave is 2.0 m.

$$f = \frac{v}{\lambda}$$

$$= \frac{10.\text{ m/s}}{2.0\text{ m}}$$

$$= 5.0\text{ Hz}$$

39. **3** Two or more waves passing simultaneously through the same area of a medium affect the medium independently but they do not affect each other. The resultant displacement of any point in the medium is the algebraic sum of the displacements of all the individual waves.
Since, in this case, one pulse is 0.05 m up and the other is 0.15 m down, the net displacement is 0.10 m down. Therefore, diagram (3) best represents the shape of the rope when both pulses are in region *AB*.

Wrong Choices Explained:

(1) Correct magnitude, but you forgot direction or made a mistake when considering it.

(2), (4) You simply added the two amplitudes with no regard for direction.

40. **1** Resonance is defined as the spontaneous vibration of an object at a frequency equal to that of the wave that initiates the resonant vibration.

Wrong Choices Explained:

(2) Doppler effect: an apparent change in frequency that results when a wave source and an observer are in relative motion with respect to each other.

(3) Reflection: when a wave travels from one medium into another, part of the energy of the wave is transmitted to the new medium and part moves back into the original medium (i.e., it is reflected) with the same frequency.

(4) Interference: the result of the combination of two or more waves that pass simultaneously through the same area of a medium.

41. **2** The distance between two successive points on a periodic wave that are in phase is equal to the wavelength of the wave. Alternatively, it can be said that one wavelength incorporates one full crest and one full trough.

Therefore, waves *A* and *C* have the same wavelength.

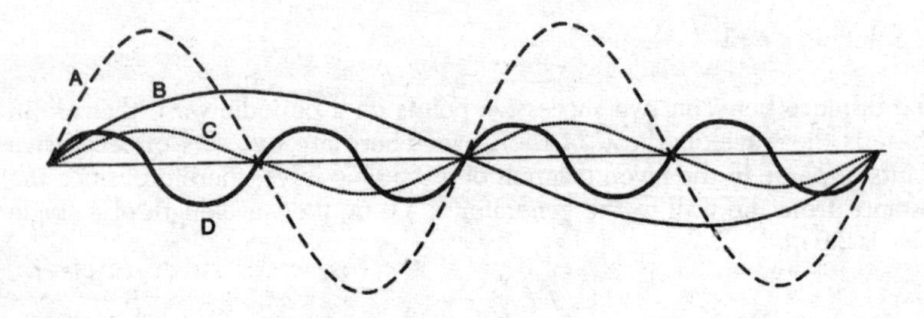

Wrong Choices Explained:

(1), (3), (4) Wave *B* has twice the wavelength of wave *A*, wave *B* has four times the wavelength of wave *D*, wave *C* has twice the wavelength of wave *D*.

42. **2** Period is defined as the time for one complete repetition of a periodic phenomenon. In this instance, a complete repetition would be the time for one whole wave (i.e., a wavelength).

In the time span represented in the diagram, wave *B* completes only one cycle while waves *A* and *C* complete two cycles and wave *D* completes four cycles. Therefore, wave *B* has the longest period.

Wrong Choices Explained:

(1), (3) Waves *A* and *C* have periods that are half as long as the period for wave *B*.

(4) Wave *D* has the shortest period; it completes four cycles in the time wave *B* takes to complete one.

43. **4** The law of reflection states that the angle of incidence is equal to the angle of reflection. The angles are measured with respect to a normal (a perpendicular line drawn to the surface).

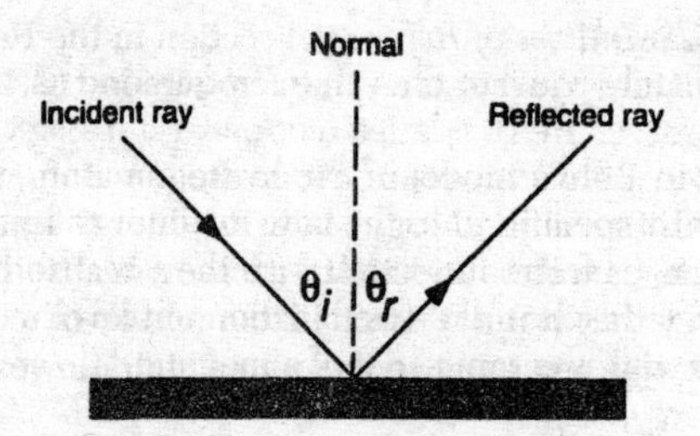

Since the angle of incidence in the question is 35°, the angle of reflection is also 35°. The angle between the incident ray and the reflected ray is the sum of their angles, or 70°.

Wrong Choices Explained:

(1) A 0° angle results only if the incident ray strikes the surface along the normal (i.e., at 0°).

(2) You simply stated the angle of reflection. Read the question carefully.

(3) You subtracted 35° from 90° and obtained the angle that the ray makes with the surface.

44. **3** Refer to the *Absolute Indices of Refraction* section in the *Reference Tables*. (Hint: The problem states that the ray has a wavelength of 5.9×10^{-7} m and mentions refraction; both items are used to identify the required reference.) Find the index of refraction for air (1.00), and identify medium *X* based on its index of refraction, which you will calculate.

Given: $\theta_{air} = 30.°$ Find: identity of medium X = ?

$n_{air} = 1.00$

$\theta_X = 12°$

Refer to the *Wave Phenomena* section in the *Reference Tables* to find an equation(s) that relates angles with absolute indices of refraction.

Solution:

$$n_1 \sin\theta_1 = n_2 \sin\theta_2$$

$$\therefore n_{air} \sin\theta_{air} = n_x \sin\theta_x$$

$$n_x = \frac{n_{air} \sin\theta_{air}}{\sin\theta_x}$$

$$= \frac{(1.00)(\sin 30.°)}{\sin 12°}$$

$$= \frac{(1.00)(0.5000)}{0.2079}$$

$$= 2.4$$

Refer to the *Absolute Indices of Refraction* section in the *Reference Tables*. This value is approximately equal to the value for diamond (2.42).

45. **1** According to Bohr's model of the hydrogen atom, an electron can be associated only with specific energy states or energy levels in an atom. When an electron changes from one state to another, it absorbs or emits specific amounts of energy that bring it exactly from one level to another, not in between. If the electron is returning to its ground state (lowest energy level), it will emit or lose energy.

Refer to the *Energy Level Diagrams for Mercury and Hydrogen* in the *Reference Tables* to find the acceptable energy levels for hydrogen atom. To go from level $n = 2$ (–3.40 eV) to level $n = 1$ (–13.60 eV), the ground state, the electron in the hydrogen atom would have to emit, or lose, 10.20 eV. Choice (1) correctly defines this change.

Wrong Choices Explained:

(2) Correct magnitude; but an electron in an atom returning to the ground state loses energy.

(3) Level $n = 2$ has an energy state of –3.40 eV, but no two levels have a difference of 3.40 eV between them.

(4) Level $n = 2$ has an energy state of –3.40 eV, but no two levels have a difference of 3.40 eV between them; also, when returning to the ground state, electrons in an atom lose energy.

46. **4** Dispersion is the separation of polychromatic light into its individual colors. Different colors of light have different frequencies, and the speed of a wave in a dispersive medium, such as glass, depends on its frequency.

When a ray of light strikes a surface between two media obliquely (at an angle), it will change direction (be refracted). The slower the wave, the smaller the angle of refraction. Different colors will therefore have different angles of refraction.

Diagram (4) correctly represents light from a polychromatic light source (sunlight) striking a glass surface obliquely, and thus this setup could be used to demonstrate the phenomenon of dispersion.

Wrong Choices Explained:

(1) This diagram shows a monochromatic light source, and the light is not striking the surface of the glass obliquely.

(2) This diagram shows a monochromatic light source.

(3) This diagram shows a ray that is not striking the surface of the glass obliquely.

47. **4** The Doppler effect is an apparent change in frequency that results when a wave source and an observer are in relative motion with respect to each other. A star moving away from Earth and emitting light waves from the

hydrogen spectrum would appear to be emitting these waves at a lower frequency; that is, the wavelengths would appear to be longer (very much like a source of sound moving away from an observer, illustrated in the diagram below).

Since all of the light waves would be emitted at the same time and would be moving away from Earth at the same speed, they would each experience the same shift. Spectrum 4 correctly shows an equal increase (0.2) in wavelength for all four lines of the hydrogen spectrum.

Wrong Choices Explained:

(1) This spectrum would be correct had the star been moving toward Earth.

(2), (3) These spectrums are incorrect because they show unequal shifts for the four lines.

48. **1** When a ray of light strikes a surface between two media obliquely (at an angle), it will change direction (be refracted). Flint glass has a greater index of refraction (1.61) than air (1.00) (refer to the *Absolute Indices of Refraction* section in the *Reference Tables*); therefore, the ray in the flint glass will bend toward the normal.

When the ray strikes the plane mirror, the law of reflection applies and the angle of incidence will be equal to the angle of reflection. When the ray strikes the glass-air surface again, it will bend away from the normal because air has a lower index of refraction than flint glass. Diagram (1) correctly represents these transitions.

Wrong Choices Explained:

(2) This diagram shows incorrect refraction between the air and the flint glass media.

(3) This diagram shows no refraction at all.

(4) This diagram shows incorrect refraction at the flint glass-air surface (the fourth arrow).

49. **4** Alpha particles are positively charged, as Rutherford had determined in earlier experiments. Like charges repel; therefore, for positively charged alpha particles to be scattered at large angles, these alpha particles would have to have been repelled by the gold's positive atomic nuclei.

50. **3** Refer to the *Modern Physics* section in the *Reference Tables* to find an equation(s) dealing with kinetic energy and photons.

$$KE_{max} = hf - W_o \qquad E_{photon} = hf$$

According to the above equations, the maximum kinetic energy of an electron ejected from a metal by a photon depends on both the photon's frequency and the metal's work function.

51. **3** The laws of conservation of energy and conservation of momentum hold for all collisions, including photon-electron collisions. This phenomenon, which was observed by Arthur Compton, a U.S. physicist, is known as the Compton effect.

52. **1** When a ray of light strikes an interface with an angle of incidence greater than the critical angle for the medium, total internal reflection takes place and the law of reflection governs the path of the reflected ray. Diagram (1) correctly represents the path of the light rays experiencing total internal reflection.

Wrong Choices Explained:

(2) This diagram shows no refraction or reflection.

(3) This diagram shows normal refraction for a flint glass-air interface for an angle of incidence less than the critical angle.

(4) This diagram shows normal refraction (90°) for a flint glass-air interface for an angle of incidence equal to the critical angle.

53. **2** The resistance of a material depends on (1) the nature of the material, (2) the geometry of the conductor, and (3) the temperature at which the resistance is measured. The resistance of conductors, such as those in a wire filament, increases with increasing temperature. This fact also makes sense because increasing the temperature of a conductor increases the vibrational

kinetic energy of its atoms, making collisions with electrons more likely and thus impeding the flow.

54. **3** According to the law of conservation of momentum, the total momentum of two objects not subjected to outside forces before they collide will be equal to the total momentum of the two objects after they collide.

Therefore, the momentum of the clay-block system after the collision is the same as the total momentum of the clay and the block before the collision.

Wrong Choice Explained:

(1) You were thinking of kinetic energy. In an inelastic collision such as the one shown in this question, kinetic energy is not conserved; some is lost to heat energy.

55. **1** When electrons are ejected from a photoemissive surface, the color of the incident light determines how energetic the electrons are, and the intensity determines the rate at which the electrons are ejected. Decreasing the intensity of the light will decrease the rate (i.e., the number of photoelectrons emitted per second).

PART II

GROUP 1—**Motion in a Plane**

56. **3** To know how far above the level ground the snowball hits the smokestack, you need to calculate how far it falls in the 0.65 s before it hits. Motions in the horizontal and vertical directions are independent of each other. Since it is the vertical direction you are interested in, you must consider that an object falling freely near the surface of Earth is subject to a constant acceleration due to Earth's gravity (refer to the *List of Physical Constants* in the *Reference Tables*) and that initially this object has no velocity in the vertical direction.

Given: $\Delta s_{yi} = 4.5$ m Find: height above level ground = ?

$v_{iy} = 0$ m/s

$\Delta t = 0.65$ s

$a_y = 9.8$ m/s^2

Refer to the *Mechanics* section in the *Reference Tables* to find an equation(s) that relates displacement with acceleration and time.

Solution:

$$\Delta s_y = v_{iy}\,\Delta t + \frac{1}{2}a_y(\Delta t)^2$$
$$= (0\text{ m/s})(0.65\text{ s}) + \frac{1}{2}(9.8\text{ m/s}^2)(0.65\text{ s})^2$$
$$= 2.1\text{ m}$$

Subtract the 2.1 m that the snowball fell from the initial 4.5 m above the ground at which it started to find that the snowball hits the smokestack 2.4 m above the level ground

57. 4 Motions in the horizontal and vertical directions are independent of each other. In this case, no external forces are acting in the horizontal direction, and therefore velocity is constant.

Given: $\Delta s_x = 15$ m
$\Delta t = 0.65$ s

Find: $v_x = ?$

Refer to the *Mechanics* section in the *Reference Tables* to find an equation(s) that relates velocity with displacement and time.

Solution:

$$\bar{v}_x = \frac{\Delta s_x}{\Delta t}$$
$$= \frac{15\text{ m}}{0.65\text{ s}}$$
$$= 23\text{ m/s}$$

58. 4 Write out the problem.

Given: $m = 2.0$ kg
constant speed
$r = 3.0$ m
$F_c = 24$ N

Find: $a_c = ?$

Refer to the *Motion in a Plane* section in the *Reference Tables* to find an equation(s) that relates centripetal acceleration with mass, radius, and centripetal force.

Solution:

$$F_c = \frac{mv^2}{r}$$

$$a_c = \frac{v^2}{r}$$

$$\therefore F_c = ma_c$$

$$a_c = \frac{F_c}{m}$$

$$= \frac{24\text{ N}}{2.0\text{ kg}}$$

$$= 12\text{ m/s}^2$$

By definition, centripetal acceleration always points toward the center of the circular path; therefore, in this case it is directed toward point *D*.

59. **2** At any point in a circular path the velocity of an object is tangent to the circle. Therefore, in the diagram the velocity is directed toward point *B*.

By definition, the centripetal force is directed toward the center of the circle; in this case that is toward point *D*.

60. **1** Write out the problem.

Given: $m = 2.0\text{ kg}$ Find: $v = ?$

constant speed

$r = 3.0\text{ m}$

$F_c = 24\text{ N}$

Refer to the *Motion in a Plane* section in the *Reference Tables* to find an equation(s) that relates speed with mass, radius, and centripetal force.

Solution:

$$F_c = \frac{mv^2}{r}$$

$$v^2 = \frac{F_c r}{m}$$

$$v = \sqrt{\frac{F_c r}{m}}$$

$$= \sqrt{\frac{(24\text{ N})(3.0\text{ m})}{2.0\text{ kg}}}$$

$$= 6.0\text{ m/s}$$

Wrong Choices Explained:

(2) You substituted the values incorrectly by placing 2.0 in the numerator and 3.0 in the denominator (you can avoid this mistake by always using units), and you forgot to take the square root and solved for v^2 instead of v.

(3) You forgot to take the square root and solved for v^2 instead of v.

(4) You substituted the values incorrectly by placing 2.0 in the numerator and 3.0 in the denominator. (You can avoid this mistake by always using units.)

61. **3** Velocity is a vector quantity, and therefore arrows can be used to represent the horizontal and vertical velocities. To find a *resultant vector*, place the given vectors head to tail, and draw the resultant vector from the tail of the first vector to the head of the second vector.

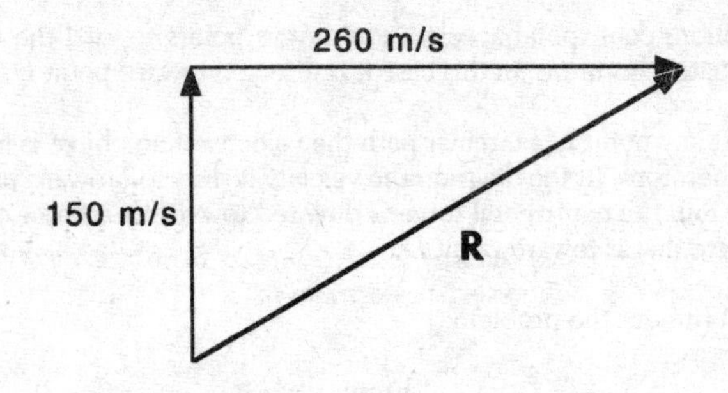

Since the result is a right triangle, use the Pythagorean theorem

$$a^2 + b^2 = c^2$$

to determine the magnitude of the resultant.

$$\begin{aligned} R^2 &= (150 \text{ m/s})^2 + (260 \text{ m/s})^2 \\ &= 22{,}500 \text{ m}^2/\text{s}^2 + 67{,}600 \text{ m}^2/\text{s}^2 \\ &= 90{,}100 \text{ m}^2/\text{s}^2 \\ R &= 300 \text{ m/s} \\ &= 3.0 \times 10^2 \text{ m/s} \end{aligned}$$

62. **4** The only acceleration that the projectile is experiencing is acceleration due to Earth's gravity. This acceleration always points downward, toward the ground.

Vector (4) best represents the acceleration of the projectile in the position shown.

63. **3** Since the planet moves from point A to point B in the same amount of time it takes to move from point C to point D, the planet must be moving faster between C and D because it has a greater distance to cover in the same amount of time. Therefore, since it moves faster between C and D than between A and B, the kinetic energy, which is proportional to the square of the speed, must be greater at point C than at point B. Refer to the *Energy* section in the *Reference Tables* to find the equation for kinetic energy:

$$\text{KE} = \frac{1}{2}mv^2$$

Potential energy is dependent on height or, in this case, the distance between the planet and the Sun. At point C the planet is closer to the Sun than at point B; therefore at point C the planet has less potential energy than at point B. Refer to the *Energy* section in the *Reference Tables* to find the equation for potential energy:

$$\text{PE} = mgh$$

According to the law of conservation of energy, this makes sense: as the planet's kinetic energy increases as it moves from point B to point C, its potential energy decreases (energy is transformed from potential to kinetic; that is, it is conserved, not created or destroyed).

64. **3** Kepler's second law states that each planet moves in such a way that an imaginary line drawn from the Sun to the planet sweeps out equal areas of space in equal periods of time. According to the information presented, the sweep from A to B takes the same time as the sweep from C to D; therefore, according to Kepler's second law the area of region ABS is the same as the to the area of region CDS.

65. **1** Kepler's third law states that the cube of a planet's average distance from the sun (r^3) divided by the square of its period (T^2) is a constant for all planets. Planet A is the farthest away from the star. Therefore, according to Kepler's third law, planet A has the greatest orbital period.

GROUP 2—**Internal Energy**

66. **2** Temperature is defined as a measure of the average kinetic energy ($\overline{\text{KE}}$) of the particles in a sample. Absolute temperature or Kelvin temperature is directly proportional to the average kinetic energy of the particles in a sample.

Graph (2) best represents a directly proportional relationship between the average kinetic energy of the random motion of the particles of an ideal gas and its absolute temperature.

Wrong Choices Explained:

(1) This graph indicates an exponential relationship between temperature and average kinetic energy.

(3) This graph indicates that an inverse proportion exists between temperature and average kinetic energy.

(4) This graph indicates no relationship between temperature and average kinetic energy.

67. **4** To determine Kelvin temperature from Celsius degrees, add 273 to the Celsius value.

$$\begin{aligned} K &= {}^\circ C + 273 \\ &= 37^\circ C + 273 \\ &= 310. \end{aligned}$$

The normal body temperature of 37° Celsius is equivalent to 310. K.

Wrong Choice Explained:

(3) You subtracted 37 from 273.

68. **3** According to the second law of thermodynamics, heat will flow spontaneously from a region of higher kinetic energy to a region of lower kinetic energy. If the respective temperatures of two objects are known, it can be determined whether a heat exchange would take place if the objects were in contact. (Heat exchange will occur if the objects are at different temperatures.)

Wrong Choices Explained:

(1), (2), (4) To calculate any of these quantities, you would at least need to know the mass of each object as well as the nature (chemical composition) of each object.

69. **1** The amount of heat required to change 1 gram of a substance from the liquid phase to the gaseous phase is known as its heat of vaporization. Refer to the *Heat Constants* section in the *Reference Tables* to find the heat of vaporization for each of the four choices. Of the four metals, aluminum has the highest heat of vaporization (10,500 kJ/kg); and since all four samples have the same mass, it therefore requires the most heat to change from the liquid to the gaseous phase.

70. **3** As heat is added, the temperature of the ice-water mixture will not increase because temperature remains constant during a phase change until all of the substance has changed phase. In this case, once the ice has melted into water and the phase change is complete, as heat is added the temperature will rise. When the temperature reaches the boiling point of water, the temperature will again remain constant, as heat is added, until all of the water has changed to steam.

Graph (3) best represents the relationship between the temperature of the mixture and the heat added.

71. **1** According to the second law of thermodynamics, heat will flow spontaneously from a region of higher temperature to a region of lower temperature. When equilibrium has been reached, the final temperature of the mixture will be somewhere between the initial temperatures of the ice (0.0°C) and of the water (25°C).

In this case, heat will flow from the 0.250 kg of water (lowering the temperature of the water), to the 0.060-kg ice cube, first melting all the ice, and then raising the temperature of the now 0.060 kg of water until the two temperatures are equal.

Choice 1 correctly states the conditions of this system when equilibrium is reached.

72. **3** Write out the problem.

Given: $m_{\text{liquid}} = 6.00 \text{ kg}$ Find: $H_{f\text{liquid}} = ?$

$Q_{\text{liquid}_{\text{freeze}}} = Q_{\text{ice}_{\text{melt}}}$

$m_{\text{ice}} = 3.00 \text{ kg}$

Refer to the *Internal Energy* section in the *Reference Tables* to find an equation(s) that relates heat of fusion with mass and heat.

Solution: $Q = mH_f$

Since $Q_{\text{liquid}} = Q_{ice}$:

$$m_{\text{liquid}} H_{f\text{liquid}} = m_{\text{ice}} H_{f\text{ice}}$$

$$H_{f\text{liquid}} = \frac{m_{\text{ice}} H_{f\text{ice}}}{m_{\text{liquid}}}$$

Refer to the *Heat Constants* section in the *Reference Tables* to find the heat of fusion for ice.

$$H_{f\text{liquid}} = \frac{(3.00 \text{ kg})(334 \text{ kJ/kg})}{6.00 \text{ kg}}$$

$$= 167 \text{ kJ/kg}$$

73. **4** The melting of lead represents a phase change. During a phase change, the energy released or, in this case, absorbed is associated with changes in the average potential energy of the molecules of a substance.

Wrong Choices Explained:

(1) Temperature remains constant during a phase change.

(2) Heat of fusion is a constant.

(3) Temperature remains constant during a phase change; and if temperature doesn't change, kinetic energy doesn't change.

74. **2** When a substance (or substances) is dissolved in water, the particles of the substance(s) interfere with the freezing and boiling process. As a consequence, the resulting solution has a lower freezing point and a higher boiling point than the pure water.

Choice 2 correctly states that ocean water (a solution) will freeze at a lower temperature and boil at a higher temperature than distilled (pure) water.

75. **3** The first law of thermodynamics states that the change in internal energy of a system (ΔU) is equal to the heat (Q) that the system absorbs (or releases) minus the work (W) the system does (or has done to it). If all the added heat is converted to work, then $Q - W$, or ΔU, the change in internal energy, will equal zero; therefore, in the case in question, the internal energy of the gas will remain the same.

GROUP 3—**Electromagnetic Applications**

76. **4** First use one of the left-hand rules to determine the direction of the magnetic field around the conductor. Point the left thumb in the direction of the electron flow; the fingers of the left hand (from wrist to fingertips) will curl in the direction of the magnetic field. At the point where the electron is located, the magnetic field is directed out of this page.

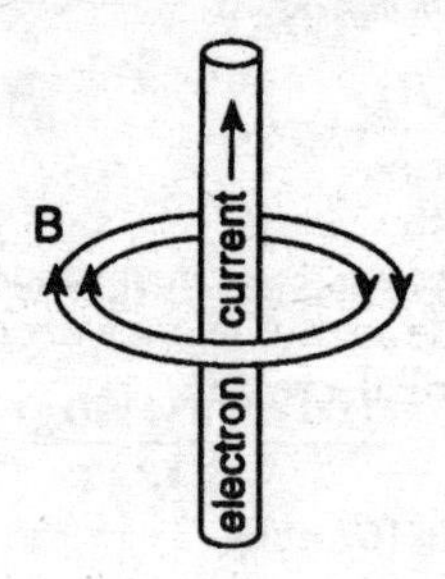

Next, determine the resulting force on the electron by using another left-hand rule. Point the thumb of the left hand in the direction of the velocity of the electron, and point the fingers in the direction of the magnetic field (at the location of the electron, the magnetic field points out of the paper). The force on the electron points away from the palm of the left hand. In this case the palm points toward the left side of the page.

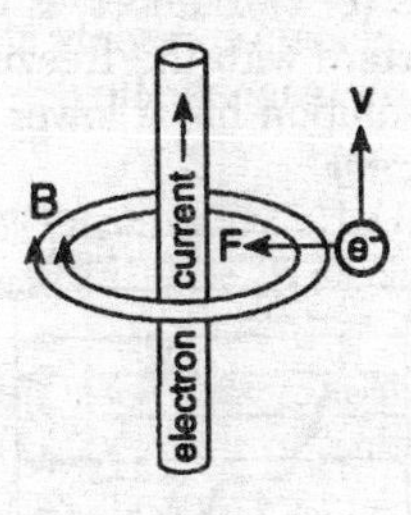

Diagram (4) best represents the resulting magnetic field (*B*) around the conductor and the direction of the magnetic force (*F*) on the free electron.

77. **1** Refer to the *Electromagnetic Applications* section in the *Reference Tables* to find an equation(s) that relates force with speed, particle charge, and magnetic field strength.

$$F = qvB$$

From this equation it can be seen that magnetic force is directly proportional to charge, speed, and flux density. Therefore, the magnitude of the magnetic force on the particle will increase if the flux density of the magnetic field increases.

Wrong Choices Explained:

(2) If time of travel of the charge increases for the same distance, the speed of the particle decreases and the force will decrease.

(3), (4) If the magnitude of the charge decreases or the speed of the charge decreases, the force will decrease.

78. **2** If a wire carrying a current is placed in a magnetic field, so that the direction of the current is perpendicular to the direction of the magnetic field, the magnetic field of the wire will interact with the magnetic field of the magnet to produce a force on the wire. If, however the current is parallel to the magnetic field, no force will be present on the wire.

A current-carrying loop of wire in an external magnetic field is simply an arrangement in which wire is both perpendicular and parallel to the external magnetic field. The diagram below represents a current-carrying rectangular loop of wire placed in a uniform magnetic field.

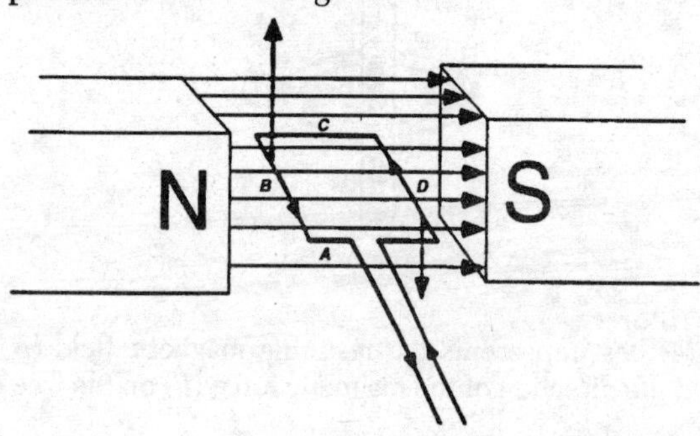

The sections of the loop in the diagram that are parallel to the magnetic field (A and C) will experience no force. If the right-hand rule is applied to section B of the loop, it experiences a force directly upward out of the page. Similarly, section D will experience a force directly downward into the page. The net result of these two forces is to rotate the loop.

Forces that produce rotational motion are called *torques*. Therefore, the torque experienced by a current-carrying loop of wire in an external magnetic field is due to the interaction of the external magnetic field and the magnetic field produced by the current in the loop.

79. **4** An ammeter is a galvanometer that has been modified in order to measure larger currents. A low-resistance device, known as a shunt, is placed in parallel with the wires of the galvanometer.

Wrong Choices Explained:

(1) The mass spectrometer is a device that uses a magnetic field to separate charged particles of differing masses.

(2) A transformer is a device that uses a changing magnetic field to increase or decrease the potential difference between a primary and a secondary circuit.

(3) A voltmeter is a device used to measure potential difference and is constructed by placing a large resistor in series with the coil of a galvanometer.

80. **1** A back emf is a potential difference that develops in a circuit that opposes the potential difference of the source. As a result, the net potential difference in the circuit is less; and the armature current, which is proportional to the potential difference, decreases.

81. **2** Thermionic emission is the emission of electrons from substances such as metallic filaments when these substances are heated. It follows, then, that as the temperature of a surface increases, electrons are emitted at a higher rate.

82. **3** The diagram below represents a mass spectrometer.

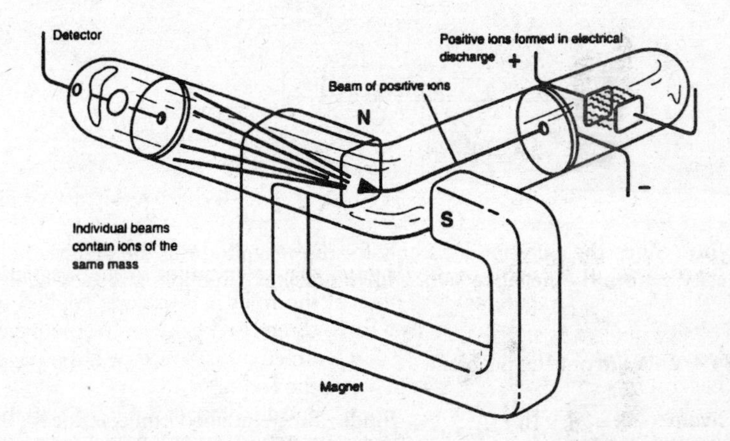

The positive ions formed by electric discharge are initially accelerated by an electric field and then deflected by a magnetic field. Therefore, in an operating mass spectrometer, the motion of the positive ion beams is influenced by both electric and magnetic fields.

83. **2** Write out the problem.

Given: $m = 3.2 \times 10^{-14}$ kg Find: $E = ?$

$q = -1.6 \times 10^{-18}$ C

$F_{\text{electric}} = F_{\text{gravity}}$ at equilibrium

F_{gravity} is the weight, w, of the oil drop

$\therefore F_{\text{electric}} = w$ at equilibrium

Refer to the *Mechanics* section in the *Reference Tables* to find an equation(s) for weight.

$$w = mg$$

Refer to the *List of Physical Constants* in the *Reference Tables* to find the value for g.

Refer to the *Electricity and Magnetism* section in the *Reference Tables* to find an equation(s) that relates electric field intensity to force and charge.

Solution:

$$E = \frac{F}{q}$$

$$F = F_{gravity} = w = mg$$

$$\therefore E = \frac{mg}{q}$$

$$= \frac{(3.2 \times 10^{-14} \text{ kg})(9.8 \text{ m/s}^2)}{-1.6 \times 10^{-18} \text{ C}}$$

$$= -2.0 \times 10^5 \text{ N/C}$$

Note: Since the question asks only for the magnitude of the electric field intensity, ignore the negative sign, which indicates direction, and pick choice (2).

84. **2** Write out the problem.

Given:

$B = 2.0 \times 10^{-2}$ T

$\ell = 2.0$ m

$v = 3.0$ m/s

Find: Since induced emf is induced potential difference, V = ?

Refer to the *Electromagnetic Applications* section in the *Reference Tables* to find an equation(s) that relates induced potential difference with flux density, length, and speed.

Solution:

$$V = B\ell v$$

$$= (2.0 \times 10^{-2} \text{ T})(2.0 \text{ m})(3.0 \text{ m/s})$$

$$= 0.12 \text{ V}$$

85. **3** Write out the problem.

Given: $N_p = 2$ Find: $I = ?$

$N_s = 4$

$V_p = 12\text{ V}$

$R = 8.0\ \Omega$

Refer to the *Electromagnetic Applications* and the *Electricity and Magnetism* sections in the *Reference Tables* to find an equation(s) that relates current with number of turns of coil, potential difference, and resistance.

Solution:

$$\frac{N_p}{N_s} = \frac{V_p}{V_s} \qquad R = \frac{V}{I}$$

$$V_s = \frac{V_p N_s}{N_p} \qquad I = \frac{V}{R}$$

$$= \frac{(12\text{ V})(4)}{2} \qquad = \frac{24\text{ V}}{8.0\ \Omega}$$

$$= 24\text{ V} \qquad = 3.0\text{ A}$$

GROUP 4—**Geometric Optics**

86. **2** Write out the problem.

Given: $S_o = 0.020\text{ m}$ Find: $f = ?$

$d_o = 0.15\text{ m}$

$d_i = 0.30\text{ m}$

(Note that, by convention, the image distance is designated as positive because the image is located in front of the mirror, on the reflecting side, and is considered a real image.)

Refer to the *Geometric Optics* section in the *Reference Tables* to find an equation(s) that relates focal length with distance and size.

Solution:

$$\frac{1}{f} = \frac{1}{d_o} + \frac{1}{d_i}$$

$$= \frac{1}{0.15\text{ m}} + \frac{1}{0.30\text{ m}}$$

$$= \frac{2}{0.30\text{ m}} + \frac{1}{0.30\text{ m}}$$

$$= \frac{3}{0.30\text{ m}}$$

$$f = \frac{0.30\text{ m}}{3}$$

$$= 0.10\text{ m}$$

(Note that the size of the object is not needed in this calculation.)

87. **2** Individual converging mirrors can produce real, inverted images or virtual, erect images. Since the image is located in front of the mirror, on the reflecting side, it is considered a real image and is therefore inverted.

88. **4** Write out the problem.

Given: $S_o = 0.020$ m Find: $S_i = ?$
$d_o = 0.15$ m
$d_i = 0.30$ m

(Note that, by convention, the image distance is designated as positive because the image is located in front of the mirror, on the reflecting side, and is considered a real image.)

Refer to the *Geometric Optics* section in the *Reference Tables* to find an equation(s) that relates image size with distance and object size.

Solution:
$$\frac{S_o}{S_i} = \frac{d_o}{d_i}$$
$$S_i = \frac{S_o d_i}{d_o}$$
$$= \frac{(0.20\text{ m})(0.30\text{ m})}{0.15\text{ m}}$$
$$= 0.040\text{ m}$$

Wrong Choice Explained:

(1) You incorrectly substituted the value of d_o for d_i and the value of d_i for d_o.

89. **3** When a ray parallel to the principal axis strikes the surface of a convex mirror, it is reflected back from the surface of the mirror in such a way that, if it is projected back behind the mirror, it passes through a virtual focus. This reflective light ray will therefore pass through point *D*. See the diagram that follows.

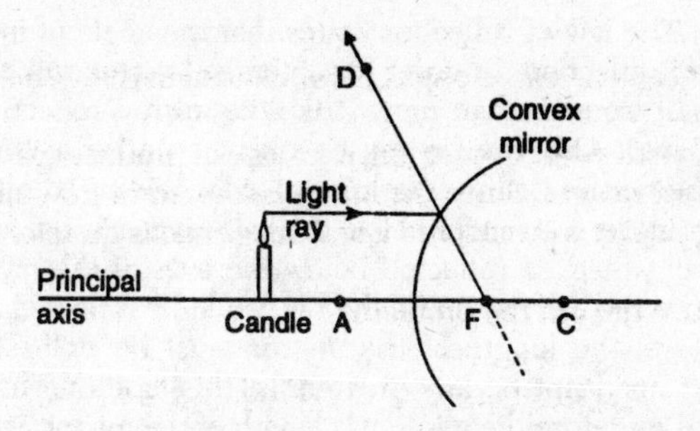

90. **4** Single convex lenses can produce real, inverted images if the object is at a distance greater than one focal length from the lens or virtual, erect images if the object is less than one focal length away from the lens. In this case, since the diagram indicates that the candle (object) is located at a distance greater than the focal length of the lens, you may conclude that the lens is being used to produce a real, inverted image of the candle.

91. **4** A ray, such as *RF*, that strikes the principal axis at one focal length from the lens will emerge parallel to the principal axis because the light is refracted as it passes through the lens.

Wrong Choices Explained:

(1) Reflection occurs in mirrors, not in lenses.

(2) Polarization is the separation of a beam of light so that the vibrations are in one plane. A convex lens is not a polarizing filter and will not polarize light.

(3) None of the three rays drawn to locate an image ever passes through point 2*F*.

92. **1** Refer to the *Geometric Optics* section in the *Reference Tables.*

$$\frac{1}{f}=\frac{1}{d_o}+\frac{1}{d_i} \qquad \frac{S_o}{S_i}=\frac{d_o}{d_i}$$

The focal length of a lens is a constant. If an object is moved further away, that is, if the object distance is increased, in order for the first equation to remain an equality the image distance must become smaller.

The second equation states that the ratio of object size to image size is equal to the ratio of object distance to image distance. If, as concluded above, the image distance decreases as the object distance increases, then, for the ratios to remain equal, the image size must decrease since the object size is not changing.

93. **1** The law of reflection states that the angle of incidence is equal to the angle of reflection. In order for a person to see his or her feet in a mirror, a ray of light from the feet must strike the mirror, reflect off the mirror, and strike the eyes. Also, the ray must strike the mirror at the level of the midpoint of the person (halfway up) in order for the ray to reach the eyes. If the ray were to strike the mirror below the halfway mark, the angle would be such that the ray would be reflected below the eyes; if the ray were to strike the mirror above the halfway mark, the ray would be reflected above the eyes.

Therefore, the length of the mirror must be at least half the person's height but need not be any greater. In this case one-half of the student's height, and therefore the minimum length of the mirror, is 1.0 m.

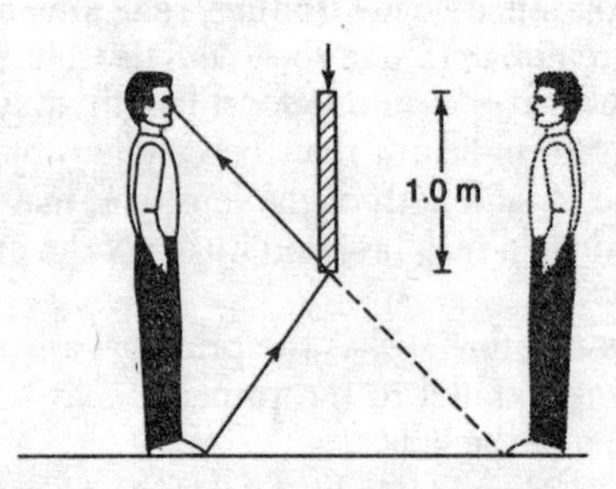

94. **2** If a ray of light passes through the principal focus of a concave mirror and strikes the surface, the reflected ray emerges parallel to the principal axis. Therefore, if the filament is placed at the mirror's principal focus, the reflected rays will produce a parallel beam of light.

Wrong Choices Explained:

(1) If the filament is placed between the mirror and the principal focus, a virtual, erect image is produced behind the mirror.

(3) If the filament is placed at the mirror's center of curvature, a real, inverted image that is the same size as the object is produced at the center of curvature.

(4) If the filament is placed beyond the mirror's center of curvature, a real, inverted image that is smaller than the object is produced between the mirror's principal focus and the center of curvature.

95. **1** Spherical mirrors are subject to a deficiency known as spherical aberration. A spherical concave mirror will not focus light to exactly one point. As a result, images appear fuzzy or out of focus. To correct this deficiency, a parabolic mirror must be used.

Wrong Choices Explained:

(2) Chromatic aberration is a lens defect in which different colors of light are focused at different points.

(3) Dispersion is the separation of polychromatic light into its component colors.

(4) Refraction is the change in direction of a wave when it passes obliquely from one medium to another in which it moves at different speed.

GROUP 5—**Solid State**

96. **3** A diode rectifier is a device that permits charge to flow in one direction only. An AC signal that passes through a diode rectifier will emerge as a pulsating DC signal. Graph (3) best represents the wave form and phase of the emerging pulsating DC signal.

Wrong Choices Explained:

(1) This diagram indicates that nothing has happened to the signal.

(2) This diagram indicates that the signal has been reversed.

(4) This diagram shows a pulsating DC signal, but the signal has been reversed.

97. **1** A forward-biased diode will show increased current readings with increased applied voltage (potential). Graph (1) best represents this relationship.

Wrong Choices Explained:

(2), (3), (4) These graphs show an inverse relationship, an inversely proportional relationship, and an uneven inverse relationship, respectively, between applied potential and current.

98. **1** The *P*-side of the diode in the diagram is represented by the arrow in the symbol, and the *N*-side of the diode by the straight, thick vertical line. When the *P*-side of the diode is connected to the negative terminal and the *N*-side is connected to the positive terminal, the electric field barrier is strengthened.

As a result, little or no current is present in the circuit because the electric field barrier severely inhibits charge flow within the diode. In this case, the diode is said to be reversed biased.

Wrong Choice Explained:

(2) For the diode to be forward biased, the *P*-side of the diode must be connected to the positive terminal and the *N*-side to the negative terminal.

99. **2** Most of the charge carriers in an *N*-type semiconductor are electrons (hence the name *N*-type, as in negative). Therefore, in this type of semiconductor the number of free electrons is greater than the number of holes.

100. **4** In a transistor, the emitter serves as the primary source of charge carriers. In a *P-N-P* transistor, these charge carriers are holes.

Wrong Choices Explained:
(2) The collector receives the charge carriers from the base.
(3) The base is the thin middle portion of the transistor.

101. **3** As the temperature of a solid conductor increases, the vibrational kinetic energy of its atom kernels also increases. As a result, more occur between electrons and atom kernels, thereby impeding the flow of current.

Wrong Choices Explained:
(1) Temperature increases usually cause the expansion of metals; therefore, the cross-sectional area would minimally increase, not decrease.
(2) Temperature is directly related to kinetic energy; higher temperature results from increased kinetic energy of the particles.
(4) Promotion of electrons occurs from the valence band to the conduction band, but this phenomenon is associated with semiconductors, not conductors.

102. **3** According to the electron band model of conduction, semiconductors conduct electricity via electrons in the conduction band and holes in the valence band.

103. **4** The semiconductor on the left is a *P*-type semiconductor, and the one on the right is an *N*-type semiconductor. Together they form a *P-N* junction diode.

Wrong Choices Explained:
(2) A transistor consists of two junction diodes that share a thin semiconductor layer between them.
(3) Doping substances, which contribute electrons to the conduction band, are referred to as donor elements.

104. **2** A doping atom such as indium, which has only three valence electrons, will accept electrons from the conduction band. For this reason it is known as an acceptor element. In this case, the indium will remain in the germanium crystal as a negative ion; therefore, an electron deficiency remains in the conduction band. Since such deficiencies, known as holes, behave as positive charges, most of the charge carriers in the semiconductor can be considered to be positive holes.

For this reason, if a very small amount of indium was added to a germanium crystal, the resulting semiconductor would be *P*-type (*P* as in positive).

Wrong Choices Explained:

(1) *N*-type semiconducting materials conduct via excess electrons in the conduction band placed there by donor elements that have five valence electrons.

(3), (4) Although there may be relative charges in the conduction and valence band, no net charge exists in the material, as all the donors and acceptors are neutral atoms.

105. **1** By definition, the emitter-base pair should be forward biased and the collector-base pair should be reverse biased regardless of whether the transistor is an *N-P-N* or a *P-N-P* transistor.

GROUP 6—**Nuclear Energy**

106. **3** All atomic nuclei (also called *nuclides*) and their component nucleons may be represented by the same general symbol:

$$\boxed{{}^{A}_{Z}X}$$

Here *X* represents the letter(s) used to identify the particle; *Z*, called the *atomic number*, indicates the number of elementary charges present (assumed to be positive unless a negative sign is written); and *A*, called the *mass number*, is equal to the sum of neutrons and protons present.

The number of neutrons (*N*) present in an atomic nucleus is given by the expression

$$N = A - Z$$

For an atom of ${}^{32}_{15}P$, the atomic number is 15 and the mass number is 32; therefore, the number of neutrons is 17 (32 – 15 = 17).

107. **1** Nuclear reactions, like any other chemical reactions, obey the laws of conservation of mass and conservation of electric charge. In the given equation, article *X*'s mass number and atomic number must be such that the sum of all the mass numbers and atomic numbers on the right side of the equation are equal to the sum of all the mass numbers and atomic numbers on the left side of the equation.

Solution: ${}^{32}_{16}S + {}^{1}_{0}n \rightarrow {}^{32}_{15}P + {}^{1}_{1}X$

The mass number and atomic number of particle *X* correspond to those of a proton, ${}^{1}_{1}H$.

Wrong Choices Explained:

(2) A neutron is represented as ${}_0^1n$.

(3) An alpha particle is represented as ${}_2^4He$.

(4) A beta particle is represented as ${}_{-1}^{0}e$.

108. **3** The stability of a nucleus is tied to the existence of two nuclear forces. These forces, called the strong and weak interactions, are much more powerful at the very small distances present between the nucleons within the nucleus than is the gravitational forces. At larger distances, however, these nuclear forces lose their effectiveness, so it can be said that they have a shorter range than the gravitational force.

109. **4** Quarks are the subatomic particles of which protons, neutrons, and other nuclear particles are composed. Quarks carry either 1/3 or 2/3 of an elementary charge and come in six "flavors": top, bottom, up, down, charm, and strange. Protons are made up of three quarks, two up quarks and one down quark.

Wrong Choices Explained:

(1) This is a fundamental particle of the baryon group. It is greater in mass than a proton or neutron.

(2) In an atom, the baryons are the particles with the largest masses. These include protons and neutrons.

(3) A positron, also known as a beta (+) particle, is the antiparticle of the electron, formed in the nucleus by the disintegration of a proton.

110. **1** Alpha decay results in the emission of an alpha particle from a nuclide. An alpha particle is a helium nucleus, which consists of two protons and two neutrons. As a result of alpha particle emission, the resultant daughter nuclide has a mass number that is 4 less and an atomic number 2 less than those of the parent nuclide.

Wrong Choices Explained:

(2), (4) Electron capture (or K-capture) and positron emission result in a daughter nuclide with the same mass number as the parent nuclide and an atomic number 1 less than that of the parent nuclide.

(3) Gamma ray emission results in a nucleus with a lower energy state; there is no change in mass or atomic number.

111. **3** Refer to the *Uranium Disintegration Series* section in the *Reference Tables*. Find the column for polonium (Po), atomic number 84, on the chart, and count the number of dots that represent the isotopes of polonium. There are three dots in this column that correspond, according to the mass numbers at the right, to isotopes Po-218, Po-214, and Po-210.

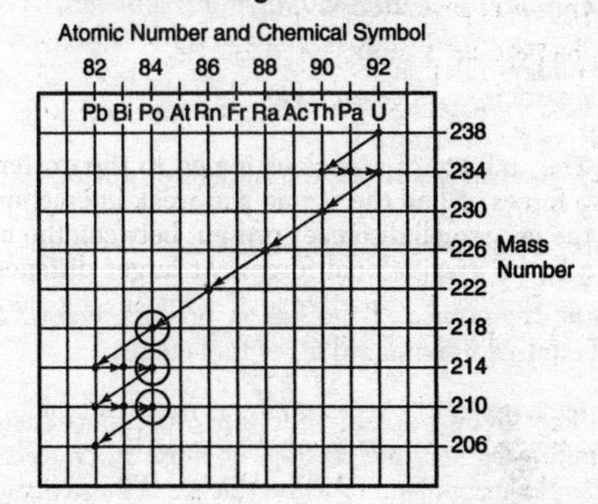

112. **1** By definition, gamma radiation is very high energy photons of electromagnetic radiation. Gamma photons have the highest frequencies in the electromagnetic spectrum.

113. **2** Write out the problem.

Given:

$$t_{\frac{1}{2}} = 3.0 \text{ days}$$

$$m_f = 2.0 \text{ kg}$$

$$\text{total time} = 9.0 \text{ days}$$

Find: $m_i = ?$

Refer to the *Nuclear Energy* section in the *Reference Tables* to find an equation(s) that relates mass with time and half-life.

Solution:

$$m_f = \frac{m_i}{2^n}$$

$$m_i = m_f \times 2^n$$

To solve this problem, you need to determine how many half-lives (n) have elapsed. You can do this by dividing the total time elapsed by the half-life time.

$$n = \frac{\text{total time}}{t_{\frac{1}{2}}} = \frac{9.0 \text{ days}}{3.0 \text{ days}} = 3.0$$

$$m_i = m_f \times 2^n = (2.0 \text{ kg})(2^3) = 16 \text{ kg}$$

114. **1** Nuclear fission is the process of splitting a heavy nucleus, such as uranium-235 or plutonium-239, into lighter fragments. Uranium-235 and plutonium-239 are used specifically as fuels in nuclear reactors because they have the ability to undergo fission.

Wrong Choices Explained:

(2) Fusion is the process of uniting lighter nuclei, such as those of deuterium, into a heavier nucleus.

(3) During the process of fission, thermal neutrons are absorbed by the fuel prior to actual splitting.

(4) During the course of the fission process, two or more neutrons are released and are free to bombard other fuel atoms.

115. **1** The given nuclear reaction represents fusion. Fusion is the process of uniting lighter nuclei, such as those of deuterium, into a heavier nucleus. Fusion is accompanied by the release of large quantities of energy.

Wrong Choices Explained:

(2) Fission is the process of splitting a heavy nucleus, such as uranium-235 or plutonium-239, into lighter fragments. Fission is accompanied by the release of a large amount of energy.

(3) Alpha decay is a natural radioactive process that results in the emission of an alpha particle from a nuclide.

(4) Beta decay is a natural radioactive process that results in the emission of a beta particle from a nuclide.

PART III

116. The three forces acting on the block are (1) the force of gravity (the block's weight), (2) the normal force, which acts perpendicular to the surface and in this situation is equal in magnitude but opposite in direction to the weight of the block, and (3) the applied horizontal force to the right.

Example of Acceptable Response

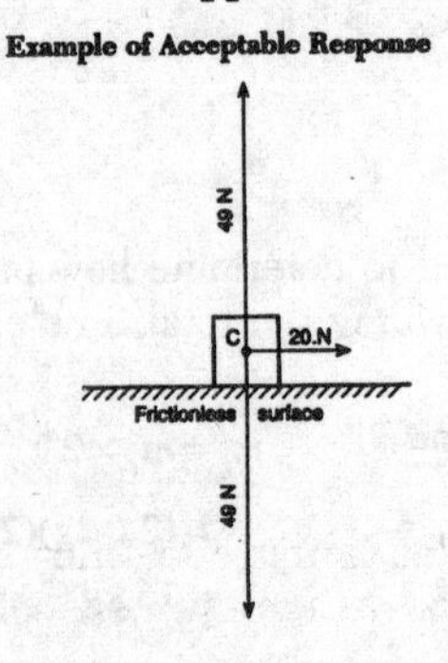

If each of the three vectors meets *all three* of the following criteria, a total of three credits is awarded.

- A line originating at point C and having an arrowhead indicating the correct direction, such as those shown in the preceding diagram. (A vector with either of its ends in contact with an edge of the block is acceptable.)
  —applied force: parallel to the horizontal
  —weight: perpendicular to the horizontal
  —normal force: perpendicular to the horizontal
- The vector, including its arrowhead, is drawn to the appropriate length.
  —applied force: 2.0 cm ± 0.2 cm long
  —weight: 4.9 cm ± 0.2 cm long
  —normal force: 2.0 cm ± 0.2 cm long
- The vectors are labeled.
  —applied force: 20. N or 20 N
  —weight: 49 N
  —normal force: 49 N

If each of the three vectors meets *at least two* of the three criteria, a total of two credits is awarded.

If each of the three vectors meets *at least one* of the three criteria, a total of one credit is awarded.

117. Write out the problem.

Given: $m = 5.0$ kg Find: $a = ?$

$w = 49$ N

$F_{applied} = 20.$ N

Refer to the *Mechanics* section in the *Reference Tables* to find an equation(s) that relates acceleration with mass, weight, and force.

Solution: $F = ma$

$$a = \frac{F}{m}$$

$$= \frac{20.\text{ N}}{5.0\text{ kg}}$$

$$= 4.0\text{ m/s}^2 \text{ or } 4.0\text{ N/kg}$$

(Note that weight was not needed in this calculation as the weight was balanced by the normal force and the net force was due solely to the applied horizontal force.)

Note: A total of two credits is allowed. One credit is awarded for writing the equation and for the substitution of values with units. If the equation and/or units are not shown, you do not receive this credit. One credit is awarded for the correct answer (number and units). If there are no units, you do not receive this credit. Significant figures and scientific notation are not required to obtain this credit.

118. Write out the problem.

Given: $q = 25 \text{ C}$ Find: $\Delta t = ?$

$V = 1.8 \times 10^6 \text{ V}$

$I = 2.0 \times 10^4 \text{ A}$

Refer to the *Electricity and Magnetism* section in the *Reference Tables* to find an equation(s) that relates time with charge, electric potential difference, and current.

Solution:

$$I = \frac{\Delta q}{\Delta t}$$

$$\Delta t = \frac{\Delta q}{I}$$

$$= \frac{25 \text{ C}}{2.0 \times 10^4 \text{ A}}$$

$$= 1.3 \times 10^{-3} \text{ s or } 1.25 \times 10^{-3} \text{ C/A}$$

(Note that electric potential difference was not needed in this calculation.)

Note: A total of two credits is allowed. One credit is awarded for writing the equation and for the substitution of values with units. If the equation and/or units are not shown, you do not receive this credit. One credit is awarded for the correct answer (number and units). If there are no units, you do not receive this credit. Significant figures and scientific notation are not required to obtain this credit.

119. Write out the problem.

Given: $q = 25 \text{ C}$ Find: $W = ?$

$V = 1.8 \times 10^6 \text{ V}$

$I = 2.0 \times 10^4 \text{ A}$

Refer to the *Electricity and Magnetism* section in the *Reference Tables* to find an equation(s) that relates energy with charge, electric potential difference, and current.

Solution:

$$V = \frac{W}{q}$$

$$W = Vq$$

$$= (1.8 \times 10^6 \text{ V})(25 \text{ C})$$

$$= 4.5 \times 10^7 \text{ J or } 45 \times 10^6 \text{ V} \cdot \text{C}$$

(Note that current was not needed in this calculation.)

Note: A total of two credits is allowed. One credit is awarded for writing the equation and for the substitution of values with units. If the equation and/or units are not shown, you do not receive this credit. One credit is awarded for the correct answer (number and units). If there are no units, you do not receive this credit. Significant figures and scientific notation are not required to obtain this credit.

120. To receive this credit the response must be written in one or more *complete* sentences.
Examples of Acceptable Responses
Light travels much faster than sound.
The speed of light is 3.0×10^8 m/s and the speed of sound is only 3.3×10^2 m/s.
(**Note:** Refer to the *List of Physical Constants* section in the *Reference Tables* to find these values.)

Note: One credit is granted for an acceptable response.

121. One credit is granted if the scale is linear and the scale divisions are appropriate. A scale of 0.10 m per division is *not* acceptable.
See the graph located after question 123 for examples of acceptable graphs.

122. One credit is granted if all points are plotted accurately (±0.3 grid space). This credit will be granted if you correctly use your response from question 121.
See the graph located after question 123 for examples of acceptable graphs.

123. One credit is granted if the best-fit line is straight. If one or more points are plotted incorrectly in question 122 but a best-fit straight line is drawn, this credit will be granted.

Examples of Acceptable Graphs

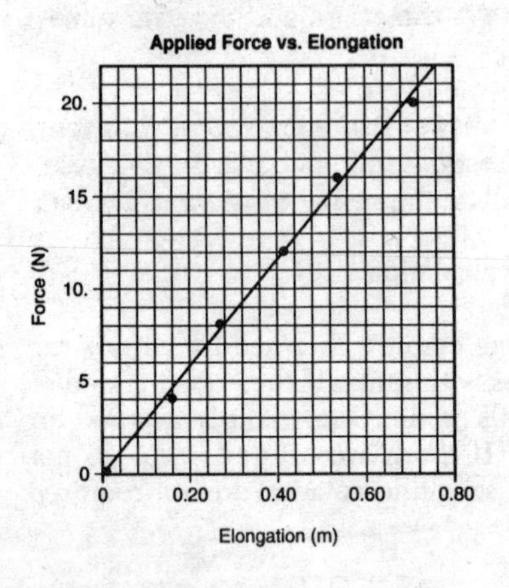

or

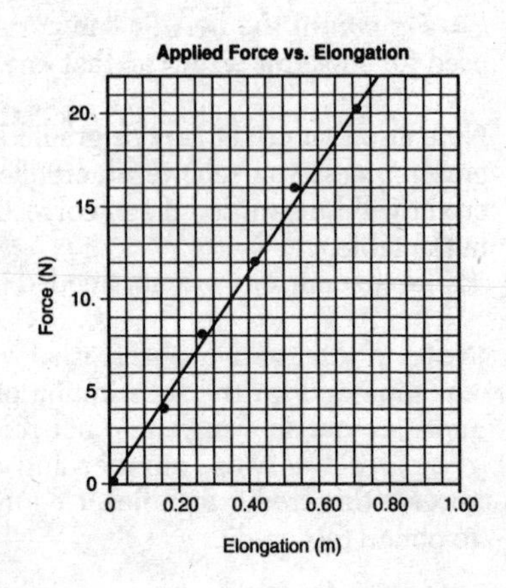

124. Write out the problem.

Given: $F = 12.0\ \text{N}$
$x = 0.42\ \text{m}$

Find: $k = ?$

Refer to the *Energy* section in the *Reference Tables* to find an equation(s) that relates the spring constant with applied force and elongation of a spring.

Solution:

$$F = kx$$

$$k = \frac{F}{x}$$

$$= \frac{12.0\ \text{N}}{0.42\ \text{m}}$$

$$= 29\ \text{N/m}\ (\pm 2\ \text{N/m})$$

(Note that this response is based on the assumption that the elongation of 0.42 m due to an applied force of 12.0 N lies on the best-fit line and that the line passes through the origin.)

Note: A total of two credits is allowed. One credit is awarded for writing the equation and for the substitution of values with units. If the equation and/or units are not shown, you do not receive this credit. One credit is awarded for the correct answer (number and units). If there are no units, you do not receive this credit. Significant figures and scientific notation are not required to obtain this credit.

Note: The slope *may* be determined by direct substitution into the equation $k = F/x$ *only* if the best-fit line passes through the *origin* and the data values used for substitution are on that line.

Note also that credit may be granted for an answer that is consistent with your graph, *unless* you receive no credits for questions 122 and 123. In that case, credit will be awarded if you correctly calculate the spring constant using data in the table.

Topic	Questions Numbers (total)	Wrong Answers (x)	Grade
Mechanics	1–16, 54, 116–117, 120: (20)		$\frac{100(20-x)}{20}=\%$
Energy	17–21, 121–124: (9)		$\frac{100(9-x)}{9}=\%$
Electricity and Magnetism	22–24, 36, 53, 83, 118–119: (17)		$\frac{100(17-x)}{17}=\%$
Wave Phenomena	36, 37–44, 46–48: (12)		$\frac{100(12-x)}{12}=\%$
Modern Physics	45, 49–52, 55: (6)		$\frac{100(6-x)}{6}=\%$
Motion in a Plane	56–65: (10)		$\frac{100(10-x)}{10}=\%$
Internal Energy	66–75: (10)		$\frac{100(10-x)}{10}=\%$
Electromagnetic Applications	76–82, 84, 85: (10)		$\frac{100(9-x)}{9}=\%$
Geometrical Optics	86–95: (10)		$\frac{100(10-x)}{10}=\%$
Solid State Physics	96–105: (10)		$\frac{100(10-x)}{10}=\%$
Nuclear Energy	106–115: (10)		$\frac{100(10-x)}{10}=\%$

To further pinpoint you weak areas, use the Topic Outline in the front of the book.